Introduction to Relativistic Statistical Mechanics

Classical and Quantum

Introduction to Relativistic Statistical Mechanics

Classical and Quantum

Rémi Hakim
Paris-Meudon Observatory, France

World Scientific
NEW JERSEY • LONDON • SINGAPORE • BEIJING • SHANGHAI • HONG KONG • TAIPEI • CHENNAI

Published by

World Scientific Publishing Co. Pte. Ltd.
5 Toh Tuck Link, Singapore 596224
USA office: 27 Warren Street, Suite 401-402, Hackensack, NJ 07601
UK office: 57 Shelton Street, Covent Garden, London WC2H 9HE

Library of Congress Cataloging-in-Publication
Hakim, Rémi, 1936–
Introduction to relativistic statistical mechanics : classical and quantum / Rémi Hakim.
p. cm.
Includes bibliographical references and index.
ISBN-13 978-981-4322-43-0 (hardcover)
1. Statistical mechanics. 2. Relativistic quantum theory. 3. Relativistic kinematics. I. Title.
QC174.86.C6H35 2011
530.13--dc22
2010054042

British Library Cataloguing-in-Publication Data
A catalogue record for this book is available from the British Library.

Typeset by Stallion Press
Email: enquiries@stallionpress.com

Printed in Singapore.

This book is dedicated to:

my friend and colleague Daniel Gerbal
1935, Paris — 2006, Paris Za"l

my colleague and friend Horacio Dario Sivak
1946, Buenos Aires — 2000, Villejuif Za"l

Contents

Preface

Relativistic statistical mechanics has long ceased to be considered as a simple matter where it is sufficient to change the expression of the energy from the Newtonian to the relativistic one and to check the Lorentz invariance of the final result. For about 30 years, this field has grown exponentially and there now exist several thousand articles devoted to it. The reasons for such an explosion are briefly presented in the introduction. They not only come from the requirements of astrophysics (white dwarfs, pulsars/neutron stars/magnetars, the early universe, etc.) and elementary particle physics (production of particles, heavy ion collisions and the search for the quark–gluon plasma), but are also increasingly in demand in condensed matter physics (a notable example is the development of the petawatt laser). The presently evolving and exploding nature of this domain explains why the subject cannot be dealt with in an exhaustive way.

This book is intended to be an *introduction* to some recent developments of relativistic statistical mechanics rather than a standard textbook. Owing to the dynamical character of the field, particularly in the quantum domain, only a few applications — or, more accurately, illustrations — of the notions presented are given, mainly in view of the comprehension of some astrophysical problems. The book may also serve as an introduction to the current literature on the subject, and it had a relatively well-furnished bibliography — albeit, unfortunately, nonexhaustive. It contains the basics of nonquantal relativistic kinetic theory, referring very often to the well-known book by S.R. de Groot, M.C.J. van Leeuwen and Ch. G. van Weert (1980), and of classical statistical mechanics. However, most applications are related to quantum systems (such as relativistic plasmas and nuclear matter), and hence slightly more than half of the book is devoted to relativistic quantum statistical mechanics.

Whereas many works rest on quantum thermal field theory — essentially the study of the partition function with the field-theoretical method — the subject is not treated along this line here and, for the sake of completeness, is only briefly outlined: there exist excellent books in this domain, such as the ones by M. Le Bellac (2000) or J. Kapusta (1989). Rather, a covariant version of the Wigner function is the central object of the formalism under consideration. This approach presents the advantage of being somewhat closer to what is generally known by astrophysicists, and also permits one to recover all expressions familiar to those working in the field of heavy ion collisions. On several occasions the covariant Wigner function formalism appears simpler than thermal field theory. This is illustrated in the case of the Walecka model (1974) of nuclear matter and in that of relativistic quantum plasmas. Whereas field-theoretical methods rely heavily on the use of Feynman diagrams and are therefore, at least in spirit, perturbative — even though well-chosen resummations can describe nonperturbative effects satisfactorily — the close kinship of the covariant Wigner formalism with standard tools of classical plasma physics allows the introduction of methods of approximation well tested in that field. Finally, the covariant Wigner operator can be expressed in terms of the central object of field-theoretical methods, viz. the Green function. On the other hand, the covariant Wigner formalism presents the disadvantage of being much less studied than, for example, finite temperature Green functions, which the present work will hopefully remedy in some measure.

The case of non-Abelian plasmas — such as the quark–gluon plasma — is not considered in this book; not only is it a domain of its own which would deserve an entire book but the subject is still in an uncertain state. Furthermore, this would drive us far away from a simple introduction.

Finally, most calculations are only outlined, especially whenever long and tedious, in favor of the basic concepts and by referring to original works.

Acknowledgments: The author is indebted to Drs. J. Diaz Alonso, M. Lemoine, L. Mornas and to Dr. H. Sivak for comments and for reading some manuscripts and making comments, respectively.

Notations and Conventions

We generally use a system of units where $\hbar = c = 1$ and a flat space–time metric $\eta^{\mu\upsilon}$ endowed with signature $(+\ -\ -\ -)$. Greek indices vary from 0 to 3, while Latin ones run from 1 to 3. Boldface symbols generally designate spatial three-vectors. x or p designates four-vectors: $x = (x^0, \mathbf{x})$, $p = (p^0, \mathbf{p})$. The Minkowski pseudoscalar product of two four-vectors a and b is designated by $a \cdot b$; $a \cdot b = \eta_{\mu\nu} a^\mu b^\nu = a^0 b^0 - a \cdot b$. The symbol

$$\Delta^{\mu\upsilon}(a) \equiv \eta^{\mu\upsilon} - \frac{a^\mu a^\upsilon}{a^2}$$

is the projector over the three-plane orthogonal to the four-vector a^μ. As usual, tensor indices placed between parentheses (resp. between brackets) indicate a full symmetrization (resp. antisymmetrization). The Levi-Civita pseudotensor is defined as

$$\begin{cases} \varepsilon^{0123} = +1, \\ \varepsilon^{\mu\nu\alpha\beta} = \begin{cases} +1 & \text{if } (\mu, \nu, \alpha, \beta) \text{ form an even permutation of } (0, 1, 2, 3), \\ -1 & \text{if } (\mu, \nu, \alpha, \beta) \text{ form an odd permutation of } (0, 1, 2, 3), \\ 0 & \text{otherwise.} \end{cases} \end{cases}$$

We use the same symbol for a mathematical notion and its Fourier transform

$$A(k) = \frac{1}{\sqrt{2\pi}} \int dx \exp(-ik \cdot x) A(x),$$

and the name of variables will allow correct identification.

The works which are quoted are according to whether they are in the bibliography of relativistic statistical mechanics or not; for instance, J.D. Walecka (1974) appears in the bibliography while G. Baym is quoted as a note — L.P. Kadanoff and G. Baym, *Quantum Statistical Mechanics*, etc.

Introduction

Relativistic statistical mechanics is nowadays a *bona fide* subject, in fields from astrophysics to heavy ion collisions, not forgetting nuclear matter, etc. After the first articles by F. Jüttner (1911) [see also W. Pauli (1921)], it did not attract much attention until the beginning of the 1960s, the more so since possible applications seemed to be quite speculative at that time.

In 1928, Jüttner generalized his 1911 ideal gas results to the case of the ideal quantum gas, which was soon applied to the theory of white dwarfs by S. Chandrasekhar (1934), with the now well-known consequence of the existence of a limiting mass for this kind of star — the so-called Chandrasekhar mass.

A lesser known work in the domain is the article by A.G. Walker (1934) where, for the first time, general relativity was introduced and kinetic theory applied to the expanding universe.

Slightly later, D. van Dantzig improved relativistic hydrodynamics and studied the ideal gas case (1939); his results were described and extended by J.L. Synge (1957). P.G. Bergmann (1951, 1962) provided various tools for use in relativistic statistical mechanics (essentially, techniques involving differential forms, well suited to such a case). At about the same time, A.E. Scheidegger and C.D. McKay (1951) and A.O. Barut (1958) devised techniques for performing "statistics of fields," still in the noninteracting case.

The interest raised by nuclear fusion, in the late 1950s, led to various studies on relativistic plasmas: S. Titeica (1956), S.T. Beliaev and G.I. Budker (1956), and Yu. L. Klimontovich (1960). While Titeica gave a covariant version of the Vlasov equation, Beliaev and Budker included a Landau-like collision term.

However, Klimontovich achieved a decisive advance — using M. Schönberg's method of second quantization in phase space[1] — and was able to provide a BBGKY hierarchy for the covariant one-, two-, etc.–particle distribution function of an electron plasma embedded in a neutralizing uniform background. From this hierarchy, he was able to derive the relativistic Landau collision term and hence the plasma Fokker–Planck equation; he also obtained the Balescu–Guernsey–Leenhardt equation, whose collision term involves the influence of the plasma modes.[2]

Although Klimontovich took a great step, the general situation — discussed in detail by P. Havas (1964) — was still unclear since, apart from plasmas, no other nonquantum physical system was known. Furthermore, it was believed that only Hamiltonian equations of motion were needed in relativistic statistical mechanics. As a matter of fact, a "no-interaction theorem" was proven by D.G. Currie, T.F. Jordan and E.C.G. Sudarshan[3] to the effect that a Hamiltonian formalism applies only to systems constituted by noninteracting particles. Therefore, the sole remaining possibility was the simultaneous statistical treatment of particles and field(s) although they were supposed to be interacting.

Such an approach was already known in the nonrelativistic case (see e.g. E.G. Harrison, I. Prigogine) and could easily be extended to relativity [see e.g. A. Mangeney (1965)] although the detailed calculations were not trivial at all. The results were not manifestly covariant and hence the proof that they actually satisfy the principle of relativity had to be given for each particular case. Accordingly, the Brussels school (Prigogine and his collaborators) imagined a formalism that provided the Lorentz transformation properties of their equations and also of the physical observables [see e.g. R. Balescu and E. Pena (1967, 1968)]. However, their formalism, although ingenious and corresponding to an implicit and quite admissible philosophical position as to relativity (space and time must be kept separated), was extremely involved and had the consequence that the Lorentz

---

[1]M. Schönberg, Application of second quantization methods to the classical statistical mechanics, *Nuovo Cimento*, **9**, 1139 (1952); A general theory of the second quantization methods, *ibid.* **10**, 697 (1953).

[2]S. Ichimaru, *Basic Principles of Plasma Physics* (Benjamin, Reading, Massachusetts, 1973).

[3]D.G. Currie, T.F. Jordan and E.C.G. Sudarshan, *Rev. Mod. Phys.* **35**, 350 (1963); see also G. Marmo, N. Mukunda and E.C.G. Sudarshan, *Phys. Rev.* **D20**, 2120 (1984).

transformation acquired a curious dynamical meaning while, according to common wisdom, it is a merely kinematical transformation.[4]

Meanwhile, N.A. Chernikov (1956–1964), G.E. Tauber and J.W. Weinberg (1961), and W. Israel (1963) studied the covariant Boltzmann equation, whether in a flat space–time case or in a curved one. These studies were taken up later by numerous authors and applied to the calculation of transport coefficients (bulk and shear viscosity, heat conduction coefficient, diffusion coefficient, etc.) via the use of approximation methods (Chapman–Enskog, moments, etc.) adapted to the case of relativity.

As to quantum systems, impulsions to their active study were provided by the so-called statistical model of multiple production of particles [L. Landau (1953)] and by its extension by R. Hagedorn (1965) to the statistical bootstrap model. In the mid-1970s, still in view of multiproduction of particles, P.A. Carruthers and F. Zachariasen (1974–1983) first used a covariant form of the Wigner function[5]; at about the same time, F. Cooper, Sharp and Feigelbaum (1976) and others worked in the same direction. This latter was then generalized to fermions, or given a gauge-invariant form [E.A. Remler, V.V. Klimov (1982), J. Winter (1984), U. Heinz (1983, 1985), H.-Th. Elze, M. Gyulassy and D. Vasak (1986a,b). The covariant Wigner function was used in the study of relativistic quantum plasmas, embedded or not in strong magnetic fields, for the derivation of the main properties of nuclear matter through the use of the J.D. Walecka's model (1974) or other phenomenological ones.

However, the QED plasma was studied from a mere theoretical point of view by several authors, beginning with Fradkin (1959) (who extended Matsubara's results to the relativistic case), Akhiezer and Peletminski (1960), Tsytovich (1961), etc., with the help of Green function methods.

The development of experimental data on the 3 K blackbody universal background radiation since 1965 led to more and more support for the big bang cosmological model and motivated theoretical works on the state of matter in the primeval universe, i.e. the universe before roughly 1 s. This required studies of quantum field theory at finite temperature[6] and/or

---

[4]The dynamical interpretation of I. Prigogine and his coworkers is perfectly admissible but it does not correspond to the general trend of physicists looking for symmetries in the laws of physics.

[5]E.P. Wigner, *Phys. Rev.* **40**, 749 (1932).

[6]S. Weinberg (1974), etc.

density that gradually became a domain in their own right. The main trend of these works was the study of phase transitions of various orders in the primeval universe and, in particular, people were looking for the restoration of broken symmetries at high temperatures [D.A. Kirzhnitz and A.D. Linde (1972)].

At about the same time, the asymptotic freedom[7] property of quantum chromodynamics, and of other gauge theories, indicated that at high density and/or temperature — which is the case in the primitive universe [see e.g. M.J. Perry and J.C. Collins (1975)] — hadron matter presumably undergoes a phase transition to a phase of quasi-free quarks.

Order-of-magnitude calculations [and also lattice calculations; see e.g. M. Creutz (1985)] then gave a critical temperature ranging from 100 MeV to 200 MeV. This is an energy which can be obtained in nucleus–nucleus collisions and therefore, in order to discover the quark–gluon phase of baryon matter, many efforts were undertaken and are still in progress. Unfortunately, there is presently no obvious signal for the manifestation of a possible quark phase. As a consequence, theoretical works in this field are exploding, allowing thereby a thorough exploration of finite temperature quantum field theory.

It was mentioned above that astrophysical objects — the interior of compact stars, the pulsar's magnetosphere, the primeval universe — resort to the use of relativistic statistical mechanics, whether classical or quantum. Therefore, we now review very briefly these objects and also the microscopic applications such as the heavy ion collisions. This is of course not intended to provide a fully developed theory but rather to specify the main applications a little bit further.

It should now be the place for an interesting tour of multiparticle production and the statistical bootstrap model, since they played an important role in the development of relativistic statistical mechanics.

In high energy collisions, one observes the emission of a great variety of particles: the ones allowed by energy–momentum and internal quantum number conservation. The higher the energy involved in the collision, the larger the number of particles produced, so that the idea of a statistical treatment gradually emerged. The first statistical model — which was not relativistic — goes back to E. Fermi and L. Landau,[8] and

---

[7]D.H. Politzer, Asymptotic freedom, an approach to strong interactions, *Phys. Rep.* **14**, 130 (1974).

[8]E. Fermi, *Prog. Theor. Phys.* **5**, 570 (1950); L. D. Landau, *Izv. Akad. Nauk SSSR* **78**, 51 (1953).

was eventually improved and made Lorentz-covariant during the 1950s and '60s.

The basic idea was to replace the transition probability involved in the S matrix by a constant, possibly some average value, while keeping the energy–momentum conservation relations

$$\frac{d^n P}{dp_1 dp_2 \ldots dp_n} = W\left(a + b \to x_1 + x_2 + \cdots + x_n\right)$$
$$\times \delta^{(4)}\left(p_a + p_b - p_1 - p_2 \cdots - p_n\right) \prod_{i=1}^{i=n} 2\theta(p_i^0)\delta\left(p_i^2 - m_i^2\right),$$

where W — the transition probability per unit of volume and time — has been approximated by a constant, avoiding thereby all dynamical complications. In the above expression, W is given by

$$W(a + b \to x_1 + x_2 + \cdots + x_n) = |M(a + b \to x_1 + x_2 + \cdots + x_n)|^2, \tag{1}$$

where M is the transition amplitude of the reaction.

This model is "statistical" in the sense of a phase space dominance and in general not with a thermodynamic meaning. One is generally interested in the probability of producing N particles of a given species whatever the X other ones, i.e. in

$$\begin{cases} P(N) = \sum_X P(N, X) \\ P(N, X) \approx \int \prod_{i=1}^{i=n+X} \frac{d^3 p_i}{p_{i0}} \delta^{(4)}\left(P - \sum_{i=1}^{i=n+X} p_i\right) \\ \qquad \times \langle W(a + b \to x_1 + x_2 + \cdots + x_n)\rangle \end{cases} \tag{2}$$

($P = p_a + p_b$), and since the details of the transition probability become less and less relevant when one integrates over the large phase space implied by a large number of particles produced in a high energy collision, W can be replaced by a constant as, for instance, its average value. Finally, $P(N)$ appears to be essentially a microcanonical probability. It has been evaluated via the use of the central limit theorem by F. Lurçat and P. Mazur (1964).

This was, however, not completely satisfactory and R. Hagedorn (1965) reintroduced some dynamics with his "statistical bootstrap," which was originally[9] built to explain the approximate constancy of the (average)

[9]See R. Hagedorn (1995) for the history of his interesting model.

transverse momenta of the produced particles in high energy hadron–hadron collisions. It was also based on the remark that the secondaries produced in such a collision result from the decay of a number of *fireballs*,

$$a + b \to \text{fireballs} \to n + X,$$

or resonances, from which the idea of the statistical bootstrap finally emerged: fireballs are made of fireballs, themselves made of fireballs, etc. In this model, one thus has to make a statistics of fireballs with a particular mass spectrum $\rho(m)$ and it is described by its partition function, roughly given by

$$Z(V,T) = \sum_{\text{all states } i} \exp\left(-\frac{E_i}{T}\right) = \int_0^\infty dE\sigma(E,V)\exp\left(-\frac{E}{T}\right), \quad (3)$$

where $\sigma(E, V)$ is the density of state of the system, connected to the mass spectrum essentially by a relation of the form $\log[\rho(m)] \approx \log[\sigma(m, V_0)]$ for m large, and where V_0 is the interaction volume. Finally, Hagedorn was led to a mass spectrum of the asymptotic form

$$\rho(m) \approx m^{-\alpha} \exp\left(-\frac{m}{T_0}\right), \quad (4)$$

where T_0, now known as *Hagedorn's temperature*, is a constant of the order of 160 MeV, which appeared as being a limiting temperature since the partition function

$$Z(V,T) \approx V_0 \int_0^\infty dm\rho(m)m^{-\alpha} \exp\left(-m\left\{\frac{1}{T} - \frac{1}{T_0}\right\}\right) \quad (5)$$

has a meaning only when $T < T_0$. Such a spectrum — which is essentially verified when plotting the various particles and their resonances as a function of their masses — led to an explanation of the constancy of the average transverse momenta of secondaries produced in high energy hadron–hadron collisions. The model, however, suffered from some obvious drawbacks: for instance, it implicitly involved only attractive interactions, the ones necessary for forming fireballs, and no repulsion[10] at all. Also, it needed many improvements, such as the conservation of some internal quantum numbers.[11] A few years later, the statistical bootstrap was used in a possible description of the quark–gluon plasma, the more so since the

[10] The use of the so-called "pressure ensemble" can be considered as a first attempt at taking repulsion into account [R. Hagedorn (1995); R. Hagedorn and J. Rafelski.

[11] K. Redlich and L. Turko (1980); L. Turko (1981, 1994); H.T. Elze and W. Greiner (1986).

phenomenological MIT bag model[12] of hadrons also provides an exponential energy spectrum.

This led to an enormous literature, which cannot be invoked here [see e.g. the references quoted in R. Hagedorn (1995)]. Numerous high energy physicists became used to thinking in terms of dense matter and hence of relativistic statistical physics; moreover, they found a natural domain of applications and/or theoretical toy models in various fields of relativistic astrophysics. The statistical bootstrap also gave an impulsion to the study of statistical mechanics of particles endowed with a mass spectrum [R. Hakim (1974), C. Barrabes (1976, 1982a,b), L. Burakovsky and L.P. Horwitz (1993, 1994)].

At the beginning of the 1970s, another line of thought, which aimed at introducing more dynamical considerations in the statistical approach to multiparticle production, arose with the works of P.A. Carruthers and F. Zachariasen (1974, 1975, 1976, 1983).

[12]A. Chodos, R.L. Jaffe, K. Johnson, C.B. Thorne and V.F. Weisskopf, *Phys. Rev.* **D9**, 3471 (1974); R.C. Tolman, *Phys. Rev.* **55**, 364 (1939); J. R. Oppenheimer and G. M. Volkoff, *Phys. Rev.* **55**, 374 (1939).

Chapter 1

The One-Particle Relativistic Distribution Function

The first works began, as could be expected, with those notions derived from kinetic theory, such as the distribution function, the Maxwell–Boltzmann distribution function, and the kinetic equations it is supposed to obey. Accordingly, the same path is followed in this first chapter. The first use of the covariant one-particle distribution function seems to have been made by A.G. Walker (1934), D. van Dantzig (1939), S. Titeica (1956) and J.L. Synge (1957). The approach presented here is due to Yu. L. Klimontovich (1960) and R. Hakim (1967) [see also N.G. van Kampen (1969)].

In this chapter, we shall briefly show how the one-particle distribution function can be defined in a simple way and on what phase space. The equilibrium distribution function — the relativistic Maxwell–Boltzmann distribution, hereafter called the Jüttner–Synge function — is then briefly derived and its main properties given.

1.1. The One-Particle Relativistic Distribution Function

Rather than elaborating on the transformation laws of the distribution function, on the phase space element, etc., it is much simpler to start from the main physical observables — i.e. the four-current and the energy–momentum tensor — to build the definition of the covariant distribution function.

Let us first consider a classical, i.e. nonquantum, relativistic particle. The numerical four-current it defines in space–time is provided by the so-called Feynman four-current:

$$J^{\mu}(x) = \int ds\, \delta^{(4)}[x - x(s)] \frac{d}{ds} x^{\mu}(s), \tag{1.1}$$

where x is the space–time point and s is an arbitrary parameter — generally taken to be the proper time — along the space–time trajectory over which the integral is extended. It can immediately be verified that $n(\mathbf{x}, t)$, its space–time density, is given by

$$n(x,t) = \sum_i \delta^{(3)}[x - x_i(t)], \tag{1.2}$$

and that the usual three-current is still

$$\mathbf{j}(x,t) = \sum_i \delta^{(3)}[x - x_i(t)]\mathbf{v}_i(t). \tag{1.3}$$

Similarly, the energy–momentum tensor of the particle is given by

$$T^{\mu\nu}(x) = \int ds\, \delta^{(4)}[x - x(s)]p^{\mu}(s)\frac{d}{ds}x_i^{v}(s). \tag{1.4}$$

For a system of N particles, the four-current and the energy–momentum tensor of the particles are then provided by

$$J^{\mu}(x) = \sum_{i=1}^{i=N} \int ds\, \delta^{(4)}[x - x_i(s)]\frac{d}{ds}x_i^{\mu}(s), \tag{1.5}$$

$$T^{\mu\nu}(x) = \sum_{i=1}^{i=N} \int ds\, \delta^{(4)}[x - x_i(s)]p_i^{\mu}(s)\frac{d}{ds}x_i^{v}(s), \tag{1.6}$$

which can be rewritten as

$$\begin{aligned} J^{\mu}(x) &= \int d^4u \int ds \sum_{i=1}^{i=N} \delta^{(4)}[p - p_i(s)]\,\delta^{(4)}[x - x_i(s)]u_i^{\mu}(s) \\ &= \int d^4u \int ds \sum_{i=1}^{i=N} \delta^{(4)}[p - p_i(s)]\,\delta^{(4)}[x - x_i(s)]u_i^{\mu}(s) \qquad (1.7) \\ &= \int d^4u\frac{p^{\mu}}{m} \int ds \sum_{i=1}^{i=N} \delta^{(4)}[p - p_i(s)]\,\delta^{(4)}[x - x_i(s)] \\ &\equiv \int d^4p\frac{p^{\mu}}{m}R(x,p) \qquad (1.8) \end{aligned}$$

for the four-current, where the properties of the δ function are used, and as

$$T^{\mu v} = \int d^4p\frac{p^{\mu}p^{v}}{m}R(x,p) \tag{1.9}$$

for the energy–momentum tensor. u^{μ} is the four-velocity of the particles, generally a function of x and p. In these last two equations we have used the definition

$$R(x,p) \equiv \int ds \sum_{i=1}^{i=N} \delta^{(4)}[p - p_i(s)]\, \delta^{(4)}[x - x_i(s)]. \tag{1.10}$$

$R(x,p)$ depends on the initial data chosen for the trajectories of the relativistic particles and thus is a *random function* in the context of a statistical ensemble where these data are known only in a statistical manner.

The covariant distribution function $f(x,p)$ is thus defined as

$$f(x,p) \equiv \langle R(x,p) \rangle, \tag{1.11}$$

where the average value $\langle \cdots \rangle$ is taken over the initial data, whatever they might be,[13] so that, by construction, it allows the calculation of any kind of average values of observable quantities whatsoever.

Therefore, it appears that the one-particle relativistic phase space, or μ space, is formally the eight-dimensional space subtended by $(x,\, p)$. As a matter of fact, the momentum p is generally constrained by a mass shell condition of the type $p^2 = m^2$ or by any other, such as

$$[p - eA(x)]^2 = m^2, \tag{1.12}$$

when one is dealing with a charged system embedded in an electromagnetic four-potential A^μ.

Let $A^{\cdots}_{\cdots}(x,p)$ be a tensor observable connected to the particles; its space–time density is given by

$$A^{\mu\cdots}_{\cdots}(x) = \int d^4p \frac{p^\mu}{m} A^{\cdots}_{\cdots}(x,p) f(x,p), \tag{1.13}$$

where the global quantity of $A^{\cdots}_{\cdots}(x,p)$ in the system is given by

$$A^{\cdots}_{\cdots} = \int_\Sigma \int d\Sigma_\mu d^4p \frac{p^\mu}{m} A^{\cdots}_{\cdots}(x,p) f(x,p), \tag{1.14}$$

where Σ is an arbitrary spacelike three-surface, i.e. $A^{\cdots}_{\cdots}$ is the flux of the four-current $A^{\mu\cdots}_{\cdots}(x)$ through Σ. In general, the average value of A depends on Σ; the only case where it is independent of Σ is the one where

$$\partial_\mu A^{\mu\cdots}_{\cdots}(x) = 0. \tag{1.15}$$

[13] In the classical relativistic context of the so-called action-at-a-distance formalism of interacting particles, the initial value problem is not yet solved and the initial data necessary for determining completely the future of the system might consist of the initial positions and velocities of the particles *and* some part of the trajectories in the past.

As an example, the entropy of the system is given by

$$S = \int_\Sigma d\Sigma_\mu \, S^\mu(x), \tag{1.16}$$

where $S^\mu(x)$ is the entropy four-current

$$S^\mu(x) = -k_B \int d^4p \frac{p^\mu}{m} f(x,p) \log f(x,p), \tag{1.17}$$

while the entropy invariant density is simply $u \cdot S$, where u^μ is the average four-vector that defines the rest frame of the gas.

From what has been discussed above, the normalization of the covariant distribution function reads

$$\begin{aligned}\int_\Sigma \int d\Sigma_\mu d^4p \frac{p^\mu}{m} f(x,p) &\equiv \int_\Sigma d\Sigma_\mu J^\mu(x) \\ &= N\end{aligned} \tag{1.18}$$

when there are N particles in the system.[14] In other words, the flux of the four-current through an arbitrary spacelike three-surface defines the normalization of the distribution function: there are as many intersections of world lines with Σ as particles in the system. In the above equation $d\Sigma_\mu$ is the differential form

$$d\Sigma_\mu = \frac{1}{3!}\varepsilon_{\mu\nu\alpha\beta}\, dx^\nu \times dx^\alpha \times dx^\beta, \tag{1.19}$$

the *surface element* on Σ. Note that, owing to the mass shell condition $p^2 = m^2$, the integration element d^4p in μ space actually reduces to a three-dimensional one,[15]

$$d^4p \to m\frac{d^3p}{p_0}, \tag{1.20}$$

where the factor m has been added so that the integration element has the dimension of a mass cube, as usual. Also of use is the variable $v = p/m$, whose integration element is just d^3v/v_0. Finally, it appears that the integration extends over a six-dimensional μ space,

$$\Sigma(x)\{p^2 = m^2\}, \tag{1.21}$$

as in the Newtonian case. Whether this last six-dimensional phase space or the covariant eight-dimensional one is called "phase space" is only a matter of definition.

---

[14]Instead of N, the normalization is often chosen to be 1, in order for f to be a probability.

[15]The use of d^4p is generally more convenient; however, it can be a source of confusion if one is not cautious enough [see e.g. B. Kursunoglu (1967)].

For an infinite system, the normalization of $f(x, p)$ occurs via the four-current or the local $n(x)$ density

$$n_{\text{eq}}(x) = [J^\mu(x) \cdot J_\mu(x)]^{1/2}, \tag{1.22}$$

i.e. via its definition or, equivalently, as

$$n_{\text{eq}}(x) = \int d^4p u_\mu(x) \frac{p^\mu}{m} f(x, p), \tag{1.23}$$

with $u^\mu(x) \equiv J^\mu(x)/n_{\text{eq}}(x)$ the average four-velocity of the system.

1.1.1. *The phase space "volume element"*

When one considers the six-dimensional phase space $\Sigma \times \mu$, its invariant "volume element" is given by

$$d\Omega(x, p) = d\Sigma_\mu(x) \times d\Sigma^\mu(p), \tag{1.24}$$

where

$$d\Sigma_\mu(p) = \frac{1}{3!} \varepsilon_{\mu\nu\alpha\beta}\, dp^\nu \times dp^\alpha \times dp^\beta \tag{1.25}$$

is the differential form "element of the three-surface." The above element of integration on phase space is, of course, written in an obvious system of coordinates adapted to its structure as a product of two three-surfaces. Let us briefly calculate $d\Sigma_\mu(p)$ restricted to the hyperboloid $p^2 = m^2$, and let us choose the coordinate system of $\{p^i\}_{i=1,2,3}$ so that $p^0 = \sqrt{\mathbf{p}^2 + m^2}$. For instance, the zeroth component of $d\Sigma_0$ immediately yields

$$d\Sigma_0 = dp^1 \times dp^2 \times dp^3 = p_0 \frac{d^3p}{p_0} \tag{1.26}$$

and, finally, one recovers

$$d\Sigma_\mu = p_\mu \frac{d^3p}{p_0}. \tag{1.27}$$

The volume element is sometimes taken to be truly d^4p and the constraint $p^2 = m^2$ occurs either explicitly,

$$d^4p 2\theta(p^0)\delta(p^2 - m^2) f(p), \tag{1.28}$$

or implicitly in the distribution function. In any case, care must always be taken when dealing with either the integration element or the distribution function: is the mass shell restriction included in the former or the latter? Or is it explicit?

1.2. The Jüttner–Synge Equilibrium Distribution

The relativistic Maxwell–Boltzmann distribution function, hereafter called the *Jüttner–Synge distribution*, was first derived by F. Jüttner in 1911 and studied in detail by J.L. Synge (1957). It can be derived in numerous possible ways: by noting that the Boltzmann factor $\exp(-\beta E)$ can be obtained from thermodynamic considerations, independently of relativity theory, and hence it is sufficient to replace E by its relativistic expression and to normalize the result; by maximizing the entropy of the system while taking account of the constraints provided by the average energy and the number of particles within the system; by solving the covariant Boltzmann equation [W. Israel (1963)]; by using a covariant formulation for the passage of a microcanonical ensemble to a canonical one [R. Hakim (1973)], as first shown by A.I. Khinchin (1956) in the nonrelativistic domain; etc.

First, the Jüttner–Synge distribution is briefly derived by maximizing the free energy of the system,

$$F = U - TS, \tag{1.29}$$

while the number (N) of particles is kept conserved; equivalently, the same can be done for densities

$$\begin{aligned} \beta F &= \beta\rho - s \equiv \beta\rho - u_\mu S^\mu, \\ \beta &\equiv (k_B T)^{-1}, \end{aligned} \tag{1.30}$$

or

$$\delta F = 0, \quad \delta N = 0. \tag{1.31}$$

Therefore, one has to maximize the free energy

$$\begin{cases} \delta(\beta F) = \delta \displaystyle\int \frac{d^3p}{p_0}\{\beta(p\cdot u)^2 - (p\cdot u)\log f_{\mathrm{eq}}(p)\} f_{\mathrm{eq}}(p) = 0, \\ \delta n_{\mathrm{eq}} = \delta \displaystyle\int \frac{d^3p}{p_0}(p\cdot u) f_{\mathrm{eq}}(p) = 0, \end{cases} \tag{1.32}$$

while N is conserved.

Introducing a Lagrange multiplier α for the constraint on N, one has

$$\delta \int \frac{d^3p}{p_0}\{\beta(p\cdot u)^2 - k_B(p\cdot u)\log f_{\mathrm{eq}}(p) + \alpha(p\cdot u)\} f_{\mathrm{eq}}(p) = 0, \tag{1.33}$$

(where k_B is the Boltzmann constant) from which one is immediately led to the following form for the equilibrium distribution function,

$$f_{\mathrm{eq}}(p) = A\exp(-\beta u^\mu p_\mu) \tag{1.34}$$

(with $p_0 \equiv \sqrt{\mathbf{p}^2 + m^2}$), where A is directly connected to the Lagrange multiplier; it is determined by the normalization condition. One gets successively

$$J^\mu \equiv n_{\rm eq} u^\mu = \int_{\substack{p^2 - m^2 = 0 \\ p^0 > 0}} m \frac{d^3 p}{p_0} \left(\frac{p^\mu}{m}\right) A \exp(-\beta u^\mu p_\mu)$$

$$= -A \frac{\partial}{\partial(\beta u_\mu)} \int_{\substack{p^2 - m^2 = 0 \\ p^0 > 0}} m \frac{d^3 p}{p_0} \exp(-\beta u^\mu p_\mu), \qquad (1.35)$$

where the "generating function" [see J.L. Synge (1957)] $A\Phi(m\beta)$, for the moments of $f_{\rm eq}(p)$, is defined by

$$\Phi(m\beta) = \int_{\substack{p^2 - m^2 = 0 \\ p^0 > 0}} m \frac{d^3 p}{p_0} \exp(-\beta u^\mu p_\mu), \qquad (1.36)$$

and is explicitly given by

$$\Phi(m\beta) = 4\pi m^3 \int_{\substack{p^2 - m^2 = 0 \\ p^0 > 0}} d\chi \, sh^2\chi \exp(-\beta m ch\chi), \qquad (1.37)$$

where use has been made of the *relativistic polar coordinates*

$$\begin{cases} p^1 = m sh\chi \sin\theta \cos\varphi, \\ p^2 = m sh\chi \sin\theta \cos\varphi, \\ p^3 = m sh\chi \cos\theta, \\ p^0 = m ch\chi. \end{cases} \qquad (1.38)$$

Finally, $\Phi(m\beta)$ turns out to be

$$\Phi(m\beta) = 4\pi m \frac{K_1(m\beta)}{\beta}, \qquad (1.39)$$

where the Kelvin functions[16] $K_n(\xi)$ are defined by

$$K_n(\xi) = \left(\frac{\xi}{2}\right)^n \frac{\Gamma\left(\frac{1}{2}\right)}{\Gamma\left(n + \frac{1}{2}\right)} \int_0^\infty d\chi \exp(-\xi ch\chi) \sinh^{2n}\chi$$

$$= \frac{(2\xi)^n n!}{(2n)!} \int_0^\infty d\chi \exp(-\xi ch\chi) \sinh^{2n}\chi \qquad (1.40)$$

$$= \int_0^\infty d\chi \exp(-\xi ch\chi) \cosh(n\chi).$$

[16] See Abramovitz and Stegun, *Handbook of Mathematical Functions* (Dover, New York, 1965).

From the generating equation $\Phi(m\beta)$, and the recurrence relations obeyed by K_n and their derivatives (see App. A), one obtains

$$A = \frac{n_{\text{eq}}\beta}{4\pi m^2 K_2(m\beta)}, \tag{1.41}$$

which is connected to the chemical potential μ through

$$A = \frac{n_{\text{eq}}\beta}{4\pi m^2 K_2(m\beta)} = \exp(\beta\mu); \tag{1.42}$$

this property can be seen by calculating the various terms of the thermodynamic relation[17]

$$s = \frac{\rho - \mu n_{\text{eq}}}{T} + k_B n_{\text{eq}}, \tag{1.43}$$

where k_B is the usual Boltzmann constant.

For the energy–momentum tensor, one obtains

$$\begin{aligned} T^{\mu\upsilon} &= A\frac{\partial^2}{\partial\xi_\mu \partial\xi_\nu}\Phi(m\beta) \\ &= \left[n_{\text{eq}} m \frac{K_3(m\beta)}{K_2(m\beta)} + \frac{n_{\text{eq}}}{\beta}\right] u^\mu u^\upsilon - \frac{n_{\text{eq}}}{\beta}\eta^{\mu\upsilon}. \end{aligned} \tag{1.44}$$

An alternative form of $T^{\mu\nu}$ can be obtained with the recursion relations obeyed by the Kelvin functions (see App. A) and reads

$$T^{\mu\nu} = \left\{ m n_{\text{eq}} \frac{K_1(\beta m)}{K_2(\beta m)} + \frac{4 n_{\text{eq}}}{\beta} \right\} u^\mu u^\nu - \frac{n_{\text{eq}}}{\beta}\eta^{\mu\nu}. \tag{1.45}$$

The Lagrange multiplier β is determined from the equation of state of the relativistic gas. A comparison of the energy–momentum tensor, which has the so-called *perfect fluid* form[18]

$$T^{\mu\upsilon} = (\rho + P)u^\mu u^\upsilon - P\eta^{\mu\upsilon} \tag{1.46}$$

(see Chap. 2) finally yields $P\beta = n_{\text{eq}}$, which is nothing but the perfect gas equation of state and hence this terminates the identification of β with $1/k_B T$(k_B is the usual Boltzmann constant). In this last equation ρ is the (invariant) energy density of the system and P is its pressure.

[17]See the details in S.R. de Groot, W.A. van Leeuwen and Ch. G. van Weert (1980).
[18]This means that the energy–momentum tensor does not contain any dissipation term which would introduce gradients of some macroscopic quantities, such as the average four-velocity or the temperature.

1.2.1. *Thermodynamics of the Jüttner–Synge gas*[19]

The covariant form of the first law of thermodynamics reads

$$\beta_\nu T^{\mu\nu} = S^\mu + \alpha J^\mu, \tag{1.47}$$

which can be rewritten, after multiplying by u^μ, as

$$\beta d\rho = ds + \alpha dn_{\text{eq}} \tag{1.48}$$

or, through multiplying by an arbitrary volume V,

$$dU = dS + \alpha dN. \tag{1.49}$$

The identification of $T^{\mu\nu}$ with its perfect fluid form, or a direct calculation, provides

$$\rho = mn_{\text{eq}}\frac{K_1(\beta m)}{K_2(\beta m)} + \frac{3n_{\text{eq}}}{\beta} = n_{\text{eq}}m\frac{K_3(m\beta)}{K_2(m\beta)} \quad \text{(energy density)}, \tag{1.50}$$

$$h = \rho + P = n_{\text{eq}}m\frac{K_3(m\beta)}{K_2(m\beta)} + \frac{n_{\text{eq}}}{\beta} \quad \text{(density of enthalpy per particle).} \tag{1.51}$$

Their first relativistic corrections are given by the asymptotic forms of the Kelvin functions (see App. A), namely

$$\begin{cases} \rho = n_{\text{eq}}\left\{m + \dfrac{3}{2}k_BT + \dfrac{15}{8}\dfrac{(k_BT)^2}{m} + \cdots\right\}, \\ h = n_{\text{eq}}\left\{m + \dfrac{5}{2}k_BT + \dfrac{15}{8}\dfrac{(k_BT)^2}{m} + \cdots\right\}. \end{cases} \tag{1.52}$$

As expected, these expressions contain the rest energy of a generic particle. The limit $\beta m \gg 1$ of the Jüttner–Synge function can easily be shown to be the ordinary Maxwell–Boltzmann distribution [J.L. Synge (1957)] with the help of the asymptotic formula given in App. A.

$$K_n(\xi) = \left(\frac{\pi}{2\xi}\right)^{1/2} \exp(-\xi)\left\{1 + \frac{4n^2-1}{1!8\xi} + \frac{(4n^2-1^2)(4n^2-3^2)}{2!(8\xi)^2} + \cdots\right\}. \tag{1.53}$$

[19] See J.L. Synge (1957), W. Israel (1976, 1981), or S.R. de Groot, W.A. van Leeuwen and Ch.G. van Weert (1980).

In such a case, of course, the rest mass contribution is eliminated *ipso facto*. From these two quantities, ρ and h, one obtains the heat capacity at constant volume and pressure through

$$\begin{cases} C_V = -\beta^2 k_B^{-1} \dfrac{\partial}{\partial \beta}\left(\dfrac{\rho}{n_{\text{eq}}}\right)\Bigg|_V, \\ C_P = -\beta^2 k_B^{-1} \dfrac{\partial}{\partial \beta}\left(\dfrac{h}{n_{\text{eq}}}\right)\Bigg|_P, \end{cases} \tag{1.54}$$

which provides the *adiabatic index* $\gamma \equiv C_P/C_V$ of the Jüttner–Synge gas through

$$\frac{\gamma}{\gamma - 1} = (\beta m)^2 + 5\beta h - (\beta h)^2. \tag{1.55}$$

From the above expansions of ρ and h, the relativistic corrections to the adiabatic index are obtained as

$$\gamma = \frac{5}{3} - \frac{5}{3}\frac{1}{\beta} + \cdots . \tag{1.56}$$

The adiabatic index plays an important role in problems of stability concerning various types of stars.

1.2.2. *Thermal velocity*

In nonrelativistic physics, the average thermal velocity of a generic particle of an ordinary Maxwellian gas is given by

$$v_{\text{th}} = \sqrt{3\frac{k_B T}{m}}, \tag{1.57}$$

and, as a matter of fact, it is often used in the relativistic context. However, J.L. Synge (1957) considers the most probable speed of a relativistic ideal gas, which appears to be a solution to the equation

$$9v^6 + [(\beta m)^2 + 3]v^4 - 8v^2 + 4 = 0; \tag{1.58}$$

when βm is close to zero, the equilibrium distribution possesses a sharp maximum so that the most probable speed is close to the thermal velocity. In this case, J.L. Synge gives

$$v^2 \approx 1 - \frac{(\beta m)^2}{25}, \tag{1.59}$$

which shows that the relativistic thermal velocity is quite different from the Newtonian one. It might seem that it would be sufficient to take the

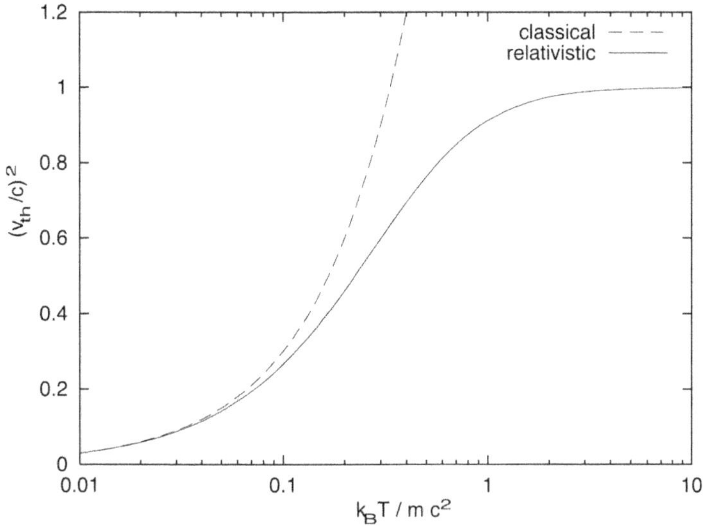

Fig. 1.1 The relativistic thermal velocity compared to the classical one. (Calculation by Dr. L. Mornas.)

relativistic average value of v^2 to obtain a coherent definition for the thermal velocity, i.e.

$$v_{\text{th}}^2 = \frac{\beta m}{K_2(\beta m)} \int d\chi \exp(-\beta m \cosh\chi) \frac{\sinh^4\chi}{\cosh\chi}. \tag{1.60}$$

However, such a definition does not involve the usual energy content included in the classical definition; this occurs because of the different relationship between energy and velocity.

In order to obtain a thermal velocity with the same energy content as in the nonrelativistic case, the following equality is considered as a definition of v_{th}:

$$\frac{m}{\sqrt{1 - v_{\text{th}}^2}} \underset{\text{def}}{\equiv} \langle E \rangle, \tag{1.61}$$

or

$$v_{\text{th}}^2 = \frac{\langle E \rangle^2 - m^2}{\langle E \rangle^2} \tag{1.62}$$

(see Fig. 1.1).

The expression for $\langle E \rangle$ can be obtained from the energy–momentum tensor, or from ρ via

$$\langle E \rangle = n_{\text{eq}}^{-1} \rho, \tag{1.63}$$

and one obtains (see also J.L. Synge (1957))

$$\langle E\rangle = m\left\{\frac{K_1(m\beta)}{K_2(m\beta)} + \frac{3}{m\beta}\right\} \tag{1.64}$$

so that

$$v_{\text{th}}^2 = 1 - \left\{\frac{K_1(m\beta)}{K_2(m\beta)} + \frac{3}{m\beta}\right\}^{-2}. \tag{1.65}$$

For large values of $m\beta$ (low temperature case) one recovers the classical value, while for small $m\beta$ (ultrarelativistic case) one obtains

$$v^2 \approx 1 - \frac{1}{9}(\beta m)^2, \tag{1.66}$$

which is, as expected, of the same order of magnitude as J.L. Synge's most probable speed.

1.2.3. *Moments of the Jüttner–Synge function*

When one is dealing with dissipative phenomena, a hierarchy of moments can be obtained from a relativistic kinetic equation (see Chap. 2) and their explicit form generally depends on the first moments of the equilibrium distribution [see e.g. S.S. Moiseev (1960), for the case of rarefied gases]. The various moments of the Jüttner–Synge distribution are obtained, as mentioned above, from the function $\Phi(m\beta)$ and are particularly useful in some approximation schemes employed in obtaining solutions to the relativistic Boltzmann equation or to other kinetic equations. Accordingly, the first few moments are explicitly given here.

They are given by

$$M_0 = \int \frac{d^3p}{p_0} f_{\text{eq}}(p) = 4\pi m A \frac{K_1(m\beta)}{\beta}, \tag{1.67}$$

$$M_1^\mu = \int \frac{d^3p}{p_0} p^\mu f_{\text{eq}}(p) = 4\pi m^2 A \frac{K_2(m\beta)}{\beta} u^\mu, \tag{1.68}$$

$$\begin{aligned} M_2^{\mu\nu} &= \int \frac{d^3p}{p_0} p^\mu p^\nu f_{\text{eq}}(p) \\ &= \frac{4\pi m^3}{\beta} A\left[K_3(m\beta)u^\mu u^\nu - \frac{1}{m\beta}K_2(m\beta)\eta^{\mu\nu}\right], \end{aligned} \tag{1.69}$$

$$M_3^{\mu\nu\alpha} = \int \frac{d^3p}{p_0} p^\mu p^\nu p^\alpha f_{\text{eq}}(p)$$
$$= \frac{4\pi m^4}{\beta} A \left[K_4(m\beta) u^\mu u^\nu u^\alpha - \frac{K_3(m\beta)}{m\beta} u^{(\mu} \eta^{\nu\alpha)} \right], \tag{1.70}$$

$$M_4^{\mu\nu\lambda\beta} = \int \frac{d^3p}{p_0} p^\mu p^\nu p^\alpha p^\beta f_{\text{eq}}(p)$$
$$= \frac{4\pi m^5}{\beta} A \left[K_5(m\beta) u^\mu u^\nu u^\alpha u^\beta - \frac{K_4(m\beta)}{m\beta} u^{(\mu} u^\nu \eta^{\alpha\beta)} + \frac{K_3(m\beta)}{(m\beta)^2} \eta^{(\mu\nu} \eta^{\alpha\beta)} \right], \tag{1.71}$$

$$M_5^{\mu\nu\lambda\beta\lambda} = \int \frac{d^3p}{p_0} p^\mu p^\nu p^\alpha p^\beta p^\lambda f_{\text{eq}}(p)$$
$$= \frac{4\pi m^6}{\beta} A \left[K_6(m\beta) u^\mu u^\nu u^\alpha u^\beta u^\lambda - \frac{K_5(m\beta)}{m\beta} u^{(\mu} u^\nu u^\alpha \eta^{\beta\lambda)} + \frac{K_4(m\beta)}{(m\beta)^2} u^{(\mu} \eta^{\nu\alpha} \eta^{\beta\gamma)} \right]. \tag{1.72}$$

In these expressions use has been made of the conventional symmetrization notations on the indices.

1.2.4. *Orthogonal polynomials*

When one is dealing with off-equilibrium processes, the distribution function has to be approximated in some way. For instance, in the non-relativistic case, the distribution function is often expanded — and next truncated at some order — as

$$f = f_{\text{eq}} \sum_{n=0}^{\infty} a_n H_n, \tag{1.73}$$

where f_{eq} is the usual Maxwell–Boltzmann distribution and H_n are the orthogonal polynomials associated with the weight defined by this function, i.e. Hermite's polynomials.

It would therefore appear to be desirable to find the family of those polynomials that are orthogonal with respect to the weight defined by the Jüttner–Synge function. Remember that this function is defined on the mass hyperboloid $p^2 = m^2$.

The first polynomials to be used in the solution of the relativistic Boltzmann equation were actually neither polynomials nor even orthogonal [N.A. Chernikov (1963, 1964)]. D.C. Kelly (1968, 1969) and Ch. Marle (1969) studied the first orthogonal polynomials from a mathematical point of view and constructed them with the well-known Schmidt orthogonalization procedure. However, these were not very easy to express explicitly except, of course, the first few ones, which are given below. Their general form was improved by J. Stewart (1971) and J.L. Anderson (1974), and finally given more specific forms by J.C. Lucquiaud (1986) on the basis of group-theoretical arguments. Finally, an improved version is presented in the book by S.R. de Groot, W.A. van Leeuwen and Ch. G. van Weert (1980). Here the first few orthogonal polynomials are defined and given and, as a matter of fact, they are those which are actually used in practice.

A general distribution function[20] f is expanded as

$$f(p) = f_{\text{eq}}(p) \sum_{n=0}^{\infty} a_{(n)\mu_1\mu_2\ldots\mu_n} H_n^{\mu_1\mu_2\ldots\mu_n}(p), \tag{1.74}$$

where H_n are mutually orthogonal:

$$\frac{1}{n_{\text{eq}}} \int \frac{d^3p}{p_0} f_{\text{eq}}(p) H_n^{\mu_1\mu_2\ldots\mu_n}(p) H_\ell^{\mu_1\mu_2\ldots\mu_\ell}(p) = 0, \tag{1.75}$$

with $n \neq \ell$.

Ch. Marle (1969) proved the following properties for H_n:

(i) These polynomials are symmetric in the indices $\{\mu_1, \mu_2, \ldots, \mu_n\}$; they obey the following relations:
(ii) $\eta_{\mu_1\mu_2} H_n^{\mu_1\mu_2\ldots\mu_n}(p) = 0$ for $p \geq 2$.
(iii) They form a complete system for those functions $g(p)$ such that $g(p)\exp\left(-\frac{1}{2}\beta u \cdot p\right)$ is square-integrable.

He gave the first few polynomials as

$$H_0 = 1, H_1^\mu(p) = p^\mu - \frac{K_2(m\beta)}{K_1(m\beta)} u^\mu, \tag{1.76}$$

$$H_2^{\mu\nu}(p) = p^\mu p^\nu - C_\lambda^{\mu\nu} H_1^\lambda(p) - C^{\mu\nu}, \tag{1.77}$$

[20]We have omitted the x dependence of f and possibly of the *local* distribution f_{eq}.

with

$$C^{\mu\nu} = \frac{K_3(m\beta)}{K_1(m\beta)} u^\mu u^\nu - \frac{K_2(m\beta)}{m\beta K_1(m\beta)} \eta^{\mu\nu}, \tag{1.78}$$

$$\begin{aligned} C_\lambda^{\mu\nu} = &-\frac{(\beta m)^2}{K_1(m\beta)} [\eta_{\lambda\alpha} + Y(m\beta) u_\lambda u_\alpha] \\ &\times \left\{ \left(\frac{K_4(m\beta)}{m\beta} - \frac{K_2(m\beta) K_3(m\beta)}{m\beta K_1(m\beta)} \right) u^\mu u^\nu u^\alpha \right. \\ &- \left(\frac{K_3(m\beta)}{(m\beta)^2} - \frac{[K_2(m\beta)]^2}{(m\beta)^2 K_1(m\beta)} \right) \eta^{\mu\nu} u^\alpha \\ &\left. - \frac{K_3(m\beta)}{(m\beta)^2} (\eta^{\mu\alpha} u^\nu + \eta^{\nu\alpha} u^\mu) \right\}, \end{aligned} \tag{1.79}$$

where the function $Y(m\beta)$ is defined by

$$[1 + Y(m\beta)]^{-1} = 1 + m\beta \left\{ \frac{K_2(m\beta)}{K_1(m\beta)} - \frac{K_3(m\beta)}{K_2(m\beta)} \right\}. \tag{1.80}$$

1.2.5. *Zero mass particles*

Let us start from the energy–momentum tensor of an ideal gas composed of zero mass particles:

$$T^{\mu\nu} = \left\{ m n_{\text{eq}} \frac{K_1(\beta m)}{K_2(\beta m)} + \frac{4 n_{\text{eq}}}{\beta} \right\} u^\mu u^\nu - \frac{n_{\text{eq}}}{\beta} \eta^{\mu\nu}, \tag{1.81}$$

and let m tends to zero. Using the properties of Kelvin's functions for small arguments (see App. A), one obtains

$$\lim_{m \to 0} T_\mu^\mu = 0, \tag{1.82}$$

as one could have expected for photons for instance, and which indicates the usual equation of state for massless particles, i.e.

$$P = \frac{1}{3} \rho, \tag{1.83}$$

so that the sound velocity[21] for such a gas is $c/\sqrt{3}$.

Similarly, when one starts from the expression of the four-current and let βm tend to zero, the normalization coefficient of the Jüttner–Synge function is found to be

$$A = \frac{n\beta^3}{8\pi m^2}, \tag{1.84}$$

---

[21] See Chap. 2.

i.e. m tends to zero and n to infinity so that A remains finite; while the integration element becomes

$$\frac{d^3p}{p_0} \to \frac{d^3p}{|\mathbf{p}|}. \tag{1.85}$$

Finally, all that is needed for the description of classical zero mass particles in thermal equilibrium is available. As an example, Stefan's law is derived as follows. The energy density of such a gas is given by

$$\begin{aligned} \rho = u_\mu u_\nu T^{\mu\nu} &= \int \frac{d^3p}{p}(p\cdot u)^2 A\exp(-\beta p\cdot u) \\ &\beta^{-4} = T^4, \end{aligned} \tag{1.86}$$

where the proportionality to T^4 appears after the elementary change $p \to \beta p$.

1.3. From the Microcanonical Distribution to the Jüttner–Synge One[22]

Now the Jüttner–Synge function will be derived from the free microcanonical distribution. Why from the "free" one and not from interaction? There are several reasons. The first one is that it is not derived in the nonrelativistic case, except for weak interactions. Next, in the relativistic case, the question of interaction is completely different from the nonrelativistic and certainly much more complex. Finally, the noninteracting case is sufficiently instructive as such.

The relativistic form of the microcanonical model for N free particles endowed with a total energy–momentum P^μ is

$$\begin{aligned} &f^{(N)}_{\mathrm{micro}}(P^\mu, \{p_i^\mu\}) \\ &\quad = \mathrm{const}\ \delta\left(P^\mu - \sum_{i=1}^{i=N} p_i^\mu\right)\prod 2m_i\theta(p_i^0)\delta(p_i^\lambda p_{i\lambda} - m_i^2), \end{aligned} \tag{1.87}$$

where $\theta(p^0)$ is the Heaviside step function and where the normalization constant depends (i) on the number of particles, (ii) on the *total* energy–momentum P^μ of the gas and (iii) on the spatial volume occupied. Note that the relativistic microcanonical model has the same content as the classical one: the particles lie on the total energy and momentum of the whole system

[22]It might be useful to read Chap. 5 first.

of particles, which obeys $P \cdot P = M^2$. Also, note that $f_{\text{micro}}^{(N)}(P^\mu, \{p_i^\mu\})$ is normalized through

$$\int \prod_{i=1}^{i=N} d^4 p_i \frac{p_i^\mu}{m_i} f_{\text{micro}}^{(N)}(P^\mu, \{p_i^\mu\}) \equiv J^{\mu_1 \mu_2 \dots \mu_N}$$
$$= N \left(\frac{N}{V}\right)^{2N} \frac{P^{\mu_1} P^{\mu_2} \dots P^{\mu_N}}{M^N}. \tag{1.88}$$

However, instead of employing $f_{\text{micro}}^{(N)}(P^\mu, \{p_i^\mu\})$, we shall use a probability, more adapted to the use of the *central limit theorem*;[23] it is, therefore, a true density of probability in energy–momentum space.

Let us, however, mention that the central limit theorem, in its simplest form, is:

Central limit theorem. Let $\{X_i\}_{i=1,2,\dots,N}$be random independent variables of densities $\{f(x_i)\}_{i=1,2,\dots,N}$. Then the law of

$$X = \frac{X_1 + X_2 + \cdots + X_N}{N}$$

is that of a Gaussian:

$$f_\Sigma(x) = \frac{1}{\sqrt{2\pi}} \exp\left(-\frac{1}{2}x^2\right) + O\left(\frac{1}{N}\right).$$

This theorem is valid for conditions that are valid in most ordinary cases and has a number of important applications. Among the assumptions necessary for the central limit theorem, one finds:

- existence of the first two moments of the distribution function;
- existence of the characteristic function (i.e. Fourier existence) of the distribution function;

and for many other cases definite complex data $f_{\text{micro}}^{(N)}(P^\mu, \{p_i^\mu\})$ reads

$$f_{\text{micro}}^{(N)}(x|P^\lambda, \{p_i^\lambda\}) = f_{\text{micro}}^{(N)}(P^\mu, \{p_i^\mu\}) \times \prod_{i=1}^{i=N} \frac{P \cdot p_i}{M m_i}. \tag{1.89}$$

However, rather than employing $f_{\text{micro}}^{(N)}(P^\mu, \{p_i^\mu\})$, we shall use a probability, more adapted to the use of the *central limit theorem*;[24] it is thus a true density of probability in energy–momentum space.

[23] A.I. Khinchin, *Mathematical Foundation of Information Theory* (Dover, New York, 1957).
[24] A.I. Khinchin, *loc. cit.*

It follows that the one-particle distribution function is

$$f(x|p) = \int \prod_{i=1}^{i=N} d^4p_i\, \delta(p^\mu - p_i^\mu) f_{\text{micro}}^{(N)}(x^\mu | P^\mu, \{p_i^\mu\}), \tag{1.90}$$

where we have made clear in the notation that the local distribution is also a conditional one. A simple calculation provides

$$f(x|p) = \frac{\Omega_{N-1}(\sqrt{[P^\lambda - p^\lambda]^2})}{\Omega_N(M)} \frac{P^\mu p_\mu}{Mm}, \tag{1.91}$$

where

$$\Omega_N(M) = \int \delta\left(P^\lambda - \sum_{i=1}^{i=N} p_i^\lambda\right) \prod_{i=1}^{i=B} 2m_i \theta(p_i^0) \delta(p_i^2 - m_i^2) \frac{P^\mu p_{i\mu}}{Mm_i} d^4p_i. \tag{1.92}$$

An expression similar to $\Omega_N(M)$ has been evaluated by F. Lurçat and P. Mazur (1964) employing the central limit theorem, which we shall use now. Note also that their expression differs from ours by the absence of the term

$$\prod_{i=1}^{i=N} \frac{P^\mu p_{i\mu}}{Mm_i}. \tag{1.93}$$

We now apply the central limit theorem to

$$P^\mu = \frac{p^{\mu_1} + p^{\mu_2} + \cdots + p^{\mu_N}}{N}; \tag{1.94}$$

the probability density is

$$g_N(\beta, P) = \frac{1}{\phi_N(\beta)} \Omega_N(P) \exp(-\beta \cdot P) \tag{1.95}$$

and, as a result, tends toward a Gaussian distribution:

$$g_N(\beta, P) = \frac{1}{(2\pi)^2} \frac{1}{(\text{Det} B_N)^{1/2}} \exp\left[-\frac{1}{2}(P - A_N)_\mu B_N^{\mu\nu} (P - A_N)_\nu\right] + O\left(\frac{1}{N}\right), \tag{1.96}$$

where A is the average value of P and B_N is its dispersion matrix. The calculation of these first two moments is easy, since they are the first two derivatives of $\log \phi_N(\beta)$. One finds that

$$A_N^\mu = -\frac{\partial \log \phi_N(\beta)}{\partial \beta}\beta^\mu, \tag{1.97}$$

$$B_N^{\mu\nu}(\beta) = \frac{\partial^2 \log \phi_N(\beta)}{\partial \beta^2}u^\mu u^\nu + \frac{1}{\beta}\frac{\partial \log \phi_N(\beta)}{\partial \beta}\Delta^{\mu\nu}(u), \tag{1.98}$$

where u^μ is a unit four-vector to be derived elsewhere but which is also parallel to β^μ. In our case, it turns out that, the generating function is

$$\begin{aligned}\phi_N(\beta) &\underset{\text{def}}{\equiv} \int d^4P\,\Omega_N(P)\exp(-\beta^\mu P_\mu)\\ &= \left(\frac{1}{Mm}\right)^N \left(\frac{\partial^2}{\partial \alpha^2} + \frac{3}{\alpha}\frac{\partial}{\partial \alpha}\right)\left(\int_0^\infty d^4P\frac{4\pi m^2}{\alpha}K_1(\alpha P)\right)^N,\end{aligned} \tag{1.99}$$

with $\alpha^2 = \alpha^\lambda \alpha_\lambda$.

However, the important point for the derivation of the Jüttner–Synge distribution is the *form* of $\Omega_N(M)$ and, more particularly, the fact that

$$\Omega_N(M) \approx \exp(\beta M). \tag{1.100}$$

Inserting now this form into the expression for $f(x|p)$, we find that

$$f(x|p) = L(\beta, M)\exp\Big(-\beta M + [M^2 + m^2 - 2P^\mu p_\mu]^{1/2}\Big)\frac{P\cdot p}{Mm} \tag{1.101}$$

and, since $N \gg 1$, $M \gg m$, we have

$$f(x|p) = L(\beta, M)\frac{P\cdot p}{Mm}\exp\left(-\frac{\beta P^\mu p_\mu}{M}\right). \tag{1.102}$$

This last form is precisely the Jüttner–Synge function, except that $L(\beta, M)$ has to be determined: actually it could be determined by looking at the limit $N \gg 1$ and $M \gg m$; in fact, it can be determined simply by a normalization condition although it is actually furnished by the limiting form of $\Omega_N(M)$. Note that P^μ/M is precisely u^μ.

Finally, the actual form of the Jüttner–Synge function has been established but it could be derived more rigorously; this involves, however, quite lengthy calculations.

1.4. Equilibrium Fluctuations

In this section the four-current equilibrium fluctuations are calculated:

$$\begin{cases} \delta J^{\mu\nu}(x, x') = \langle \delta J^\mu(x)\delta J^\nu(x')\rangle \\ \qquad\qquad\quad \equiv \langle (J^\mu_{\text{micro}}(x) - \langle J^\mu(x)\rangle)(J^\nu_{\text{micro}}(x') - \langle J^\nu(x')\rangle)\rangle. \end{cases} \tag{1.103}$$

They are required in the use of the fluctuation–dissipation theorem in order to obtain, for instance, the modes of oscillation of a plasma.[25] Other kinds of fluctuations, like those of the energy–momentum tensor, can be calculated in a similar manner.

The starting point is the (random) four-current $J^\mu(x)$ of free particles: their trajectories are straight lines. The microscopic four-current then reads

$$J^\mu_{\text{micro}}(x) = \sum_i \int_{-\infty}^{+\infty} ds_i \frac{p^\mu}{m} \delta^{(4)} \left[x - x_i - \left(\frac{p_i}{m_i} \right) s_i \right] \tag{1.104}$$

between collisions, and its equilibrium average value is, of course, $\langle J^\mu(x) \rangle = nu^\mu$. In this last equation x_i and p_i/m_i are the initial four-positions and four-velocities of the particles of the system. With the definition of relativistic average values

$$A^{\cdots}_{\cdots} = \int_\Sigma \int d\Sigma_\mu \, d^4 p \frac{p^\mu}{m} A^{\cdots}_{\cdots}(x,p) f(x,p), \tag{1.105}$$

we get

$$\delta J^{\mu\nu}(x, x') = \sum_i \int_{-\infty}^{+\infty} ds_i' \, ds_i \int_\Sigma \int d\Sigma_\alpha \, d^4 p \frac{p^\alpha}{m} f_{\text{eq}}(p) \\ \times p_i^\mu p_i^\nu \delta^{(4)} \left[x - x_i - \left(\frac{p_i}{m_i} \right) s_i \right] \delta^{(4)} \left[x' - x_i - \left(\frac{p_i}{m_i} \right) s_i' \right] \tag{1.106}$$

or

$$\delta J^{\mu\nu}(x, x') = \sum_i \int_{-\infty}^{+\infty} ds_i' \, ds_i \int_\Sigma \int d\Sigma_\alpha \, d^4 p \frac{p^\alpha}{m} f_{\text{eq}}(p) \\ \times p_i^\mu p_i^\nu \delta^{(4)} \left[x - x' - \left(\frac{p_i}{m_i} \right) (s_i - s_i') \right] \delta^{(4)} \left[x' - x_i - \left(\frac{p_i}{m_i} \right) s_i' \right], \tag{1.107}$$

and with extensive use of the properties of the δ function and of the consequent relation

$$d\Sigma_\alpha \, dx^\alpha = d^4 x \tag{1.108}$$

and from the fact that

$$\frac{dx_i^\alpha}{ds_i} = \frac{p_i^\alpha}{m}, \tag{1.109}$$

we find that

$$\delta J^{\mu\nu}(x - x') = \int d\tau \int_{\substack{p^2 = m^2 \\ p^0 > 0}} d^4 p \frac{p^\mu p^\nu}{m^2} f_{\text{eq}}(p) \delta((x - x') - u\tau). \tag{1.110}$$

[25] See e.g. A.G. Sitenko, *Electromagnetic Fluctuations in Plasmas* (Wiley, New York, 1959); for the relativistic calculation see R. Hakim (1974).

The integrations are easily performed and we obtain

$$\delta J^{\mu\nu}(X) = \frac{m\beta n_{\rm eq}}{4\pi K_2(m\beta)} \frac{X^\mu X^\nu}{(X\cdot X)^{5/2}} \exp\left(-m\beta \frac{u\cdot X}{(X\cdot X)^{1/2}}\right), \tag{1.111}$$

where we have set $X \equiv x - x'$, in agreement with the expression given by A.G. Sitenko (1959) for the numerical density fluctuations. Obviously, this expression makes sense only when the separation X is timelike. However, it should be noted that, in general, for an arbitrary physical quantity, the situation is not necessarily of this type: a given phenomenon in the past can influence two events separated by a spacelike distance.

1.5. One-Particle Liouville Theorem

A kinetic equation is often needed in order to obtain the distribution function $f(x, p)$. The general scheme for a kinetic equation is constituted by three elements. First, the Liouville equation gives the general flow of the particles in phase space. It is then coupled to a second element, the collision term, which renders possible deviations from this general flow. Finally, collective effects, which affect both the flow in phase space and the collision term, must be taken into account. Several examples will be given in subsequent chapters.

Before studying the kinetic equations, we first indicate briefly how the one-particle Liouville theorem occurs in μ space. Since the number of particles in the system is assumed to be conserved,[26] the eight-current in μ space is necessarily conserved and its "continuity equation" then reads

$$\partial_\mu \left[\frac{dx^\mu}{d\tau} f(x,p)\right] + \frac{\partial}{\partial p^\mu}\left[\frac{dp^\mu}{d\tau} f(x,p)\right] = 0, \tag{1.112}$$

where τ is the proper time; or, equivalently, after it is noted that the "velocity" in this μ space is given by

$$\begin{aligned} u^\mu &\equiv \frac{dx^\mu}{d\tau} \quad \text{(4-velocity)}, \\ F^\mu(x,p) &= \frac{dp^\mu}{d\tau} \quad \text{(4-force)}, \end{aligned} \tag{1.113}$$

it reads

$$\left[u^\mu \partial_\mu + \frac{\partial}{\partial p^\mu}(F^\mu(x,p))\right] \cdot f(x,p) = 0. \tag{1.114}$$

[26]See e.g. Ch. Marle (1969) for the case of decaying or mutually transforming particles.

This is *not* the Liouville equation , which is

$$\frac{d}{d\tau}f(x,p) \equiv \left\{u^{\mu}\partial_{\mu} + F^{\mu}(x,p)\frac{\partial}{\partial p^{\mu}}\right\} f(x,p) = 0. \tag{1.115}$$

The Liouville equation is obeyed *only* by those four-forces that satisfy the condition

$$\frac{\partial}{\partial p^{\mu}}(F^{\mu}(x,p)) = 0. \tag{1.116}$$

For instance, this is the case of a system composed of charged particles submitted to an external electromagnetic field $F^{\mu\nu}$ where $F^{\mu} = (e/m)p_{\nu}F^{\mu\nu}$: indeed, one has

$$\frac{\partial}{\partial p^{\mu}}(F^{\mu}(x,p)) = \frac{e}{m}F^{\mu\nu}(x)\eta_{\mu\nu} = 0. \tag{1.117}$$

Note that when this condition is not satisfied, the (one-particle) Liouville theorem is no longer valid but should be considered as it is.

1.5.1. *Relativistic Liouville equation from the Hamiltonian equations of motion*

When the dynamical equations can be cast into a Hamiltonian form,

$$\begin{cases} \dfrac{dx^{\mu}}{d\tau} = \dfrac{\partial H(x,p)}{\partial p_{\mu}}, \\ \dfrac{dp_{\mu}}{d\tau} = -\dfrac{\partial H(x,p)}{\partial x^{\mu}}, \end{cases} \tag{1.118}$$

the relativistic Liouville equation is recovered as usual and reads[27]

$$\frac{\partial H(x,p)}{\partial p_{\mu}}\partial_{\mu}f(x,p) - \frac{\partial H(x,p)}{\partial x^{\mu}}\frac{\partial f(x,p)}{\partial p_{\mu}} \equiv \{H,f\} = 0, \tag{1.119}$$

where $\{H,f\}$ is the relativistic Poisson bracket. An example of such a Hamiltonian system is that of the charged particle embedded in an electromagnetic four-potential $A^{\mu}(x)$, for which one has

$$H(x,p) = \frac{[p - eA(x)]^2}{2m}. \tag{1.120}$$

It should be emphasized that although the equations of motion can be *formally* recovered, this Hamiltonian is purely "technical" and has not the

[27]G. Kalman, *Phys. Rev.* **123**, 384 (1961); G. Schay Jr., *Nuovo Cimento. Suppl.* **26**, 291 (1962).

meaning of an energy; remember, for instance, that actually one also has to impose the constraint

$$[p - eA(x)]^2 = m^2, \tag{1.121}$$

and not too much importance should be attached to this pseudo-Hamiltonian character. Here we have to insist that, in a covariant context, there is no Hamiltonian with the meaning of an energy; for instance, in the case of a free particle, a formal Hamiltonian is

$$H = \frac{p^2 - m^2}{2m}, \tag{1.122}$$

which cannot in any way lead to an energy.

In the example considered, and when the four-potential is invariant along timelike lines parallel to a four-vector u^μ, one has

$$A^\mu(x + \chi u) = A^\mu(x) \quad \text{(for all } \chi) \tag{1.123}$$

so that it depends on x through the combination $\Delta^{\mu\nu}(u)x_\mu x_\nu$. A first integral of the motion is

$$u \cdot [p - eA(x)], \tag{1.124}$$

which is the energy in the rest frame of the system, so that the equilibrium distribution is

$$f_{\text{eq}}(p) = A \exp[-\beta u \cdot p]; \tag{1.125}$$

in other words, the equilibrium distribution function of free charged particles embedded in an electromagnetic field is identical to the Jüttner–Synge function *except* that the proper numerical density is changed as[28]

$$n_{\text{eq}} \to n_{\text{eq}} \exp[-\beta_\mu e A^\mu(x)], \tag{1.126}$$

since

$$p^\mu = m\frac{dx^\mu}{d\tau} + eA^\mu(x); \tag{1.127}$$

otherwise, the equilibrium distribution function would not obey the Liouville equation . Then n, the invariant density, reduces to

$$n_{\text{eq}} \exp[\beta e V(x)], \tag{1.128}$$

in the local rest frame where $u^\mu = (1, \mathbf{0})$; this is the usual relation.

---

[28]See an equivalent derivation in S.R. de Groot, W.A. van Leeuwen and Ch. G. van Weert (1980).

1.5.2. *Conditions for the Jüttner–Synge functions to be an equilibrium*

The necessary condition for a distribution function to represent an equilibrium is, of course, that the equilibrium distribution be a solution to the Liouville equation . Therefore, let us see what conditions the local Jüttner–Synge function should obey; by 'local' it is meant that the macroscopic variables of this function do depend on x. Also, we assume that there is no external force present.

To this end, let us introduce this function into the Liouville equation, and let us first write the Jüttner–Synge distribution as

$$f_{\rm eq}(p) = A(x)\exp(-\beta_\mu(x)p^\mu). \tag{1.129}$$

It turns out that we should have the equation

$$\begin{aligned} p\cdot\partial f_{\rm eq}(p) &= 0 \\ &= [p\cdot\partial A\cdot\exp(-\beta_\mu p^\mu) - A\exp(-\beta_\mu p^\mu)p_\mu p\cdot\partial\beta^\mu]. \end{aligned} \tag{1.130}$$

In other words, this implies that (i) $\partial_\mu A = 0$ and (ii) that the coefficient of $p^\mu p^\nu$ is zero whatever the coefficient. Explicitly, one should have

$$\partial_\mu\beta_\nu + \partial_\nu\beta_\mu = 0, \tag{1.131}$$

or, in arbitrary coordinates or in the case of gravitation (see Chap. 4),

$$\nabla_\mu\beta_\nu + \nabla_\nu\beta_\mu = 0. \tag{1.132}$$

Such an equation for β^μ is said to be a Killing equation. It shows that the local distribution function cannot be arbitrary. We see this in the case of the relativistic rotating gas. Note that when the particles are massless the Liouville equation is obeyed by a less stringent equation,

$$\partial_\mu\beta_\nu + \partial_\nu\beta_\mu = \chi(x)\cdot\eta_{\mu\nu}, \tag{1.133}$$

or, in arbitrary coordinates,

$$\nabla_\mu\beta_\nu + \nabla_\nu\beta_\mu = \chi(x)\cdot\eta_{\mu\nu}. \tag{1.134}$$

In such a case the vector β_μ is said to be conformal Killing.

1.6. The Relativistic Rotating Gas

As is well known, rigid rotation of a physical system is not possible in a relativistic context since it would imply velocities higher than that of light. However, differential rotation is still possible with a vanishing rotation velocity at the light cylinder, that is, the cylinder where the velocity is the one of light. Therefore, a *local* thermal equilibrium for a relativistic gas in rotation can be found, and it will now be investigated.

The projection of the kinetic momentum over the axis $\mathbf{n}$ — i.e. $\mathbf{n}\cdot\mathbf{L}$ — is an additive first integral of the motion and thus must be integrated in the equilibrium distribution function. Let $L_{\mu\nu}$ be the tensor

$$L_{\mu\nu} = p_\mu x_\nu - p_\nu x_\mu, \tag{1.135}$$

and let n^μ be a unitary spacelike four-vector orthogonal to the average local four-velocity u^μ of the gas:

$$n\cdot n = -1, \quad n\cdot u = 0, \quad u\cdot u = 1. \tag{1.136}$$

The tensor

$$L^{\mu\nu}_{\text{spat}} = \Delta^\mu_\alpha(u)\Delta^\nu_\beta(u)L^{\alpha\beta} \tag{1.137}$$

reduces to the usual kinetic momentum in the local frame $u^\mu = (1, \mathbf{0})$. The scalar

$$\Theta = u_\mu n_\nu \varepsilon^{\mu\nu\alpha\beta} L_{\alpha\beta} \tag{1.138}$$

reduces to $\mathbf{n}\cdot\mathbf{L}$ in the local frame of the gas and is an additive constant of the motion of a generic particle. Consequently, the equilibrium distribution function, which is a linear function of the additive first integrals of the motion, reads

$$f_{\text{eq}}(p) = \tilde{A}\exp(-\beta_\mu p^\mu + \lambda\Theta), \tag{1.139}$$

where λ is an appropriate Lagrange multiplier. Let now ω be the constant and uniform angular velocity of the gas: the Lagrange multiplier λ can be expressed in terms of ω and the equilibrium distribution can be rewritten as

$$\begin{aligned} f_{\text{eq}}(p) &= \tilde{A}\exp(-\beta_\mu p^\mu + \beta\omega\Theta) \\ &= \tilde{A}\exp(-\beta_\mu p^\mu + \beta_\mu \omega n_\nu \varepsilon^{\mu\nu\alpha\beta}[x_\alpha p_\beta - x_\beta p_\alpha]) \\ &= \tilde{A}\exp(-\beta_\mu[p^\mu - \omega n_\nu \varepsilon^{\mu\nu\alpha\beta}(x_\alpha p_\beta - x_\beta p_\alpha)]), \end{aligned} \tag{1.140}$$

where $\tilde{A}$ is the normalization constant, to be calculated from the equilibrium four-current. This distribution function must be normalizable and, accordingly, the following constraint must be satisfied:

$$\begin{cases} \beta_\mu[p^\mu - \omega n_\nu \varepsilon^{\mu\nu\alpha\beta}(x_\alpha p_\beta - x_\beta p_\alpha)] > 0, \\ \text{for all } p\text{'s} \end{cases} \tag{1.141}$$

which, in the rest frame of the system, reduces to

$$\omega^2 r^2 < 1. \tag{1.142}$$

This means that the velocity of a rotating piece of gas should not exceed the light velocity. $f_{\text{eq}}(p)$ must also obey the one-particle Liouville equation and, introduced in the latter, it yields a constraint involving the numerical invariant density, the four-velocity and the rotation velocity. Note that the "rotating" Jüttner function can be rewritten as

$$\begin{aligned} f_{\text{eq}}(p) &= \tilde{A}\exp(-\beta_\mu[p^\mu - \omega n_\nu \varepsilon^{\mu\nu\alpha\beta}(x_\alpha p_\beta - x_\beta p_\alpha)]) \\ &= \tilde{A}\exp(-B_\mu p^\mu), \end{aligned} \tag{1.143}$$

with

$$B_\lambda = \beta^\mu(\eta_{\mu\lambda} - 2\omega n^\nu \varepsilon_{\mu\nu\alpha\lambda} x^\alpha), \tag{1.144}$$

so that, in order for f_{eq} to obey the relativistic Liouville equation , the four-vector B^μ must be a Killing vector (or conformal Killing when the particles are massless)

$$\nabla_\mu B_\nu + \nabla_\nu B_\mu = 0, \tag{1.145}$$

which imposes an r-dependent temperature [N.A. Chernikov (1964)]:

$$T(r) = \frac{T(0)}{(1 - \omega^2 r^2)^{1/2}}. \tag{1.146}$$

It should be noted that in practice, i.e. in astrophysical situations, we do not have to face objects with a rigid rotation, but rather the rotating gas is in a *differential rotation* where not only is the temperature r-dependent but also the rotation velocity itself. In such a case, we have to deal with a local equilibrium and not with a global one as studied above.

Chapter 2

Relativistic Kinetic Theory and the BGK Equation

Once the relativistic distribution function is defined and the Liouville equation derived, the next step is obtaining a kinetic equation that the covariant distribution function is supposed to obey. From a theoretical point of view, such an equation has been used in many instances: derivation of covariant and dissipative hydrodynamics, technical developments in the calculation of transport coefficients, mathematical theorems (existence and uniqueness of the solutions) as to the Boltzmann–Einstein system, the gravitational case (see Chap. 4), propagation of sound and related phenomena (dispersion and absorption), description of transient effects in nonequilibrium thermodynamics [W. Israel and J.M. Stewart (1976, 1979), J. Stewart (1977), M. Kranys (1976, 1977), etc.], and so on.

The natural candidate for a relativistic kinetic theory was a relativistic generalization of the usual Boltzmann equation. This was first performed by A.G. Walker (1934) and A. Lichnérowicz and R. Marrot (1940) in a nonmanifestly covariant form, while it was later studied in a fully relativistic way[1] by many authors: G. Tauber and J.W. Weinberg (1961), W. Israel (1963), N.A. Chernikov (1956*ff*), Ch. Marle (1969), H. Akama (1970), K. Bitcheler (1965, 1967), etc. This equation has been found to have the form [S.R. de Groot, W.A. van Leuwen and Ch. Van Weert (1980)]

$$\left[p^\mu \partial_\mu + F^\mu(x,p)\frac{\partial}{\partial p^\mu}\right] f(x,p) = C\{f(x,p)\}, \tag{2.1}$$

where $C\{f(x,p)\}$ is the Boltzmann collision term, which we give in terms of the collision differential cross-section $\sigma(\Omega)$ of the particles within

[1]See the excellent review by W. Israel (1972).

the system:

$$C\{f(x,p)\} = \frac{1}{2}\int \frac{d^3p'}{p'_0} d\Omega \, [f(x,p')f(x,p'') - f(x,p)f(x,\bar{p} = p' + p'')]\sqrt{(p\cdot p')^2 - m^4\sigma(\Omega)} \qquad (2.2)$$

[S.R. de Groot, W.A. van Leuwen and Ch. Van Weert (1980)]; $F^\mu(x,p)$ is an external force in which the system is possibly embedded. In terms of the transition probability *per* unit of time, the relativistic Boltzmann equation reads

$$p\cdot\partial f(x,p) + F^\mu(x,p)\frac{\partial}{\partial p^\mu} f(x,p) = \frac{1}{2}\int \frac{d^3p'}{p'_0}\frac{d^3p''}{p''_0}\frac{d^3\bar{p}}{\bar{p}_0} W(p',p'' \to p,\bar{p})\, \delta^{(4)}(p + p'' - p' - \bar{p}) \times [f(x,p')f(x,p'') - f(x,p)f(x,\bar{p})]. \qquad (2.3)$$

This equation has been studied mainly by the above-mentioned authors and, in particular, by W. Israel (1963). The nonrelativistic approximation methods have been extended by various authors to the special relativity case and applied in several formal situations, such as the derivation of the various relativistic forms of hydrodynamics (see below).

More generally, the collision term of any valid relativistic kinetic equation should be such that the conservation laws (energy–momentum and particle number) are obeyed; this leads, after successively multiplying both sides of the kinetic equation by 1 and p^μ and integrating over p, to

$$\begin{cases} \partial_v J^v(x) = 0 = \int d^4p \; C\{f(x,p)\} = 0, \\ \partial_v T^{\mu v}(x) = 0 = \int d^4p \; p^\mu C\{f(x,p)\} = 0. \end{cases} \qquad (2.4)$$

Furthermore, the collision term must be such that

$$C[f_{\text{eq}}(p)] = 0, \qquad (2.5)$$

since $f_{\text{eq}}(p)$ has to obey the Liouville equation. A last important condition that a fully valid collision term should satisfy is that there should exist an H theorem, or

$$\partial_\mu S^\mu(x) \geq 0. \qquad (2.6)$$

Indeed, given two spacelike three-surfaces Σ_1 and Σ_2, the latter being in the future of the former, the above inequality for the entropy four-current implies that

$$S(\Sigma_1) \leq S(\Sigma_2), \tag{2.7}$$

which expresses the growth of entropy in the course of time. This condition is, however, not sufficient since $f_{\text{eq}}(p)$ must also be a solution at infinity in time. We shall see (Chap. 5) that the condition (2.7) is indeed not sufficient for getting an equilibrium.

The relativistic Boltzmann equation obeys all the above conditions and does constitute the archetype of all relativistic kinetic equations as it is in the Newtonian case. Unfortunately, it seems to be only an interesting conceptual tool, useful for the evaluation of those differences existing between the relativistic and classical cases that do not apply to any real physical situation.[2] In particular, it requires point-particles collisions while in relativistic dense matter one deals with collisions of collective (extended) modes. Therefore, in physical situations, it seems preferable to use a merely *phenomenological* kinetic equation, which should contain the main features of the physical system under study. This can be achieved with the relaxation time approximation studied by J.L. Anderson and H.R. Witting (1974a,b). Of course, such a phenomenological equation does not apply to all physical cases, but it is sufficient in many interesting situations.

Finally, we shall no longer continue with the Boltzmann equation or, more particularly, with its different approximation: there exist excellent books on the subject, to which the reader is referred — an older one is that of S.R. de Groot, W.A. van Leeuwen and Ch. G. van Weert (1980), while a more recent one is that of C. Cercignani and G.M. Kremer (2000).

2.1. Relativistic Hydrodynamics

For reasons clearly put forward by relativistic kinetic theory [see W. Israel (1963) and Ch. Marle (1969)],[3] there exist several equivalent forms of relativistic hydrodynamics. The basic reason for this particular feature of relativity lies in the fact that, while timelike four-vectors have generally different directions in the Minkowski space–time, they are all parallel in the

[2]For instance, it has been used to explain the matter/antimatter asymmetry in the primeval universe [see e.g. R. Omnès, *Phys. Rev. Lett.* **23**, 38 (1969); *ibid.*, *Phys. Rep.* **C3**, 1 (1970)]. However, the situation prevailing at the time considered [E. Alvarez (1982)] is one of high densities and/or temperatures, where collective effects are predominant and not collisions.

[3]See also S.R. de Groot, W.A. van Leeuwen and Ch. G. van Weert (1980).

Galilean one. The best-known forms of relativistic hydrodynamics are those given by C. Eckart (1940) and by L. Landau and E. Lifschitz (1959). As in the classical Newtonian case, relativistic hydrodynamics allow the propagation of signals at velocities higher than that of light, and this problem has been addressed by and investigated with different interesting methods by several authors [W. Israel (1976, 1981, 1979), B. Carter (1989), etc.].

The starting-point of relativistic hydrodynamics is the perfect fluid form of the energy–momentum tensor

$$T^{\mu\nu}_{\text{perf}} = (\rho + P)u^\mu u^\nu - P\eta^{\mu\nu}, \tag{2.8}$$

where ρ is the invariant energy density and P the pressure of the fluid under consideration, while u^μ is its four-velocity. In addition to this general form, the basic conservation relations have to be obeyed:

$$\begin{cases} \partial_\mu T^{\mu\nu} = 0, \\ \partial_\mu (n u^\mu) = 0, \end{cases} \tag{2.9}$$

so that the fluid is described by *four* equations while it possesses *five* unknown data: ρ, P, n, u^μ. The description of the system must then be supplemented by one more equation, generally the equation of state $P = P(\rho)$ or often in a parametric form such as

$$P = P(n), \quad \rho = \rho(n). \tag{2.10}$$

When the system contains other macroscopic quantities — such as the local temperature — other equations must be provided in order that the system may be complete [see e.g. A. Lichnérowicz (1967)]; for instance, when the system possesses a field of local temperature, a relativistic temperature equation should generally be provided for the full determination of the system (see below).

When the fluid is *dissipative*, its energy–momentum tensor contains gradients of one or several macroscopic quantities; the presence of gradients of the four-velocity gives rise to *viscosity*, the gradient of particle density n provides *diffusion* (when there exist several species of particles or external fields), the gradient of temperature T leads to *heat diffusion*, etc.

Historically, there exist two approachs to relativistic hydrodynamics: the one given by C. Eckart (1940) and that connected with the names of L. Landau and E. Lifschitz (1959). From the point of view of relativistic kinetic theory, the two are equivalent "up to higher order terms": as shown, essentially by W. Israel (1963), Ch. Marle (1969) and others, there exist an infinity of possible forms of relativistic hydrodynamics that differ by terms of order $O(\tau_0^2)$, τ_0 being a small parameters (see below).

Landau's and Eckart's forms do admit the same basic perfect fluid form but differ in their choice of the off-equilibrium part of the energy–momentum tensor and of the four-current,

$$T^{\mu\nu}_{\text{off}} \underset{\text{def}}{\equiv} T^{\mu\nu} - T^{\mu\nu}_{\text{perf}}, \tag{2.11}$$

$$J^{\mu\nu}_{\text{off}} \underset{\text{def}}{\equiv} J^{\mu\nu} - J^{\mu\nu}_{\text{perf}}, \tag{2.12}$$

and also in the choice of the hydrodynamic four-velocity u^μ. A detailed comparison of these two approaches is discussed in the book by S. R. de Groot, W.A. van Leeuwen and Ch. G. van Weert (1980). Clear connections between the two formalisms can also be found in the article by J.L. Anderson and H.R. Witting (1974) and also in C. Cercignani and G.M. Kremer (2000).

2.1.1. *Sound velocity*

In the case of an adiabatic fluid, an infinitesimal variation of the energy–momentum tensor reads

$$\delta T^{\mu\nu} = (\delta\rho + \delta P)u^\mu u^\nu - \delta P \eta^{\mu\nu} + (\rho + P)u^{(\mu}\delta u^{\nu)}, \tag{2.13}$$

where δu^μ is orthogonal to u^μ. The equations of motion are then

$$\begin{aligned} \partial_\mu \delta T^{\mu\nu} &= 0, \\ k_\mu \delta T^{\mu\nu} &= 0, \end{aligned} \tag{2.14}$$

where we have given the same name for $T^{\mu\nu}$ and its Fourier transform. The last equation explicitly reads

$$k_\mu[(\delta\rho + \delta P)u^\mu u^\nu - \delta P \eta^{\mu\nu} + (\rho + P)u^{(\mu}\delta u^{\nu)}] = 0, \tag{2.15}$$

while the other equation of motion, $\partial_\mu J^\mu = 0$, is

$$k_\mu(n\delta u^\mu + u^\mu \delta n) = 0, \tag{2.16}$$

and is of no use here. Let us multiply Eq. (2.15) by k^μ and u^μ, respectively; after setting $\varpi \equiv u^\mu k_\mu$, we get

$$\begin{cases} k_\nu k_\mu[(\delta\rho + \delta P)u^\nu u^\mu - k^2\delta P + (\rho + P)u^{(\mu}\delta u^{\nu)}] = 0, \\ u_\nu k_\mu[(\delta\rho + \delta P)u^\nu u^\mu - \delta P\varpi + (\rho + P)u^{(\mu}\delta u^{\nu)}] = 0. \end{cases} \tag{2.17}$$

We obtain

$$\begin{cases} [(\delta\rho + \delta P)\varpi^2 - \delta P k^2 + 2(\rho + P)\varpi\delta\varpi] = 0, \\ [(\delta\rho + \delta P)\varpi - \delta P\varpi + (\rho + P)\delta\varpi] = 0. \end{cases} \tag{2.18}$$

Multiplying the last equation by 2ϖ and substracting both equations, we get

$$\omega^2 - |\mathbf{k}|^2 \frac{\partial P}{\partial \rho} = 0, \tag{2.19}$$

which is the equation obeyed by usual longitudinal propagation; thus, we have

$$v_{\text{sound}}^2 = \left(\frac{\partial P}{\partial \rho}\right) \tag{2.20}$$

after a comparison with the classical equation. Note that it is not $\partial P/\partial n$ that occurs but $\partial P/\partial \rho$, and while $\rho = mn$ in the classical case, it is not so in relativity.

In general, one should have

$$v_{\text{sound}}^2 = \frac{\partial P}{\partial \rho} \leq 1; \tag{2.21}$$

indeed, the velocity of sound must be less than (or equal to) that of light. This is an important point; indeed, the velocity of sound is to be calculated from the equation of state and therefore it must *a priori* be rejected when $v_{\text{sound}} > 1$. An acceptable equation of state, from this point of view, is for example the one provided by the equation for nuclear matter calculated by S.A. Chin and J.D. Walecka (1974a,b).

2.1.2. *The Eckart approach*

Whereas the energy–momentum tensor has the perfect gas law form, no problem occurs while one is choosing a four-velocity. However, this is not so when gradients become important, since there exist new tensors in the theory, and therefore new "velocities" cannot be imposed but *defined.* We shall first look at the choice of C. Eckart (1940).

C. Eckart chose a natural timelike four-vector as the hydrodynamic velocity, the one defined by the particle flow $J^\mu(x)$, or

$$u^\mu(x) = \frac{J^\mu(x)}{n(x)} = \frac{J^\mu(x)}{\sqrt{J_\alpha(x)J^\alpha(x)}}. \tag{2.22}$$

The heat flow four-vector q^μ is introduced as a spacelike four-vector orthogonal to the hydrodynamic local four-velocity,

$$q^\mu : \begin{cases} q_\mu q^\mu < 0, \\ q_\mu u^\mu = 0, \end{cases} \tag{2.23}$$

and its contribution to the energy–momentum tensor as

$$T_Q^{\mu\nu} = q^\mu u^\nu + q^\nu u^\mu; \tag{2.24}$$

the viscous stress tensor is defined as

$$\theta^{\mu\nu} = -\chi\Delta^{\mu\nu}(u)\partial_\alpha u^\alpha - \eta\Delta^{\mu\alpha}(u)\Delta^{\nu\beta}(u)(\partial_\alpha u_\beta + \partial_\beta u_\alpha), \tag{2.25}$$

where χ is the *bulk* viscosity coefficient and η the *shear* viscosity. Finally, the *dissipative* part of the energy–momentum tensor is such that

$$T_{\text{off}}^{\mu\nu} = T_Q^{\mu\nu} + \theta^{\mu\nu}. \tag{2.26}$$

Note that

$$u_\mu T_{\text{off}}^{\mu\nu} = u_\nu T_{\text{off}}^{\mu\nu} = 0. \tag{2.27}$$

It remains for one to connect the heat flow four-vector q^μ to the gradient of temperature. C. Eckart (1940) uses the following:

$$q^\mu = -\lambda\Delta^{\mu\nu}(u)\left\{\frac{1}{T}u^\alpha\partial_\alpha u_\nu - \partial_\nu\left(\frac{1}{T}\right)\right\}, \tag{2.28}$$

where λ is the *heat conductivity* coefficient.

A few words as to the off-equilibrium of the energy–momentum tensor are in order. Firstly, the off-part of $T^{\mu\nu}$ is necessarily proportional to $\Delta^{\alpha\beta}(u)$, since it lies in the three-plane orthogonal to u. Secondly, there are two viscosity coefficients since $T_{\text{off}}^{\mu\nu}$ contains two independent tensors, $\partial_\alpha u^\alpha$ and $\partial_\alpha u_\beta + \partial_\beta u_\alpha$, multiplied of course by $\Delta^{\mu\alpha}(u)\Delta^{\beta\nu}(u)$.

As to the form of the flow of heat q^μ, it contains one term that does not exist in the classical theory. This is the term

$$-\lambda\Delta^{\mu\nu}(u)\frac{1}{T}u^\alpha\partial_\alpha u_\nu, \tag{2.29}$$

called sometimes "the inertia of heat" which is not proportional to the gradient of the temperature; it has no equivalent to the Newtonian analog and is only an isothermal flow of heat opposite to the acceleration of matter.

Finally, the matching conditions in the case of C. Eckart read

$$\begin{cases} J^{\mu}_{\text{off}}(x) = 0, \\ u_{\mu} T^{\mu\nu}_{\text{off}}(x) = 0. \end{cases} \tag{2.30}$$

See also S. Weinberg (1971) for a discussion.

2.1.3. *The Landau–Lifschitz approach*

L. Landau and E. Lifschitz have a more sophisticated — but equivalent — choice as to the local hydrodynamic velocity; they choose the flow of matter. Therefore, while in Eckart's local rest frame of the fluid there is no particle flow, in the Landau–Lifschitz one there is no matter flow. The Landau–Lifschitz hydrodynamical four-velocity is thus the timelike eigen-four-vector of the energy–momentum tensor

$$T^{\mu\nu} u_{\text{LL}\nu} = \rho u^{\mu}_{\text{LL}}, \tag{2.31}$$

where the index LL indicates the Landau–Lifschitz choice and where ρ is the energy density; this can also read

$$u^{\mu}_{\text{LL}} = \frac{T^{\mu\nu} u_{\text{LL}\nu}}{T_{\alpha\beta} u^{\alpha}_{\text{LL}} u^{\beta}_{\text{LL}}}. \tag{2.32}$$

This implies that the dissipative part of $T^{\mu\nu}$ is orthogonal to u^{μ}_{LL}, as is the case in Eckart's form. The four-current has thus a different expression than Eckart's and is given by

$$J^{\mu}_{\text{LL}} = n u^{\mu}_{\text{LL}} - q^{\mu}_{\text{LL}}, \tag{2.33}$$

where q^{μ}_{LL} is the Landau–Lifschitz heat flow four-vector. Note the relation

$$q^{\mu} = \frac{\rho + P}{n} q^{\mu}_{\text{LL}}, \tag{2.34}$$

given by J.L. Anderson and H.R. Witting (1974), shown together with

$$\begin{cases} n_{\text{E}} = n_{\text{LL}} = n, \\ \rho_{\text{E}} = \rho_{\text{LL}} = \rho, \\ P_{\text{E}} = P_{\text{LL}} = P, \end{cases} \tag{2.35}$$

and the matching conditions for Landau and Lifschitz read

$$\begin{cases} u_{\mu} J^{\mu}(x) = u_{\mu} J^{\mu}_{\text{eq}}(x), \\ u_{\mu} T^{\mu\nu}(x) = u_{\mu} T^{\mu\nu}_{\text{eq}}(x). \end{cases} \tag{2.36}$$

2.2. The Relaxation Time Approximation

The simplest of the various relativistic kinetic equations is the covariant version of the Bathnagar–Gross–Krook (BGK) equation[4] studied in detail by J.L. Anderson and H.R. Witting (1974):

$$\left[p^\mu \partial_\mu + F^\mu(x,p)\frac{\partial}{\partial p^\mu}\right] f(x,p) = C\left\{f(x,p)\right\} \equiv -p\cdot u\frac{f(x,p)-f_{\rm eq}(p)}{\tau_0}, \tag{2.37}$$

where τ_0 is a *relaxation time* to be evaluated with the help of other considerations and which might possibly be a function of x and p. τ_0 can be *roughly* evaluated as

$$\tau_0 \approx \frac{1}{n v_{\rm th}\sigma}, \tag{2.38}$$

where n is the numerical density of particles, $v_{\rm th}$ the thermal velocity and σ the total collision cross-section; however, more sophisticated evaluations can be obtained, like

$$\begin{aligned}\frac{1}{\tau_0} = \frac{1}{2}\int \frac{d^3p}{p_0}\frac{d^3p'}{p'_0}\frac{d^3p''}{p''_0}\frac{d^3\bar{p}}{\bar{p}_0} W(p',p'' \to p,\bar{p}) \\ \times\, \delta^{(4)}\left(p+p''-p'-\bar{p}\right) f_{\rm eq}(p') f_{\rm eq}(p'')\end{aligned} \tag{2.39}$$

[D. Gerbal (1972)]. Since the left hand side of the BGK equation represents essentially the flow along the trajectories in μ space, it can *formally* be rewritten as

$$\frac{d}{d\tau}f = -p\cdot u\frac{f-f_{\rm eq}}{m\tau_0}, \tag{2.40}$$

whose formal "solution" reads

$$f \approx f_{\rm eq} + \left[f(t=0) - f_{\rm eq}\right]\exp\left(-\frac{\tau p\cdot u}{m\tau_0}\right), \tag{2.41}$$

showing thereby that the nonequilibrium distribution f relaxes toward the equilibrium one, $f_{\rm eq}$, exponentially in "time." It can be seen that the quantity $\tau p\cdot u/m$ is essentially the usual time t in the local rest frame of the fluid, and hence τ_0 appears to be a true relaxation time. Note that in the quantum case one has a similar equation with the difference that $f_{\rm eq}$ is a quantum distribution of the Fermi–Dirac or Bose–Einstein type [see J.L. Anderson and H.R. Witting (1974b)].

[4] D. Bathnagar, D. Gross and M. Krook, *Phys. Rev.* **94**, 511 (1954).

The conservation laws — four-current and energy–momentum tensor — yield the equation

$$\begin{cases} J^{\mu}u_{\mu} = J^{\mu}_{\text{eq}}u_{\mu}, \\ T^{\mu\upsilon}u_{\mu} = T^{\mu\upsilon}_{\text{eq}}u_{\mu}, \end{cases} \tag{2.42}$$

obtained by integrating the BGK equation over p, after it has been multiplied by 1 or p^{μ}. These conditions are the Landau–Lifschitz (1959) *matching conditions* (see next section). They express the constancy of the particle and energy–momentum densities in the rest frame of the equilibrium system. It should be noted that the covariant BGK equation does depend on the so-called matching.

2.3. The Relativistic Kinetic Theory Approach to Hydrodynamics

Although the first derivations of relativistic hydrodynamices were made with the Boltzmann equation [W. Israel (1963), N.A. Chernikov (1964)], it is much simpler to use the BGK equation although the transport coefficients obtained have a different value. To this end, the Chapman–Enskog method — among many others — is the simplest to use and the deduction of J.L. Anderson and H.R. Witting (1974) is followed in this section.

The Chapman–Enskog method is expressed, in this case, by an expansion in powers of the small parameter

$$\varepsilon \equiv \frac{\tau}{\tau_{\text{macro}}}, \tag{2.43}$$

where τ is the relaxation time and τ_{macro} a macroscopic hydrodynamical time. Furthermore, assumptions of weak spatial gradients of the macroscopic quantities like $\{T, n, u^{\mu}\}$ are made. Then one expands the solution to the BGK equation into the small parameter ε (for convenience, we choose $\tau_{\text{macro}} = 1$):

$$f(x,p) = \sum_{\ell} \tau^{\ell} f_{\ell}(x,p), \tag{2.44}$$

with

$$f_0(x,p) \equiv f_{\text{eq}}(p). \tag{2.45}$$

At the lowest order, once f_{eq} is introduced into the conservation equations

$$\begin{cases} \partial_{\mu}T^{\mu\nu}_{\text{eq}}(x) \approx 0, \\ \partial_{\mu}J^{\mu}_{\text{eq}}(x) \approx 0, \end{cases} \tag{2.46}$$

one arrives at the following relations between the macroscopic quantities:

$$\begin{cases} \dot{n} + n\theta = 0, \\ \dot{\rho} + (\rho + P)\theta = 0, \\ (\rho + P)\Delta^{\mu\lambda}(u)\dot{u}_\mu = \Delta^{\mu\lambda}(u)\partial_\mu P, \end{cases} \tag{2.47}$$

where use has been made of the notations

$$\theta \equiv \partial_\alpha u^\alpha, \qquad \begin{bmatrix} \tilde{n} \\ \tilde{\rho} \end{bmatrix} \equiv u \cdot \partial \begin{bmatrix} n \\ \rho \end{bmatrix}, \tag{2.48}$$

to be employed later.

Once the expansion for f is introduced into the BGK equation, it immediately provides the solution at order 1 in τ:

$$f_1(x,p) = -\tau\{p \cdot \partial f_{\text{eq}}(\beta(x), n(x), u^\mu(x); p)\} + O(\tau^2). \tag{2.49}$$

From this solution, one can calculate the first order correction to the four-current and to the energy–momentum tensor

$$\begin{cases} J_1^\mu(x) = \displaystyle\int \frac{d^3p}{p_0} p^\mu f_1(x,p), \\ T_1^{\mu\nu}(x) = \displaystyle\int \frac{d^3p}{p_0} p^\mu p^\nu f_1(x,p), \end{cases} \tag{2.50}$$

which must also satisfy the Landau–Lifschitz matching conditions

$$\begin{cases} u_\mu J_1^\mu(x) = 0, \\ u_\mu T_1^{\mu\nu}(x) = 0, \end{cases} \tag{2.51}$$

in order to obey both the conservation equations and the relativistic BGK equation. These two first order quantities are obtained as

$$\begin{aligned} J_1^\mu(x) &= \int \frac{d^3p}{p_0} \frac{p^\mu p^\sigma}{u \cdot p} \{-\partial_\sigma \alpha + \partial_\sigma(\beta u_\gamma) p^\gamma\} f_{\text{eq}}(p) \\ &= -\partial_\sigma \alpha \int \frac{d^3p}{p_0} \frac{p^\mu p^\sigma}{u \cdot p} f_{\text{eq}}(p) + \partial_\sigma(\beta u_\gamma) \int \frac{d^3p}{p_0} \frac{p^\mu p^\sigma p^\gamma}{u \cdot p} f_{\text{eq}}(p), \end{aligned} \tag{2.52}$$

$$T_1^{\mu\nu}(x) = -\partial_\sigma \alpha \int \frac{d^3p}{p_0} \frac{p^\mu p^\nu p^\sigma}{u \cdot p} f_{\text{eq}}(p) + \partial_\sigma(\beta u_\gamma) \int \frac{d^3p}{p_0} \frac{p^\mu p^\nu p^\sigma p^\gamma}{u \cdot p} f_{\text{eq}}(p), \tag{2.53}$$

where we have set

$$\alpha \equiv \ln\left\{\frac{n_{\text{eq}}\beta}{4\pi m^2 K_2(m\beta)}\right\}. \tag{2.54}$$

The three integrals appearing in these last expressions can be decomposed on the various possible tensors that can be constructed from u^μ and $\eta^{\mu\nu}$, and using the notations $I_2^{\mu\nu}$, $I_3^{\mu\nu\sigma}$, $I_4^{\mu\nu\sigma\gamma}$ for them, one can write

$$I_2^{\mu\nu} = \mathbf{T}_1 u^\mu u^\nu - \mathbf{T}_2 \eta^{\mu\nu}, \tag{2.55}$$

$$I_3^{\mu\nu\sigma} = \mathbf{S}_1 u^\mu u^\nu u^\sigma - \mathbf{S}_2(u^\mu \eta^{\nu\sigma} + u^\nu \eta^{\sigma\mu} + u^\sigma \eta^{\mu\nu}), \tag{2.56}$$

$$\begin{aligned} I_4^{\mu\nu\sigma\gamma} &= +\mathbf{Q}_1 u^\mu u^\nu u^\sigma u^\gamma \\ &\quad -\mathbf{Q}_2(u^\mu u^\nu \eta^{\sigma\gamma} + u^\mu u^\sigma \eta^{\nu\gamma} + u^\nu u^\sigma \eta^{\mu\gamma} + u^\nu u^\gamma \eta^{\mu\sigma} + u^\mu u^\gamma \eta^{\nu\sigma} + u^\sigma u^\gamma \eta^{\mu\nu}) \\ &\quad + \mathbf{Q}_3(\eta^{\mu\nu}\eta^{\sigma\gamma} + \eta^{\mu\sigma}\eta^{\nu\gamma} + \eta^{\mu\gamma}\eta^{\nu\sigma}), \end{aligned} \tag{2.57}$$

and the various coefficients $\mathbf{T}_1, \mathbf{T}_2, \ldots, \mathbf{Q}_3$ are obtained from the scalars formed with the various integrals and the final result is[5]

$$\begin{aligned} T_1 &= \frac{1}{3}[4n - 4\pi m^3 \exp(\beta\mu)(K_1 - K_{i1})], \\ T_2 &= \frac{1}{3}[n - 4\pi m^3 \exp(\beta\mu)(K_1 - K_{i1})], \\ S_1 &= 2\rho - m^2 A_0, \\ S_2 &= \frac{1}{3}(\rho - m^2 A_0), \\ Q_1 &= \frac{1}{5}[16S_1 - 48S_2 - 12m^2 n + 4\pi m^5 \exp(\beta\mu)(K_1 - K_{i1})], \\ Q_2 &= \frac{1}{15}[6S_1 - 18S_2 - 7m^2 n + 4\pi m^5 \exp(\beta\mu)(K_1 - K_{i1})], \\ Q_3 &= \frac{1}{15}[S_1 - 3S_2 - 2m^2 n + 4\pi m^5 \exp(\beta\mu)(K_1 - K_{i1})], \end{aligned} \tag{2.58}$$

where they are given by

$$\begin{aligned} n &= 4\pi \frac{m^2}{\beta} \exp(\beta\mu) K_2(\beta m), \\ \rho &= 4\pi m^4 \exp(\beta\mu) \left[\frac{K_3(\beta m)}{\beta m} - \frac{K_2(\beta m)}{(\beta m)^2} \right], \\ P &= 4\pi m^4 \exp(\beta\mu) \frac{K_2(\beta m)}{(\beta m)^2}, \end{aligned} \tag{2.59}$$

[5]We follow the article by J.L. Anderson and H.R. Witting (1974a).

$$\begin{aligned} S_1 &= 4\pi m^5 \exp(\beta\mu)\frac{K_4(\beta m)}{\beta m}, \\ S_2 &= 4\pi m^5 \exp(\beta\mu)\frac{K_3(\beta m)}{(\beta m)^2}, \end{aligned} \tag{2.60}$$

$$\begin{aligned} Q_1 &= 4\pi m^6 \exp(\beta\mu)\frac{K_5(\beta m)}{\beta m}, \\ Q_2 &= 4\pi m^6 \exp(\beta\mu)\frac{K_4(\beta m)}{(\beta m)^2}, \end{aligned} \tag{2.61}$$

$$Q_3 = 4\pi m^6 \exp(\beta\mu)\frac{K_3(\beta m)}{(\beta m)^3}. \tag{2.62}$$

One then gets

$$J_1^\mu = \tau\beta^2\left[\frac{T_2 S_2}{P} - S_2\right]\Delta^{\mu\nu}(u)\left[\dot{u}_\nu + \frac{\partial_\nu\beta}{\beta}\right], \tag{2.63}$$

$$\begin{aligned} T_1^{\mu\nu} &= \tau u^\mu u^\nu\left[(Q_2 - Q_3)\dot{\beta} - m\dot{\beta}S_2 - \frac{5}{3}Q_3\beta\theta\right] \\ &\quad - \tau\eta^{\mu\nu}\left[m^{-1}(Q_2 - Q_3)\dot{\beta} - (\dot{\beta\mu})S_2 - \frac{5}{3}Q_3\beta\theta\right] + 2\tau\beta Q_3\sigma^{\mu\nu} \end{aligned} \tag{2.64}$$

for the off-equilibrium parts of the four-current and of the energy–momentum tensor. Let us specify that these tensors can hardly be obtained without the use of (local) equilibrium.

In order to obtain the heat conductivity and the viscosity coefficients, we have to go back to Eckart's choice and, to this end, to use Eq. (2.28). Accordingly, the various dissipative coefficients are

$$\lambda = \tau\frac{\rho + P}{n}\beta^2 m\left(\frac{T_2 S_2}{P} - S_2\right), \tag{2.65}$$

$$\eta = \tau\beta Q_3, \tag{2.66}$$

$$\zeta = -\frac{\tau}{\theta}\left[\beta^2 m(Q_2 - Q_3)u\cdot\partial\left(\frac{1}{\beta m}\right) + S_2\, u\cdot\partial(\beta\mu) + \frac{8}{3}Q_3\beta\theta\right]. \tag{2.67}$$

After simplifying by eliminating θ, etc., and using the exact values of (Q_i, S_i), one obtains

$$\lambda = \frac{\tau}{3}\beta^2 m^6 4\pi\exp(\beta\mu)\left[h\left(\frac{K_2}{\beta m} + K_{i1} - K_1\right) - 3\frac{K_2}{(\beta m)^2}\right], \tag{2.68}$$

the coefficient of heat conductivity,

$$\eta = \frac{\tau}{15}\beta m^5 4\pi \exp(\beta\mu)\left[3\frac{K_3}{(\beta m)^2} - \frac{K_2}{\beta m} + K_1 - K_{i1}\right], \tag{2.69}$$

the coefficient of shear viscosity, and

$$\chi = -\tau m^4 4\pi \exp(\beta\mu)\left[\frac{K_2}{(m\beta)^2}\frac{(\beta m)^2 h'(\beta m) + \beta m h(\beta m)}{(\beta m)^2 h'(\beta m) + 1}\right.$$
$$\left. - \frac{K_3}{\beta m}\frac{1}{(\beta m)^2 h'(\beta m) + 1} - \frac{\beta m}{9}\left(\frac{3K_3}{(\beta m)^2} - \frac{K_2}{\beta m} + K_1 - K_{i1}\right)\right], \tag{2.70}$$

the coefficient of bulk viscosity.

In this expression, h is the enthalpy per particle and per unit of mass of the system, given by

$$h(\beta m) = \frac{K_3(\beta m)}{K_2(\beta m)}. \tag{2.71}$$

This shows that, in relativity, there exist two kinds of viscosity. In the Newtonian case the bulk viscosity does not appear in the usual relativistic case. Furthermore, it is clear that, for instance, a $\tau(p)$ can elaborate more precisely a specific case.

2.4. The Static Conductivity Tensor

In order to calculate the conductivity of the system, let us assume that it is embedded in a small electric field, $F^{\mu\nu}$, and let us assume further that the system is static and homogeneous. Then the relativistic BGK equation just reads

$$eF^{\mu\nu}p_\nu \frac{\partial}{\partial p^\mu} f(p) = -p\cdot u\frac{f(p) - f_{\text{eq}}(p)}{\tau_0}, \tag{2.72}$$

and the answer to the electric field is the four-current

$$J^\mu_{(1)} = \int \frac{d^3p}{p_0} p^\mu f_{(1)}(p), \tag{2.73}$$

where $f_{(1)}(p)$ is only $f(p) - f_{\text{eq}}(p)$. Finally, at the simplest order in the Chapman–Enskog expansion, we have

$$f_{(1)}(p) = -\tau_0 \frac{eF^{\mu\nu}p_\nu}{p\cdot u}\frac{\partial}{\partial p^\mu} f_{\text{eq}}(p) \tag{2.74}$$

so that the response four-current is

$$J^{\lambda}_{(1)} = -\tau_0 e F^{\mu\nu} \int \frac{d^3p}{p_0} p^{\lambda} \frac{p_\nu}{p \cdot u} \frac{\partial}{\partial p^{\mu}} f_{\text{eq}}(p). \tag{2.75}$$

Finally, since the conductivity tensor is defined as (J^{λ} and $F^{\mu\nu}$ being "small")

$$J^{\lambda} = \sigma^{\lambda\mu\nu} F_{\mu\nu}, \tag{2.76}$$

it is given by

$$\sigma^{\lambda\mu\nu} = -\tau_0 e \int \frac{d^3p}{p_0} \frac{p^{\lambda} p^{\nu}}{p \cdot u} \frac{\partial}{\partial p_{\mu}} f_{\text{eq}}(p). \tag{2.77}$$

It is simple to count the number of independent components of this tensor. *A priori* $F^{\mu\nu}$ has six independent components — which reduce to three when one is looking at the electric character of $F^{\mu\nu}$ — and there remain four indices for λ, which means 12 components. However, taking into account the fact that the electric field is homogeneous, it possesses only one component while the index λ is that of the current, or only one component. Finally, σ has exactly one *independent* component, as should be the case. It should be borne in mind that this result rests only on the particularly simple character of the problem.

2.5. Approximation Methods for the Relativistic Boltzmann Equation and Other Kinetic Equations

The relativistic Boltzmann equation was first studied by A.G. Walker (1934) and A. Lichnérowicz and R. Marrot (1940) and in covariant form by several other authors [G. Tauber and J.W. Weinberg (1961); N.A. Chernikov (1956*ff*); W. Israel (1963)], and was rediscovered by many others. It is found essentially by investigating the balance between those particles entering a given phase space volume element and those which leave it, exactly the same way as in the nonrelativistic case. A large number of approximate methods have been devised for its solution which also apply to other kinetic equations. A simple version of the Chapman–Enskog method has been applied above to the BGK equation, and in this section other useful methods are very briefly explained. Details and more serious analyses can be found in the book by S.R. de Groot, W.A. van Leeuwen and Ch. G. van Weert (1980) [see also C. Cercignani and G.M. Kremer (2000)]. Note that these methods have numerous variations and can be interconnected.

2.5.1. *A simple Chapman–Enskog approximation*

We begin with the simplest version of the Chapman–Enskog method; more formal approximations are studied in the book by S.R. de Groot, W.A. van Leeuwen and Ch. G. van Weert (1980), to which we refer for other methods and more serious studies.

First, we set the relativistic Boltzmann equation into an equation written in terms of the decomposition

$$p \cdot u u^{\nu} \cdot \partial f + p^{\mu} \Delta_{\mu\nu}(u) \partial^{\nu} f = C(f), \tag{2.78}$$

which represents a splitting into a temporal term p (along u) and a spatial one, represented by the projection of p over the spatial direction $p\Delta$. Next, this equation is written in terms of the dimensionless quantities

$$X_T{}^{\mu} = \tau_0 x \cdot u^{\mu} u, \quad X_S^{\mu} = r_0 \Delta^{\mu}{}_{\nu}(u) x^{\nu}, \tag{2.79}$$

as

$$\frac{1}{\tau_0} p \cdot \frac{\partial f}{\partial X_T} + \frac{1}{r_0} p^{\mu} \Delta_{\mu\nu}(u) \frac{\partial f}{\partial X_{S\nu}} = \frac{1}{\tau} \bar{C}(f), \tag{2.80}$$

where τ^{-1} denotes the number of collisions per unit of time in the system and $\bar{C}(f)$ indicates what is laid after the dimensionless system has been made. The meaning of (τ_0, r_0) is the following: they are macroscopic quantities such as hydrodynamic ones; they are to be compared with τ. Therefore, there are *a priori* two dimensionless parameters in the problem $\tau\tau_0^{-1}$ and τr_0^{-1} and combinations. In the Chapman–Enskog expansion we use only the parameter τr_0^{-1}. However, there exist two others — those connected with the *external* force field and an eventual *collective* field, which we have taken to be zero here. These two supplementary parameters correspond to, for example, the length on which these force fields vary.

Therefore, the basis of the method is an expansion in the parameter

$$\varepsilon = \frac{\tau}{r_0}, \tag{2.81}$$

where r_0 is a typical hydrodynamical scale, on which the system varies appreciably, and where τ *also* designates the typical mean free path, i.e. the average distance between two successive collisions. Such a parameter is supposed to be small enough for keeping only the first order. Of course, higher orders can be considered, but we shall not be concerned with them here. Thus, we have

$$f(x,p) = f_0(x,p)[1 + \phi(x,p)] + 0(\varepsilon^2), \tag{2.82}$$

where the term $f_1 \equiv f_0\phi$ is of order 1 in ε, and f_0 is generally the Jüttner–Synge distribution. Note that f_0 is not necessarily the Jüttner–Synge distribution; it could be a *normal solution*, which is a solution taking account of x via the data of ρ, u^μ and τ in smooth ways. The concept of a normal solution was introduced by D. Hilbert (1912).[6]

Inserting this last expression of $f(x,p)$ into the relativistic Boltzmann equation and equating equal powers of the parameter ε, we get

$$p\cdot\partial f_0(x,p) = \frac{1}{2}\int \frac{d^3p'}{p'_0}\frac{d^3p''}{p''_0}\frac{d^3\bar{p}}{\bar{p}_0}W(p',p''\to p,\bar{p})\delta^{(4)}(p+p''-p'-\bar{p}) \\ \times f_0(x,p')f_0(x,p'')[\phi(x,p')+\phi(x,p'')-\phi(x,p)-\phi(x,\bar{p})], \tag{2.83}$$

where use has been made of the relationship

$$f_0(x,p')f_0(x,p'') = f_0(x,p_*)f_0(x,p). \tag{2.84}$$

At this point several remarks have to be made.

The first one deals with the function f_0 inserted in the Boltzmann equation. It should be the Jüttner–Synge distribution; however, the various parameters entering this function may *not* necessarily be u^μ, n and ρ. The main reason why this is so is that the four-vector u^μ is arbitrary, unlike the classical case where all four-vectors are parallel. We are thus obliged to define these five quantities with either the C. Eckart (1940) or the L. Landau and E. Lifschitz (1959) ones; these conditions are not the only possible ones, of course. They are briefly studied in the section "Relativistic hydrodynamics."

The next remark deals with the solution of Eq. (2.83). We could first solve this equation by expanding $f_{\text{off}}(x,p)$ in orthogonal polynomials and obtaining the coefficients. But this is not the only possibility. From another point of view, one can find in S.R. de Groot (1973) a succint but complementarity solution for solving the Chapman–Enskog approximation. Also, they can be found in J. Stewart (1974).

2.6. Transport Coefficients for a System Embedded in a Magnetic Field

Consider a system composed of charged particles embedded in a static and homogeneous magnetic field; the particles, i.e. the electrons, are immersed in an opposite charge whose only role is to have neutrality. The equivalent

---

[6]D. Hilbert, *Grundzüg einer allgemeinen Theorie der linearen Integralgleischungen* (Teubener, 1912).

relativistic BGK equation then reads

$$\left[p^\mu \partial_\mu + eF^{\mu\nu} p_\nu \frac{\partial}{\partial p^\mu}\right] f(x,p) = -p \cdot u \frac{f(x,p) - f_{\text{eq}}(p)}{\tau_0}. \tag{2.85}$$

Let us now introduce into this equation the hydrodynamical length (or times) so as to write it in terms of dimensionless variables (note that, owing to the linearity of the equation, it is not necessary to change f and f_{eq}). To this end, we introduce the hydrodynamical length ℓ (or time) and the Larmor radius of the electron,

$$r_0 = \frac{eB}{m}, \tag{2.86}$$

and m^{-1}, the Compton wavelength of the electron. With the change of variables

$$x = \ell \bar{x}, \quad p = m\bar{p}, \quad F^{\mu\nu} = mr_0 \bar{F}^{\mu\nu}, \tag{2.87}$$

the kinetic equation reads

$$\tau_0 \left[\bar{p}^\mu \partial_\mu + r_0 \bar{F}^{\mu\nu} \bar{p}_\nu \frac{\partial}{\partial \bar{p}^\mu}\right] f(x,p) = -\bar{p} \cdot u[f(x,p) - f_{\text{eq}}(p)]. \tag{2.88}$$

Note that we have set $L = 1$, since L does not play an explicit role at this stage. Therefore, in order to have a Chapman–Enskog expansion, we have to discuss the effect of different terms that appear in this last equation. Let us expand f as

$$f = f_{\text{eq}} + f_{(1)} + f_{(2)} + \cdots \tag{2.89}$$

so that, from the kinetic equation, we have

$$f_{(1)} = \frac{\tau_0}{\bar{p} \cdot u}\left[\bar{p} \cdot \partial f_{\text{eq}} + \tau r_0 \bar{F}^{\mu\nu} \bar{p}_\nu \frac{\partial f_{\text{eq}}}{\partial \bar{p}^\mu} + \tau r_0 \bar{F}^{\mu\nu} \bar{p}_\nu \frac{\partial f_{(1)}}{\partial \bar{p}^\mu} + 0(\tau_0^2)\right]. \tag{2.90}$$

Therefore, either $\tau_0 r_0$ is of the order of $\tau_0 r_0 \leq \tau_0^2$ or it is not. In the first case, one has the relatively simple situation [R. Domingez Tenreiro and R. Hakim (1977b)]

$$f_{(1)} = \frac{\tau_0}{\bar{p} \cdot u} f_{\text{eq}} + \tau_0 r_0 \bar{F}^{\mu\nu} \bar{p}_\nu \frac{\partial f_{\text{eq}}}{\partial \bar{p}^\mu}, \tag{2.91}$$

in which it is sufficient to use the equilibrium distribution. Note that this distribution is either the Jüttner–Synge function (classical case) or a more complex Wigner function (quantum case; Chap. 11); it appears that the quantum case has the same kinetic equation when spin effects are neglected. The other possibility is much more complex and has been performed in the quantum case (Chap. 12).

It remains for us to define the viscosity coefficients, and also the heat coefficients. As to the viscosity, they have been introduced by S.I. Braginskii (1965) in the nonrelativistic domain. In the relativistic case, they are given by

$$\psi^{\alpha\beta} = -\eta_0 W_0^{\alpha\beta} - \eta_1 W_1^{\alpha\beta} - \eta_2 W_2^{\alpha\beta} - \eta_3 W_3^{\alpha\beta} - \eta_4 W_4^{\alpha\beta} - \bar{\eta} \bar{W}^{\alpha\beta}, \tag{2.92}$$

where the η's are the viscosity coefficients and the tensors $W^{\alpha\beta}$ are defined by

$$\begin{aligned}
W_0^{\alpha\beta} &= \frac{3}{2}\left[\left(n^\alpha n^\beta + \frac{1}{3}\Delta^{\alpha\beta}(u)\right)\left(n^\mu n^\nu + \frac{1}{3}\Delta^{\mu\nu}(u)\right)\right]\sigma_{\mu\mu},\\
W_1^{\alpha\beta} &= \left(\pi^{\alpha\mu}\pi^{\beta\nu} - \frac{1}{2}\pi^{\alpha\beta} n^\mu n^\nu\right)\sigma_{\mu\nu},\\
W_2^{\alpha\beta} &= -\left(\pi^{\alpha\mu} n^\beta n^\rho + \pi^{\beta\rho} n^\alpha n^\mu\right)\sigma_{\mu\rho},\\
W_3^{\alpha\beta} &= \frac{1}{2}\left(\pi^{\alpha\mu}\varepsilon^{\beta\gamma\rho} + \pi^{\beta\rho}\varepsilon^{\alpha\gamma\mu}\right) n_\gamma \sigma_{\mu\rho},\\
W_4^{\alpha\beta} &= -\left(n^\alpha n^\mu \varepsilon^{\beta\gamma\rho} + n^\beta n^\rho \varepsilon^{\alpha\gamma\mu}\right) n_\gamma \sigma_{\mu\rho},\\
W^{\alpha\beta} &= \left(\pi^{\alpha\beta} + 2n^\alpha n^\beta\right)\theta,
\end{aligned} \tag{2.93}$$

where

$$\varepsilon^{\alpha\gamma\mu} \underset{\text{def}}{\equiv} \varepsilon^{\alpha\gamma\mu\lambda} u_\lambda. \tag{2.94}$$

The n's and $\pi^{\alpha\beta}$ are defined through the electromagnetic field via

$$\begin{gathered}
F^{\alpha\beta} = H\varepsilon^{\mu\nu\alpha\beta} n_\alpha u_\beta, \quad {}^*F^{\mu\nu} = H u^{[\mu} n^{\nu]},\\
u_\mu u^\mu = +1, \quad n_\mu u^\mu = 0, \quad n_\mu n^\mu = -1, \quad \pi^{\mu\nu} = \Delta^{\mu\nu}(u) + n^\mu n^\nu.
\end{gathered} \tag{2.95}$$

Similarly, the heat tensor is defined as

$$\begin{aligned}
-\tau_0 J^\mu_{(1)} &= q^\mu_\perp + q^\mu_p,\\
q^\alpha_\perp &= \pi^{\alpha\beta} q_\beta,\\
q^\mu_p &= -n^\alpha n^\beta q_\beta,
\end{aligned} \tag{2.96}$$

and it is all that is relevant. Actually, if we insist that the heat coefficients be defined as

$$Q^\alpha_\perp = \lambda_\perp \pi^{\alpha\nu}(\beta^{-1}\dot{u}_\nu - \partial_\nu \beta^{-1}), \quad Q^\alpha_p = -\lambda_\perp n^\alpha n^\nu (\beta^{-1}\dot{u}_\nu - \partial_\nu \beta^{-1}), \tag{2.97}$$

it remains for us to interpret the two coefficients that subsist: they probably are diffusion parts, which we do not study here.

Let us now compute the dissipative part of J^μ and $T^{\mu\nu}$, in the simplest case,

$$f_{(1)} = -\tau_0 \left[\frac{1}{p \cdot u} p \cdot \partial f_{\text{eq}} + \frac{e}{m} F^{\mu\nu} p_\nu \frac{\partial f_{\text{eq}}}{\partial p^\mu} \right], \tag{2.98}$$

where we have kept the nondimensional quantity. The first term in the brackets is exactly the same as in the nonmagnetic case. Let us examine the second term. The derivative with respect to p^μ gives rise to a factor β^μ. This term is proportional to u^μ, which, once contracted with $F^{\mu\nu}$, is exactly zero. The consequence of this is the identity of the dissipative process of heat with the nonmagnetic case. The same thing arises equally with the $T^{\mu\nu}_{(1)}$ dissipation part of the energy–momentum tensor. It should be noted that this is true only in the simple case we have considered.

Chapter 3

Relativistic Plasmas

Relativistic plasmas are objects encountered in many astrophysical situations. For instance, they occur in the magnetosphere of pulsars where they are strongly magnetized, or in quasar jets. In the case of white dwarfs, the electron plasma is both relativistic and degenerate and could possibly be magnetized. The first relativistic Vlasov equation — an equation used to describe the collective behavior of plasma — was given by S. Titeica (1956) and by numerous other authors; let us, however, mention V.N. Tsytovich (1961) and A.G. Sitenko.[1] In this chapter the electromagnetic excitations — dispersion relations — are mainly studied after some general and simple relations are provided as to various data about electromagnetism. Also, we are concerned only with the classical plasmas here; the quantum ones are dealt with in Chap. 15.

3.1. Electromagnetic Quantities in Covariant Form

Let us first look at the conductivity tensor, which relates the electromagnetic field $F^{\mu\nu}$ and the four-current J^μ through

$$J^\mu(k) = \Lambda^{\mu\alpha\beta}(k)F_{\alpha\beta}(k), \tag{3.1}$$

where use has been made of Fourier transforms. Owing to the conserved character of the four-current, $k_\mu J^\mu = 0$, whatever the electromagnetic field, the *conductivity tensor* must obey

$$k_\mu \Lambda^{\mu\alpha\beta}(k) = 0. \tag{3.2}$$

[1] A.G. Sitenko, *Electromagnetic Fluctuations in Plasmas* (Academic Press, New York, 1967).

This tensor is antisymmetrical with respect to the indices (α, β) and thus possesses *a priori* $24 - 6 = 18$ independent components. When the only macroscopic four-vector present in the system is the average four-velocity u^μ, it has the general form

$$\Lambda^{\mu\alpha\beta}(k) = [a(k)u^\mu + b(k)k^\mu](k^\alpha u^\beta - k^\beta u^\alpha) + [c(k)u^{(\alpha} + d(k)k^{(\alpha}]\eta^{\beta)\mu}, \tag{3.3}$$

where the various coefficients are connected through the relation

$$a(k)k \cdot u + b(k)k^2 = c(k), \tag{3.4}$$

which expresses the four-current conservation constraint.

The next important quantity is the *polarization tensor*, defined by

$$J^\mu(k) = \Pi^\mu{}_\nu(k)A^\nu(k), \tag{3.5}$$

where $A^\nu(k)$ is the *total* electromagnetic field:

$$\begin{cases} A^\nu(k) = A^\nu_{\text{int}}(k) + A^\nu_{\text{ext}}(k), \\ A^\nu_{\text{int}}(k)\text{: internal field of the system,} \\ A^\nu_{\text{ext}}(k)\text{: external (applied) field.} \end{cases} \tag{3.6}$$

The following relation connects it to the conductivity tensor,

$$\Pi^{\mu\nu}(k) = 2k_\alpha\Lambda^{\mu\alpha\nu}(k), \tag{3.7}$$

and obeys

$$k_\mu\Pi^{\mu\nu}(k) = k_\nu\Pi^{\mu\nu}(k) = 0, \tag{3.8}$$

which insures the gauge invariance of the definition; indeed, when one is performing a gauge transformation on $A^\mu(k)$,

$$\begin{cases} A^\mu(k) \to A^\mu(k) - ik^\mu\Lambda(k), \\ k^2\Lambda(k) = 0, \end{cases} \tag{3.9}$$

the relation $J^\mu(k) = \Pi^\mu{}_\nu(k)A^\nu(k)$ remains invariant.

As a consequence of Maxwell's equations,

$$\begin{cases} \partial_\nu F^{\mu\nu}(x) = 4\pi J^\mu(x), \\ \partial^*_\nu F^{\mu\nu}(x) = 0, \end{cases} \tag{3.10}$$

the equations obeyed by the four-potential $A^\mu(k)$ are then written as[2]

$$[k^2\eta^{\mu\nu} - k^\mu k^\nu]A_\nu(k) = -4\pi J^\mu(k) = -4\pi[J^\mu_{\text{ext}}(k) + J^\mu_{\text{int}}(k)], \tag{3.11}$$

[2]For similar considerations, see e.g. D.B. Melrose (1973).

where $J^\mu_{\rm int}(k)$ generates the internal field while $J^\mu_{\rm ext}(k)$ is responsible for the external one. Accordingly, one has

$$\{[k^2\eta^{\mu\nu} - k^\mu k^\nu] + 4\pi\Pi^{\mu\nu}(k)\}A_\nu(k) = -4\pi J^\mu_{\rm ext}(k). \tag{3.12}$$

When one switches off the external field, the internal field can oscillate according to

$$\{[k^2\eta^{\mu\nu} - k^\mu k^\nu] + 4\pi\Pi^{\mu\nu}(k)\}A_\nu(k) = 0, \tag{3.13}$$

which implies the relation

$$\mathrm{Det}\{[k^2\eta^{\mu\nu} - k^\mu k^\nu] + 4\pi\Pi^{\mu\nu}(k)\} = 0. \tag{3.14}$$

However, such a relation provides the proper modes of oscillation of the plasma only after a gauge condition has been chosen and when one is working in the three-space allowed by the rank 3 character of the above 4×4 matrix.

The gauge invariance of the final result can be checked by choosing a gauge-fixing parameter λ (see Chap. 7), so that the above equation reads

$$\mathrm{Det}\{[k^2\eta^{\mu\nu} - (1-\lambda)k^\mu k^\nu] + 4\pi\Pi^{\mu\nu}(k)\} = 0; \tag{3.15}$$

the choice of λ corresponds to different possible gauges — for instance, $\lambda = 1$ implies the use of the Feynman gauge. The gauge invariance of the dispersion relation is then expressed by its independence from the gauge-fixing parameter. Actually, in the nonquantum case, the question of the gauge invariance is almost irrelevant, essentially because they can also be derived directly from the fields.

Some deeper insight into the plasma modes can be obtained after the polarization tensor $\Pi^{\mu\nu}(k)$ has been decomposed on the two orthogonal projectors

$$\begin{cases} P^{\mu\nu}(k) = \Delta^{\mu\nu}(u) + \dfrac{(k^\mu - k\cdot u\, u^\mu)(k^\nu - k\cdot u\, u^\nu)}{\Delta^{\alpha\beta}(u)k_\alpha k_\beta}, \\[2ex] Q^{\mu\nu}(k) = \dfrac{(k^2 u^\mu - k\cdot u\, k^\mu)(k^2 u^\nu - k\cdot u\, k^\nu)}{k^2\Delta^{\alpha\beta}(u)k_\alpha k_\beta}, \end{cases} \tag{3.16}$$

as

$$\Pi^{\mu\nu}(k) = \pi_T(k)P^{\mu\nu}(k) + \pi_L(k)Q^{\mu\nu}(k). \tag{3.17}$$

Note the completeness relation of these projectors,

$$\Delta^{\mu\nu}(k) = P^{\mu\nu}(k) + Q^{\mu\nu}(k), \tag{3.18}$$

whose meaning is that they span the three-space orthogonal to the four-vector k^μ.

Let us now come back to the above dispersion condition. For a polarization vector $e^\mu(k)$,

$$\begin{cases} e_T^\mu = P^{\mu\nu} e_\nu, \\ e_L^\mu = Q^{\mu\nu} e_\nu, \\ e_\lambda^\mu = \dfrac{k^\mu k^\nu}{k^2}, \end{cases} \tag{3.19}$$

the dispersion relation is split into

$$\begin{cases} [k^2 - \pi_T(k)] e_T^\mu(k) = 0, \\ [k^2 - \pi_L(k)] e_L^\mu(k) = 0, \\ \lambda k^2 k^\mu k_\nu e_\lambda^\nu(k) = 0 \end{cases} \tag{3.20}$$

for the transverse (twice-degenerated) and longitudinal polarization four-vectors. The third one corresponds to a gauge degree of freedom, which is not physical. The transverse modes are related to the propagation of photons in a dispersive medium, while the longitudinal one is the *plasmon* mode.

3.2. The Static Conductivity Tensor

Let us consider a neutral plasma in thermal equilibrium and let us apply a constant external field, $F_{\text{ext}}^{\mu\nu}$. We want to evaluate the effects of collisions on the electrical conductivity or, more specifically, the conductivity tensor has to be calculated. To this end, it is still assumed that the collisions of the electrons with either the ions, the other electrons or possible plasmons are described adequately by the relativistic BGK equation written as

$$p \cdot \partial f(x,p) + e p_\nu F_{\text{ext}}^{\mu\nu} \frac{\partial}{\partial p^\mu} f(x,p) = -p \cdot u \frac{f(x,p) - f_{\text{eq}}(p)}{\tau}. \tag{3.21}$$

The *linear* response to the external field $F_{\text{ext}}^{\mu\nu}$ is a four-current $J_1^\lambda(x)$, given by

$$J_1^\lambda(x) = \int d^4p \frac{p^\lambda}{m} f_1(x,p), \tag{3.22}$$

where $f_1(x,p)$ is given by

$$\begin{aligned} f_1(x,p) &\approx -\frac{\tau}{p \cdot u} \left(p \cdot \partial + e p_\nu F_{\text{ext}}^{\mu\nu} \frac{\partial}{\partial p^\mu} \right) f_{\text{eq}}(p) \\ &= -\frac{\tau}{p \cdot u} e p_\nu F_{\text{ext}}^{\mu\nu} \frac{\partial}{\partial p^\mu} f_{\text{eq}}(p). \end{aligned} \tag{3.23}$$

The four-current J_1^λ is thus

$$J_1^\lambda = -\tau e F_{\text{ext}}^{\mu\nu} \int d^4p \frac{p^\lambda p_\nu}{p \cdot u} \frac{\partial}{\partial p^\mu} f_{\text{eq}}(p), \tag{3.24}$$

so that the static conductivity tensor is given by

$$\Lambda_{\nu\mu}^\lambda = \tau e \int d^4p f_{\text{eq}}(p) \frac{\partial}{\partial p^\mu} \left(\frac{p^\lambda p_\nu}{p \cdot u} \right) \tag{3.25}$$

and hence the static conductivity tensor is obtained.

3.3. Debye–Hückel Law

The general form of the relativistic Debye–Hückel law can be obtained from the expression of the conductivity tensor (see below) or, more simply, by looking at the four-potential experienced by an extra particle embedded at rest in a plasma. Let us now look at this simple approach [R. Hakim (1967); S.R. de Groot, W.A. van Leuwen and Ch. G. van Weert (1980)].

Let A^μ be the four-potential to which is subjected a test particle whose charge is Q, at rest within the plasma, and let $J_{\text{plasma}}(x)$ be the plasma four-current. $A^\mu(x)$ obeys Maxwell's equations written in the form

$$\Box A^\mu(x) = 4\pi [J_{\text{plasma}}^\mu(x) + J_{\text{test}}^\mu(x)] \tag{3.26}$$

with the Lorentz condition

$$\partial_\mu A^\mu(x) = 0, \tag{3.27}$$

and where

$$\begin{cases} J_{\text{test}}^\mu(x) = Q \displaystyle\int_{-\infty}^{+\infty} d\tau\, u^\mu \delta^{(4)}(x - u\tau), \\ J_{\text{plasma}}^\mu(x) = e \left\{ \displaystyle\int dp\, p^\mu f_{\text{electrons}}(x,p) - \int dp\, p^\mu f_{\text{ions}}(x,p) \right\}. \end{cases} \tag{3.28}$$

In these last equations u^μ is the average four-velocity both of the plasma and of the test particle (remember that the latter is at rest with respect to the plasma). In the last equation, one has

$$\underset{\text{electrons}}{f_{\text{ions}}}(x,p) = \alpha \exp[-\beta \cdot (p \mp eA)], \tag{3.29}$$

which is the Jüttner–Synge equilibrium function in the presence of the electromagnetic field brought by the perturbation constituted by the test particle. The plasma proper charge density is then (see Chap. 1)

$$n_{\text{plasma}} = en\{\exp(-e\beta \cdot A) - \exp(e\beta \cdot A)\} \approx 2e^2 n \beta_\mu A^\mu(x) \tag{3.30}$$

with $|\beta eu \cdot A|^2 \ll |\beta eu \cdot A|$, while the plasma four-current is $n_{\text{plasma}} u^\mu$. It follows that the equation for A^μ reduces to

$$\Box A^\mu(x) + 8\pi n e^2 \beta u \cdot A u^\mu = 4\pi J^\mu_{\text{test}}(x), \tag{3.31}$$

whose Fourier transform is

$$\{-k^2 \eta^{\mu\nu} + (8\pi n e^2 \beta) u^\mu u^\nu\} A_\nu(k) = 4\pi J^\mu_{\text{test}}(k). \tag{3.32}$$

This algebraic equation can easily be solved as[3]

$$A^\mu(k) = -\frac{1}{k^2} 4\pi J^\mu_{\text{test}}(k) + \frac{8\pi n e^2 \beta}{k^2} \frac{u_\nu u^\mu}{8\pi n e^2 \beta - k^2} 4\pi J^\nu_{\text{test}}(k). \tag{3.33}$$

The first term on the right hand side of this relation is obviously the contribution of the test particle, while the second one represents the screening of the test charge by the electron of the plasma. In the extreme relativistic limit, $\beta \to 0$, the field experienced by the test particle is not screened; no screening is possible since the electrons and the ions suffer a violent thermal agitation. In the opposite case, $\beta \to \infty$ (or T tends to zero), the above formula yields

$$A^\mu(k) \to -\frac{1}{k^2} \Delta^\mu{}_\nu(u) J^\nu_{\text{test}}(k) = 0, \tag{3.34}$$

since $J^\nu_{\text{test}}(k) \propto u^\mu$. Finally, in a reference frame where $u^\mu = (1, \mathbf{0})$, A^0 obeys the equation

$$\Box A^0(x) + 8\pi n e^2 \beta A^0(x) = 4\pi J^0_{\text{test}}, \tag{3.35}$$

which in the static limit is the well-known equation

$$\Delta V(x) - 8\pi n e^2 \beta V(x) = 4\pi n_{\text{test}}, \tag{3.36}$$

whose solution is the usual Debye–Hückel law.

3.4. Derivation of the Plasma Modes

In this section, the covariant Vlasov equation is given and the dispersion equation obeyed by the collective modes of an electron plasma embedded in a positively charged neutralizing background is briefly derived. The latter equation is nothing but the Liouville equation for electrons subjected to

[3]The reference R. Hakim (1967) contains some misprints and the correct expression is the one given here.

the average electromagnetic field of the plasma coupled to the Maxwell's equations, or

$$\left[p^{\mu}\partial_{\mu} + ep_{\upsilon}F^{\mu\upsilon}(x)\frac{\partial}{\partial p^{\mu}}\right] f(x,p) = 0, \tag{3.37}$$

$$\begin{cases} \partial_{\mu}F^{\mu\upsilon}(x) = 4\pi J^{\upsilon}(x), \\ \partial_{\mu}^{*}F^{\mu\upsilon}(x) = 0, \end{cases} \tag{3.38}$$

where $J^{\upsilon}(x)$ is the four-current that generates the electromagnetic field $F^{\mu\upsilon}(x)$:

$$J^{\upsilon}(x) = e\int d^4p\frac{p^{\upsilon}}{m}f(x,p). \tag{3.39}$$

The dispersion relations obeyed by the normal modes of the plasma are derived as usual by looking at the propagation of small electron perturbations within the plasma while retaining first order effects only:

$$\begin{cases} f(x,p) = f_{\mathrm{eq}}(p) + f_{(1)}(x,p), \\ F^{\mu\upsilon}(x) = F^{\mu\upsilon}_{(\mathrm{eq})}(x) + F^{\mu\upsilon}_{(1)}(x), \end{cases} \tag{3.40}$$

with $F^{\mu\upsilon}_{(\mathrm{eq})}(x) \equiv 0$, since the plasma is neutral i.e. $J^{\upsilon}_{(\mathrm{eq})} \equiv 0$ and

$$[f_{(1)}(x,p)]^2 \ll f_{(1)}(x,p). \tag{3.41}$$

The equation for the off-equilibrium quantities, indexed by 1, is thus

$$J^{\nu}_{(1)}(x) = e\int d^4p\frac{p^{\upsilon}}{m}f_{(1)}(x,p), \tag{3.42}$$

$$\partial_{\mu}F^{\mu\nu}_{(1)}(x) = 4\pi J^{\nu}_{(1)}(x), \tag{3.43}$$

$$ep_{\upsilon}F^{\mu\nu}_{(1)}(x)\frac{\partial}{\partial p^{\mu}}f_{\mathrm{eq}}(p) = -p\cdot\partial f_{(1)}(x,p). \tag{3.44}$$

The last equation — the linearized Vlasov equation — after performing a Fourier transform and using the definition of $J^{\lambda}(k)$ yields

$$\begin{aligned} J^{\lambda}(k) &= \int d^4p\, p^{\nu} f_{(1)}(k,p) = -eF^{\mu\nu}_{(1)}(k)\int d^4p\frac{p_{\nu}p^{\lambda}}{k\cdot p - i\varepsilon}\frac{\partial}{\partial p^{\mu}}f_{\mathrm{eq}}(p) \\ &= -ie\big(k^{\mu}A^{\nu}_{(1)} - k^{\nu}A^{\mu}_{(1)}\big)\int d^4p\frac{p_{\nu}p^{\lambda}}{k\cdot p - i\varepsilon}\frac{\partial}{\partial p^{\mu}}f_{\mathrm{eq}}(p). \end{aligned} \tag{3.45}$$

Then, Maxwell's equations written in the form

$$k^2A^{\mu}_{(1)}(k) = -4\pi J^{\mu}_{(1)}(k), \tag{3.46}$$

together with the Lorentz gauge condition

$$k_{\mu}A^{\mu}_{(1)}(k) = 0, \tag{3.47}$$

yield the following homogeneous system for $A^{\mu}_{(1)}(k)$:

$$\left\{\eta^{\mu\alpha}k^2 + 4\pi i e^2\left(k^{\mu}\eta^{\nu\alpha} - k^{\nu}\eta^{\mu\alpha}\right)\int d^4p \frac{p^{\nu}}{k\cdot p - i\varepsilon}\frac{\partial}{\partial p^{\mu}} f_{\mathrm{eq}}(p)\right\} A_{\alpha}(k) = 0. \tag{3.48}$$

It possesses solutions only when its determinant vanishes, i.e. when

$$\mathrm{Det}\left\{\eta^{\beta}_{\lambda}\left(1 - \frac{k^{\alpha}I_{\alpha}}{k^2}\right) + \omega_P^2 I^{\beta}_{\lambda} + \omega_P^2 k^{\beta}\frac{I_{\lambda}}{k^2}\right\} = 0, \tag{3.49}$$

and, of course, the gauge condition $\partial_{\mu}A^{\mu} = 0$.

In this expression the various integrals I are defined by

$$\begin{cases} I_{\nu} = \dfrac{m\beta}{4\pi K_2(m\beta)} \displaystyle\int d^4\varsigma \frac{\varsigma_{\nu}}{k_{\alpha}\varsigma^{\alpha}} \exp(-m\beta_{\mu}\varsigma^{\mu}), \\ I^{\nu}_{\mu} = \dfrac{m\beta}{4\pi K_2(m\beta)} \displaystyle\int d^4\varsigma \frac{\varsigma_{\mu}\varsigma^{\nu}}{(k_{\alpha}\varsigma^{\alpha})^2} \exp(-m\beta_{\mu}\varsigma^{\mu}), \end{cases} \tag{3.50}$$

while the plasma frequency is denoted, as usual, by

$$\omega_P^2 = \frac{4\pi n e^2}{m}. \tag{3.51}$$

In the local frame of reference of the plasma, i.e. the one where $u^{\mu} = (1, \mathbf{0})$, and choosing the third axis as the direction of propagation of the plasma waves,

$$k^{\mu} = (k^0, 0, 0, k^3) \equiv (\omega, 0, 0, k), \tag{3.52}$$

the above determinant is split into the two equations

$$\left\{1 - \frac{1}{k^{\alpha}k_{\alpha}}\omega_P^2 \frac{K_1(m\beta)}{K_2(m\beta)}\right\} + \omega_P^2 I^1_1 = 0 \qquad \text{(for transverse modes)}$$

$$\left[\left\{1 - \frac{1}{k^{\alpha}k_{\alpha}}\omega_P^2 \frac{K_1(m\beta)}{K_2(m\beta)}\right\} + \omega_P^2 I^0_0 + \frac{\omega_P^2\omega}{k^{\alpha}k_{\alpha}} I_0\right]$$
$$\times \left[\left\{1 - \frac{1}{k^{\alpha}k_{\alpha}}\omega_P^2 \frac{K_1(m\beta)}{K_2(m\beta)}\right\} + \omega_P^2 I^3_3 + \frac{\omega_P^2 k}{k^{\alpha}k_{\alpha}} I_3\right]$$
$$= \omega_P^4\left\{\frac{k}{k^{\alpha}k_{\alpha}} I_0 + I^3_0\right\} \times \left\{I^3_0 + \frac{\omega I_3}{k^{\alpha}k_{\alpha}}\right\} \qquad \text{(for longitudinal modes).} \tag{3.53}$$

This equation for the longitudinal modes can also be cast into the form

$$\frac{\Omega_P^2 + k^2}{\omega^2 - k^2} = 2I_0 \frac{\omega_P^2 \omega}{\omega^2 - k^2} + \omega_P^2 \frac{\partial}{\partial \omega} I_0, \tag{3.54}$$

where we have set

$$\Omega_P^2 = \omega_P^2 \frac{K_1(m\beta)}{K_2(m\beta)}, \tag{3.55}$$

which will appear as an *effective* relativistic plasma frequency. For transverse modes, we have also

$$\left\{1 - \frac{\Omega_P^2}{\omega^2 - k^2}\right\} + \omega_P^2 I_1^1 = 0. \tag{3.56}$$

The various properties of the Kelvin functions indicate that Ω_P^2 is an increasing function of the parameter $m\beta$, which vanishes for $m\beta = 0$ and tends, as expected, toward the ordinary plasma frequency ω_P^2 for $m\beta \to \infty$.

Finally, let us note that these dispersion relations do not agree with those obtained by other authors. For instance, they do not agree with B. Kursunoglu's (1961) results since this author uses an incorrect equilibrium distribution function because of a confusion in his notation. As to the results of F. Santini and G. Szamosi (1965) or of K.-K. Tam (1968), they are different from ours because of an incorrect normalization of their distribution function.

3.4.1. *Evaluation of the various integrals*

In this section the various integrals appearing in the above dispersion relations are briefly evaluated.[4] First, notice that the relation

$$k^\nu I_{\mu\nu} = -I_\mu \tag{3.57}$$

reduces the calculation to that of I_μ only. This integral is a function of β, $k \cdot k$ and $k \cdot u$. It has the general form

$$I^\mu = c_1 \beta^\mu + c_2 k^\mu, \tag{3.58}$$

since β^μ and k^μ are the only four-vectors in the theory. Multiplying now this expression of I^μ by β^μ and by k^μ successively, one gets a linear system for c_1 and c_2, whose solution is

$$c_1 = \frac{k \cdot \beta \varphi - k \cdot k \psi}{(k \cdot k)(\beta \cdot \beta) - (k \cdot \beta)^2}, \tag{3.59}$$

$$c_1 = \frac{\beta \cdot \beta \varphi - k \cdot \beta \psi}{(k \cdot k)(\beta \cdot \beta) - (k \cdot \beta)^2}, \tag{3.60}$$

[4]For more details, see R. Hakim and A. Mangeney (1968).

where the following notations have been used:

$$\begin{cases} \varphi \equiv k_\mu I^\mu = \dfrac{K_1(m\beta)}{K_2(m\beta)}, \\ \psi \equiv \beta_\mu I^\mu. \end{cases} \tag{3.61}$$

It remains for one to calculate the invariant ψ, and one finds that

$$\begin{aligned} \mathrm{Im}\psi &= \frac{\pi}{2k}\frac{\beta}{K_2(m\beta)}\exp\left(-\frac{m\beta}{\sqrt{1-(\omega/k)^2}}\right) \\ &\quad \times \left[\frac{1}{m\beta}+\frac{1}{\sqrt{1-(\omega/k)^2}}\right] \ (\text{for } \omega\coth\chi \leq k), \end{aligned} \tag{3.62}$$

$$\begin{aligned} \mathrm{Re}\psi &= -\frac{m\beta^2\omega}{k^2K_2(m\beta)}\cdot\frac{d}{d(m\beta)} \\ &\quad \times \left\{\frac{1}{m\beta}\int_0^{+\infty} d\chi \frac{\exp(-m\beta ch\chi)}{((\omega/k)^2-1)\cosh^2\chi+1}\right\}. \end{aligned} \tag{3.63}$$

Details can be found elsewhere (see Ref. 3).

3.4.2. *Collective modes in extreme cases*

In this subsection, two limiting cases are examined; the zero temperature limit and the infinite temperature case.

(1) At absolute zero, the Jüttner–Synge equilibrium distribution becomes

$$f_{\mathrm{eq}}(p) = nm\delta^{(4)}(p-mu) \tag{3.64}$$

and the various integrals reduce to

$$\begin{cases} I_\nu = \dfrac{u_\nu}{k\cdot u}, \\ I_\mu^\nu = \dfrac{u_\mu u^\nu}{(k\cdot u)^2}, \end{cases} \tag{3.65}$$

so that the dispersion relations for the transverse modes read

$$k\cdot k = \omega_P^2 \tag{3.66}$$

or

$$\omega^2 = \mathbf{k}^2 + \omega_P^2, \tag{3.67}$$

which is the usual result, in the local reference frame and for waves propagating along the third axis. For longitudinal oscillations, we obtain

$$\begin{cases} \omega_1^2 = \omega_P^2, \\ \omega_2^2 = \mathbf{k}^2 + \omega_P^2. \end{cases} \tag{3.68}$$

While the first solution is the ordinary one, the second seems to be due to gauge properties' modes and it disappears at the Newtonian limit.

(2) Let us now investigate briefly the extreme relativistic case (infinite temperature, or $m\beta \to 0$). In this extreme relativistic limit, the dispersion relations reduce to $k^2 = 0$ and the medium becomes "transparent" for plasma waves. However, as we shall show in Chap. 5, such waves are strongly damped by the emission of radiation.

The plasma modes have been studied elsewhere [R. Hakim and A. Mangeney (1971)] and the main results will only be outlined here. There are two most interesting cases. First, for transverse waves, it is the case of *supraluminous* waves, i.e. waves having a phase velocity greater than the speed of light in a vacuum. Second, for longitudinal waves, two cases are of particular interest, supraluminous waves and suprathermal waves. We shall limit ourselves to these two cases.

3.5. Brief Discussion of the Plasma Modes

Supraluminous transverse waves. From the above dispersion relation for transverse waves and the evaluation of the integral I_1^1 at order 2 in the "parameter" k/ω, one finds that

$$\omega^4 - [k^2 + \Omega_P^2 - \omega_P^2\phi_1(\beta m)]\omega^2 + k^2\omega_P^2[\phi_2(\beta m) - \phi_1(\beta m)] = 0, \tag{3.69}$$

where ϕ_1 and ϕ_2 are the following functions [R. Hakim and A. Mangeney (1971)]:

$$\phi_\ell(x) = \frac{x}{K_2(x)}\frac{1}{2\ell+1}\int_0^\infty d\chi\,(\tanh\chi)^{2\ell+1}\cosh^2\chi\exp(-x\cosh\chi). \tag{3.70}$$

The dispersion relation for transverse waves is now solved by iteration and the resulting equation has always a (physical) positive and a (unphysical) negative solution.[5] For k^2 very small, one finds that

$$\omega_0^2 \approx \Omega_P^2 - \omega_P^2\phi_1(\beta m) \approx \omega_P^2[\phi_0(\beta m) - \phi_1(\beta m)], \tag{3.71}$$

[5]This is a consequence of the fact that $\phi_1 - \phi_2 > 0$, as can easily be checked.

whereas, in the same case (small values of k^2), the classical case is such that $\omega_0^2 \approx \omega_P^2$. Therefore, one of the relativistic effects consists in a shift of the plasma frequency. This frequency shift can be evaluated via an asymptotic expansion of ϕ_i, and one obtains

$$\omega_0^2(k^2 \sim 0) \approx \omega_P^2 \left[1 - \frac{5}{2\beta m}\right] \tag{3.72}$$

so that

$$\left|\frac{\Delta\omega_0^2}{\omega_0^2}\right| \approx \frac{5}{2\beta m}, \tag{3.73}$$

and a shift of 10% could be expected in plasmas whose temperatures are of the order of a few hundred thousand degrees.

Supraluminous longitudinal waves. First, the dispersion relation is rewritten as

$$\Omega_P^2 + k^2 = \left\{\sum_{\ell=0}^{\infty}\left(\frac{k}{\omega}\right)^{2\ell}\phi_\ell(\beta m)\right\} - \omega_P^2\left(1 - \frac{k^2}{\omega^2}\right) \times \left\{\sum_{\ell=0}^{\infty}\left(\frac{k}{\omega}\right)^{2\ell}(2\ell+1)\phi_\ell(\beta m)\right\}, \tag{3.74}$$

where the real part of the integral I_0 has been expanded in powers of the inverse phase velocity. At order 2 a straightforward calculation provides

$$\omega_0^2 = \left(\frac{K_1(m\beta)}{K_2(m\beta)} - \phi_1(m\beta)\right)\omega_P^2 + 3k^2\frac{\phi_1(m\beta) - \phi_2(m\beta)}{\frac{K_1(m\beta)}{K_2(m\beta)} - \phi_1(m\beta)}, \tag{3.75}$$

and, as in the case of transverse waves, the plasma frequency is shifted from its nonrelativistic value. Using the asymptotic expansions of the various functions K's and ϕ's, this dispersion relations yields the usual nonrelativistic relation:

$$\omega_0^2 = \omega_P^2 + 3\frac{k_BT}{m}k^2. \tag{3.76}$$

The first relativistic correction turns out to be

$$\omega_0^2 = \omega_P^2 + 3\frac{k_BT}{m}k^2 - \frac{5}{2}\omega_P^2\frac{k_BT}{mc^2} - \frac{33}{2}\left(\frac{k_BT}{mc^2}\right)^2 k^2c^2, \tag{3.77}$$

where the factors c have been re-established and which is nothing but an expansion in powers of the (usual) thermal velocity over c^2. A deviation of 10% of the coefficient of k^2 occurs at relatively moderate temperatures, of the order of 10^8 K.

Suprathermal longitudinal waves. While in the case of supraluminous phase velocities there is no possible resonance between plasma waves and electrons, such an effect is possible for suprathermal velocities, inferior to that of light, i.e. when

$$v_{\text{thermal}} < \frac{\omega}{k} < 1. \tag{3.78}$$

Accordingly, the "small parameter"

$$\begin{cases} \eta^2 = \frac{1-\frac{\omega^2}{k^2}}{(\beta m)^2}, \\ \omega < k, \end{cases} \tag{3.79}$$

is introduced and we restrict ourselves to those μ's (or those β's) such that $\eta^2 \ll 1$. It follows that if ω is to be considered as a given quantity, k should satisfy the condition[6]

$$k^2 < \frac{\omega^2}{1-\alpha^2}, \quad \alpha < 1. \tag{3.80}$$

Note that the term $\cosh\chi$, which occurs in the exponential of the various integrals of the dispersion relations is at most of order $(\beta m)^{-1}$ and hence the velocity of a typical electron, namely $\tanh\chi$, is at most of order $1-(\beta m)^2$. It follows that $\eta^2 < 1$ represents $v_{\text{thermal}} < \omega/k$. At order 1 in the "small parameter" η^2, the dispersion relation for transverse waves turns out to be

$$\omega^2 - k^2 = \frac{\Omega_P^2 k^2}{k^2 + \Omega_P^2 + \omega_P^2\left[\frac{1}{\beta m} - 2\frac{K_1(m\beta)}{K_2(m\beta)}\right]}. \tag{3.81}$$

The right hand side of this equation may be shown to be always positive, so that $\omega^2/k^2 > 0$, which is not compatible with our assumption $\omega < k$. Therefore, as in the nonrelativistic case, there is no propagation[7] of infraluminous transverse waves[8] and we concentrate on the longitudinal case.

We first neglect the damping term, since its weakness is expected and is evaluated at the end of the calculation. The dispersion equation then reads

$$\Omega_P^2 + k^2 = 2\omega\omega_P^2 I_0 + \omega_P^2(\omega^2 - k^2)\frac{\partial}{\partial\omega} I_0. \tag{3.82}$$

[6]When $\alpha > 1$, we always have $\eta^2 < 1$ since $\omega/k < 1$.
[7]This result is also valid at order 2 in η^2.
[8]See e.g. S. Ichimaru, *Basic Principles of Plasma Physics* (Benjamin, Reading, 1973).

In order to obtain an explicit relation valid at order 1 in the expansion parameter η^2, we set

$$\operatorname{Re} I_0 = \frac{A}{\omega} + \frac{B}{\omega}\eta^2 + O(\eta^4), \tag{3.83}$$

where A and B are functions of βm. The expansion of I_0 and various recursion relations between the Kelvin functions provide

$$\begin{cases} A(\beta m) = \dfrac{K_1(m\beta)}{K_2(m\beta)}\left\{1 - \dfrac{2}{(m\beta)^2}\right\} + \dfrac{1}{m\beta}, \\ B(\beta m) = 3\dfrac{K_1(m\beta)}{K_2(m\beta)} - m\beta. \end{cases} \tag{3.84}$$

Using the fact that

$$\omega^2 - k^2 = -\omega^2\eta^2(\beta m)^2 + O(\eta^4), \tag{3.85}$$

the dispersion relation can be cast into the form

$$\omega^2 = [2\omega_P^2 A(\beta m) - \Omega_P^2] + \eta^2[2\omega_P^2 B(\beta m) + \Omega_P^2(\beta m)^2 - \omega_P^2(\beta m)^2 A(\beta m)], \tag{3.86}$$

which itself leads to

$$\begin{aligned} \omega^2 = \omega_P^2 \left(A(m\beta) + \frac{B(m\beta)}{(m\beta)^2}\right) - \frac{1}{k^2}(2\omega_P^2 A(m\beta) - \Omega_P^2) \\ \times \left(\Omega_P^2 - \omega_P^2 A(m\beta) + \frac{2\omega_P^2 B(m\beta)}{(m\beta)^2}\right). \end{aligned} \tag{3.87}$$

This last equation is obtained from the preceding one by replacing $\eta^2(\omega, k)$ with its value for $\omega = \omega_0$, where ω_0 is evaluated at order 0 in $\eta^2 \cdot \omega_0$ is given by

$$\omega_0^2 = 2\omega_P^2 A(\beta m) - \Omega_P^2; \tag{3.88}$$

it is a wave propagating at velocity c since, in this approximation, $\eta^2 = 0$ and hence $\omega = k$.

The dispersion curves are most easily plotted with the dimensionless variables

$$\begin{cases} y = \omega^2\omega_P^{-2}, \\ x = k^2\omega_P^{-2}, \end{cases} \tag{3.89}$$

and are depicted in Fig. 3.1. They must, of course, be suprathermal and infraluminous, two properties expressed by

$$\begin{cases} y < x, \\ yx^{-1} > 1 - (\beta m)^2, \end{cases} \tag{3.90}$$

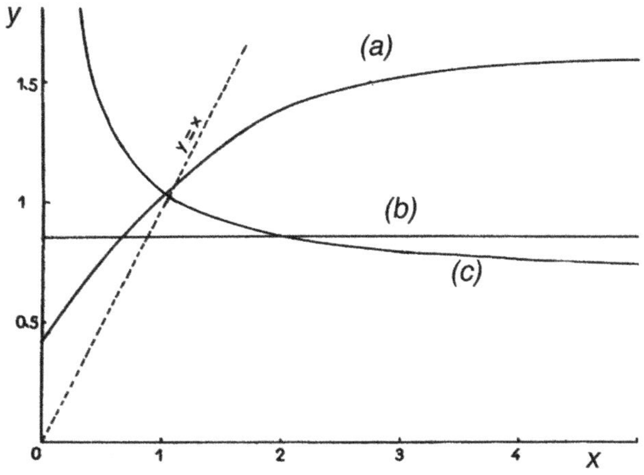

Fig. 3.1 Typical dispersion curves for longitudinal suprathermal waves [after R. Hakim and A. Mangeney, *Phys. Fluids* **14**, 2751 (1971)].

and $\beta m < 1$. Three typical dispersion curves have been plotted with qualitatively different behavior according to the sign of x^{-1}.

Different behaviors. These differences occur according to the sign of the coefficient of x^{-1}. The dispersion curves show that the phase and group velocities for waves such as $\beta m > \beta_0 m$ have opposite directions, while they have the same direction in the opposite case. For $\beta = \beta_0$, only stationary waves are possible; β_0 is a solution of

$$\frac{K_1(m\beta)}{K_2(m\beta)} = \frac{3m\beta}{8} \tag{3.91}$$

and is found to be of order 1.

Let us now turn to the nonrelativistic approximation of the dispersion equation. As already indicated, this approximation is obtained for $\beta m \to \infty$. However, the expansions in $\eta^2\alpha^2$ used in the derivation of the approximate dispersion relation are convergent only when $\eta^2\alpha^2 < 1$. Therefore, as βm increases, η^2 must go to zero in such a way that this condition is satisfied. It follows that the nonrelativistic approximation is valid only in a small neighborhood of

$$k^2 = \omega_P^2 \left[1 + O\left(\frac{1}{(\beta m)^2}\right)\right]. \tag{3.92}$$

In this neighborhood, the dispersion equation is equivalent to the classical dispersion relation and the first relativistic corrections are exactly the same as those already obtained above.

Let us now evaluate the damping decrement γ and let us still assume that $\gamma \ll \omega_0$, so that

$$\omega^2 \approx \omega_0^2 + 2i\gamma\omega_0. \tag{3.93}$$

Inserting this relation into the dispersion equation and using

$$I_0(\omega + i\gamma) \approx I_0(\omega) + i\gamma \frac{\partial}{\partial\omega} I_0 \bigg|_{\omega=\omega_0}, \tag{3.94}$$

we find that

$$\gamma = \frac{\pi}{4} \frac{\beta m}{K_2(\beta m)} \frac{\omega_0^2 \omega_P^2}{k^2} \frac{1}{\eta^2(\beta m)^2} \exp\left(-\frac{1}{\eta}\right). \tag{3.95}$$

This is the relativistic expression for the *Landau damping*, to which it reduces at the Newtonian limit. This expression is valid for $\omega = k$, since it corresponds to the lowest approximation in η^2, and when $\beta m \gg 1$, i.e. at the ultrarelativistic limit.

3.6. The Conductivity Tensor

From the expression of the small deviation from equilibrium

$$f_{(1)}(k,p) = \frac{1}{k \cdot p - i\varepsilon} e p_\beta F_{(1)}^{\alpha\beta}(k) \frac{\partial}{\partial p^\alpha} f_{\text{eq}}(p), \tag{3.96}$$

a result of the linearized Vlasov equation, one obtains the expression for the off-equilibrium four-current

$$J^\mu(k) = \int d^4p\, p^\mu f_{(1)}(k,p) = -eF_{(1)}^{\alpha\beta}(k) \int d^4p \frac{p_\alpha p^\mu}{k \cdot p - i\varepsilon} \frac{\partial}{\partial p^\beta} f_{\text{eq}}(p), \tag{3.97}$$

from which one immediately reads the expression for the conductivity tensor:

$$\Lambda^{\mu\alpha\beta}(k) = -e \int d^4p \frac{p^\mu p^\alpha}{k \cdot p - i\varepsilon} \frac{\partial}{\partial p_\beta} f_{\text{eq}}(p). \tag{3.98}$$

In terms of our various integrals, it can be written as

$$\Lambda^{\mu\alpha\beta}(k) = \frac{\omega_P^2}{8\pi i} \left\{ (\eta^{\mu\beta} I^\alpha - \eta^{\mu\alpha} I^\beta) + (k^\beta I^{\alpha\mu} - k^\alpha I^{\beta\mu}) \right\}. \tag{3.99}$$

At zero temperature, this tensor has the explicit form

$$\Lambda^{\mu\alpha\beta}(k) = \frac{\omega_P^2}{8\pi i} \left\{ \frac{(\eta^{\mu\beta} u^\alpha - \eta^{\mu\alpha} u^\beta)}{k \cdot u} + \frac{u^\mu (k^\alpha u^\beta - k^\beta u^\alpha)}{(k \cdot u)^2} \right\}, \tag{3.100}$$

which reduces to its usual form[9] in the frame of reference where $u^\mu = (1, \mathbf{0})$. At infinite temperature, one may show that $\Lambda^{\mu\alpha\beta}(k)$ vanishes.

3.7. Plasma–Beam Instability

Let us now consider the problem of a relativistic plasma into which a beam of neutralized relativistic electrons is injected. Such a problem can be described by two distribution functions: a Jüttner–Synge distribution for the plasma and another one, of a different type, for the beam. Denoting by u^μ_{beam} the four-velocity of the beam, the latter could be described by

$$f_{\text{beam}}(p) = n_{\text{beam}} \delta^{(4)}(p - m u_{\text{beam}}). \tag{3.101}$$

This distribution is, however, much too singular and a dispersion in momentum should be allowed. Therefore, we are led to choose another Jüttner–Synge distribution for the beam, which is the simplest but not the only possible choice:

$$f_{\text{beam}}(p) = \frac{n_{\text{beam}} \beta^*}{4\pi m^2 K_2(m\beta^*)} \exp(-\beta^* u_{\text{beam}} \cdot p). \tag{3.102}$$

Furthermore, the beam is assumed to be cold, i.e. $m\beta^* \gg 1$, and constitutes a weak perturbation of the basic plasma: $n_{\text{beam}} \ll n$.

3.7.1. *Perturbed dispersion relations for the plasma–beam system*

Let us limit the present discussion to longitudinal modes. Writing the dispersion relations for these plasma waves as

$$D(\omega, k) = 0, \tag{3.103}$$

the perturbation brought about by the beam yields

$$D(\omega, k) + \delta D(\omega, k) = 0, \tag{3.104}$$

where $\delta D(\omega, k)$ is the modification introduced by the presence of the beam. Solutions to the perturbed dispersion relations will now be sought, close to those of the unperturbed system, as

$$\omega = \omega_0 + \delta\omega. \tag{3.105}$$

---

[9] See e.g. S. Gartenhaus, *Elements of Plasma Physics* (Holt, Rinehart and Winston, New York, 1964).

From the perturbed dispersion relation, we immediately obtain

$$\delta D(\omega,k)+\delta\omega\left.\frac{\partial D(\omega,k)}{\partial\omega}\right|_{\omega=\omega_0}=0, \tag{3.106}$$

where terms of order $(m\beta^*)^2$ are implicitly neglected. Thus, we have

$$\delta\omega=-\left.\frac{\delta D(\omega,k)}{\frac{\partial D(\omega,k)}{\partial\omega}}\right|_{\omega=\omega_0}. \tag{3.107}$$

The denominator occurring in this equation is easily calculated at order 0 in η^2 and is found to be

$$\left.\frac{\partial D(\omega,k)}{\partial\omega}\right|_{\omega=\omega_0}=\frac{1}{2\omega_0}, \tag{3.108}$$

where ω_0 is the solution to the longitudinal dispersion relation, given above. Let us now concentrate on the calculation of $\delta D(\omega,k)$. From the longitudinal dispersion relation

$$\frac{\Omega_P^2+k^2}{\omega^2-k^2}=2I_0\frac{\omega_P^2\omega}{\omega^2-k^2}+\omega_P^2\frac{\partial}{\partial\omega}I_0, \tag{3.109}$$

we obtain

$$\delta D(\omega,k)=\delta\Omega_P^2-2\omega_P^2\omega_0\delta I_0-(\omega_0^2-k^2)\omega_P^2\frac{\partial}{\partial\omega}\delta I_0, \tag{3.110}$$

with

$$\begin{cases}\delta\Omega_P^2=\dfrac{n_{\text{beam}}}{n}\omega_P^2\dfrac{K_1(m\beta^*)}{K_2(m\beta^*)},\\[2ex] \delta I_\mu=\dfrac{1}{nm^2}\displaystyle\int\frac{d^3p}{p_0}\frac{p_\mu}{\omega p_0-kp_3}f_{\text{beam}}(p).\end{cases} \tag{3.111}$$

Despite the formal similarity between I_0 and δI_0, some care is needed when one is evaluating δI_0; indeed, I_0 has been calculated in the rest frame of the background plasma where δI_0 must also be evaluated, and *not* in the rest frame of the beam.

3.7.2. *Stability of the beam–plasma system*

Since we are mainly interested in the stability of the system, we deal with only the imaginary part of $\delta\omega$, and we find that

$$\begin{aligned}\text{Im}\delta\omega=\delta\gamma&=\frac{\pi}{4}\frac{n_{\text{beam}}}{n_{\text{plasma}}}\frac{\beta^*m}{K_2(\beta^*m)}\frac{\omega_P^2\omega_0}{k^2}|\Delta|^{-1/2}\\&\quad\times(\omega u^0_{\text{beam}}-ku^3_{\text{beam}})\frac{1}{(\beta m)^2\eta^2}\exp\left(-\beta^*m\left[\frac{\Delta}{s}\right]^{1/2}\right).\end{aligned} \tag{3.112}$$

where we have set

$$\begin{cases} \Delta = \Delta_{\mu\nu}(u_{\text{beam}})k^\mu k^\nu, \\ s = k^2. \end{cases} \tag{3.113}$$

This expression reduces to the usual nonrelativistic one and this can be checked with the use of the following data:

$$\begin{cases} \beta m \gg 1 \quad \text{(nonrelativistic background plasma)}, \\ u^0_{\text{beam}} \approx 1, \ u_{\text{beam}} \approx v \quad \text{(nonrelativistic beam)}, \\ |\Delta|^{1/2} \approx k \ (\text{since } \Delta_{\mu\nu} \approx -\delta_{ij}). \end{cases} \tag{3.114}$$

Also, use is made of the asymptotic form of the Kelvin function K_2 and of the relation

$$\begin{cases} \exp\left\{\beta^* m \left[1 - \left(\dfrac{\Delta}{s}\right)^{1/2}\right]\right\} \approx \exp\left(-\dfrac{1}{2}\beta^* m \left[\dfrac{\omega}{k} - v\right]^2\right), \\ \text{with } \omega^2 \ll k^2 \text{ or, equivalently, } \beta^2\eta^2 \approx 1. \end{cases} \tag{3.115}$$

Let us now compare $\delta\gamma$ with the Landau damping decrement γ_L, obtained above. We have

$$\frac{\delta\gamma}{\gamma_L} = \frac{n_{\text{beam}}}{n_{\text{plasma}}} \frac{\beta^*}{\beta} \frac{K_2(\beta m)}{K_2(\beta^* m)} \frac{k}{|\Delta|^{1/2}} \frac{\omega u^0_{\text{beam}} - k u^3_{\text{beam}}}{\omega_0} \times \exp\left\{\frac{1}{\eta} - \beta^* m \left(\frac{\Delta}{s}\right)^{1/2}\right\}, \tag{3.116}$$

which may be rewritten as

$$\frac{\delta\gamma}{\gamma_L} = \frac{n_{\text{beam}}}{n_{\text{plasma}}} \frac{\beta^*}{\beta} \frac{K_2(\beta m)}{K_2(\beta^* m)} \frac{k}{|\Delta|^{1/2}} \frac{\omega u^0_{\text{beam}} - k u^3_{\text{beam}}}{\omega_0} \times \exp\left\{\frac{1}{\eta}\left(1 - \frac{\beta^*}{\beta}|u^3_{\text{beam}} - u^0_{\text{beam}}|\right)\right\}, \tag{3.117}$$

where the expression of Δ which occurs in the exponential has been evaluated at order 0 in η^2. Note that this equation is not valid when $v \to 1$. This equation is quite similar to the nonrelativistic one,[10] and hence similar conclusions can be drawn.

Let us examine the case of a highly relativistic beam, i.e. such that its ordinary velocity $w \equiv |\mathbf{w}|$ is close to 1. We have

$$u^3_{\text{beam}} = u^0_{\text{beam}} w \cos\theta, \tag{3.118}$$

[10] See A.G. Sitenko, *loc. cit.*

where θ is the angle between the spatial direction of the wave and that of the beam, i.e. between $\mathbf{k}$ and $\mathbf{w}$. A necessary condition for an unstable behavior is

$$\omega u^0_{\text{beam}} - k u^3_{\text{beam}} < 0 \tag{3.119}$$

or, equivalently,

$$\cos\theta > \frac{\omega}{kw}. \tag{3.120}$$

This last inequality implies that $\omega/k - w < 0$; furthermore, it is necessary that

$$\left|\frac{\delta\gamma}{\gamma_L}\right| > 1. \tag{3.121}$$

In fact, since we are interested in those ω/k's and w's which are close to 1, the angle θ is itself close to 0. For a highly relativistic beam, we have

$$\Delta \approx -(\omega u^0_{\text{beam}} - k u^3_{\text{beam}})^2 \tag{3.122}$$

so that, for waves propagating normally to the beam, we get

$$\delta\gamma \approx \frac{1}{(\beta m)^2\eta^2} \exp\left(-\frac{\beta^*}{\beta\eta}\frac{\omega}{k} u^0_{\text{beam}}\right), \tag{3.123}$$

while for waves propagating along the beam we obtain

$$\delta\gamma \approx -\frac{1}{(\beta m)^2\eta^2} \exp\left(-\frac{\beta^*}{\beta\eta}\left[\frac{\omega}{k} - w\right] u^0_{\text{beam}}\right). \tag{3.124}$$

The first equation shows that waves propagating perpendicularly to the beam are less and less damped as $w \to 1$, while those propagating along the beam are less and less unstable as the beam becomes more and more relativistic. This conclusion agrees with earlier results obtained by K.M. Watson, S.A. Bludman and M.N. Rosenbluth (1960).

Chapter 4

Curved Space–Time and Cosmology

In the presence of gravitation, i.e. in the case of a curved space–time, numerous authors have extended most of the foregoing results. The one-particle phase space has a particular mathematical structure since the configuration space is curved, while the energy–momentum space is cotangent to the space–time manifold

$$\mu = H^4(x) \times V^4, \tag{4.1}$$

where $H^4(x)$ is the energy–momentum space, generally characterized by

$$H^4(x) : g_{\mu\nu}(x)p^\mu p^\nu = m^2. \tag{4.2}$$

This μ space is thus endowed with a *fiber bundle* structure, whose fiber is nothing but the above x-dependent hyperboloid and whose invariance group is the Lorentz group. While the definition of the one-particle density is identical in the curved and flat cases, there exist some minor modifications as to the Liouville (or the kinetic) equation(s) and the Jüttner–Synge equilibrium distribution function.

Note that in this chapter we deal with relativistic kinetic theory within a given gravitational background. When the gravitational field is itself random, the situation is far more involved, and a brief outline of the problems met in such a case is given in Chap. 6.

One should also note that in general relativity one is mainly interested in *local* quantities, or four-currents of various physical quantities, and hence only the latter actually make sense, such as the energy–momentum tensor, the four-current of a given physical quantity (charge, number of particles, etc.), the entropy four-current and other thermodynamic quantities. Another point to be noted is that gravitation being a long range and very weak force, its gradients are generally negligible on the distance of two *short range* colliding particles.[1] It follows that the flat space–time collision term is still valid in the case of general relativity. Of course, this remark is valid as long as ordinary pointlike particles are considered and it should be clear that when galaxies, for instance, are dealt with and considered as "point particles," the gradients of the gravitational field do play a role and the situation is that of a *gravitational plasma* [see H.E. Kandrup (1980*ff*)].

4.1. Basic Modifications

The one-particle distribution function is now a little bit modified via the invariant (under coordinate changes) element of integration on the energy–momentum space $H^4(x)$,

$$\frac{d^3p}{p_0} \to \sqrt{|g(x)|}\,\frac{d^3p}{p_0}, \tag{4.3}$$

where $g(x)$ is the determinant of the metric tensor; of course, in a local frame of reference one has $|g(x)| = 1$. The volume element invariant under arbitrary changes of coordinates is written as $\sqrt{|g(x)|}d^4\,p$ and hence

$$\begin{aligned}\sqrt{|g(x)|}d^3p \int dp^0\; 2\theta(p^0)\delta[g_{\mu\nu}(x)p^\mu p^\nu - m^2]\\ = \sqrt{|g(x)|}\frac{d^3p}{p_0} = \sqrt{|g(x)|}\frac{d^3p}{\sqrt{g_{ij}(x)p^ip^j + m^2}}\end{aligned} \tag{4.4}$$

for a metric whose time and space coordinates are separated.

From the mass shell equation, the zeroth covariant component of p^μ is easily found to be

$$p_0 = \{m^2 g_{00}(x) + [g_{0i}(x)g_{0k}(x) - g_{00}(x)g_{ik}(x)]p^ip^k\}^{1/2}. \tag{4.5}$$

Consequently, the particles four-current and the energy–momentum tensors read, respectively,

$$J^\mu = \int_{g_{\mu\nu}(x)p^\mu p^\nu = m^2} \sqrt{|g(x)|}\frac{d^3p}{p_0(\mathbf{p})}p^\mu, \tag{4.6}$$

$$T^{\mu\nu}(x) = \int_{g_{\mu\nu}(x)p^\mu p^\nu = m^2} \sqrt{|g(x)|}\frac{d^3p}{p_0(\mathbf{p})}p^\mu p^\nu. \tag{4.7}$$

[1] This remark would be invalid for possible collisions of collective states, or quasi-particles, with a spatial extension of the order of the typical length variation of the gravitational gradients.

Let us now briefly derive the Liouville equation when an *external* gravitational field is present and in the absence of other forces, since the latter can be added without any particular difficulty, provided it does not vary too much on the typical variation of $g^{\mu\nu}(x)$. The equations of motion for one particle are the usual geodesic equation

$$\frac{dp^\mu}{d\tau} + \Gamma^\mu_{\alpha\beta} p^\alpha p^\beta = 0, \tag{4.8}$$

which can be obtained from the *formal* Hamiltonian[2]

$$H(x,p) = \frac{1}{2m} g_{\mu\nu}(x) p^\mu p^\nu, \tag{4.9}$$

as

$$\begin{cases} \dfrac{d}{d\tau} x^\mu = \dfrac{\partial}{\partial p_\mu} H(x,p) = \dfrac{p^\mu}{m}, \\ \dfrac{d}{d\tau} p_\mu = -\dfrac{\partial}{\partial x^\mu} H(x,p) = -\dfrac{1}{2m} \partial_\mu g_{\alpha\beta}(x) p_\alpha p_\beta. \end{cases} \tag{4.10}$$

Finally, from the continuity equation in phase space

$$\partial_\mu \left(\frac{dx^\mu}{d\tau} f(x,p) \right) + \frac{\partial}{\partial p_\mu} \left(\frac{dp_\mu}{d\tau} f(x,p) \right) = 0, \tag{4.11}$$

the Liouville equation follows in a straightforward way as

$$p \cdot \partial f(x,p) - \frac{1}{2} \partial^\mu g_{\alpha\beta}(x) p^\alpha p^\beta \frac{\partial}{\partial p^\mu} f(x,p) = 0. \tag{4.12}$$

Many other derivations of this equation have been obtained and it is a simple matter to show that the equation can be rewritten as

$$p \cdot \partial f(x,p) + \Gamma^\mu_{\alpha\beta} p^\alpha p^\beta \frac{\partial}{\partial p^\mu} f(x,p) = 0. \tag{4.13}$$

When the Liouville equation in curved space–time is coupled to the gravitation field via the Einstein equations

$$\begin{aligned} R_{\mu\nu}(x) - \frac{1}{2} R(x) g_{\mu\nu} &= 4\pi G T_{\mu\nu}(x) \\ &= 4\pi G \int_{\substack{g_{\mu\nu}(x) p^\mu p^\nu = m^2 \\ p^0 > 0}} \sqrt{|g(x)|} \frac{d^3p}{p_0} p_\mu \, p_\nu f(p), \end{aligned} \tag{4.14}$$

one obtains the gravitational equivalent of the usual Vlasov equation for an electromagnetic plasma [Ph. Droz-Vincent and R. Hakim (1968)]. This

[2]This means only that the geodesic equation can be recovered as a Hamiltonian equation, and *not* that H is the energy of the particle.

system is often referred to as the Einstein–Liouville equations. In particular, it has been used to find the normal modes of a gravitational plasma [E. Asseo, D. Gerbal, J. Heyvaerts and M. Signore (1978)].

4.2. Thermal Equilibrium in a Gravitational Field

Let us now consider the Jüttner–Synge equilibrium distribution in the presence of gravitation. It constitutes always a *local* equilibrium and not a global one as in the flat space–time case. This is a consequence of the equivalence principle. The gravitational field enters the equilibrium distribution $f_{\text{eq}} = A\exp[-\beta \cdot p]$ through the scalar product

$$\beta \cdot p = g_{\mu\nu}(x)\beta^\mu p^\nu. \tag{4.15}$$

Therefore, the distribution depends on g, through

$$f_{\text{eq}}(p) = A\exp(-g_{\mu\nu}\beta^\mu p^\nu). \tag{4.16}$$

The equilibrium distribution must obey the equation of motion written in the form of the Liouville equation; introducing f_{eq} in the latter, one gets

$$p \cdot \partial A - \frac{1}{2}[\nabla_\mu \beta_\nu + \nabla_\nu \beta_\mu] A p^\mu p^\nu = 0, \tag{4.17}$$

which must be satisfied whatever p is. This implies that, in general, the four-vector β_μ has to satisfy the Killing conditions

$$[\nabla_\mu \beta_\nu + \nabla_\nu \beta_\mu] = 0, \tag{4.18}$$

and A = const. This latter condition yields, for instance, n = const and T = const. These conditions can also be interpreted as conditions for rigid motion.[3] There is, however, another case — the one in which β_μ is Killing conformal:

$$\frac{1}{2}[\nabla_\mu \beta_\nu + \nabla_\nu \beta_\mu] = \phi(x) g^{\mu\nu}(x), \tag{4.19}$$

where $\phi(x)$ is an arbitrary function, and $p^2 = 0$.

A first conclusion is that the one-particle equilibrium does not always exist; for massive particles, the four-vector βu^μ has to be a Killing field while for zero mass particles it is sufficient for it to be Killing conformal.

[3]See e.g. J.L. Synge, *Relativity: The General Theory* (North–Holland, Amsterdam, 1960).

4.2.1. *Thermal equilibrium in a static isotropic metric*

N.A. Chernikov (1964) has given several examples of possible thermal distributions in an external gravitational field and we give only one of them. The metric tensor for a static, spherically symmetric gravitational field has the form

$$ds^2 = \exp[\nu(r)]dt^2 - \exp[\lambda(r)]dr^2 - r^2(d\theta^2 + \sin^2\theta d\varphi^2), \qquad (4.20)$$

written in standard K. Schwarzschild (1916) coordinates. The distribution function

$$f_{\text{eq}}(p) = A\exp(-\beta^0 p_0) \qquad (4.21)$$

is a possible solution provided that the temperature satisfies the relation

$$T(r) = T(\infty) \cdot \exp(1 - \nu[r]). \qquad (4.22)$$

This can be seen by replacing βu^μ with its expression in the Killing equation (4.18).

For the Schwarzschild metric

$$\begin{aligned} \exp[\nu(r)] &= \left(1 - 2\frac{GM}{r}\right), \\ \exp[\lambda(r)] &= \frac{1}{(1 - 2\frac{GM}{r})}, \end{aligned} \qquad (4.23)$$

the temperature should vary as

$$T(r) = T(\infty) \cdot \exp\left(\frac{GM}{r}\right). \qquad (4.24)$$

For a rotating gas (see Chap. 1) subject to the same static spherical symmetric gravitational field, one finds that

$$f_{\text{eq}}(p) = A\exp(-\beta^0 p_0 + \beta^0 \omega r^2 \sin^2\theta p^3) \qquad (4.25)$$

and

$$T(r) = T(\infty) \cdot (\exp[2\nu(r) - 2] - \omega^2 r^2 \sin^2\theta)^{-1/2}. \qquad (4.26)$$

4.3. Einstein–Vlasov Equation

We now study the archetype of a kinetic equation — that of the Vlasov equation where correlations are zero. This system was first studied by Ph. Droz-Vincent and R. Hakim (1968) and was developed by others [E. Asséo, D. Gerbal, J. Heyvaerts and M. Signore (1978)]; later it was followed and developed by H.E. Kandrup (1980*ff*) to many other cases. It is constituted by the two equations

$$\begin{cases} p^\mu \partial_\mu f + \Gamma^\alpha_{\mu\nu} p_\mu p_\nu \dfrac{\partial f}{\partial p_\alpha} = 0, \\ R_{\mu\nu} - \dfrac{1}{2} R g_{\mu\nu} - \lambda g_{\mu\nu} = \chi T_{\mu\nu}, \end{cases} \tag{4.27}$$

where the energy–momentum tensor is given by

$$T^{\eta\nu} = \frac{1}{m}\sqrt{|g|}\int d^3p f(x,p) p^\mu p^\nu, \tag{4.28}$$

and is of course conservative: $\nabla_\mu T^{\mu\nu} = 0$.

In order to be specific, we linearize this system around background data which we do not specify as yet; or some background

$$\begin{cases} f(x,p) = f_0(x,p) + f_1(x,p), \\ g_{\mu\nu}(x) = g_{0\mu\nu}(x) + h_{\mu\nu}(x), \end{cases} \tag{4.29}$$

where all quadratic quantities such as hf_1 are negligible and are neglected. It remains for us to linearize the Einstein equations.

4.3.1. *Linearization of Einstein's equation*

An arbitrary (although preserving the Riemannian property) metric disturbance h induces the first order variation [A. Lichnérowicz (1967)]

$$\delta\Gamma^\gamma_{\alpha\beta} = \frac{1}{2}\{\nabla_\alpha h^\gamma_\beta + \nabla_\beta h^\gamma_\alpha - \nabla^\gamma h_{\alpha\beta}\}. \tag{4.30}$$

It is important to note that the covariant differentiation is defined with the help of the background Riemannian affinity (Christoffel symbols corresponding to $g_{\mu\nu}$) while indices are raised or lowered with the background metric only. For instance, one has

$$h^\alpha_\mu = g^{\alpha\beta} h_{\mu\beta}. \tag{4.31}$$

The change in the Ricci tensor follows from Eq. (4.30) in a straightforward way:

$$\delta R_{\mu\nu} = -\frac{1}{2}\Delta h_{\mu\nu} + \frac{1}{2}(\nabla_\mu I_\nu + \nabla_\nu I_\mu), \tag{4.32}$$

where I_α is defined by

$$\begin{aligned} I_\alpha &= \nabla_\lambda h^\lambda_\alpha - \frac{1}{2}\nabla_\alpha h, \\ h &\equiv h^\alpha_\alpha, \end{aligned} \tag{4.33}$$

and is, up to the sign, de Rham's Laplacian operator extended to symmetrical tensors by A. Lichnnérowicz (1967). With our notations, it reads

$$\Delta h_{\mu\nu} = \nabla_\lambda \nabla^\lambda h_{\mu\nu} - R_{\mu\alpha} h^\alpha_\nu - R_{\nu\alpha} h^\alpha_\mu + 2R_{\mu\nu\rho\sigma} h^{\rho\sigma}. \tag{4.34}$$

From this definition, it follows that Δ reduces to the usual D'Alembertian operator when the background manifold is flat, i.e. when $R_{\mu\nu\alpha\beta} = 0$.

Let us now return to the linearization of Einstein's equation and let us set

$$S_{\mu\nu} \equiv R_{\mu\nu} - \frac{1}{2} g_{\mu\nu} R - \lambda g_{\mu\nu} \tag{4.35}$$

so that Einstein's equation reads

$$R_{\mu\nu} = \chi \left\{ T_{\mu\nu} - \frac{1}{2} g_{\mu\nu} T \right\} + \lambda g_{\mu\nu}. \tag{4.36}$$

Since we have

$$\delta T = -h^{\mu\nu} T_{\mu\nu} + g^{\mu\nu} \delta T_{\mu\nu}, \tag{4.37}$$

where we have used

$$\delta g^{\alpha\beta} = -h^{\alpha\beta}, \quad \delta g_{\alpha\beta} = h_{\alpha\beta}, \tag{4.38}$$

variation of Eq. (4.36) yields

$$\delta R_{\mu\nu} = -\frac{1}{2} h_{\mu\nu} S + \frac{1}{2} g_{\mu\nu} h^{\alpha\beta} S_{\alpha\beta} + \lambda h_{\mu\nu} + \left(j_{\mu\nu} - \frac{1}{2} g_{\mu\nu} j \right), \tag{4.39}$$

where we have set

$$j \equiv g^{\alpha\beta} j_{\alpha\beta}. \tag{4.40}$$

In Eq. (4.39), $\delta R_{\mu\nu}$ is as given by Eq. (4.32). Therefore, Eqs. (4.33) and (4.39) provide a linear partial differential equation for $h_{\alpha\beta}$, whose source term is

$$\left\{ j_{\mu\nu} - \frac{1}{2} g_{\mu\nu} j \right\}. \tag{4.41}$$

Alternatively, we might as well consider that the actual source is just the contribution of the disturbance in the contravariant components of the energy–momentum tensor. Therefore, the "new" source term would involve

$$k^{\mu\nu} = \chi \delta T^{\mu\nu}, \tag{4.42}$$

rather than j terms. Note that $j_{\mu\nu}$ and $k^{\mu\nu}$ are interrelated through

$$j_{\mu\nu} = k_{\mu\nu} + h_{\mu\alpha} S^\alpha{}_\nu + h_{\nu\alpha} S^\alpha{}_\nu. \tag{4.43}$$

Anyway, once source terms have been separated, we are left with second order partial differential equations which have to be solved with various techniques and, more particularly, with the help of Green function methods.

4.3.2. *The formal solution to the linearized Einstein equation*

Starting from Eqs. (4.39) and (4.32), we get

$$\Delta h_{\mu\nu} = -2\left\{-\frac{1}{2}h_{\mu\nu}S + \frac{1}{2}g_{\mu\nu}h_{\alpha\beta}S^{\alpha\beta} + \lambda h_{\mu\nu} + \left[j_{\mu\nu} - \frac{1}{2}g_{\mu\nu}j\right]\right\} \tag{4.44}$$

or, equivalently,

$$\begin{aligned} Lh_{\mu\nu} \equiv{} & \Delta h_{\mu\nu} + 2\lambda h_{\mu\nu} - h_{\mu\nu}S + g_{\mu\nu}h_{\alpha\beta}S^{\alpha\beta} \\ & + 2[h_{\mu\alpha}S^{\alpha}_{\ \nu} + h_{\nu\alpha}S^{\alpha}_{\ \mu}] - 2g_{\mu\nu}g^{\alpha\rho}S^{\alpha}_{\ \rho}h_{\sigma\alpha} \\ ={} & -2[k_{\mu\nu} - \frac{1}{2}g_{\mu\nu}k], \end{aligned} \tag{4.45}$$

where use has been made of Eq. (4.45). Note that Eq. (4.39) may be simplified further by imposing the usual gauge $I_\mu = 0$, or

$$\nabla_\mu h^\mu_{\ \nu} - \frac{1}{2}\nabla_\nu h = 0. \tag{4.46}$$

This gauge condition can be cast into a form similar to the common gauge Lorentz condition of electromagnetism.

The formal solution to Eq. (4.45) may be written as

$$h_{\mu\nu} = \int \eta' H_{\mu\nu\alpha'\beta'}k^{\alpha'\beta'}, \tag{4.47}$$

where the "Green function" $H_{\mu\nu\alpha'\beta'}$ is a bitensor distribution which has to be specified further by giving conditions on its support (retarded, advanced conditions, etc.). The primed indices $H_{\mu\nu\alpha'\beta'}$ are related to the variables x'^α which occur implicitly: $H_{\mu\nu\alpha'\beta'} = H_{\mu\nu\alpha'\beta'}(x, x')$. In fact, we have

$$H_{\mu\nu\alpha'\beta'} = K_{\mu\nu\alpha'\beta'} - \frac{1}{2}g^{\mu'\sigma'}K_{\mu\rho'\alpha'\beta'}g_{\alpha'\beta'}, \tag{4.48}$$

where $H_{\mu\nu\alpha'\beta'}$ is a Green function of the operator L acting on $h_{\mu\nu}$ in Eq. (4.45), i.e. we have

$$LK_{\mu\nu\alpha'\beta'} = (\tau_{\mu\alpha'}\tau_{\nu\beta'} + \tau_{\mu\beta'}\tau_{\nu\alpha'})\delta(x, x'). \tag{4.49}$$

In the case where the background is empty, $S_{\mu\nu} = 0$, Eq. (4.46) reduces to a simple Klein–Gordon-like equation. Accordingly, the Green function $K_{\mu\nu\alpha'\beta'}$ reduces to the Lichnérowicz propagators (up to the sign).

Let us now evaluate the source term $k^{\alpha\beta}$ which occur in Eq. (4.47). It is a functional of the distribution given by

$$k^{\alpha\beta} = \chi\delta \int \sqrt{|g|}\ d^4pp^\alpha p^\beta f(x,p). \tag{4.50}$$

Accordingly, Eq. (4.50) reduces to

$$k^{\alpha\beta} = \chi\delta \int \sqrt{|g|}\ d^4pp^\alpha p^\beta \left\{ \delta f(x,p) + f(x,p)\frac{\delta\sqrt{|g|}}{\sqrt{|g|}} \right\}. \tag{4.51}$$

Let us now consider the second term on the right hand side of Eq. (4.50). The well-known formula of Riemannian geometry

$$d\{\log g\} \equiv g^{\alpha\beta} dg_{\alpha\beta} \tag{4.52}$$

provides immediately

$$\delta\sqrt{|g|} = \frac{1}{2}h\sqrt{|g|}. \tag{4.53}$$

Consequently, Eq. (4.51) becomes

$$\begin{aligned} k^{\alpha\beta}(x) = \chi &\iint \eta' \sqrt{|g|}\ d^4p\ H_{\mu\nu\alpha'\beta'} p^{\alpha'} p^{\beta'} \delta f(x,p) \\ &+ \frac{\chi}{2} \iint \eta' \sqrt{|g|}\ d^4p\ H_{\mu\nu\alpha'\beta'} p^{\alpha'} p^{\beta'} h(x) f(x,p). \end{aligned} \tag{4.54}$$

It follows that $h_{\mu v}$ no longer appears to be an explicit solution to Eq. (4.43) but rather to the integral equation

$$\begin{aligned} k_{\mu\nu}(x) = \chi &\iint \eta' \sqrt{|g|}\ d^4p\ H_{\mu\nu\alpha'\beta'} p^{\alpha'} p^{\beta'} \delta f(x,p) \\ &+ \frac{\chi}{2} \iint \eta' \sqrt{|g|}\ d^4p\ H_{\mu\nu\alpha'\beta'} p^{\alpha'} p^{\beta'} h(x) f(x,p). \end{aligned} \tag{4.55}$$

Setting now

$$F = f + \frac{1}{2}hf, \tag{4.56}$$

one may rewrite Eq. (4.55):

$$h_{\mu\nu} = \chi \iint \eta' \sqrt{|g|}\ d^4p' p^{\alpha'} p^{\beta'} H_{\mu\nu\alpha'\beta'} F. \tag{4.57}$$

In the next subsection, we shall see that we do not need the explicit solution to Eq. (4.55) for $h_{\mu\nu}$ and that the change of function (4.57) is extremely useful.

4.3.3. *The self-consistent kinetic equation for the gravitating gas*

Let us now turn to Eq. (4.43) taking into account the linearization procedure. Assuming that f is a solution to

$$p^\mu \partial_\mu f + \Gamma_\alpha{}^\mu{}_\beta p^\alpha p^\beta \frac{\partial}{\partial p^\eta} f = 0, \tag{4.58}$$

i.e. that f is a background quantity, we can rewrite this equation as

$$p^\mu \partial_\mu F + \Gamma_\alpha{}^\mu{}_\beta p^\alpha p^\beta \frac{\partial}{\partial p^\mu} F + X_\alpha{}^\mu{}_\beta p^\alpha p^\beta \frac{\partial}{\partial p^\mu} f + \frac{1}{2} f p^\mu \partial_\mu h = 0. \tag{4.59}$$

Using now Eq. (4.56), we find that

$$p^\mu \partial_\mu F + \Gamma_\alpha{}^\mu{}_\beta p^\alpha p^\beta \frac{\partial}{\partial p^\mu} F + X_\alpha{}^\mu{}_\beta p^\alpha p^\beta \frac{\partial}{\partial p^\mu} f + \frac{1}{2} p^\mu \partial_\mu h\ f = 0. \tag{4.60}$$

With the help of Eqs. (4.30) and (4.57), the explicit expression for $X^\mu_{\alpha\beta}$ may be obtained:

$$X_\alpha{}^\mu{}_\beta = \frac{\chi}{2} \iint \eta' \sqrt{|g|}\ d^4 p \{\nabla_\alpha H^\mu_{\beta\rho'\sigma'} + \nabla_\beta H^\mu_{\alpha\rho'\sigma'} - \nabla^\mu H_{\alpha\beta\rho'\sigma'}\} p^{\rho'} p^{\sigma'} F'. \tag{4.61}$$

Finally, the kinetic equation looked for is

$$\begin{aligned} p^\mu \partial_\mu F &+ \Gamma_\alpha{}^\mu{}_\beta p^\alpha p^\beta \frac{\partial}{\partial p^\mu} F \\ &= \frac{\chi}{2} \iint \eta' d^4 p' p^{\rho'} p^{\sigma'} F' \\ &\quad \times \left[\{\nabla_\alpha H^\mu_{\beta\rho'\sigma'} + \nabla_\beta H^\mu_{\alpha\rho'\sigma'} - \nabla^\mu H_{\alpha\beta\rho'\sigma'}\} p^{\alpha'} p^{\beta'} \frac{\partial}{\partial p^\mu} f \right. \\ &\quad \left. + f p^\mu \partial_\mu (g^{\alpha\beta} H_{\alpha\beta\rho\sigma}) \right]. \end{aligned} \tag{4.62}$$

This equation is an integrodifferential linear equation, as we expected from the beginning. It may be called a linearized Vlasov equation for the gravitational plasma.

4.4. An Illustration in Cosmology[4]

An illustration in relativistic cosmology has been done by E. Asseo, D. Gerbal, J. Heyvaerts and M. Signore (1978) in the domain of dispersion of

[4]The reader who is not familiar with usual notion of cosmology should first go to the next section and then come back to this section.

gravitational radiation in a universe. They *mutatis mutandis* used the same technique as that in conventional plasma physics. We use a flat Robertson–Walker metric, $k = 0$,

$$ds^2 = dt^2 - R^2(t)\frac{dx^2 + dy^2 + dz^2}{1 + \frac{1}{4}kr^2}, \tag{4.63}$$

although the cases $k = \pm 1$ are also to be considered. In this equation, $R(t)$ is the so-called "radius of the universe"; it is better called the "scale factor." In the case we consider — the one where the wavelength of the wave that propagates according to dispersion relations is much smaller than the radius of the universe — this is a very good approximation, which we shall accept here.

In this case the only nonvanishing Christoffel symbols which are not zero are

$$\begin{cases} \Gamma^i_{0j} = \dfrac{\dot{R}(t)}{R(t)}\delta^j_i & (i, j = 1, 2, 3), \\ \Gamma^0_{ij} = \dot{R}(t)R(t)\delta^j_i, \end{cases} \tag{4.64}$$

and the perturbations of the Christoffel symbols are

$$\delta\Gamma^0_{00} = \frac{1}{2}\dot{h}_{00}, \tag{4.65}$$

$$\delta\Gamma^0_{i0} = \frac{1}{2}\left(\partial_i h_{00} - 2\frac{\dot{R}(t)}{R(t)}h_{0i}\right), \tag{4.66}$$

$$\delta\Gamma^0_{ij} = \frac{1}{2}(\partial_i h_{0j} + \partial_j h_{0i} - \bar{h}_{ij} - 2\dot{R}(t)R(t)h_{00}\delta_{ij}). \tag{4.67}$$

These are the data that take account of the background Robertson–Walker metric. The statistical data are then summed up in the linearized kinetic equation

$$p \cdot \partial Z + \Gamma^\alpha_{\beta\sigma}p^\beta p^\sigma \frac{\partial Z}{\partial p^\alpha} + \delta\Gamma^\alpha_{\beta\sigma}\frac{\partial N}{\partial p^\alpha}p^\beta p^\alpha = 0, \tag{4.68}$$

while the linearized Einstein equations do read

$$(Lh)_{\mu\nu} = -\left(K_{\mu\nu} - \frac{1}{2}g_{\mu\nu}K^\lambda_\lambda\right), \tag{4.69a}$$

$$K_{\mu\nu} = 8\pi G \int \frac{d^2p}{p_0}p_\mu p_\nu \left(Z + \frac{1}{2}h^\lambda_\lambda\right), \tag{4.69b}$$

where the left hand side of the equation for the gravitational waves is

$$\begin{aligned} (Lh)_{\mu\nu} = {} & \Delta h_{\mu\nu} + (\nabla_\mu I_\nu + \nabla_\nu I_\mu) - h_{\mu\nu}R^\lambda_\lambda \\ & - g_{\mu\nu}h_{\alpha\beta}R^{\alpha\beta} + 2(h_{\mu\alpha}R_\nu{}^\alpha + h_{\nu\alpha}R_\mu{}^\alpha). \end{aligned} \tag{4.70}$$

$R_{\alpha\beta}$ is the usual Einstein tensor and $\Delta h_{\alpha\beta}$ is the de Rham–Lichnérowicz tensor for a symmetric tensor; furthermore, we take $I_\beta = 0$ as the de Donder gauge.

4.4.1. *The two-time scale approximation*

In such a case, one is confronted with another timescale,

$$t_H = \frac{R(t)}{\dot{R}(t)} = \left(\frac{3}{8\pi G\rho}\right)^{1/2}, \tag{4.71}$$

the Hubble time that characterizes the evolution of the universe. It is clear that the dispersion and the expansion effects occur simultaneously, and thus no effects can be neglected. However, when $\omega^{-1} = t_H$, the universe is essentially stationary and remains — at least during a number of periods — only as acting as a very slow change to the dispersion characteristics.[5]

In order to present the two-time scale method, we shall use the following equation which looks like our linearized system:

$$[\partial_0^2 + \omega^2(t)]\varphi(t) = \frac{\dot{R}(t)}{R(t)}\partial_0\varphi(t). \tag{4.72}$$

$R(t)$ varies slowly on the same scale of time as $\omega(t)$. We obviously must introduce two times in order to treat the problem; and, for instance,

$$[\partial_0^2 + \omega^2(t_H)]\varphi(t) = \frac{\dot{R}(t_H)}{R(t_H)}\partial_0\varphi(t) \tag{4.73}$$

results from the above equation by making the time inside the two terms $\omega^2(t)$ and $\dot{R}(t)/R(t)$ a "constant." Thus, we introduce two times in the problem; a short time t_c and a long one t_H. The former applies to phenomena occurring in times of the order ω_k^{-1}, while the latter occurs in times of the order of the expansion.

Let us now specify how the various times are intermingled. We have

$$\frac{d}{dt} = \frac{d\tau}{dt}\frac{\partial}{\partial\tau} + \frac{dt_H}{dt}\frac{\partial}{\partial t_H}, \tag{4.74}$$

where

$$d\tau = \omega_k(t)dt, \quad \tau = \int_0^t dt'\ \omega_k(t'). \tag{4.75}$$

τ is the short time to be used in the following. The second term on the right hand side of the above derivative definition involves the long-time scale: if

[5]It was introduced by H. Poincaré when he studied the slightly nonlinear oscillations; see e.g. J. Cole, *Perturbation Methods in Applied Mathematics* (Ginn Blaisdell, Waltham, Massachusetts, 1968).

it is applied on a function φ, it is of a smaller contribution than the first term, and this from an order of $(\omega_k \tau_H)^{-1}$. Thus, we consider the function as depending on the two variables τ and t_H. Substituting the last equation into Eq. (4.73), we obtain the expansion of the operators

$$\frac{\partial}{\partial t} = \omega_k(t_H)\frac{\partial}{\partial \tau} + \varepsilon\frac{\partial}{\partial t_H}, \tag{4.76}$$

$$\frac{\partial^2}{\partial t^2} = \omega_k^2(t_H)\frac{\partial^2}{\partial \tau^2} + \varepsilon\left[2\omega_k(t_H)\frac{\partial}{\partial t_H}\frac{\partial}{\partial \tau} + \frac{\partial \omega_k(t_H)}{\partial t_H}\right] + \varepsilon^2\partial^2/\partial t_H^2, \quad (4.76)$$

where we have set $\varepsilon = dt_H/dt$ in order to conserve the size of the magnitude of the different terms.

Inserting this formal expansion into Eq. (4.73), we get

$$\left[\frac{\partial^2}{\partial \tau^2} + \frac{2\varepsilon}{\omega_k(t_H)}\frac{\partial^2}{\partial t_H \partial \tau} + \frac{\varepsilon}{\omega_k^2(t_H)}\left(\frac{\partial \omega_k(t_H)}{\partial t_H}\right)\frac{\partial}{\partial \tau} + \frac{\varepsilon^2}{\omega_k^2(t_H)}\frac{\partial^2}{\partial t_H^2} + 1\right]\varphi$$
$$= \varepsilon\frac{\dot{R}(t)}{R(t)}(t_H)\frac{1}{\omega_k^2}\left[\omega_k(t_H)\frac{\partial}{\partial \tau} + \varepsilon\frac{\partial}{\partial t_H}\right]\varphi. \tag{4.77}$$

We now decompose φ in the expansion

$$\varphi = \varphi^{(0)} + \varepsilon\varphi^{(1)} + \varepsilon^2\varphi^{(2)} + \cdots, \tag{4.78}$$

so that the last equation is now split into the following hierarchy:

$$\frac{\partial^2}{\partial \tau^2}\varphi^{(0)} + \varphi^{(0)} = 0, \tag{4.79}$$

$$\varepsilon\left(\frac{\partial^2}{\partial \tau^2}\varphi^{(1)} + \varphi^{(1)}\right) + \left[\frac{2\varepsilon}{\omega_k(t_H)}\frac{\partial^2}{\partial t_H \partial \tau} + \frac{\varepsilon}{\omega_k^2(t_H)}\left(\frac{\partial \omega_k(t_H)}{\partial t_H}\right)\frac{\partial}{\partial \tau}\right]\varphi^{(0)}$$
$$= \varepsilon\frac{\dot{R}(t)}{R(t)}(t_H)\frac{1}{\omega_k(t_H)}\frac{\partial}{\partial \tau}\varphi^{(0)} \tag{4.80}$$

We thus have an infinite equation to be solved immediately as

$$\varphi^{(0)} = \phi^{(0)}(t_H)\exp(i\tau), \tag{4.81}$$

where $\phi^{(0)}(t_H)$ is as yet undetermined. Setting this first solution into the second equation of the hierarchy, we get

$$\frac{\partial^2\varphi^{(1)}}{\partial \tau^2} + \varphi^{(1)} = e^{i\tau}\left[i\left(\frac{\dot{R}(t_H)}{R(t_H)}\frac{1}{\omega_k} - \frac{\dot{\omega}_k}{\omega_k}\right)\phi^{(0)}(t_H) - \frac{2i}{\omega_k}\frac{\partial \phi^{(0)}(t_H)}{\partial t_H}\right]. \tag{4.82}$$

Let us try to solve this last equation as

$$\varphi^{(1)}(\tau) = A(\tau)e^{i\tau}; \tag{4.83}$$

the immediate result is a linear growth of $A(t)$. We can now take advantage of the unknown part of $\phi^{(1)}$, and let us set the bracket on the right hand side of the preceding equation equal to zero. Finally, we find the equations at order 1:

$$\begin{cases} \dfrac{d^2\varphi^{(1)}}{d\tau^2} + \varphi^{(1)} = 0, \\ \dfrac{\dot{R}}{R} - \dfrac{\dot{\omega}_k}{\omega_k} = 2\dfrac{\dot{\phi}^{(0)}}{\varphi^{(1)}}. \end{cases} \tag{4.84}$$

It should be emphasized that this procedure, in order to prevent the linear growth of $\varphi^{(1)}(\tau)$, is the only possible one to allow the convergence of the series on an order of an interval of τ_H (E. Asseo *et al.*, whom we follow closely).

4.4.2. *Derivation of the dispersion relations (a rough outline)*

We are now in a position to derive the dispersion relation for the cosmological gravitational plasma and for that we cast our equation in the form of Eq. (4.60).

First, we set the Liouville equation in the form

$$\begin{cases} p^\mu \partial_\mu Z = P(t), \\ P(t) \equiv \dfrac{1}{2}(\partial_\rho h_\sigma^{\ \alpha} + \partial_\sigma h_\rho^{\ \alpha} - \partial^\alpha h_{\rho\sigma})p^\rho p^\alpha \dfrac{\partial N}{\partial p^\alpha}, \end{cases} \tag{4.85}$$

because of the term $\partial Z/\partial u$, which is much smaller than the others. Then the equation is Fourier-transformed in three dimensions and the result is

$$Z = \int_0^\infty d\tau \frac{P(t-\tau)}{p_0} \exp\left(-i\frac{\mathbf{q}\cdot\mathbf{p}}{p_0}\tau\right). \tag{4.86}$$

Solutions of the form

$$\begin{bmatrix} Z \\ P \end{bmatrix} = \begin{bmatrix} \tilde{Z} \\ \tilde{P} \end{bmatrix} \exp\left[-i\int_0^t d\tau' \omega(\tau')\right] \tag{4.87}$$

are looked for, with $(\tilde{Z}, \tilde{P})$ varying slowly as a function of time $[(\tilde{Z}, \tilde{P})$ are the "Fourier transforms" of $(Z, P)]$. While keeping only the lowest order terms, one obtains

$$\tilde{Z} = \tilde{P}(t) \int_0^{\infty} \frac{d\tau}{p_0} \exp\left[i\tau\left(\omega - \frac{\mathbf{q}\cdot\mathbf{p}}{p_0}\right)\right]$$
$$= i\frac{\tilde{P}(t)}{q_\lambda p^\lambda}. \quad (4.88)$$

As it is substituted into $\Sigma_{\mu\nu}$ and $K_{\mu\nu}$, it yields an equation of the form

$$\Sigma_{\mu\nu} = O^{\alpha\beta}_{\mu\nu} h_{\alpha\beta}, \quad (4.89)$$

whereas the equation $(Lh)_{\mu\nu} = \Sigma_{\mu\nu}$ reduces to a homogeneous equation for $\tilde{h}_{\mu\nu}$:

$$\mathsf{D}^{\mu\nu}_{\alpha\beta}\tilde{h}_{\mu\nu} = \mathsf{E}^{\mu\nu}_{\alpha\beta}\tilde{h}_{\mu\nu}. \quad (4.90)$$

Then one can show that the operator $\mathsf{D}^{\alpha\beta}_{\mu\nu}$ has the form

$$\mathsf{D}^{\alpha\beta}_{\mu\nu} = \left[-\omega^2 + \frac{q^2}{R^2} + 2i\omega\frac{\partial}{\partial t} + i\frac{\partial\omega}{\partial t} + \frac{\partial^2}{\partial t^2}\right]\eta^\alpha_\mu\eta^\beta_\nu - O^{\alpha\beta}_{\mu\nu}, \quad (4.91)$$

while its zeroth order is

$$\left[\left(-\omega^2 + \frac{q^2}{R^2}\right)\eta^\alpha_\mu\eta^\beta_\nu - O^{\alpha\beta}_{\mu\nu}\right]\tilde{h}_{\alpha\beta} = 0. \quad (4.92)$$

Therefore, the equation satisfied by $\tilde{h}^E_{\alpha\beta}$ [Eq. (4.90)] is

$$\left[\frac{\partial^2}{\partial t^2} + \omega_k^2(t)\right]\eta^\alpha_\mu\eta^\beta_\nu\tilde{h}^E_{\alpha\beta}(t) = \mathsf{E}^{\alpha\beta}_{\eta\nu}\tilde{h}^E_{\alpha\beta}(t), \quad (4.93)$$

which is appropriate to the two-time scale approximations. h^E represents the various polarizations of a gravitational wave.

It should be stressed that this is a very short summary whose aim is only to give an idea of the way the two-time scale method will apply. We used extensively the article of E. Asseo, D. Gerbal, J. Heyvaerts and M. Signore (1978) and it remains for us to show how it applies to the waves in a cosmological background: this is partially the fact of the article.

4.5. Cosmology and Relativistic Kinetic Theory

The first work on the subject was an article by A.G. Walker (1936) which was of a purely theoretical nature since, at that time, cosmology was considered as a merely speculative subject, owing to the lack of observations other than the "nebula recession." Moreover, cosmology was still halfway between science and metaphysics and thus did not yet have the consideration it could deserve. Furthermore, general relativity was considered as very speculative in spite of the observation of the bending of light rays grazing the Sun.

4.5.1. *Cosmology: a very brief overview*[6]

Modern cosmology is essentially based on three observational facts *and* their theoretical interpretation in the light of the physical concepts of our century: (i) the expansion of the universe, (ii) the background blackbody radiation at 2.7 K, and (iii) the abundance of light elements. All other data are connected with at least one of these three basic facts.

From a theoretical point of view, the basic assumptions are: (i) the cosmological principle, (ii) the existence of a universal time, (iii) the existence of local observers characterized by a timelike four-velocity field u^μ, and (iv) the validity of general relativity.

(i) *The expansion of the universe.* The first fact is the "recession of the nebulae," discovered by E. Hubble (1929) from an observational point of view, and interpreted as expressing the expansion of the universe with the help of the first evolutionary models by A. Friedmann (1922, 1924). This expansion was implemented in *Hubble's law*:

$$v = H_0 d, \tag{4.94}$$

where ν is the velocity of recession of an observed galaxy, d its distance to the observer and H_0 the so-called Hubble constant.

A. Friedmanns models were based on the cosmological principle, whose mathematical traduction is that space–time is essentially homogeneous and isotropic at very large scales, and on Einstein's equations,

$$R_{\mu\nu}(x) - \frac{1}{2}g_{\mu\nu}(x)R(x) - \lambda g_{\mu\nu}(x) = 8\pi G T_{\mu\nu}(x), \tag{4.95}$$

where $R_{\mu\nu}(x)$ is the Ricci tensor, $R(x)$ the curvature scalar and $T^{\mu\nu}_{mn}(x)$ the energy–momentum tensor of matter present in the universe; and G is the constant of gravitation and l the cosmological constant. In standard cosmology, T^{λ}_{mn} is taken to have the perfect fluid form; this property implies that there is no dissipation and hence that the entropy is conserved. Also, matter is supposed to form a fluid whose "molecules" are essentially galaxies. The cosmological principle implies that the space–time metric is

[6]For more details, see e.g. S. Weinberg, *Gravitation and Cosmology* (Wiley, New York, 1972).

the Friedmann–Robertson–Walker metric; in isotropic coordinates, it reads

$$ds^2 = dt^2 - R^2(t)\frac{d\mathbf{x}^2}{\left(1 + \frac{1}{4}kr^2\right)^2}, \tag{4.96}$$

where k $(= 0, \pm 1)$ is the spatial curvature index and $R(t)$ is the scale factor (sometimes improperly called the "radius of the universe"). From the conservation equations obeyed by the energy–momentum tensor and Einstein's equations, it can be shown that $R(t)$ satisfies the two equations[7]

$$\begin{cases} \left(\dfrac{\dot{R}(t)}{R(t)}\right)^2 + \dfrac{k}{R^2(t)} = \dfrac{8\pi}{3}G\rho, \\ \dfrac{\ddot{R}(t)}{R(t)} - \dfrac{1}{2}\lambda = -\dfrac{4\pi}{3}G(\rho + 3P). \end{cases} \tag{4.97}$$

It follows that, in order to obtain the solution to this system, one more equation is needed, such as the equation of state of the matter present in the universe at a given time t, i.e. $P = P(\rho)$.

The scale factor $R(t)$ expresses the dilatation of lengths with time and, in first approximation, leads to Hubble's law and provides the following relation for Hubble's constant, H_0:

$$H_0 = \left.\frac{\dot{R}(t)}{R(t)}\right|_{t=t_0} \tag{4.98}$$

where t_0 is the present time. The expansion of the universe then implies that $\dot{R}(t_0) > 0$. Furthermore, since all distances undergo a dilatation as time flows, roughly as

$$l = \frac{R(t)}{R(t_0)}l_0, \tag{4.99}$$

densities go as l^{-3} and hence vary as

$$n(t) = n(t_0)\left(\frac{R(t_0)}{R(t)}\right)^3. \tag{4.100}$$

This allows considering the universe to be denser in the past provided that it had a continuous expansion — a property of the standard model.

[7]Details can be found in many books, such as: P.J.E. Peebles, *Principles of Physical Cosmology* (Princeton University Press, 1993); S. Dodelson, *Modern Cosmology* (Elsevier, Amsterdam, 2003).

(ii) *The background blackbody radiation.* This was discovered in 1965 by A.A. Penzias and R.W. Wilson and its existence was assumed by G. Gamow in 1948 to explain the abundance of the elements.[8] In the universe there exists an electromagnetic bath of highly isotropic thermal radiation whose distribution function is

$$f(k) = \frac{1}{(2\pi)^3}\frac{1}{\exp(\beta_0 k^0) - 1}, \tag{4.101}$$

where k^0 is the frequency of a photon and β_0 is the inverse temperature at the present time. As is shown below, the temperature at time t is connected with the present time temperature through

$$T(t) = T(t_0)\frac{R(t_0)}{R(t)}. \tag{4.102}$$

This means that, going backward in time, the temperature of the cosmological blackbody radiation was higher: since the universe is expanding,

$$\begin{cases} R(t) < R(t_0), \\ \quad t < t_0. \end{cases} \tag{4.103}$$

Accordingly, this provides a *thermal history* of the universe where, for each temperature and density regime, specific physical phenomena do occur. One can then speculate on the various states through which our universe passed and whose consequences could now be observable or not. For instance, the very tiny deviations of the blackbody radiation isotropy can be interpreted in terms of the presently admitted theories.

Let us now show how the above relation between the background radiation temperature and the cosmological scale factor can be obtained. Starting from Friedmann's equations, one obtains after a few algebraic manipulations

$$d(\rho R^3) + Pd(R^3) = 0, \tag{4.105}$$

which can also be derived from the conservation of the perfect fluid form of the energy–momentum tensor. On the other hand, the equation of state of thermal radiation is given by

$$\begin{cases} \rho = \sigma T^4, \\ P = \dfrac{1}{3}\rho \end{cases} \tag{4.106}$$

(where σ is the Stefan constant), which, after it has been inserted in the preceding equation, provides the expected result. Note also that the

[8] S. Weinberg, *The First Three Minutes: A Modern View of the Origin of the Universe* (Basic Books, 1977).

above blackbody distribution function provides directly the same result since one can equivalently use the equation of states to calculate the time evolution of T.

(iii) *The abundance of light elements.* G. Gamow (1948) had the wonderful idea that the abundance of *all* elements could result from a chemical equilibrium of all possible nuclear reactions in a high temperature thermal bath. However, this idea did not work satisfactorily and only the abundance of the lightest elements — i.e. He^4, He^3, D and Li^7 — was explained in that way and, today, the heavy elements formation is attributed to nuclear processes in stars, while their dispersion in the Galaxy is thought to result from supernova explosions.

The process of light elements formation (and abundance) occurs during the "first three minutes" [S. Weinberg (1993)]. At its very beginning, in the standard model, the temperature was of the order of 3 MeV and matter was a mixture of neutrons, protons, electrons, neutrinos (and antineutrinos) and photons. Among these particles, only the last three species were relativistic and they were essentially free so that relativistic kinetic theory is of almost no use in this case. Finally, this kind of calculations sets more or less stringent constraints on the various phenomena which could possibly have existed before, say, 1 s.

4.5.2. *Kinetic theory and cosmology*

Let us consider the "cosmological fluid," whether composed of galaxies, elementary particles or possibly primordial stars. It should obey the Einstein–Liouville system, i.e.

$$\begin{cases} p\cdot\partial f(x,p)+\Gamma^{\mu}_{\alpha\beta}p^{\alpha}p^{\beta}\dfrac{\partial}{\partial p^{\mu}}f(x,p)=0, \\ R_{\mu\nu}(x)-\dfrac{1}{2}R(x)g_{\mu\nu}=4\pi G T_{\mu\nu}(x) \\ \qquad\qquad =4\pi G\displaystyle\int_{\substack{g_{\mu\nu}(x)p^{\mu}p^{\nu}=m^2 \\ p^0>0}}\sqrt{|g(x)|}\frac{d^3p}{p_0}p_{\mu}p_{\nu}f(p), \end{cases}$$

where collisions have been neglected.[9] The Einstein–Liouville system has been studied by several authors, such as J. Ehlers, P. Gehren and R.K. Sachs (1968), who investigated its possible isotropic solutions, or G.F.R. Ellis,

[9]This is not always valid as is the case in some theories of matter/antimatter separation [see e.g. E.W. Kolb and M.S. Turner, Grand unified theories and the origin of the baryon asymmetry, *Annu. Rev. Nucl. Sci.* **33**, 645 (1983)].

R. Treciokas and D.R. Matravers (1983) for anisotropic but homogeneous cosmological solutions. This system can explicitly be rewritten [R. Hakim (1968)] as

$$\begin{cases} p^0 \dfrac{\partial}{\partial t} f(t, p^0) - \dfrac{\dot{R}(t)}{R(t)} \left\{ \left[\left(p^0\right)^2 - m^2 \right] \dfrac{\partial}{\partial p^0} f(t, p^0) \right\} = 0 \\ + \text{Friedmann's equations,} \end{cases} \tag{4.107}$$

where the distribution function does not depend on $\mathbf{x}$ because of the cosmological principle: our universe is supposed to be homogeneous. The formal solution to the above cosmological Liouville equation is then easily found to be

$$f(t, p^0) = f_0 \left(\frac{\sqrt{(p^0)^2 - m^2}}{m} R(t) \right), \tag{4.108}$$

where f_0 is an arbitrary function. Note that the argument of f_0 is nothing but the well-known constant of the motion of a generic particle[10]:

$$vR(t) \equiv \frac{\sqrt{(p^0)^2 - m^2}}{m} R(t). \tag{4.109}$$

This means that the equilibrium Jüttner–Synge distribution *cannot* be a possible solution for *massive* particles since p^0 is not a constant of the motion. This is of course a consequence of the fact that, in cosmology, u^μ is not a Killing four-vector but only a Killing conformal one. As a consequence of this last property, for zero rest mass particles — e.g. photons — $p^0 R(t)$ is a first integral and the Jüttner–Synge function

$$f(t, p^0) = A \exp[-\beta p^0 R(t)] \tag{4.110}$$

is still a solution to the Einstein–Liouville system; actually, it is a pseudoequilibrium distribution function whose temperature varies exactly as indicated above.

The fact that massive particles cannot be in thermal equilibrium — except for brief periods of time — has led L. Bel (1969) to suggest another equilibrium function, of the form

$$f(t, \mathbf{p}) \approx \exp[-\beta |\mathbf{p}| R(t)]. \tag{4.111}$$

However, the above first integral is not an additive one and hence is not suitable for statistical thermodynamics as it is presently understood:

[10]See e.g. L. Landau and E. Lifschitz, *Classical Field Theory* (Addison-Wesley, Reading, 1962).

a thermodynamic system, when it is separated into smaller macroscopic pieces, should be separated from the point of view of the additive integrals.

Finally, note that in the observed universe the pressure is practically negligible, and Friedmann's equations are quite easy to solve and they yield

$$\frac{R(t)}{R(t_0)} = \left(\frac{t}{t_0}\right)^{2/3}, \tag{4.112}$$

where for simplicity $\lambda = 0$ has been chosen; but this is not an essential restriction.

4.5.3. *Kinetic theory of the observed universe*

Besides pure theory, most works about kinetic theory in the presence of gravitation (including gravitational interactions) deal with stellar clusters, galaxies or galaxy clusters, and are generally not relativistic. One of the first tasks of observational cosmology was the verification of Hubble's law, in particular with farther and farther galaxies; and the magnitude/redshift diagram, which expresses this law, was essentially in accordance with Hubble's results. However, the quasars, discovered in the early 1960s, were objects with high redshifts and, accordingly, were immediately inserted into the magnitude/redshift diagram. Unfortunately, instead of being more or less aligned on the theoretical curve, they form a non-clearly-interpretable cloud.

An interesting idea by L. Bel (1969) to explain this fact was the assumption that quasars have a great dispersion in their velocities (with respect to the cosmological flow) and could thus be described by a distribution function. Their redshift is given by

$$1 + z \equiv \frac{\lambda_0}{\lambda} = \frac{R(t_0)}{R(t)} \frac{\sqrt{v^2+1} - v\cos\theta}{\sqrt{v_0^2+1} - v_0\cos\theta'}, \tag{4.113}$$

where λ is the proper wavelength of the light emitted and λ_0 the observed wavelength; ν is the quasar velocity and ν_0 the observer's velocity; θ is the angle between the quasar's velocity and the direction of the emitted radiation, and θ_0 is the angle between the observer's velocity and the direction of the observed radiation. This has been systematically confronted with the observational data [E. Alvarez and L. Bel (1973); E. Alvarez and J.M. Gracia-Bondia (1974, 1975)] but without any completely conclusive issue whatsoever. Today, astrophysicists rather think that the redshift dispersion of quasars probably occurs because of evolution effects.

4.5.4. *Statistical mechanics in the primeval universe*

In the primeval universe — roughly before 1 s in the standard model — the temperature is so high,

$$k_B T \gg mc^2, \tag{4.114}$$

that most particles are ultrarelativistic and hence behave as massless objects. Accordingly, there exists a thermal equilibrium for these "massless" particles since the local cosmological velocity u^μ is a Killing conformal field,

$$\nabla_\mu \beta_\nu + \nabla_\nu \beta_\mu = 2 g_{\mu\nu} \phi, \tag{4.115}$$

where $g_{\mu\nu}$ is the Robertson–Walker metric and ϕ is

$$\nabla_\mu \beta_\nu + \nabla_\nu \beta_\mu = 4 g_{\mu\nu} \frac{\dot{R}}{R} \frac{1}{T(R)}, \tag{4.115}$$

with $T(R) = T_0 R_0 / R$.

First, we recall (see Chap. 7) the main properties of the blackbody radiation. It is represented by a one-particle distribution function of the type

$$f_{\text{eq}}(p) = \frac{1}{\exp(\beta p \cdot u) - 1} \tag{4.116}$$

and is such that

$$n = \int \frac{d^3p}{p_0} p^\mu u_\mu f_{\text{eq}}(p), \tag{4.117}$$

$$\rho = \int \frac{d^3p}{p_0} p^\mu p^\nu u_\mu u_\nu f_{\text{eq}}(p), \tag{4.118}$$

$$P = \frac{1}{3}\rho. \tag{4.119}$$

Explicitly, these data read

$$n = 2\frac{\varsigma(3)}{\pi^2} T^3, \tag{4.120}$$

$$\rho = \frac{\pi^2}{15} T^4, \tag{4.121}$$

$$P = \frac{\pi^2}{45} T^4, \tag{4.122}$$

where $\varsigma(3)$ is the Riemann function $\varsigma(x)$: $\varsigma(3) = 1.202$.

When the blackbody radiation contains other particles than photons, there are two possibilities. In the first case, they are ultrarelativistic. In such a case, they are essentially free and are such that $k_B T \gg mc^2$. Then the various data concerning the blackbody radiation are essentially true except that the degeneracy level has an effective value:

$$d^* = \sum_{\text{bosons}} d_i + \frac{7}{8} \sum_{\text{fermions}} d_i. \tag{4.123}$$

In this last expression, d's are degeneracies of the particle i while d^* is the total degeneracy. For instance, $d = 2$ for a photon, $d = 4$ for $e^{\pm}$ and $d = 12$ for the system $q\bar{q}$. Finally, we have

$$\rho = d^* \frac{\pi^2}{30} T^4 \tag{4.124}$$

and similar formulae for n or P, etc.

Let us at this point look more precisely at Fig. 4.1, where the various particles are shown with their effective degeneracy. We see that they are

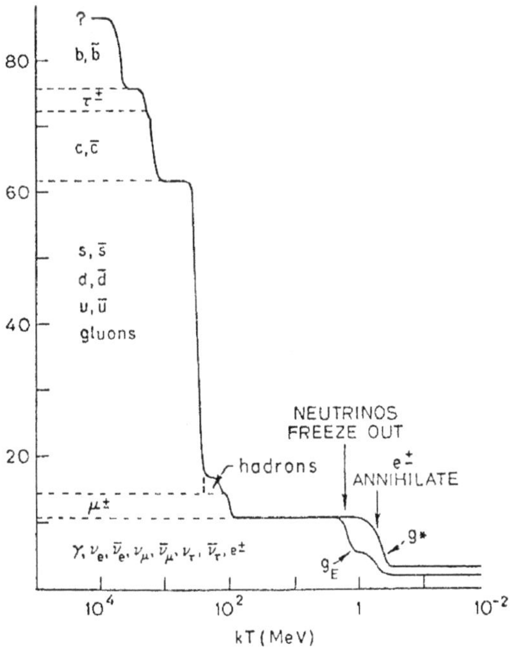

Fig. 4.1 [after R.V. Wagoner[11] (1979)]

[11]In *The Early Universe*, eds. R. Balian, J. Audouze and D. Schramm (North–Holland, Amsterdam, 1979).

placed on plateaus of the curve; and each plateau is separated from the other by a rounded part which represents a non-ultrarelativistic particle and even a nonequilibrium state of the system. However, it is generally omitted, the more so since the fact is that the "roundup" is quite small and therefore is not considered.

4.5.5. *Particle survival*

In the cosmic blackbody radiation, particles are in equilibrium,

$$m + \bar{m} \rightleftarrows n\gamma,$$

at least as long as the temperature does not fall off too much. When the particles are in equilibrium, their distribution function is roughly

$$N_{\text{eq}} \propto (m\beta)^{3/2} \exp(-m\beta), \tag{4.125}$$

at least close enough to the end of the equilibrium since their kinetic energy is in the neighborhood of zero.

If it goes down so that it is in an off-equilibrium state, a few particles would not find any partner to annihilate and thus they would survive to the blackbody. This is of particular importance when one looks, at particles produced in a symmetric manner and it would be important to evaluate the particles which survive, i.e. which do not collide with an antiparticle.

Suppose that the number of particles and of antiparticles are produced equally in the thermal blackbody radiation. Suppose further that they annihilate with a given cross-section $\sigma(m,T)$; then what is the density of the particles $n(m)$ actual at the present moment? What is the mass density $\rho(m)$ at the same density? This is called the problem of "particle survival-Particle survival,"[12] and we follow the course of J.D. Barrow (1983). The problem will be handled with a relativistic Boltzmann-equation-like; more exactly, one for the particle and the other for the antiparticle. If $f(x,p)$ and $\bar{f}(x,p)$ denote the distribution of these cases, they obey the equations

$$\begin{cases} p \cdot \partial f + \Gamma^{\mu}_{\alpha\beta} \dfrac{\partial}{\partial p^{\mu}} f = C(f, \bar{f}), \\ p \cdot \partial \bar{f} + \Gamma^{\mu}_{\alpha\beta} \dfrac{\partial}{\partial p^{\mu}} \bar{f} = C(\bar{f}, f), \end{cases} \tag{4.126}$$

[12]See e.g. J.D. Barrow, *Cosmology and Elementary Particles* (Gordon and Breach, 1983). The first articles on the subject are: H.Y. Chiu, *Phys. Rev. Lett.* **17**, 712 (1965); Y.B. Zakharov, *Adv. Astron. Astrophys.* **3**, 242 (1965); G. Steigman, *Annu. Rev. Astron. Astrophys* **29**, 313 (1979).

where $C(f, \bar{f})$ is the collision term which contains (i) the particle–particle collisions, (ii) the antiparticle–antiparticle collisions, (iii) the particle–antiparticle collisions and (iv) the production rate of particles.

Before looking at the collision term, let us simplify the streaming term — the left hand side of these equations — by noting that the standard cosmology is homogeneous and hence one has $\nabla_{\mathbf{x}} f = \nabla_{\mathbf{x}} \bar{f} = 0$. Also, the explicit calculation of the Christoffel symbols occuring in the streaming term yields

$$p^0 \frac{\partial f}{\partial t} - (p^{02} - m^2) \frac{\dot{R}}{R} \frac{\partial f}{\partial p^0} = C(f, \bar{f}) \tag{4.127}$$

and another analogous equation for $\bar{f}$, where $C(f, \bar{f})$ is the collision term. We now integrate the two equations for f and $\bar{f}$. It turns out that

$$\begin{cases} \dfrac{1}{\sqrt{-g}} \partial_\mu (\sqrt{-g}\, n u^\mu) = \displaystyle\int d^4p\; C(f, \bar{f}), \\ \dfrac{1}{\sqrt{-g}} \partial_\mu (\sqrt{-g}\, \bar{n} u^\mu) = \displaystyle\int d^4p\; C(\bar{f}, f), \end{cases} \tag{4.128}$$

or, because of the symmetry properties of the background and from the fact that we are in a frame of reference where $\mathbf{u} = 0$,

$$\frac{\partial n}{\partial t} + 3 \frac{\dot{R}}{R} n = -g_s n \bar{n}\, \langle W\delta \rangle + P(t) \tag{4.129}$$

and the other equation for $\bar{n}$.

In this equation, we made the simplification that

$$\begin{aligned} &W(p', p'' \to p, \bar{p}) \delta(p + p'' - p' - p) \\ &\qquad \approx \langle W(p', p'' \to p, \bar{p}) \delta(p + p'' - p' - p) \rangle; \end{aligned}$$

this is the basis of the statistical assumption made in elementary particle production. This hypothesis is valid insofar as the brackets, $\langle \cdots \rangle$ vary slowly. Then the various terms of $C(f, \bar{f})$ all vanish and the only surviving one is $g_s \langle \cdots \rangle$, where g_s is the degeneracy of the particles. There also subsists the production rate of the particles $P(t)$; finally, we get

$$\frac{\partial n}{\partial t} + 3 \frac{\dot{R}}{R} n = -g_s n \bar{n}\, \langle W\delta \rangle + P(t) \tag{4.130}$$

and a similar equation for $\bar{n}$. Taking account of the fact that $P(t) \equiv \bar{P}(t)$, by substracting the second equation from the former, we obtain

$$\frac{\partial (n - \bar{n})}{\partial t} = -3 \frac{\dot{R}}{R} (n - \bar{n}) \tag{4.131}$$

or

$$(n - \bar{n})R^3 = \text{const}; \tag{4.132}$$

in other words, the number of particles is a first integral of the system. We use now the detailed balance principle, which indicates that $\dot{R}/R = 0$, in equilibrium, and $\bar{N} = 0$. It follows that

$$P_{\text{eq}} = \langle W\delta\rangle \bar{n}_{\text{eq}} n_{\text{eq}} = \langle W\delta\rangle n_{\text{eq}}^2, \tag{4.133}$$

which is introduced in the equation of the particle number

$$N = nR^3, \tag{4.134}$$

i.e. Eq. (4.130), to obtain

$$\begin{aligned} \frac{dN}{dt} &= \langle W\delta\rangle (n_{\text{eq}} + n)(N_{\text{eq}} - N) \\ &= \langle W\delta\rangle n_{\text{eq}}(N_{\text{eq}}^2 - N^2), \end{aligned} \tag{4.135}$$

which shows that $N \approx N_{\text{eq}}$ and therefore

$$N \approx N_{\text{eq}} \propto (m\beta)^{3/2} \exp(-m\beta). \tag{4.136}$$

However, when the temperature falls off, N is no longer a solution to Eq. (4.136) and it occurs with a deviation from equilibrium:

$$\Delta = \frac{N - N_{\text{eq}}}{N_{\text{eq}}}. \tag{4.137}$$

Thus, for temperatures lower than that where there is no longer any production of the massive particles, we have

$$\frac{dN}{dt} = -\langle W\delta\rangle n^2 V = -\langle W\delta\rangle N^2 V^{-1}. \tag{4.138}$$

This occurs when the expansion rate equals the interaction rate; we call T^* the characteristic temperature for one type of particles. This last equation should be considered with the boundary condition that $N(T^*)$ deviates from $N_{\text{eq}}(T^*)$ by $\Delta(T^*)$. This equation, which gives rise to a large deviation Δ,

$$\left(\frac{m}{T^*}\right)^{1/2} \exp\left(\frac{m}{T^*}\right) = \text{const} \ \times g_i m \langle W\delta\rangle g^{-1/2}, \tag{4.139}$$

is solved numerously, and the result is shown in Fig. 4.2.

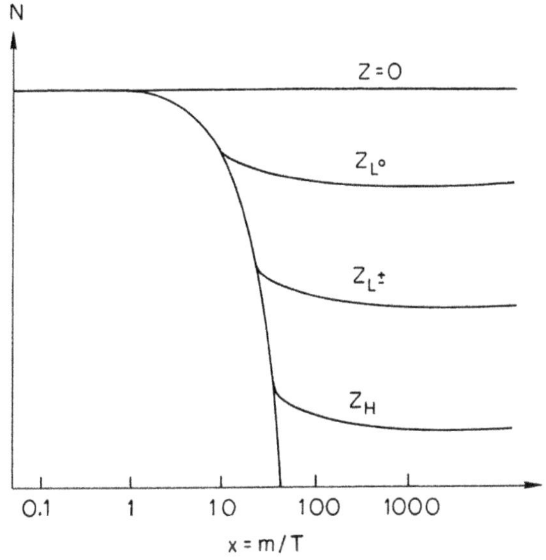

Fig. 4.2 Number of particles that survive from the cosmological radiation [after G. Steigman (1979)]. Leptons are Z_L^0 and $Z_L^{\pm}$, while hadrons are designated by Z_H. Z is proportional to the mass of the particles.

It shows the equilibrium and off-equilibrium parts (one for each kind of particles). The equilibrium part for photons, or zero rest mass particles, is the $Z = 0$ part, while the other part is composed of two parts: (i) an equilibrium part and (ii) the off-equilibrium one "flowing" out of the equilibrium.

Chapter 5

Relativistic Statistical Mechanics

From a theoretical point of view, relativistic statistical mechanics has long been a puzzling question and has finally been erected after relativistic kinetic theory has been clarified, after some tools proposed by P.G. Bergmann (1951) have been presented and after Yu. L. Klimontovich's articles (1960) on relativistic plasmas have appeared. Also, P. Havas (1965) thoroughly discussed the dynamical problems raised by relativity.

5.1. The Dynamical Problem

From a dynamical point of view, there are also deep differences between relativistic and Newtonian mechanics, and let us briefly discuss this important point. First, the known equations of motion are not of the standard Hamiltonian form and it seems that some people think that this rules out the possibility of constructing a relativistic statistical mechanics. Of course, in many cases, equations of motion, whether relativistic or not, can be cast into a Hamiltonian form; however, such a possibility is merely formal, since the Hamiltonian has not the meaning of an energy and hence rules out the possibility of defining, for example, thermal equilibrium. However, this absence of a physical Hamiltonian does not mean that a statistical mechanics is impossible to build. Statistical mechanics, as it is presently understood after J.W. Gibbs, is constructed from (i) equations of motion and (ii) random initial data whose probability distribution is supposed to be known or, at least, replaced by another assumption, such as thermal equilibrium.

Moreover, it has been shown that relativity and the use of Hamiltonian equations of motion lead to the *no-interaction theorem*[1] of D.G. Currie, T.F. Jordan and E.C.G. Sudarshan (1963), i.e. the only Hamiltonian with an energy interaction and covariant has no manifestly covariant content. This prompted a number of authors to build a statistics of both fields (essentially the electromagnetic field) and particles, following work already performed in the Newtonian case.[2] While this can be done in a manifestly covariant way — as outlined below — most authors preferred to decompose the field into elementary harmonic oscillators, breaking thereby the Lorentz invariance of the theory, but with the advantage of an easier interpretation of various elementary processes [see e.g. A. Mangeney (1965)]. The Hamiltonian used in such an approach essentially reads

$$H = \sum_{i=1}^{i=N} \sqrt{(\mathbf{p}_i - e\mathbf{A}(\mathbf{x}_i, t))^2 + m^2} + \frac{1}{8\pi} \int d^3x\ (\mathbf{E}^2 + \mathbf{B}^2), \tag{5.1}$$

while the electric and magnetic fields $\mathbf{E}$ and $\mathbf{B}$ are expanded into field oscillators. With such a Hamiltonian, a Liouville-like equation can be written and the usual methods of perturbation theory applied. This has been extensively studied by R. Balescu and his collaborators, and by I. Prigogine and his coworkers.

An interesting attempt to avoid the difficulties occurring because of the absence of a Hamiltonian was made by L.P. Horwitz, S. Shashoua and W.C. Schieve (1989).

The only known *classical* physical system[3] consists of charged particles interacting via electromagnetism; its basic equations of motion are

$$\begin{cases} \dfrac{dp^{\mu}_{(i)}}{d\tau} = \dfrac{e}{m} F^{\mu\nu}(x_{(i)}) p_{(i)\nu}, \\[2ex] \Box A^{\mu}(x) - \partial^{\mu}\partial_{\nu}A^{\nu}(x) = \dfrac{4\pi e}{m} \displaystyle\sum_{i=1}^{i=N} p^{\mu}_{(i)}(\tau_{(i)}) \delta^{(4)}[x - x_{(i)}(\tau_{(i)})]. \end{cases} \tag{5.2a}$$

Solving the second of these equations for the electromagnetic four-potential $A^{\mu}(x)$, with the Lorentz gauge condition

$$\partial_{\nu}A^{\nu}(x) = 0, \tag{5.2b}$$

[1] D.G. Currie, T.F. Jordan and E.C.G. Sudarshan, *Rev. Mod. Phys.* **35**, 350 (1963); see also G. Marmo, N. Mukunda and E.C.G. Sudarshan, *Phys. Rev.* **D20**, 2120 (1984).

[2] See e.g. E.G. Harris and A. Simon, *Phys. Fluids* **3**, 255 (1960).

[3] In the past, interactions of particles via a scalar field were considered as a possible description of nuclear matter [G. Marx, *Nucl. Phys.* **1**, 660 (1956)]; for its statistical description, see R. Hakim (1967b).

and by using retarded solutions, one obtains equations of motion which can also be derived from the so-called Fokker action principle,[4]

$$\delta I = \delta \left\{ \sum_{i=1}^{i=N} \int p_i^{\mu} p_{i\mu} \, d\tau + \frac{1}{2} \sum_{i,j} e^2 \iint p_i^{\mu} p_{j\mu} D(x_i - x_j) d\tau_i \, d\tau_j \right\} = 0, \tag{5.3}$$

where D is the retarded elementary solution to the wave equation

$$\begin{cases} \Box D(x) = \delta^{(4)}(x), \\ D(x) = \theta(x^0)\delta(x^2). \end{cases} \tag{5.4}$$

This variational principle exhibits the nonlocal nature of the equations of motion. The two viewpoints — field-*plus*-particles or particles only — are then equivalent as far as the motions of the particles are concerned. The equations of motion can thus be rewritten in terms of the particle variables only:

$$\begin{cases} \dfrac{d}{d\tau_i} p_i^{\mu}(\tau_i) = \dfrac{e}{m} F^{\mu v}[x(\tau_i)] p_{iv}(\tau_i), \\ F^{\mu v}(x) = \partial^{\mu} A^{v}(x) - \partial^{v} A^{\mu}(x), \\ i = 1, 2, \ldots, N, \end{cases} \tag{5.5}$$

with

$$A^{\mu}[x_i(\tau_i)] = \frac{4\pi e}{m} \sum_{i=1}^{i=N} \int d^4x' \, d\tau_i \, p_i^{\mu}(\tau_i) \, D[x'] \, \delta^{(4)}[x' - x_i(\tau_i)]. \tag{5.6}$$

This is equivalent to solving the field equations in terms of the particle data. There was another attempt, also based on action-at-a-distance or on direct interactions [Ph. Droz-Vincent (1996)].

5.2. Statement of the Main Statistical Problems

Basic problems in relativistic statistical mechanics are of several sorts. Firstly, as was outlined above, they are of dynamical order: What are the initial data corresponding to the equations of motion under consideration?

[4] A.D. Fokker, *Z. Phys.* **58**, 386 (1929). See also: A.O. Barut, *Electrodynamics and Classical Theory of Fields and Particles*, p. 122*ff* (MacMillan, New York, 1964); J. Rzewuski, *Field Theory*, Part I (PWN, Warsaw, 1964).

Secondly, problems of statistical order have to be considered: What may be called a "relativistic Gibbs ensemble"? How does one treat in a covariant way the random character of the possibly existing fields? Finally, they are also relevant to the possibility of actually measuring the initial data, at least via some *gedanken* experiment. In this section, those problems are addressed and their incidence is discussed.

5.2.1. *The initial value problem: observations and measures*

The equations of Newtonian mechanics are second order differential equations and the initial conditions, which are necessary for obtaining solutions, are most generally considered to be the initial positions and velocities (or momenta) of the particles that constitute the system in question. In the (nonquantum) relativistic context, the two possible points of view, i.e. field or action-at-a-distance, the statuses of the initial data are not analogous.

Let us begin with the action-at-a-distance point of view. The equations of motion are integrodifferential equations and so far the status of their initial conditions is still unclear. In order to be similar to the Newtonian ones, they should be such that, given a spacelike three-surface Σ considered as the "initial time," only the initial positions and momenta of the particles on Σ are necessary for specifying the motion of the system in the future of this three-surface. However, in the action-at-a-distance formalism, the precise nature of the initial data corresponding to the equation of motion is totally unknown. Moreover, in the case of only two particles, it has been shown that not only the usual (positions and momenta) initial data have to be known on Σ, but also the motion can be specified only if their past is partly known (i.e. a finite part of the past trajectories).

As to the field point of view, it is generally implicitly *assumed* that the initial data on Σ that allow knowledge of the future of the system are (i) the initial positions and momenta of the particles and (ii) the usual initial data of the (electromagnetic) field, which are the field on Σ and its normal derivative. This rests on the loose idea that, separately, the particles obey second order differential equations while the fields satisfy second order partial differential equations. Although natural, not only is such an assumption not proven but also it could well be incorrect if what has been shown in the case of the two-particle problem in the action-at-a-distance formalism is confirmed.

To be specific, let us consider the case of particles interacting through a scalar field, the equations of motion being

$$\frac{dp^{\mu}_{(i)}}{d\tau_{(i)}} = g\Delta^{\mu\nu}(p_{(i)})\partial_{\nu}\varphi(x_{(i)}), \quad i = 1, 2, \ldots, N, \tag{5.7}$$

for the particles and

$$\Box\varphi(x) + M^2\varphi(x) = g\sum_{i=1}^{i=N}\int_{-\infty}^{+\infty} d\tau\ \delta^{(4)}[x - x_{(i)}(\tau)] \tag{5.8}$$

for the scalar field.[5] The equations for the particles are obviously second order differential equations, and *if* the scalar field were given, the initial data would be of the form $\{x_{(i)0},\ p_{(i)0}\}_{i=1,2,\ldots,N}$. As to the field, *if* the motion of the particles were known, this second order partial differential equation would be solved as

$$\begin{aligned}\varphi(x) = g\sum_{i=1}^{i=N}\int_{0}^{+\infty} d\tau \int d^{(4)}x'\Delta(x - x')\delta^{(4)}\left[x - x_{(i)}(\tau)\right] \\ + \int_{\Sigma} d\Sigma_{\nu}\left[\Delta(x - x')\partial^{\nu}\varphi_{(0)}(x') - \partial^{\nu}\Delta(x - x')\varphi_{(0)}(x')\right],\end{aligned} \tag{5.9}$$

where the first term comes from the right hand side of the field equation and the second one makes the initial data of the field apparent. In this last equation, $\Delta(x)$ is an elementary solution of φ.

It should also be noted that the elimination of the self-field in favor of a set of supplementary variables (e.g. accelerations) for the particles renders the situation even worse and conditions of a new kind have to be imposed. F. Rohlich (1965), probably by analogy with quantum field theory, chooses to impose the condition that the particles of the system (in the absence of external fields) are free (see below).

Let us now make a brief parenthesis on the possible measurability of the particles' initial data. In the Newtonian context, the initial positions and momenta of the particles of the system can, in principle, be measured with an arbitrary accuracy and, furthermore, instantaneously. Within the relativity framework, an observer using electromagnetic signals can only "see"

[5] If one insists on a physical interpretation, the particles are classical nucleons interacting through (classical) scalar mesons.

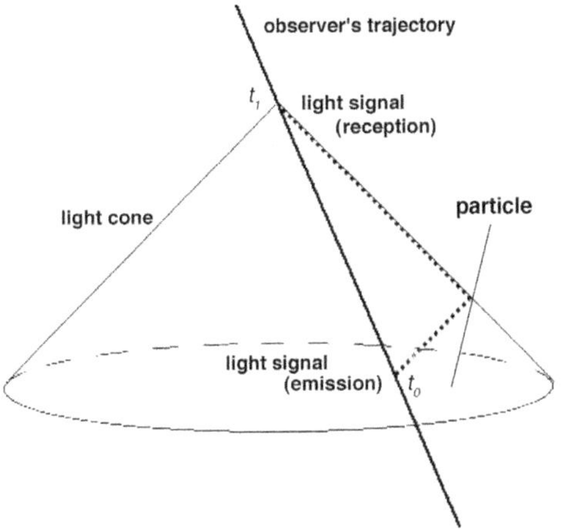

Fig. 5.1 Measure of the space–time position of a particle: an observer emits a radar signal at a given time t_0 and absorbs the signal reflected by the particle at time t_1, measuring thereby its space–time position.

what is on his past light cone[6] and the initial data accessible to his measures lie entirely on this three-surface, and not on a spacelike three-surface. Furthermore, the position measurements can never be instantaneous and always need an interval of time: the one between the emission of a radar signal and its reception (see Fig. 5.1). As to the measure of the field data, the situation is far more involved from an "operational" point of view. However (besides the particles' initial data), if with some experimental device the electromagnetic field and its normal derivative on the past light cone are measured (or known), then one faces a new problem of a mathematical nature: the Cauchy problem for the electromagnetic field is not well set on its characteristic surfaces, i.e. on the light cone. Loosely speaking, this can be physically understood as follows: since this field propagates with the speed of light, the field's initial data give rise to a loss of information reaching the point of "observation" owing to the fact that some of field data do propagate perpendicularly to the light cone.

The usual Cauchy problem — finding the solutions to the dynamical equations with given initial conditions — is such that initial data are

[6]The situation can, of course, be more involved when the background medium is dispersive. In such a case, the initial positions lie on a timelike conoid. If the observer uses signals propagating with velocities smaller than the velocity of light, then the "initial data" are dispersed in his entire past light cone.

provided *on* a spacelike three-surface, and this does not correspond to what is measured, at least in principle. In fact, this is probably a reminder of the Newtonian case, where initial data are known at a given specific time; the spacelike three-surface is the analog of the Newtonian spacelike three-plane $t = \text{const}$. If one insists on a physical interpretation, one might elaborate on the fact that they correspond to a "preparation" of the system during part of the past of the "initial" three-surface.

5.2.2. *Phase space and the Gibbs ensemble*

In the Newtonian context, a system of N particles is described by a trajectory in a $6N$-dimensional phase space,

$$\Gamma^{(6N)} \equiv \mu^{(6)} \times \mu^{(6)} \cdots \times \mu^{(6)}, \tag{5.10}$$

where $\mu^{(6)}$ is the one-particle classical phase space: $\mu^{(6)} \equiv \{x\}^{(3)} \times \{p\}^{(3)}$. In special relativity, the natural generalization of the Newtonian phase space that contains no arbitrary object, such as spacelike three-surfaces, is the $8N$-dimensional Γ space:

$$\Gamma \equiv \mu \times \mu \cdots \times \mu. \tag{5.11}$$

However, in such a space, a point representing the state of the system does not lie on a one-dimensional trajectory, as in the classical case, but rather on an N-dimensional manifold [R. Hakim (1967a)]. This can easily be seen from the fact that such a point,

$$\begin{aligned}\{x^A\}_{A=1,2,\dots,8N} = \{&[x_{(1)}(\tau_1), p_{(1)}(\tau_1)],\ [x_{(2)}(\tau_2), p_{(2)}(\tau_2)], \dots,\\ &\times [x_{(N)}(\tau_N), p_{(N)}(\tau_N)]\},\end{aligned} \tag{5.12}$$

does depend on the N proper times of the particles within the system. It follows that the statistical description of a relativistic system of point particles deeply differs from the usual one although the former gives rise to the latter in appropriate conditions.

While there is at present strong agreement on such a relativistic phase space [see e.g. Ph. Droz-Vincent (1996); L.P. Horwitz, S. Shashoua and W.C. Schieve (1989)], this is not always the case and some authors prefer working in the conventional six-dimensional phase space, at the expense of manifest Lorentz covariance and with subsequent complications [see e.g. R. Balescu, T. Kotera and E. Pina (1967)].

In Hamiltonian mechanics, phase space is the fiber bundle cotangent to the configuration space, i.e. the space whose coordinates are

$\{q_i, p_i\}_{i=1,2,\ldots,N}$; and, at the same time, it is also the space constituted by the possible initial conditions. In the relativistic domain, where the dynamics is definitely not Hamiltonian, phase space does not enjoy the properties of the Newtonian phase space; in particular, it cannot be the space of the (unknown!) initial conditions. We are thus left with the $8N$-dimensional state space (defined above), which allows a description *sub specie eternitatis* of the system; a space that, for brevity, we shall call "phase space."

Newtonian statistical mechanics was a statistics of points, whereas relativistic kinetic theory appeared as a statistics of curves (of space–time trajectories), as was remarked by J.L. Synge (1957); relativistic statistical mechanics is now a statistics of N-dimensional manifolds.

It remains for us to define what can be called a relativistic "Gibbs ensemble." In Newtonian physics, it is defined by the data of an ensemble of analogous (i.e. obeying the same dynamical laws) and noninteracting systems, which differ by their initial conditions, the latter being characterized by a probability measure. In the relativistic case, the initial conditions $\{\omega\}$ are unknown but one can still assume that they are random and still characterized by a probability measure $d\mu(\omega)$, so that an observable $K(X(\{\omega\}))$ possesses the probability density

$$\rho(K) = \int d\mu(\omega)\ \delta[K - K(X(\{\omega\}))], \tag{5.13}$$

where X is a phase space point. Finally, since in the last analysis every physical quantity does implicitly depend on the initial data $\{\omega\}$, it will be convenient to assume the existence of an averaging operation $\langle K\rangle$ for all physical observables and obeying natural mathematical conditions such as

$$\langle \alpha A + \beta B\rangle = \alpha\langle A\rangle + \beta\langle B\rangle, \quad \partial\langle A\rangle = \langle \partial A\rangle, \quad \left\langle \int A \right\rangle = \int \langle A\rangle.$$

In particular, this will be applied for functions like

$$R(x, p) \equiv \int ds \sum_{i=1}^{i=N} \delta^{(4)}[p - p_i(s)]\ \delta^{(4)}[x - x_i(s)], \tag{5.14}$$

used in Chap. 1, or in this chapter to more complex objects.

5.3. Many-Particle Distribution Functions

On the relativistic $8k$-dimensional reduced phase space, the k-particle *random* densities are first defined as

$$R_k\left[x_{\mu_1}, p_{\mu_1}; x_{\mu_2}, p_{\mu_2}; \ldots; x_{\mu_k}, p_{\mu_k}\right] = \int \cdots \int d\tau_1\, d\tau_2 \ldots d\tau_k$$

$$\times \left\{ \sum_{\substack{i_1, i_2, \ldots, i_k \\ \text{all differents}}} \prod_{j=1}^{j=k} \delta^{(4)}[x_j - x_{i_j}(\tau_j)]\delta^{(4)}[p_j - p_{i_j}(\tau_j)] \right\}. \quad (5.15)$$

From this definition the usual (multitime) k-particle distribution function

$$f_k\left[x_{\mu_1}, p_{\mu_1}; x_{\mu_2}, p_{\mu_2}; \ldots; x_{\mu_k}, p_{\mu_k}\right]$$

is defined as

$$f_k\left[x_{\mu_1}, p_{\mu_1}; x_{\mu_2}, p_{\mu_2}; \ldots; x_{\mu_k}, p_{\mu_k}\right]$$
$$= \left\langle R_k\left[x_{\mu_1}, p_{\mu_1}; x_{\mu_2}, p_{\mu_2}; \ldots; x_{\mu_k}, p_{\mu_k}\right]\right\rangle. \quad (5.16)$$

The brackets represent an average value over the initial conditions whatever they might be: it is sufficient to assume its existence with its usual properties, like linearity. To be more specific, we first note that the one-particle distribution function is of the above type. Explicitly, the two-particle distribution reads

$$f_2\left[x_1,\, p_1;\, x_2,\, p_2\right] = \left\langle \int\int d\tau_1\, d\tau_2 \sum_{i \neq j} \delta^{(4)}\left[x_1 - x_i(\tau_1)\right] \delta^{(4)}\left[p_1 - p_i(\tau_2)\right] \right\rangle \quad (5.17)$$

and is normalized as

$$N(N-1) = \int_\Sigma \int_{\Sigma'} \iint d\Sigma_\mu\, d\Sigma_\upsilon\, d^4p_1\, d^4p_2\, p_1^\mu\, p_2^\upsilon f_2\left[x_1,\, p_1;\, x_2,\, p_2\right], \quad (5.18)$$

where $\sum$ and $\sum'$ are two arbitrary spacelike three-surfaces. The normalization of the f_k's are quite similar except that they are normalized to $k!C_N^k$.

However, unlike the Newtonian case, other kinds of distribution functions must also be introduced in order to get a complete system of equations. Some of them are exhibited in the next subsection.

5.3.1. *Statistics of the particles' manifolds**

In this subsection the geometrical definitions necessary for performing a statistics of the N-manifolds representing a Gibbs ensemble in the above relativistic phase space are outlined. As a matter of fact, they can be avoided when one uses coordinates adapted to the very nature of the problem. These notions are inspired by P.G. Bergmann's "generalized statistical mechanics" (1951).

In order to make such a statistics, it is necessary to dispose of an $8N$-current $J^{A_1 A_2 \ldots A_N}(x^A)$, where the indices A run from 1 to $8N$ and are normalized through

$$\int_{\Omega} J^{A_1 A_2 \ldots A_N}(x^A) d\Sigma_{A_1 A_2 \ldots A_N} = N^N, \tag{5.19}$$

where x^A are the coordinates of a point in the relativistic phase space. Ω is a $7N$-dimensional manifold that crosses all the N-dimensional manifolds of the relativistic Gibbs ensemble. To visualize this situation, one can think to a spacelike three-surface crossed by the world lines of the independent particles of a system obeying a given force law. In this equation, $d\Sigma_{A_1 A_2 \ldots A_N}$ is the differential form "element of surface" of the $7N$-dimensional manifold Ω, embedded in an $8N$-dimensional space.

The differential form $J^{A_1 A_2 \ldots A_N}(x^A) d\Sigma_{A_1 A_2 \ldots A_N}$ is a closed form:

$$d\left(J^{A_1 A_2 \ldots A_N}(x^A) d\Sigma_{A_1 A_2 \ldots A_N}\right) = 0. \tag{5.20}$$

This property implies that

$$\nabla_{A_i} \hat{J}^{A_1 A_2 \ldots A_N}(x^A) = 0, \quad i = 1, 2, \ldots, N, \tag{5.21}$$

where $\hat{J}^{A_1 A_2 \ldots A_N}(x^A)$ is the antisymmetrical part of $J^{A_1 A_2 \ldots A_N}(x^A)$. The proof is quite similar to the one used for the usual relativistic continuity equation [see R. Hakim (1967a)].

These conditions constitute a relativistic form of the particle conservation in phase space. Note that, instead of there being only one equation in the Newtonian case — which gives rise to the usual Liouville equation, together with the equations of motion — there exist now N such equations.

The above N-particle distribution function is recovered by setting

$$J^{A_1 A_2 \ldots A_N}(x^A) \equiv f_N(x^A) \xi^{A_1 A_2 \ldots A_N}(x^A), \tag{5.22}$$

*In a first reading this subsection can be omitted.

where $\xi^{A_1A_2\ldots A_N}(x^A)$ is similar to a "four-velocity" and in coordinates adapted to the structure of the manifold Ω as the form

$$\xi^{A_1A_2\ldots A_N}(x^A) = \xi^{A_1} \otimes \xi^{A_2} \otimes \cdots \otimes \xi^{A_N}, \tag{5.23}$$

where ξ_i^A is the Ath component of the $8N$-vector

$$\xi_i = 0 \oplus 0 \oplus \cdots \oplus \eta_i \oplus 0 \cdots \oplus 0 \tag{5.24}$$

with

$$\eta_i^A \equiv \{p_{i\mu},\, F_i^\mu\} \tag{5.25}$$

in the usual coordinate system.

With these definitions, one can also write

$$J^{A_1A_2\cdots A_N}(x^A) = f_N(x^A)\xi_1^{A_1} \otimes \xi_2^{A_2} \otimes \cdots \otimes \xi_N^{A_N} \tag{5.26}$$

and the above conservation equations reduce to

$$\nabla_{A_i}\{f_N(x^A)\xi_i^{A_i}\} = 0, \quad i = 1, 2, \ldots, N,$$

or, in the usual coordinates $\{x_i^\mu, p_{i\mu}\}_{i=1,2,\ldots,N}$, read

$$\nabla_{A_i}\{f_N(x^A)\eta_i^{A_i}\} = 0, \quad i = 1, 2, \ldots, N. \tag{5.27}$$

When the dynamics is such that

$$\nabla_{A_i}\{\eta_i^{A_i}\} = 0, \quad A_i = 1, 2, \ldots, 8;\; i = 1, 2, \ldots, N, \tag{5.28}$$

one obtains the N Liouville equations

$$\eta_i^{A_i}\nabla_{A_i}\{f_N(x^A)\} = 0, \quad i = 1, 2, \ldots, N. \tag{5.29}$$

When this is true, a Liouville theorem is valid although f_N does not satisfy an N-particle Liouville equation, and one has

$$df_N(x^A) = \sum_{i=1}^{i=N} f_N(x^A)\eta_i^{A_i} d\tau_i = 0. \tag{5.30}$$

This is the case for electromagnetic interactions, for instance.

All that has been said about f_N can be repeated *mutatis mutandis* for all other densities, which can be met in what follows. We shall not pursue this geometrical approach, which can be useful only in highly particular cases.

5.4. The Relativistic BBGKY[7] Hierarchy

In order to obtain a BBGKY hierarchy for the various densities, a generating equation for the random distribution R_1 is first derived as [Yu. L. Klimontovich (1960); R. Hakim (1967b)]

$$p^\mu \partial_\mu R_1(x,p) + 4\pi e^2 \nabla_\mu \left\{ p_\upsilon \int d\tau\, d\tau'\, d^4x'\, d^4p' [p'^\mu \partial^\upsilon - p'^\upsilon \partial^\mu] D(x-x') \right.$$
$$\left. \times \sum_{i,j} \delta^{(4)}[x - x_i(\tau)]\delta^{(4)}[p - p_i(\tau)]\delta^{(4)}[x' - x_j(\tau')]\delta^{(4)}[p' - p_j(\tau')] \right\} = 0, \tag{5.31}$$

where the elementary properties of the δ distributions have been used together with the replacement of the various expressions of $F^{\mu\nu}$, A^μ by their explicit forms. The expression between the brackets represents nothing but the electromagnetic field $F_{\mu\nu}$ and its source J^μ. The double sum in this last equation can be split as

$$\sum_{i,j} = \sum_{i \neq j} + \sum_{i=j}. \tag{5.32}$$

The first term on the right hand side ($i \neq j$) is simply what has been defined as R_2 after it has been integrated over τ and τ', while the first one deserves a brief explanation. It is explicitly written as

$$W_2(x,p;x',p') \equiv \int d\tau\, d\tau' \sum_i \delta^{(4)}[x - x_i(\tau)]\delta^{(4)}[p - p_i(\tau)]$$
$$\times\, \delta^{(4)}[x' - x_i(\tau')]\delta^{(4)}[p' - p_i(\tau')] \tag{5.33}$$

and it represents essentially the probability density for a *given* particle to be in the state (x, p) and next to undergo a transition to the state (x', p'). While W_2 must vanish out of the null cone $(x-x')^2 = 0$ (x or x' being in the future of x' or x, respectively, causality is implied by the timelike character of the trajectory of the particle), this is *a priori* not the case for R_2, which refers to *different* particles. From a dynamical viewpoint, W_2 expresses the back-reaction of the particle on itself; accordingly, since one deals with the electromagnetic interaction, it is an infinite term, which is often discarded *a priori*. However, besides an infinite term, it also gives rise to a finite — albeit small — contribution (see the next subsection). Finally, setting

$$P_2 = \langle W_2 \rangle, \tag{5.34}$$

---

[7]Bogoliubov, Born, Green, Kirkwood and Yvon.

and taking the average value of the generating equation of the hierarchy, it turns out that the first equation of the hierarchy reads

$$p^\mu \partial_\mu f_1(x,p) + 4\pi e^2 \nabla_\mu \left\{ p_v \int d^4x'\ d^4p' [p'^\mu \partial^v - p'^v \partial^\mu] D(x-x') \right.$$
$$\left. \times [f_2(x,p;x'p') + P_2(x,p;x'p')] \right\} = 0. \qquad (5.35)$$

The next equation of the relativistic BBGKY hierarchy is obtained by multiplying the generating equation by R_1 and averaging; after a little algebra, one gets

$$p^\mu \partial_\mu f_2(x,p;x'p')$$
$$+ 4\pi e^2 \nabla_\mu \left\{ p_v \int d^4x'\ d^4p' [p''^\mu \partial^v - p''^v \partial^\mu] D(x-x'') \right.$$
$$\times [f_3(x,p;x',p';x''p'') + F_3^2[(x,p) \to (x',p');(x'',p'')]$$
$$\left. + F_3^2[(x'',p'') \to (x',p');(x,p)]] \right\} = 0. \qquad (5.36)$$

A few comments are now in order. While in the Newtonian case the first equation of the hierarchy is an equation that needs the knowledge of f_2 in order to evaluate f_1, in the relativistic case one also needs the knowledge of one more function — say, P_2. A glance at the next equation of the hierarchy shows that the knowledge of one more distribution — say, $F_3^2[(x,p) \to (x',p');(x'',p'')]$ — is needed too. This distribution is the distribution of one particle undergoing the transition $(x,p) \to (x',p')$, while another particle is present in the state (x'',p''). It is not useful to give the third equation of the hierarchy, since in actual practice only the first two are needed. This second equation is also supplemented by a symmetric equation on the primed variables, unlike the nonrelativistic case.[8] However, one also needs, at least in principle, an equation for P_2. It can easily be obtained from the generating equation of the relativistic hierarchy, or from an equation for W_2, as

$$p^\mu \partial_\mu P_2(x,p;x'p') + 4\pi e^2 \nabla_\mu \left\{ p_v \int d^4x''\ d^4p'' [p''^\mu \partial^v - p''^v \partial^\mu] D(x-x'') \right.$$
$$\left. \times [P_3(x,p;x',p';x''p'') + F_3^2[(x,p) \to (x',p');x'',p'')]] \right\} = 0, \qquad (5.37)$$

[8] Note, however, that in the classical "multitime" hierarchy, one gets similar distributions.

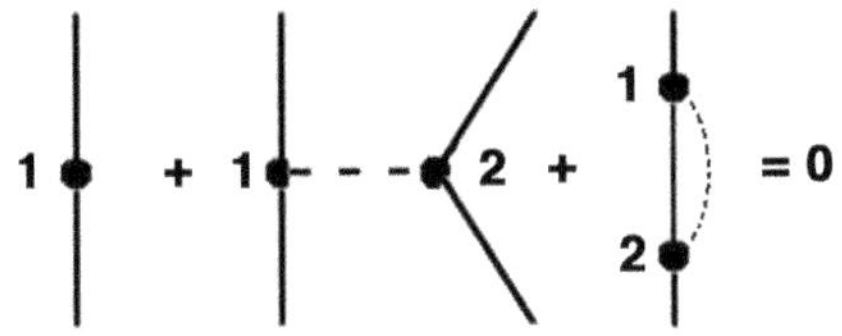

Fig. 5.2 Diagrammatic representation of the first equation of the relativistic BBGKY hierarchy: f_1 is expressed in terms of f_2 and P_2; the dotted line represents the nonlocal electromagnetic interaction, including a self-interaction on the term P_2. This distribution does not appear in the Newtonian theory.

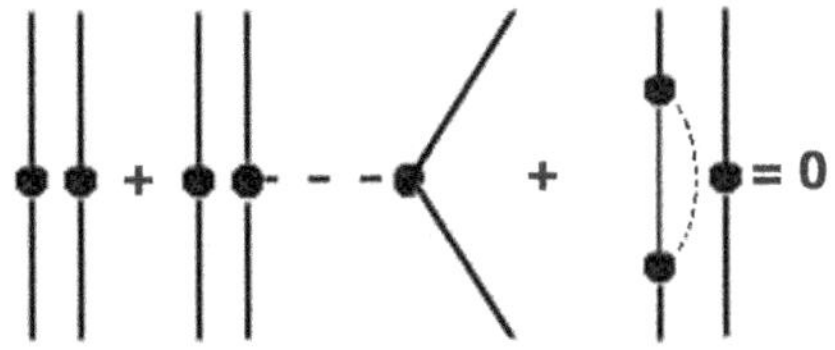

Fig. 5.3 Representation of one of the second equations of the BBGKY relativistic hierarchy — it can be obtained from the preceding one by adding a line to Fig. 5.2. Once more, this shows the appearance of new kinds of distribution functions, which are not encountered in the Newtonian framework.

where P_3 is the distribution function for a given particle to undergo the transitions

$$(x,p) \to (x',p') \to (x'',p''). \tag{5.38}$$

The generating equation of the relativistic hierarchy can be represented by Fig. 5.2, from which all the equations of the hierarchy can be represented and given an explicit form. In this diagram f_1 is represented by one vertex and two external lines, f_2 by two vertices and four external lines, and P_2 by two vertices, one internal line and two external lines.

In order to obtain the second equation of the hierarchy, it is sufficient to add to both diagrams a new vertex with two external lines.

From the above considerations, one can find a few rules allowing the diagrammatic expression of any equation of the hierarchy; this is, however, useless since only the first few equations are actually used.

5.4.1. *Cluster decomposition of the relativistic distribution functions*

The various multiparticle distribution functions should generally be decomposed into cluster decompositions in order to close the hierarchy with some ansatz like the assumed vanishing of a particular correlation function. The representations of the distribution functions by appropriate diagrams show the way the new functions occurring in the relativistic BBGKY hierarchy should be decomposed. First, it should be noted that f_2 can be decomposed as

$$f_2 = f_1 \times f_1 + g_2, \tag{5.39}$$

where g_2 is the two-body correlation function or diagrammatically (see Fig. 5.4). Similarly, f_3 is cluster-decomposed as

$$\begin{aligned} f_3 = {} & f_1 \times f_1 \times f_1 \\ & + g_2 \times f_1 + g_2 \times f_1 + g_2 \times f_1 \\ & + g_3, \end{aligned} \tag{5.40}$$

which decomposition is represented by the following diagram:

Obviously, distribution functions like P_2 or P_3 cannot be decomposed; this is, however, not the case for mixed distributions like $F_3^2[(x,p) \to (x',p');(x'',p'')]$. A glance at its representative diagram shows that it must be decomposed in a way similar to f_2, i.e. as

$$F_3^2 = P_1 \times f_1 + g_{(3)2}, \tag{5.41}$$

diagrammatically represented in Fig. 5.6, where $g_{(3)2}$ is a three-point two-body correlation.

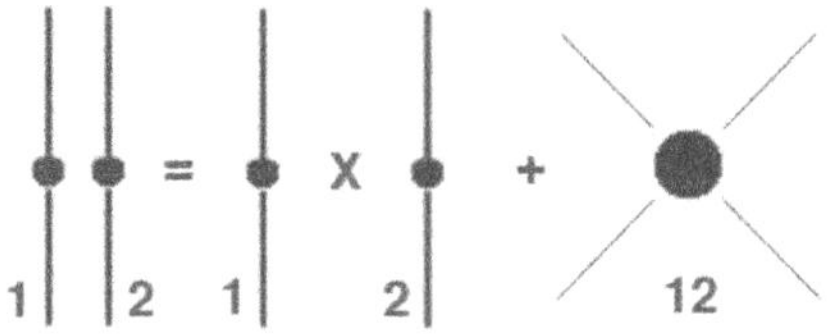

Fig. 5.4 Cluster decomposition of f_2. This does not present any new particularity with respect to the usual case; however, it constitutes a model for the cluster decomposition of more involved distribution functions, suggesting that the connected parts of the representative diagram of a distribution function have to be factored.

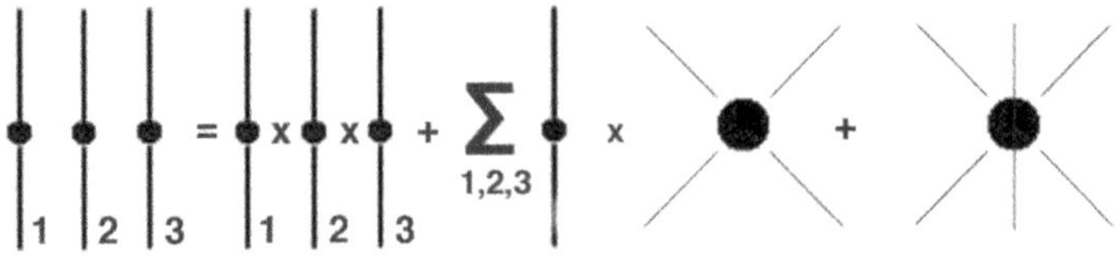

Fig. 5.5 Illustration of the cluster decomposition — the case of f_3.

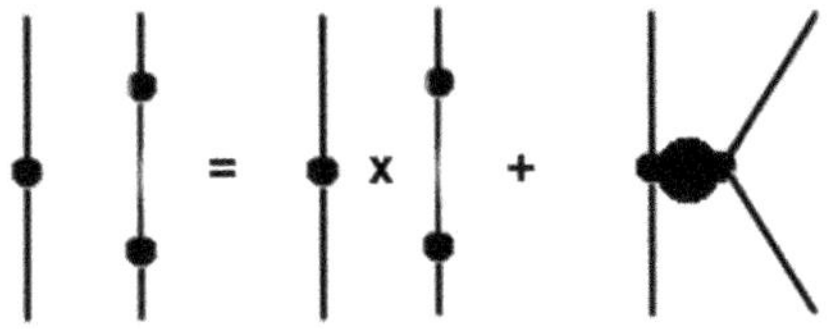

Fig. 5.6 Cluster decomposition of F_3^2. The diagram representing this distribution function is composed of two disconnected lines and hence its cluster decomposition contains the product of the two distribution functions represented by these lines.

More generally, the rule for a cluster decomposition of a given distribution function — when needed — is to decompose disconnected parts of the representative diagram exactly as for f_k.

5.5. Self-interaction and Radiation

In a previous section, it was mentioned that the interaction term involving P_2 is connected with a *self-interaction* of the particle. This has to be elaborated a little bit further. When we go back to the original δ terms from which it results, we can realize that this self-interaction is a consequence of the action of the retarded electromagnetic field $F_{\text{ret}}^{\mu v}$ emitted by a given particle and acting on the same one. When this retarded field is split as[9]

$$F_{\text{ret}}^{\mu v} = \frac{1}{2}[F_{\text{ret}}^{\mu v} + F_{\text{adv}}^{\mu v}] + \frac{1}{2}[F_{\text{ret}}^{\mu v} - F_{\text{adv}}^{\mu v}], \tag{5.42}$$

it can be shown that the self-interaction of a particle contains two parts: the first one is *infinite* and of the general form

$$\infty \times \frac{dx^\mu(\tau)}{d\tau}, \tag{5.43}$$

while the second one is finite and takes account of the back-reaction of the radiation emitted on the particle. The infinite term can be absorbed in a formal mass renormalization. Finally, the equations of motion obeyed by a system of electrons, embedded within a uniformly charged neutralizing background, read

$$\frac{dp_i^\mu(\tau_i)}{d\tau_i} = eF_{\text{ret}}^{(i)\mu v}(x_i)p_{iv}(\tau_i) + \frac{2}{3}e^2[\dot{\gamma}_i^\mu(\tau_i) + \gamma(\tau_i)\cdot\gamma(\tau_i)u_i^\mu(\tau_i)], \tag{5.44}$$

[9] A.O. Barut, *Electrodynamics and Classical Theory of Fields and Particles* (MacMillan, New York, 1965); F. Rohrlich, *Classical Charged Particles* (Addison-Wesley, Reading, Massachusetts, 1965).

$$\partial_\mu F_{\text{ret}}^{(i)\mu\upsilon*} = 0,$$

$$\partial_\mu F_{\text{ret}}^{(i)\mu\upsilon}(x_i) = \frac{e}{m}\int \sum_{j\neq i} \delta^{(4)}[x_i - x_j(\tau_j)]\delta^{(4)}[p_i - p_j(\tau_j)]p_j^\upsilon(\tau_j),$$

$$i = 1, 2, \ldots, N, \quad j = 1, 2, \ldots \neq i, \ldots, N. \tag{5.45}$$

These equations must be supplemented by the asymptotic condition

$$\lim_{\tau\to\infty} \gamma^\mu(\tau) = 0, \tag{5.46}$$

which expresses the fact that the electrons are free at infinity. This condition[10] allows the elimination of the so-called "runaway solutions" [see F. Rohrlich (1965)].

These equations are discussed in detail in the books by A.O. Barut (1965) and F. Rohrlich (1965). The presence of the proper time derivative of the four-acceleration in the equations of motion requires a modification of the above statistical treatment of the system. In particular, phase space needs to be enlarged so as to take account of the acceleration variables as

$$\Gamma_{\text{new}} = \Gamma \times \{\gamma\}^{(4N)}, \tag{5.47}$$

which necessitates the introduction of new distribution functions on this new phase space. Accordingly, we introduce a random distribution R_1, (as in a preceding section),

$$R_1[x, p, \gamma] = \int d\tau \sum_i \delta^{(4)}[x - x_i(\tau)]\, \delta^{(4)}[p - p_i(\tau)]\, \delta^{(4)}[\gamma - \gamma_i(\tau)], \tag{5.48}$$

and also its average value over the initial conditions $f_1 = \langle R_1\rangle$. Note that this new f_1 is normalized through

$$\int_\Sigma d\Sigma_\mu \int d^{(4)}p \int d^{(4)}\gamma\ \delta(p\cdot\gamma)\frac{p^\mu}{m} f_1(x, p, \gamma) = N, \tag{5.49}$$

since the integration must obey the constraint $p \cdot \gamma = 0$.

The continuity equation in the new one-particle phase space,

$$\partial_\mu(p^\mu R_1) + \nabla_\mu(\gamma^\mu R_1) + \frac{\partial}{\partial\gamma^\mu}(\bar{\gamma}^\mu R_1) = 0, \tag{5.50}$$

[10] According to A. Lichnérowicz (private communication), the runaway solutions can be eliminated on the ground of some analyticity assumptions, at least for the one-particle problem.

gives rise to a new generating equation for the relativistic BBGKY hierarchy with radiation effects; or, explicitly, one obtains

$$
\begin{aligned}
& p^\mu \partial_\mu R_1(x,p) + \gamma^\mu \nabla_\mu R_1 + \frac{\partial}{\partial \gamma^\mu}\left(\left[\frac{\gamma^\mu}{m\tau_0} - \gamma\cdot\gamma\frac{p^\mu}{m}\right] R_1\right) \\
& \quad + \frac{4\pi e^2}{m\tau_0}\frac{\partial}{\partial\gamma^\mu}\left\{ p_\upsilon \int d^4x'\, d^4p'\, d^{(4)}\gamma' [p'^\mu \partial^\upsilon - p'^\upsilon \partial^\mu] \right. \\
& \quad \left. \times D(x-x')R_2(x,p;x'p') \right\} = 0, \\
& \nabla_\mu \equiv \frac{\partial}{\partial p^\mu}, \\
& \tau_0 = \frac{2}{3}\frac{e^2}{mc^3}.
\end{aligned} \tag{5.51}
$$

The term P_2 has disappeared from this generating equation: the self-interaction has been eliminated in favor of a mass renormalization and of the finite γ terms. This equation, like the nonrelativistic one, is an equation for f_1 as a function of f_2. However, in the higher order equations "mixed" distributions still appear. This equation looks quite different from the former generating equation and it should be cast into a more useful form.

5.5.1. *An alternative treatment of radiation reaction*

As a matter of fact, there exists an alternative phase space and hence alternative distribution functions, more suitable for a perturbation in powers of the small parameter τ_0. This quantity is actually much smaller than any physically meaningful times in the system. This stems from the remark that the equations of motion of the particles can be cast into the form

$$
\frac{dp_i^\mu(\tau_i)}{d\tau_i} = \frac{e}{m}F_{\text{ret}}^{(i)\mu\upsilon}(x_i)p_{i\upsilon}(\tau_i) + m\tau_0\Delta^{\mu\upsilon}\left(\frac{p_i}{m}\right)\dot{\gamma}_{i\upsilon}, \tag{5.52}
$$

which shows clearly that the acceleration γ can be expressed in terms of p and $\dot{\gamma}$. This means that another phase space can be used, namely

$$
\Gamma_{\text{new}'} = \Gamma \times \{\dot{\gamma}\}^{(4N)}, \tag{5.53}
$$

with, of course, distribution functions depending on the variables $(x, p, \dot{\gamma})$. Similarly, from the continuity equation in this new phase space,

$$
\partial_\mu(p^\mu R_1) + \nabla_\mu[\dot{\gamma}^\mu(x,p,\gamma)R_1] + \frac{\partial}{\partial\dot{\gamma}^\mu}(\ddot{\gamma}^\mu R_1) = 0, \tag{5.54}
$$

after some algebra, one obtains a generating equation of the form

$$p^\mu \partial_\mu R_1(x,p,\dot\gamma) + 4\pi e^2 \nabla_\mu \Big\{ p_\upsilon \int d^4x'\, d^4p'\, d^4\dot\gamma' [p'^\mu \partial^\upsilon - p'^\upsilon \partial^\mu] \times D(x-x') R_2(x,p,\dot\gamma;x',p',\dot\gamma') \Big\} = -\tau_0 \nabla_\mu \left\{ \Delta^{\mu\upsilon}(p) \int d^4\dot\gamma \dot\gamma_\upsilon R_1(x,p,\dot\gamma) \right\}. \tag{5.55}$$

The right hand side of this equation is nothing but the ordinary generating equation, without self-interacting terms, while its left hand side represents the effects of the back-reaction of the radiation on the electrons.

In view of obtaining the first corrections due to radiation, it is sufficient to evaluate the right hand side at order 1 in τ_0. At this order, the equations of motion for the particles read

$$\frac{dp_i^\mu}{d\tau_i} = \frac{e}{m} F^{\mu\upsilon(i)} p_{i\upsilon} + m\tau_0 \Delta^{\mu\upsilon}(p) \dot\gamma_{i\upsilon}^{[0]}, \tag{5.56}$$

where $\dot\gamma$ is evaluated at order 0 in τ_0, i.e.

$$\dot\gamma_{i\nu}^{[0]} = \frac{d}{d\tau_i}\gamma_{i\nu}^{[0]} = \frac{d}{d\tau_i}\left\{ \frac{e}{m^2} F_\nu^{(i)\mu} p_{i\mu} \right\}^{[0]} = \frac{e}{m^2}\{ F_{\nu\alpha}^{(i)} F^{\alpha\beta(i)} p_{i\beta} + p_i^\rho p_i^\alpha \partial_{i\rho} F_{\nu\alpha}^{(i)} \}. \tag{5.57}$$

Finally, inserting this expression into the equation for R_1 and integrating over the variable $\dot\gamma$, the generating equation for a relativistic BBGKY hierarchy at order 1 in τ_0 reads

$$p^\mu \partial_\mu R_1(x,p) + 4\pi e^2 \nabla_\mu \Big\{ p_\upsilon \int d^4x'\, d^4p' [p'^\mu \partial^\upsilon - p'^\upsilon \partial^\mu] \times D(x-x') R_2(x,p;x',p') \Big\} = \text{radiation term} = -\tau_0 \nabla_\mu \left\{ \Delta^{\mu\upsilon}(p) \int d^4\dot\gamma \dot\gamma_\upsilon^{[0]}(x,p) R_1(x,p) \right\}. \tag{5.58}$$

More explicitly, the radiation term reads

$$\text{Radiation term} = -\tau_0 \nabla_\mu \Big\{ \Delta^{\mu\upsilon}(p) \Big[\Big(\frac{4\pi e^2}{m} p^\rho\, p^\alpha \int d^4x'\, d^4p' \times [p'_\alpha \partial_\nu - p'_\nu \partial_\alpha] \partial_\rho D(x-x') R_2(x,p;x',p') \Big)$$

$$+\left[\left(\frac{4\pi e^2}{m}\right)^2 p_\alpha \int\int d^4x'\, d^4p'\, d^4x''\, d^4p''\right.$$
$$\times (p'_\nu\partial_\beta - p'_\beta\partial_\nu)D(x-x')(p''^\alpha\partial_\nu - p''_\nu\partial^\alpha)D(x-x')$$
$$\left.\times [R_3(x,p;x',p';x'',p'') + W_3^2(x',p';x'',p'';\{x,p\})]\Big]\Big]\Big\}. \tag{5.59}$$

The radiation term can easily be taken for a Vlasov approximation. However, there exist two parameters with which one can sum. The first one is the usual plasma parameter; the second one is proportional to τ_0. The term in $e^2\tau_0$ is the one which remains in the lowest approximation and the first equation in order of radiation is studied briefly later on.

5.5.2. *Remarks on irreversibility*

In the case where radiation reaction is directly taken into account, one expects an irreversible behavior of the system since radiation is emitted and never absorbed. Therefore, one might expect some kind of an H theorem,

$$\partial_\mu S^\mu(x) > 0, \tag{5.60}$$

even though the approximation made (e.g. the Vlasov approximation) does not usually provide any kind of irreversible behavior of the system. This is indeed what is actually found (see below).

However, a few words of caution are in order. Irreversibility is a complex phenomenon, connected with time and length scales, to what is observed, to the strength of interactions, to the instabilities of trajectories in phase space, etc. The irreversibility reached with the emission of radiation has little to do with the latter problems. Rather, it is connected to the absence of a Liouville theorem: radiation emission gives rise to an exponential growth of the phase space element, corresponding partly to the so-called runaway solutions[11] of the Abraham–Lorentz–Dirac equations.

Moreover, in spite of the fact that the particle entropy is growing in the course of time — this is an immediate consequence of the above inequality — the system never achieves an equilibrium state, as it should if there would exist a true H theorem.

Let us now show briefly how the peculiar nature of the equations of motion leads to an exploding phase space element. We shall show this in

---

[11] See F. Rohrlich, *loc. cit.*

the case of noninteracting particles; they obey the equation

$$\partial_\mu(p^\mu f) + \nabla_\mu(\gamma^\mu f) + \frac{\partial}{\partial\gamma^\mu}(\dot{\gamma}^\mu f) = 0, \tag{5.61}$$

or, equivalently, after the explicit calculation of its last term,

$$\frac{df}{d\tau} + \frac{6m}{e^2} f = 0, \tag{5.62}$$

which indicates that

$$f(\tau) \approx \exp\left[-\left(\frac{6m}{e^2}\right)\tau\right] f(0). \tag{5.63}$$

On the other hand, the conservation of the particle number reads

$$f(\tau)\delta\mu(\tau) = f(0)\delta\mu(0), \tag{5.64}$$

so that one also has

$$\delta\mu(\tau) \approx \exp\left[\left(\frac{6m}{e^2}\right)\tau\right] \delta\mu(0). \tag{5.65}$$

5.5.3. *Remarks on thermal equilibrium*

From the basic principles of statistical thermodynamics, the grand-canonical ensemble could be defined as

$$f_{\mathrm{GC}} = \frac{1}{Z}\exp[-\beta(u\cdot P - \mu N)], \tag{5.66}$$

where

$$\begin{cases} P^\lambda = \int d\Sigma_\nu T^{\lambda\nu}, \\ N \ = \int d\Sigma_\nu J^\nu, \end{cases} \tag{5.67}$$

and where Z is the partition function.

This means that if the interactions between two macroscopic parts of the system are switched off, then the two subsequent subsystems are still in thermal equilibrium with the implicitly existing heat bath. Let us now look at the above equations a bit closer and let us point out the differences from the nonrelativistic case. In the nonrelativistic case, the energy $E = P^0$ depends on the microscopic variable $\{x_i, p_i\}_{i=1,2,\ldots,N}$ only. In the relativistic case, P^0 not only depends on these latter quantities but also on the initial data themselves, at least implicitly. Moreover, the nonlocal character of the interactions is quite peculiar to relativity. One might try to remedy this situation by treating both the particles' degrees of freedom and the fields responsible for the interaction between them; however, such an idea does not yield the desired result. Furthermore, the only data of f_{GC} do not seem to provide a full description of thermal equilibrium; indeed, as we have seen earlier when writing the relativistic BBGKY hierarchy, a complete description of the system involves not only f_N but also an infinity of other types of distribution functions, such as $P_2, P_3, \ldots; W_3^2, \ldots, W_n^p, \ldots$.

In order to be more specific and get some insight into the problem of equilibrium, let us consider the case of a noninteracting gas. In this case, with the expression for $T^{\mu\nu}$ and J^ν given in Chap. 1, $f_{\rm GC}$ can be written as

$$f_{\rm GC} = \frac{1}{Z}\exp\Bigg(-\beta u_\mu \sum_{i=1}^{i=N}\int_\Sigma d\Sigma_\nu \int d^4p \int ds\; \delta^{(4)}[x - x_i(s)] \times \delta^{(4)}[p - p_i(s)] p_i^\mu \frac{d}{ds} x_i^\nu(s)\Bigg) \times \exp\Bigg(\beta\mu \sum_{i=1}^{i=N}\int_\Sigma d\Sigma_\nu \int d^4p \int ds\; \delta^{(4)}[x - x_i(s)] \times \delta^{(4)}[p - p_i(s)] \frac{d}{ds} x_i^\nu(s)\Bigg), \quad (5.68)$$

which, after use of the ordinary properties of the δ functions and of the fact that[12]

$$d\Sigma_\nu\, dx^\nu = d^4x, \quad (5.69)$$

can be cast into the form

$$f_{\rm GC} = \frac{1}{Z}\prod_{i=1}^{i=N} \exp(-\beta[u\cdot p_i - \mu]); \quad (5.70)$$

this is merely a product of N Jüttner–Synge distribution, as it should be. This result is satisfactory in this sense that, despite the differences existing between the relativistic and the Newtonian cases, the same sensible physical consequence is provided. Unfortunately, the situation is so intricate with nonlocal interactions that it is quite difficult to elaborate on this definition of $f_{\rm GC}$.

Let us now take a glance at a possible simultaneous treatment of a system composed of charged particles and of the present electromagnetic fields.

The "canonical distribution," for the particles and the classical electromagnetic field, then reads

$$f_{\rm canonical} = \frac{1}{Z}\exp(-\beta u\cdot P + \beta\mu N)\exp\left(-\beta u_\mu \int_\Sigma d\Sigma_\nu [T^{\mu\nu}_{\rm em} + \eta^{\mu\nu} J\cdot A]\right), \quad (5.71)$$

[12]This can be justified by using adapted coordinates.

where the term $J{\cdot}A$ represents the interaction between the electromagnetic field and the particles, whereas

$$T^{\mu\nu}_{\text{em}} = \frac{1}{4\pi}\left\{F^{\mu\alpha}F^{\nu}_{\alpha} - \frac{1}{4}F_{\alpha\beta}F^{\alpha\beta}\right\} \tag{5.72}$$

is the electromagnetic energy–momentum tensor.

To look at this expression in another way, let us consider the canonical case in the Vlasov approximation. In such a case one has

$$\langle A^{\mu}\rangle = 0, \tag{5.73}$$

since the plasma is neutral and also

$$\langle A^{\mu}A^{\nu}\rangle \approx \langle A^{\mu}\rangle\langle A^{\nu}\rangle. \tag{5.74}$$

Finally, the canonical distribution in the case of the Vlasov approximation reduces to a product of Jüttner–Synge distribution, as it should.

The expression for $f_{\text{canonical}}$ is quadratic in the four-potential A^{μ}. This is quite satisfactory for an equilibrium state of the electromagnetic field [see e.g. T.W. Marshall (1963, 1965)], since in this case the canonical distribution of the field is a Gaussian.

However, even though the field part of $f_{\text{canonical}}$ can easily be functionally integrated, the remaining expression is quite involved and nothing can be done with it; in particular, Z cannot be evaluated. On the other hand, particles and fields are fully dealt with in the relativistic quantum case and it is useless to spend so much effort on the classical relativistic distribution. It might happen, however, that some particular equilibrium quantities — such as some correlation functions — are needed in a given problem. In this case, the simplest way to solve the problem is to start from the lowest equations of the BBGKY hierarchy and to cut it at some order and use approximations.

5.6. Radiation Quantities

The analysis of the electromagnetic field within a system of relativistic charged particles has been made by A.O. Barut (1964) and by F. Rohrlich (1965).[13] They found that the radiation field, i.e. the *far field* part of $F^{\mu\nu}$, is given by

$$F^{\mu\nu}_{\text{rad}\,\infty}(x) = e\left\{\frac{(X^{\mu}\gamma^{\nu} - X^{\nu}\gamma^{\mu})R - (X^{\mu}u^{\nu} - X^{\nu}u^{\mu})Q}{R^3}\right\}, \tag{5.75}$$

[13] See also the references quoted in these books.

where we have set

$$\begin{cases} X^\mu \equiv x^\mu - z^\mu, & R^2 = R^\mu R_\mu, \\ Q \equiv X^\mu \gamma_\mu. \end{cases} \tag{5.76}$$

In these relations, z^μ is the position of the emitting particle. $F^{\mu\nu}_{\text{rad}}(x)$ is a quantity which enjoys all the expected properties of radiation, i.e.

$$\begin{cases} \partial_\mu F^{\mu\nu}_{\text{rad}}(x) = 0, \\ \partial^*_\mu F^{\mu\nu}_{\text{rad}}(x) = 0, \end{cases} \qquad \begin{cases} F^{\mu\nu}_{\text{rad}}(x) F_{\text{rad}\,\mu\nu}(x) = 0, \\ {}^*F^{\mu\nu}_{\text{rad}}(x) F_{\text{rad}\,\mu\nu}(x) = 0; \end{cases} \tag{5.77}$$

in other words, $F^{\mu\nu}_{\text{rad}}(x)$ is a free field whose magnetic field is orthogonal to the electric field and their energy densities are equal.

On the other hand, the radiation field (which obeys the above equations), whether far or not, is given by [A.O. Barut (1964)]

$$F^{\mu\nu}_{\text{rad}}(x) = -\frac{2}{3}\frac{e}{m}\{p^\mu \dot{\gamma}^\nu - p^\nu \dot{\gamma}^\mu\}, \tag{5.78}$$

and thus its average value or correlation, etc., can be calculated without difficulty; for example, one has

$$\langle F^{\mu\nu}_{\text{rad}}(x)\rangle = -\frac{2}{3}\frac{e}{m}\int d^4p\, d^4\dot{\gamma}\{p^\mu \dot{\gamma}^\nu - p^\nu \dot{\gamma}^\mu\}\delta[p\cdot\gamma(\dot{\gamma})]f(x,p,\dot{\gamma}). \tag{5.79}$$

However, quantities like $F^{\mu\nu}_{\text{rad}\,\infty}(x)$ are much more difficult to evaluate. Indeed, they require the knowledge of, for example, the absorption of radiation during its course, but also the data of the hyperboloid-like surface where the photons do propagate.

Another quantity of physical interest is the energy and momentum radiated per unit of proper time, i.e.

$$\frac{dP^\mu}{d\tau} = \frac{2}{3}\frac{e^2}{m}\gamma^\lambda \gamma_\lambda p^\mu. \tag{5.80}$$

From the above expression of the radiation field, one can calculate important quantities such as the (local) average at point x:

$$\langle F^{\mu\nu}_{\text{rad}}(x)\rangle = -\frac{2}{3}\frac{e}{m}\frac{1}{n}\int d^4p\, d^4\dot{\gamma}\frac{p\cdot u}{m}\{p^\mu \dot{\gamma}^\nu - p^\nu \dot{\gamma}^\mu\}f(x,p,\gamma). \tag{5.81}$$

The spectrum of the radiation field, when defined as usual from the Fourier transform of the intensity, has neither an invariant meaning nor a covariant one. The quantity which does possess such a property is the correlation tensor of the radiation field

$$\langle F^{\mu\nu}_{\text{rad}}(x) F^{\alpha\beta}_{\text{rad}}(x')\rangle - \langle F^{\mu\nu}_{\text{rad}}(x)\rangle\langle F^{\alpha\beta}_{\text{rad}}(x')\rangle \tag{5.82}$$

and, as remarked by T.W. Marshall (1963, 1965), what is of interest as to the energy–momentum spectrum of radiation is rather the contraction

$$\Gamma_{\mu\nu}(x, x+y) = \langle F_{\mathrm{rad}\,\mu\alpha}(x) F^{\alpha}_{\mathrm{rad}\,\nu}(x+y)\rangle - \frac{1}{4}\eta_{\mu\nu}\langle F^{\alpha\beta}_{\mathrm{rad}}(x)\rangle\langle F_{\mathrm{rad}\,\alpha\beta}(x+y)\rangle, \tag{5.83}$$

from which the spectrum can be obtained.

5.7. A Few Relativistic Kinetic Equations

The first relativistic kinetic equation is, of course, the Vlasov one [S. Titeica (1956)], and it is obtained from the relativistic hierarchy from the lowest equation in which the two-body correlation function has been considered to be negligible. It leads to a large number of applications to relativistic plasmas (see the bibliography) and, as shown above, to the relativistic normal modes of plasma. Here, we briefly outline the derivation by Yu. L. Klimontovich (1961) of the relativistic version of the Landau equation, later extended by W. Thomson (1968) to a covariant Lenard–Balescu equation. This section is then concluded with simple relativistic equations of the Vlasov kind that include the effect of radiation.

5.7.1. *Derivation of the covariant Landau equation*[14]

This equation is based on several assumptions, such as the absence of three-body correlations, $g_3 \sim 0$, or

$$f_3(1,2,3) \approx f_3(1)f_3(2)f_3(3) + f_1(1)f_2(2,3) + f_1(2)f_2(1,3) + f_1(3)f_2(1,2), \tag{5.84}$$

where we have used the notation $1 \equiv (x_1, p_1)$, etc. Also, this equation is valid at order 4 in a small parameter proportional to e while the electron gas is sufficiently diluted to allow small energy–momenta transfers only during collisions; the system is spatially homogeneous. Finally, there exist two timescales, the correlation time being the smallest one. Of course, as in the classical case, these approximations can be justified rigorously and it can be shown that they are not completely independent.

[14]See Yu. L. Klimontovich (1960) and R. Hakim (1967).

As usual,[15] one starts from the first two equations of the relativistic BBGKY hierarchy, dropping the radiation terms and the infinite self-mass term. Using the cluster expansion of f_2 and f_3, these equations may be rewritten in a symbolic way as

$$p \cdot \partial f_1 + e^2 \int G[f_1 \otimes f_1 + g_2] = 0, \tag{5.85}$$

$$p \cdot \partial g_2 + e^2 \int G[W_3^2 + f_1 \otimes g_2 + f_1 \otimes g_2 + g_3] = 0, \tag{5.86}$$

where G is an operator, which can easily be obtained from the exact equations. Note that there exists a second equation, in the primed variables.

Let us now use the assumptions indicated at the beginning of this section. First, $g_3 \sim 0$. Second, since we look for a kinetic equation at order e^4, g_2 must be calculated at order e^2, i.e. at its lowest order. It follows that the second equation of the hierarchy reduces to

$$p \cdot \partial g_2 + e^2 \int G W_3^{2[0]} = 0, \tag{5.87}$$

where $W_3^{2[0]}$ is nothing but W_3^2 evaluated at order 0 in e^2. At this order W_3^2 is given by

$$\begin{aligned} &W_3^{2[0]}[x'', p''; x', p'; \{x, p\}] \\ &\quad \approx f_1(x,p) f_1(x',p') \int d\tau'' \delta^{(4)} \left[x'' - x' - p'' \frac{(\tau'' - \tau')}{m} \right] \end{aligned} \tag{5.88}$$

This means that the electrons move practically along straight world lines or, equivalently, that the field acting on particle 1 is the field produced by particle 2 moving along a straight world line.

The equation for g_2 is now solved by taking account of the so-called "adiabatic" hypothesis made above, namely that the time in f_1 is "frozen" as compared to the one in g_2.

The next assumption is used in solving the inhomogeneous equations (5.87) and (5.88); it amounts to neglecting the arbitrary solution to this equation. Furthermore, the densities f_1 which occur in the same equations are to be "frozen" (adiabatic hypothesis)[16] in the calculation of g_2.

---

[15]D.C. Montgomery and D.A. Tidman, *Plasma Kinetic Theory* (McGraw–Hill, New York, 1964).
[16]D.C. Montgomery and D.A. Tidman, *loc. cit.*

Equation (5.87) may be solved either with the use of Fourier transformation or, more simply, by using the "causal" Green function

$$K(x-x';u-u') = \int d\tau\, d\tau' \theta(\tau-\tau')\delta[x-x'-u\cdot(\tau-\tau')]\delta(u-u') \tag{5.89}$$

and letting the "initial" proper time tends to minus infinity: this is legitimate because of the existence of two timescales; "initial" correlations are destroyed. The expression obtained for g_2, after tedious calculations, is the one given by Klimontovich.

Finally, the covariant Landau equation, written in a symbolic way, reads

$$p\cdot\partial f_1 = e^4\nabla_\beta \int \varepsilon^{\alpha\beta}(p',p)\{f_1 \otimes \nabla_\alpha f_1 - f_1 \otimes \nabla_\alpha f_1\}, \tag{5.90}$$

where the tensor $\varepsilon^{\alpha\beta}(p',p)$ is given by

$$\varepsilon_{\alpha\beta} = \frac{e^2 n}{32\pi^4}\int d^4k\, k_\alpha k_\beta (A^\mu(k)\cdot k_\mu)^2 \tag{5.91}$$

[Yu. L. Klimontovich (1960b)], where $A^\mu(k)$ is the four-potential of the charged electron in the independent motion approximation. This equation is the one previously derived by S.T. Belyaev and G.I. Budker (1956). Integration over k can also be performed and one gets

$$\begin{aligned}\varepsilon_{\alpha\beta} = 2\pi e^4 nL\cdot(u^\mu u'_\mu)^2[(u^\mu u'_\mu)^2-1]^{3/2}\{[(u^\mu u'_\mu)^2-1]\eta_{\alpha\beta} \\ -(u_\alpha u_\beta + u'_\alpha u'_\beta) - u_\mu u'^\mu(u_\alpha u_\beta + u'_\alpha u'_\beta)\},\end{aligned} \tag{5.92}$$

where L is the Coulomb logarithm,

$$L = \int \frac{dk}{k}, \tag{5.93}$$

which is infinite and has to be treated by plasma techniques. Of course, from such an equation a covariant Fokker–Planck equation can easily be derived.

Let us finally note that the above calculation provides an evaluation of the correlation function g_2 as a functional of f_1 at order e^2 so that, when f_1 is specialized to be the equilibrium Jüttner–Synge distribution, one gets the equilibrium correlation function needed in the derivation of a relativistic Guernsey kinetic equation.

5.7.2. *The relativistic Vlasov equation with radiation effects*[17]

The first equation of the renormalized hierarchy is now used with the simplest assumption — the neglect of correlations. In this approximation, it is not necessary to enter into the full machinery of the BBGKY hierarchy, since only the average (collective) electromagnetic field is dealt with. It then reads

$$\begin{aligned} &p^\mu \partial_\mu f(x,p) + 4\pi e^2 \nabla_\mu f(x,p) \\ &\quad = -\tau_0 \nabla_\mu \{\Delta^{\mu\upsilon}(p) \int d^4\dot{\gamma}\, \dot{\gamma}_\upsilon^{[0]}(x,p) f_1(x,p,\dot{\gamma})\}, \end{aligned} \tag{5.94}$$

or, after replacing $\dot{\gamma}_\upsilon^{[0]}$ by its expression,

$$\dot{\gamma}_\upsilon^{[0]} = \frac{e}{m^3}\{eF_{\nu\alpha}F^{\alpha\beta}p_\beta + p^\rho p^\alpha \partial_\rho F_{\nu\alpha}\}, \tag{5.95}$$

one obtains

$$\begin{aligned} &p^\mu \partial_\mu f(x,p) + 4\pi e^2 p_\nu F^{\mu\nu} \nabla_\mu f(x,p) \\ &\quad = -\frac{e\tau_0}{m^3} \nabla_\mu \{\Delta^{\mu\upsilon}(p)\{eF_{\nu\alpha}F^{\alpha\beta}p_\beta + p^\rho p^\alpha \partial_\rho F_{\nu\alpha}\}\} \\ &\quad = -\frac{e\tau_0}{m^3} f(x,p) \Big\{ \Delta^{\mu\upsilon}(p)\{eF_{\nu\alpha}F^\alpha_{\ \mu} + p^\alpha \partial_\mu F_{\nu\alpha} + p^\rho \partial_\rho F_{\nu\mu}\} \\ &\quad\quad - \frac{5p^\nu}{m^2} eF_{\nu\alpha}F^{\alpha\beta}p_\beta \Big\}. \end{aligned} \tag{5.96}$$

As mentioned above, one can easily prove [R. Hakim and A. Mangeney (1968)] that this equation *does* possess an *irreversible* behavior in the sense that the entropy of the system is always growing:

$$\partial_\mu S^\mu(x) > 0. \tag{5.97}$$

This is due to the irreversible emission of radiation. It can be understood in another way, by going back to the equations of motion, which indicate that the phase space element is not conserved during the motion but instead "explodes," as has been shown above.

In order to evaluate the importance of the right hand side of this equation, i.e. of the radiation terms, let us make a brief order-of-magnitude analysis. To this end, dimensionless quantities are first defined with the

---

[17]See R. Hakim and A. Mangeney (1968, 1971).

substitutions

$$\begin{cases} \tau = \tilde{\tau}\hat{\tau}, \\ x = l_0\hat{x}, \\ p = l_0\tilde{\tau}^{-1}\hat{p} \equiv \varepsilon\hat{p}, \\ f(x,p) = \varepsilon^{-4} l_0^{-4} \hat{f}(\hat{x},\hat{p}), \end{cases} \tag{5.98}$$

where l_0 and $\tilde{\tau}$ are some characteristic length and time of the electron plasma, respectively. Note that these substitutions preserve the light cone and hence causality. Finally, the above kinetic equation for the reduced quantities contains (i) an interactive collective term on its left hand side that is of the order of

$$\approx \frac{4\pi e^2}{m}\tilde{\tau}\, l_0^{-1}, \tag{5.99}$$

and (ii) the radiation term on the right hand side, which is of the order of $\tau_0\tilde{\tau}^{-1}$. They are of the same order of magnitude when $l_0 \approx \tilde{\tau}$. Choosing now the Debye length for l_0,

$$l_0 \approx e^{-1} n^{-1/2} m^{-1/2}, \tag{5.100}$$

and the plasma frequency ω_P for $\tilde{\tau}^{-1}$,

$$\omega_P \approx e n^{1/2} m^{-1/2}, \tag{5.101}$$

one finds that the radiation term is of the same order of magnitude as the collective interaction term when

$$k_B T \approx mc^2, \tag{5.102}$$

i.e. when the plasma is relativistic.

From the above kinetic equation, one can derive dispersion relations for the collective mode of the plasma and, by using the same techniques as in the usual relativistic equation, one gets [R. Hakim and A. Mangeney (1968, 1971)]

$$\left(1 - \frac{\omega_P^2}{k\cdot k}G\right) + \omega_P^2[I_1^1 + i\tau_0 J_1^1] = 0 \quad \text{(transverse modes)},$$

$$\left\{\left(1 - \frac{\omega_P^2}{k\cdot k}G\right) + \omega_P^2[I_0^0 - i\tau_0 J_0^0] + \frac{\omega_P^2\omega}{k\cdot k}[I_0 + i\tau_0 J_0]\right\}$$
$$\times \left\{\left(1 - \frac{\omega_P^2}{k\cdot k}G\right) + \omega_P^2[I_3^3 - i\tau_0 J_3^3] + \frac{\omega_P^2 k^3}{k\cdot k}I_3\right\}$$

$$= \omega_P^4 \left\{ \frac{k^3}{k \cdot k} [I_0 + i\tau_0 J_0] + [I_3^3 + i\tau_0 J_3^3] \right\}$$

$$\times \left\{ [I_3^3 - i\tau_0 J_3^3] + \frac{\omega I_3}{k \cdot k} \right\} \quad \text{(longitudinal modes)},$$

which have been studied in detail elsewhere [R. Hakim and A. Mangeney (1971)]. In these equations, use was made of the notations

$$G = k_\nu I^\nu + i\tau_0 k \cdot u, \tag{5.103}$$

$$I_\mu = \frac{1}{n} \int d^4 p \frac{p_\mu}{k \cdot p} f_{\text{eq}}(p), \tag{5.104}$$

$$I_\mu^\nu = -\frac{1}{n} \int d^4 p \frac{p_\mu p^\nu}{(k \cdot p)^2} f_{\text{eq}}(p), \tag{5.105}$$

$$J_\mu^\nu = \frac{1}{n} \int d^4 p \frac{p_\mu p^\nu}{k \cdot p} f_{\text{eq}}(p). \tag{5.106}$$

Note that the τ_0 factor in G is generally negligible since it would imply plasma frequencies of the order of 10^{23} cycles/s. Also, the propagation of the plasma waves was chosen to be along the third axis while these dispersion relations were given in the rest frame of the plasma $u = (1, \mathbf{0})$.

Finally, the radiation terms lead, as expected, to a damping of the plasma waves.

To conclude this subsection, let us briefly outline the extension of the above kinetic equation to the case of the relativistic Landau equation with radiation effects. It is then sufficient to evaluate the correlation function $g_2(k,\ p,\ p')$ at order 1 in τ_0, while the order 0 was given by [Yu. L. Klimontovich (1960b)]

$$g_2^{(0)}(k; p, p') \approx \frac{8\pi^2 e^2}{m^2} \frac{k^{[\mu} p^{\nu]}}{(k \cdot p)(k \cdot k)} \delta(k \cdot p') f_1(k, p') p_{[\mu} \nabla_{\nu]} f_1(k, p) \tag{5.107}$$

and a similar expression with p and p' exchanged. The first order in τ_0 turns out to be

$$g_2^{(1)}(k; p, p') \approx -i\tau_0 \frac{8\pi^2 e^2}{m^2} \frac{1}{k \cdot u} \nabla_\mu$$

$$\times \left(\Delta_{\mu\nu}(p) p^\rho p^\alpha \frac{p'_{[\alpha} k_{\nu]} k_\rho}{k \cdot k} \delta(k \cdot p') f_1(k, p) f_1(k, p') \right). \tag{5.108}$$

With these elements, the relativistic Landau equation including radiation effects can easily be obtained and questions such as that of a Fokker–Planck equation for a plasma with emission of radiation can be addressed [Yu. L. Klimontovich (1960b)].

5.7.3. *Radiation effects for a relativistic plasma in a magnetic field*[18]

A kinetic equation for such a system can be obtained once the substitution

$$F^{\mu\nu} \to F^{\mu\nu} + F^{\mu\nu}_{\text{ext}} \tag{5.109}$$

is performed in the equations of motion. Accordingly, one has

$$\begin{aligned} &p^\mu \partial_\mu f(x,p) + 4\pi e^2 p_\nu F^{\mu\nu} \nabla_\mu f(x,p) \\ &= -\frac{e\tau_0}{m^3} \nabla_\mu [f(x,p) \Delta^{\mu\upsilon}(p) \{eF_{\nu\alpha} F^{\alpha\beta} p_\beta + p^\alpha \; p^\rho \partial_\rho F_{\nu\alpha}\}], \end{aligned} \tag{5.110}$$

where the external field $F^{\mu\nu}_{\text{ext}}$ has been assumed to be homogeneous and stationary. In order to evaluate the various terms of this equation through an order-of-magnitude analysis, one must note that, in the presence of an external magnetic field, one more dimensioned quantity now occurs, namely the Larmor frequency

$$\omega_L = \frac{eB}{m}, \tag{5.111}$$

where B is the strength of the external magnetic field. Therefore, a large number of scales of length and of time are available, giving rise to numerous possible regimes (for practical purposes, there are almost an infinite number of different regimes!). Typical lengths are

$n^{-1/3}$: average interparticle distance

$e^{-1}n^{-1/2}(k_B T)^{1/2}$: Debye length

$c\omega_P^{-1}$: wavelength of a plasma oscillation

$\nu_T \omega_L^{-1}$: ν_T, thermal velocity

etc.

Typical times are

$$\omega_P^{-1} \omega_L^{-1}, n^{-1/3}/v_T.$$

The various regimes are governed by the interplay of the various available energies: $mc^2, k_B T, e^2/l_0$ and $B^2 l_0^3$, where l_0 is one of the above lengths.

Note that the above considerations can be used as the basis of a relativistic magnetohydrodynamics of a radiating system.

---

[18]See F. Grassi, R. Hakim and H. Sivak (1986).

Finally, let us note that V.I. Berezhiani, R.D. Hazeltine and S.M. Mahajan (2004), still starting from the Abraham–Lorentz–Dirac equations, obtained the fluid equations of a relativistic magnetized plasma with radiation reaction. They wrote the equations obeyed by the first few moments and used a closure ansatz through a Jüttner–Synge local equilibrium distribution.

5.8. Statistics of Fields and Particles

As was mentioned at the beginning of this chapter, instead of treating particle variables only, one might also deal with both fields and particle data. This allows the use of a Hamiltonian theory, although at the expense of manifest Lorentz invariance. In this section, we do not deal with this Hamiltonian character and preserve the invariance properties of the theory and only outline this possibility, the more so since only quantum statistics fully account for both fields and particles, the latter being degrees of excitation of the former. Furthermore, only the case of charged particles interacting via an electromagnetic field is considered below since this is the only case known in classical physics.

The starting point is, of course, the equation of motions for the particles

$$\frac{dp_i^\mu}{d\tau} = \frac{e}{m} F^{\mu\nu}(x_i) p_{i\nu}, \quad i = 1, 2, \ldots, N, \tag{5.112}$$

and for the electromagnetic field

$$\begin{cases} \partial_\nu F^{\mu\nu}(x) = e \displaystyle\int d^4p \frac{p^\mu}{m} R_1(x,p), \\ \partial_\nu {}^*F^{\mu\nu}(x) = 0. \end{cases} \tag{5.113}$$

The equations of motion then give rise to the generating equation of the BBGKY hierarchy, which still reads

$$p \cdot \partial R(x,p) + \frac{e}{m} F^{\mu\nu}(x) p_\nu \frac{\partial R(x,p)}{\partial p^\mu} = 0. \tag{5.114}$$

Finally, the first equations of the hierarchy are written as

$$\begin{cases} p \cdot \partial f_1(x,p) + \dfrac{e}{m} p_\nu \dfrac{\partial}{\partial p^\mu} \langle R(x,p) F^{\mu\nu}(x) \rangle = 0, \\ \begin{cases} \partial_\nu \langle F^{\mu\nu}(x) \rangle \displaystyle\int d^4p \frac{p^\mu}{m} f_1(x,p), \\ \partial_\nu \langle {}^*F^{\mu\nu}(x) \rangle, = 0, \end{cases} \end{cases} \tag{5.115}$$

while the second group is obtained from the generating (random) equations by multiplying by either $R(x',p')$ or $F_{\mu\nu}(x')$ and reads

$$\begin{cases} p\cdot\partial\langle R(x,p)R(x',p')\rangle + \dfrac{e}{m}p_\nu\dfrac{\partial}{\partial p^\mu}\langle R(x,p)R(x',p')F^{\mu\nu}(x)\rangle = 0, \\ \qquad\begin{cases} \partial_\nu\langle R(x',p')F^{\mu\nu}(x)\rangle = \displaystyle\int d^4p\frac{p^\mu}{m}\langle R(x,p)R(x',p')\rangle, \\ \partial_\nu\langle R(x',p')^*F^{\mu\nu}(x)\rangle = 0, \end{cases} \end{cases} \tag{5.116}$$

$$\begin{cases} p\cdot\partial\langle R(x,p)F^{\alpha\beta}(x')\rangle + \dfrac{e}{m}p_\nu\dfrac{\partial}{\partial p^\mu}\langle R(x,p)F^{\mu\nu}(x)F^{\alpha\beta}(x')\rangle = 0, \\ \qquad\begin{cases} \partial_\nu\langle F^{\mu\nu}(x)F^{\alpha\beta}(x')\rangle = \displaystyle\int d^4p\frac{p^\mu}{m}\langle R_1(x,p)F^{\alpha\beta}(x')\rangle, \\ \partial_\nu\langle {}^*F^{\mu\nu}(x)\; F^{\alpha\beta}(x')\rangle = 0. \end{cases} \end{cases} \tag{5.117}$$

One also obtains nonwritten equations after multiplication by ${}^*F^{\alpha\beta}(x')$ and averaging. These equations can, of course, be given alternative forms by introducing $f_1,\ f_2,\ldots$ or correlation functions. As usual, this hierarchy can be truncated with various assumptions, the simplest being the Vlasov one

$$\langle R_1(x,p)F^{\mu\nu}(x)\rangle \approx f_1(x,p)\langle F^{\mu\nu}(x)\rangle,$$

which yields the relativistic Vlasov equation with a few subtleties, which we now discuss a little further.

First, a closer inspection of the generating equation of this BBGK hierarchy indicates that the self-field of the particles is included in $\langle F^{\mu\nu}(x)\rangle$: the random four-current occurring on the right hand side of Maxwell's equations does contain the contribution of each particle of the system. This means that when using the above Vlasov ansatz (i) one gives up the radiation reaction terms and (ii) the mass appearing in the subsequent Vlasov equation should be the finite physical mass, an implicit mass renormalization being performed. Next, it should be clear, from these considerations, that the mass occurring in the generating equations is the bare mass, which must always be renormalized, at least implicitly.

Finally, it must be noted that the radiation reaction contributions in the various equations of this hierarchy are not at all easy to separate from interaction terms, for instance. To some extent this has been done by I. Prigogine and F. Henin (1962, 1963), starting from equivalent equations; this is, however, not quite simple.

As a first conclusion, one can assert that the field-*plus*-particle viewpoint, in the nonquantum domain, is much more involved than the action-at-a-distance one, for instance.

Chapter 6

Relativistic Stochastic Processes and Related Questions

Besides their intrinsic interest, relativistic stochastic processes may be involved in a series of semiphenomenological theories. For instance, they permit the establishment of relativistic irreversible processes and hence relativistic Onsager relations. They also permit one to give a probabilistic interpretation of various Fokker–Planck equations considered when one is dealing with relativistic plasmas.

The problem of the relativistic Brownian motion has attracted much attention not only because of its own interest, but also owing to the resemblance of ordinary quantum mechanics to such a stochastic process. Such a formal similarity was noted long ago and was the object of many investigations.[1] Along this line of thought, it was thus quite natural to look for a relativistic generalization of the ordinary Brownian motion. There exist at least two main lines for achieving such a program: one is more mathematical in essence, while the other is more physical. In the first case, it is the stochastic process aspect which is generalized to relativity, while in the second one, the physical problem of a heavy particle embedded in a thermal *substratum* of light particles is dealt with in the relativistic context. The first approach is interesting in itself but it often contains (physically) arbitrary assumptions of a mathematical nature and leads to several indiscriminate possibilities. The physical approach seems to be more natural and rests only on the validity of the approximations performed to obtain the theory.

[1]See, for instance, E. Nelson, *Phys. Rev.* **150**, 1079 (1966); *Dynamical Theories of the Brownian Motion* (Princeton University Press, 1967). U. Ben-Yaacov (1981) has made an interesting attempt at the relativistic Brownian motion; J. Dunkel and P. Haenggi, *Phys. Rep.* **471**, 1 (2009) gives an important review of what was done recently on relativistic Brownian motion.

The main problem when one is dealing with stochastic processes, or Brownian motion in relativity, is that space and time have to be considered on the same footing: time cannot play a particular role, unless one encounters many difficulties in proving the Lorentz invariance of the theory. Moreover, such a separation between space and time is definitely outside of the essence of relativity. Finally, let us also add that a theory in which space and time would *a priori* be separated would be extremely difficult to extend to general relativity. One could think of an indexation of stochastic processes A_Σ by spacelike three-surfaces, which would themselves be partially ordered by causality:

$$\Sigma_1 \prec \Sigma_2 \Leftrightarrow \Sigma_2 \text{ is in the future of } \Sigma_1. \tag{6.1}$$

Although this is not impossible, it appears more efficient to proceed in another way, as can be seen below.

Also, the theory of stochastic quantization gives rise to such problems.

6.1. Stochastic Processes in Minkowski Space–Time

Let $A(x)$ be a physical quantity defined as a tensor (or spinor) field on Minkowski space–time. It is a random field whenever it depends on a measurable (and measured) sample set Ω. Whether one deals with relativity or not, the random process $A(x)$ can be apprehended and used in any physical problem when its moments are given,[2] i.e.

$$\langle A(x)\rangle, \langle A(x_1)\otimes A(x_2)\rangle, \ldots, \langle A(x_1)\otimes A(x_2)\otimes\cdots\otimes A(x_n)\rangle, \ldots. \tag{6.2}$$

While this represents a stochastic process on space–time, it does not present any supplementary difficulties and can be treated with the help of turbulence methods,[3] for instance.

For example, when one is dealing with plasmas and radiation, another possible treatment consists in dealing with the electromagnetic field by such methods, as has briefly been outlined in Chap. 5.

[2]See e.g. R.L. Stratonovich, *Topics in the Theory of Random Noise* (Gordon and Breach, New York, 1963).

[3]G.K. Batchelor, *The Theory of Homogeneous Turbulence* (Cambridge University Press, 1982).

6.1.1. Basic definitions

In Minkowski space–time, a stochastic process $X^\mu(\omega)$ will be characterized, in a statistical sense, by the data of the currents

$$J^{\mu_1}(x_1), J^{\mu_1\mu_2}(x_1, x_2), \ldots, J^{\mu_1\mu_2\ldots\mu_n}(x_1, x_2, \ldots, x_n), \ldots, \tag{6.3}$$

normalized through

$$\int d\Sigma_{\mu_1\mu_2\ldots\mu_n} J^{\mu_1\mu_2\ldots\mu_n}(x_1, x_2, \ldots, x_n) \\ = \int_{\Sigma_1\times\Sigma_2\times\cdots\times\Sigma_n} d\Sigma_{\mu_1} d\Sigma_{\mu_2} \ldots d\Sigma_{\mu_n} J^{\mu_1\mu_2\ldots\mu_n}(x_1, x_2, \ldots, x_n) = 1, \tag{6.4}$$

where $\Sigma_1, \Sigma_2, \ldots, \Sigma_n$ are arbitrary spacelike three-surfaces. The probability that X^μ is in the domain Δ_1 of Σ_1, Δ_2 of Σ_2, etc. is defined as

$$\text{Prob}\{\Delta_1 \subset \Sigma_1, \Delta_2 \subset \Sigma_2, \ldots, \Delta_n \subset \Sigma_n\} \\ = \int_{\Delta_1\subset\Sigma_1\times\Delta_2\subset\Sigma_2\times\cdots\times\Delta_n\subset\Sigma_n} d\Sigma_{\mu_1} d\Sigma_{\mu_2} \ldots d\Sigma_{\mu_n} J^{\mu_1\mu_2\ldots\mu_n}(x_1, x_2, \ldots, x_n). \tag{6.5}$$

A more intrinsic definition could be given [R. Hakim (1968)] but is useless for a practical purpose. The above definition reduces to the usual one when one specializes the spacelike three-surfaces to three-planes $t = \text{const}$; with coordinates adapted to these latter surfaces, one has

$$J^{\mu_1\mu_2\ldots\mu_n}(x_1, x_2, \ldots, x_n) d\Sigma_{\mu_1} d\Sigma_{\mu_2} \ldots d\Sigma_{\mu_n} \\ = J^{00\ldots0}(x_1, x_2, \ldots, x_n) d^3x_1\, d^3x_2 \ldots d^3x_n, \tag{6.6}$$

which shows that the zeroth components of the various currents play the role of the usual probability densities.

These currents must be consistent in the following sense:

$$\int_{\Sigma_1\times\Sigma_2\times\cdots\times\Sigma_\ell} d\Sigma_{\mu_1} d\Sigma_{\mu_2} \ldots d\Sigma_{\mu_\ell} J^{\mu_1\mu_2\ldots\mu_n}(x_1, x_2, \ldots, x_n) \\ = J^{\mu_{\ell+1}\mu_2\ldots\mu_n}(x_{\ell+1}, x_{\ell+2}, \ldots, x_n), \tag{6.7}$$

for all n and ℓ with $\ell \leq n$. Furthermore, they must not depend on the arbitrary surfaces Σ_1, Σ_2, etc.; these latter conditions imply the conservation relations

$$\begin{cases} \partial_{\mu_\ell} J^{\mu_1\mu_2\ldots\mu_n}(x_1, x_2, \ldots, x_n) = 0, \\ \text{for all } \ell \leq n = 1, 2, \ldots. \end{cases} \tag{6.8}$$

6.1.2. *Conditional currents*

In view of the definition of Markovian processes in Minkowski space–time, let us now define the current of transition probability as being

$$J^{\mu}(x_0 \to x_1) \underset{\text{def}}{\equiv} \frac{J_{\nu}(x_0)J^{\mu\nu}(x_0, x_1)}{J_{\lambda}(x_0)J^{\lambda}(x_0)}, \tag{6.9}$$

where the event x_1 is in the future of the event x_0. $J^{\mu}(x_0 \to x_1)$ is thus directly connected with the probability density that X, being in x_0, will be in x_1, and this can easily be seen by specializing three-surfaces to three-planes and looking at the zeroth components of the various currents. The conditional current $J^{\mu}(x_0 \to x_1)$ is normalized by

$$\int_{\Sigma} d\Sigma_{\mu}(x_1)J^{\mu}(x_0 \to x_1) = 1, \tag{6.10}$$

and this also implies the conservation relation

$$\partial_{\mu(x_1)}J^{\mu}(x_0 \to x_1) = 0. \tag{6.11}$$

Conversely, given the currents $J^{\mu}(x_1)$ and $J^{\mu}(x_0 \to x_1)$, one has

$$J^{\mu\nu}(x_0, x_1) = J^{\mu}(x_0)J^{\nu}(x_0 \to x_1), \tag{6.12}$$

which may be found either from the probabilistic interpretation of the various currents or from the definition of $J^{\mu}(x_0 \to x_1)$ considered in a local coordinate system where $J^{\mu}(x_1)$ reduces to its zeroth component.

The transition current $J^{\mu}(x_0 \to x_1)$ must also satisfy some causality requirements, specific to relativity. For instance, velocities higher than the speed of light should be forbidden and, accordingly, one should have

$$\begin{cases} J^{\mu}(x_0 \to x_1) \equiv 0, \\ \text{for } (x_1 - x_0)^2 \leq 0; \end{cases} \tag{6.13}$$

the transition four-current should thus vanish outside the null cone.

6.1.3. *Markovian processes in space–time*

We pursue our *physicist* approach of stochastic processes in Minkowski space by looking at what could be a Markovian process. Such a process is defined as usual except that we have to deal with currents rather than densities. A Markovian process is thus completely characterized when the first two currents, or the first and the transition currents, are given while all other currents are expressed in terms of these currents. For instance, $J^{\mu\nu\lambda}(x_1, x_2, x_3)$ is of the form

$$J^{\mu\nu\lambda}(x_1, x_2, x_3) = J^{\mu}(x_1)J^{\nu}(x_1 \to x_2)J^{\lambda}(x_2 \to x_3), \tag{6.14}$$

where x_3 is in the future of x_2, itself in the future of x_1. Integrating this equation over x_2 and taking the consistency relation between currents, one obtains

$$J^\mu(x_1 \to x_3) = \int_\Sigma d\Sigma_{\nu(2)} J^\nu(x_1 \to x_2) J^\mu(x_2 \to x_3), \tag{6.15}$$

which is nothing but the relativistic Chapman–Kolmogorov equation. The independence of this relation from the three-surface Σ implies the conservation equation

$$\partial_{\nu(2)} \left(J^\nu(x_1 \to x_2) J^\mu(x_2 \to x_3) \right) = 0, \tag{6.16}$$

which reduces to

$$J^\nu(x_1 \to x_2) \partial_{\nu(2)} J^\mu(x_2 \to x_3) = 0, \tag{6.17}$$

after one uses the conservation relation obeyed by J^ν. This relativistic Chapman–Enskog equation has already been given by J. Lopuszanski (1953), though without proof. He has also shown that the Fokker–Planck equation obtained from it reduces to this last conservation equation. This absence of a second order Fokker–Planck equation shows that a theory of relativistic Brownian motion cannot be erected on the assumption that such a relativistic Markovian process may represent it.

Let us examine the consequence of the strict Lorentz–Poincaré invariance on Markovian processes. For the first order four-current $J^\mu(x)$, one has

$$J^\mu(x) = J^\mu(0) = 0, \tag{6.18}$$

since there is no four-vector in the theory. The second order current, $J^{\mu\nu}(x_1, x_2)$, depends on the difference $x_1 - x_2$ and so does the transition four-current

$$J^\nu(x_1 \to x_2) = J^\nu(x_2 - x_1). \tag{6.19}$$

From the continuity equation and the fact that the Lorentz–Poincaré invariance requires that it should possess the form

$$\begin{cases} J^\nu(x_2 - x_1) = (x_2^\nu - x_1^\nu) f(\tau), \\ \tau^2 = x \cdot x, \end{cases} \tag{6.20}$$

it follows that the function $f(\tau)$ obeys

$$\tau \frac{d}{d\tau} f(\tau) + 4 f(\tau) = 0, \tag{6.21}$$

whose solution is

$$f(\tau) = \frac{\text{const}}{\tau^4}. \tag{6.22}$$

Consequently, the transition four-current does not satisfy either the continuity equation (it is too singular on the light cone) or the causality requirements and hence there does not exist any relativistic causal Markovian process obeying strictly the full Lorentz–Poincaré covariance.

Therefore, it seems *a priori* difficult to generalize the interesting results of E. Nelson (1967) in a relativistic framework unless the causality requirements are relaxed. This is certainly possible, since quantum mechanics is neither local nor causal (think that in its path integral formulation, sums over all kinds of trajectories, including spacelike ones, are performed).

Let us now examine a few consequences of the existence of a macroscopic four-vector, namely u^μ, and of Lorentz–Poincaré covariance. Then we have

$$J^\mu(x) = J^\mu(0) = nu^\mu, \tag{6.23}$$

which needs a brief comment. Since n is a constant — because of space–time homogeneity — this four-current cannot be normalize to unity; and this corresponds to a uniform spatial probability density. The transition four-current has the form

$$\begin{aligned} J^\nu(x_1 \to x_2) &= J^\nu(x_2 - x_1) \\ &\equiv J^\nu(x) = x^\nu f_1 + u^\nu f_2, \end{aligned} \tag{6.24}$$

where f_1 and f_2 are functions of the available invariants, namely $u \cdot x$, $\Delta(u) \cdot x \cdot x$ and $x \cdot x$. The "Fokker–Planck" equation written under the form

$$J^\nu(x)\partial_\nu J^\mu(y) = 0 \tag{6.25}$$

yields

$$[x^\nu f_1(x) + u^\nu f_2(x)]\, \partial_\nu \left[y^\mu f_1(y) + u^\mu f_2(y)\right] = 0, \tag{6.26}$$

while the continuity equation gives rise to

$$\partial_\nu \left[x^\nu f_1(x) + u^\nu f_2(x)\right] = 0. \tag{6.27}$$

6.2. Stochastic Processes in μ Space

Elsewhere, a "geometrical" approach was given for stochastic processes in μ space [R. Hakim (1968)]; however, in this chapter we shall restrict ourselves to a more intuitive way to deal with such problems, by using a proper-time-dependent formalism which is equivalent to the "geometrical" one, insofar as equivalent statistical assumptions are used.

6.2.1. *An overview*

Let us consider a random point in μ space, $x^A \equiv \{x^\mu, p_\nu\}$. It may be considered as being proper-time-dependent $x^A(\tau)$ and thus as a "true" stochastic process in μ space. However, the proper time characterization of the trajectory is only one among an infinity of other possible ones. Nevertheless, it is useful owing to the physical meaning of τ and also because in the end it does not appear in the results. Therefore, we first extend the methods used in the study of stochastic processes in Minkowski space–time.

First, the probability that the state of a random process is in a domain A of μ space is defined by

$$\mathrm{Prob}\{(x,p) \subset A\} = \int_{A\subset\Sigma\times P^4} d\Sigma_\lambda \, d^4p \; P_1^\lambda(x^\mu, p^\mu) \tag{6.28}$$

and is normalized as

$$\int_\Sigma d\Sigma_\lambda \, d^4p \; P_1^\lambda(x^\mu, p^\mu) = 1, \tag{6.29}$$

implying the following continuity equation in order to insure its independence from the arbitrary spacelike three-surface $\sum$:

$$\partial_\lambda P_1^\lambda(x^\mu, p^\mu) = 0. \tag{6.30}$$

Note that all this has also the form

$$\mathrm{Prob}\{(x,p) \subset A\} = \int_{A\subset\Sigma\times P^4} d\Sigma_B P^B(X), \tag{6.31}$$

where $X = (x,p)$ and when $\{p\}$ is on a mass shell different from that of $p^2 - m^2 = 0$, as is the case in Chap. 13. Therefore, the equation obeyed by P is now

$$\partial_B P^B = 0, \tag{6.32}$$

or, rather, in adapted coordinates,

$$\partial_\mu(u^\mu f) + \frac{\partial[F^\mu(x,p)f]}{\partial p^\mu} = 0. \tag{6.33}$$

More generally, the probability that the state of a random process is in a domain A_1 of μ space, and $A_2, \ldots, A_n$, is defined by

$$\mathrm{Prob}\{(x_1,p_1) \subset A_1; (x_2,p_2) \subset A_2; \ldots; (x_n,p_n) \subset A_n\}$$
$$= \int_{A_1\times A_2\times\cdots\times A_n\subset\Sigma^{\times n}} \prod_{\ell=1}^{\ell=n} d\Sigma_{\lambda_\ell} \, d^4p_\ell \; P_n^{\lambda_1\lambda_2\ldots\lambda_n}(x_1^\mu, p_1^\mu; \ldots; x_n^\mu, p_n^\mu), \tag{6.34}$$

which is normalized through

$$\int_{\Sigma^{\times n}\times P^{\times 4n}} \prod_{\ell=1}^{\ell=n} d\Sigma_{\lambda_\ell}\, d^4p_\ell\, P_n^{\lambda_1\lambda_2\ldots\lambda_n}(x_1^\mu,p_1^\mu;\ldots;x_n^\mu,p_n^\mu) = 1, \tag{6.35}$$

which implies the continuity equations

$$\partial_{\lambda_\ell} P_n^{\lambda_1\lambda_2\ldots\lambda_n}(x_1^\mu,p_1^\mu;\ldots;x_n^\mu,p_n^\mu) = 0, \quad \ell = 1,2,\ldots,n, \tag{6.36}$$

which still express the conservation of probabilities.

The probability currents $P_n^{\lambda_1\lambda_2\ldots\lambda_n}(x_1^\mu,p_1^\mu;\ldots;x_n^\mu,p_n^\mu)$, besides being positive, must satisfy the consistency relations

$$\int \prod_{\ell=k+1}^{\ell=n} (d\Sigma_{\mu_\ell} d^4p_\ell) P_n^{\lambda_1\lambda_2\ldots\lambda_n}(x_1^\mu,p_1^\mu;\ldots;x_n^\mu,p_n^\mu)$$
$$= P_k^{\lambda_1\lambda_2\ldots\lambda_k}(x_1^\mu,p_1^\mu;\ldots;x_k^\mu,p_k^\mu). \tag{6.37}$$

Finally, the stochastic process in μ space is defined, in a statistical sense, when the probability currents $P_n^{\lambda_1\lambda_2\ldots\lambda_n}(x_1^\mu,p_1^\mu;\ldots;x_n^\mu,p_n^\mu)$ are all known.

An important remark is now in order concerning these probability currents. If one thinks of the stochastic process at hand as representing some statistical system of massive point particles, then necessarily [see Chap. 5] the various indices λ_i's refer to p_i^λ's, and one should have, for instance,

$$P_n^{\lambda_1\lambda_2\ldots\lambda_n}(x_1^\mu,p_1^\mu;\ldots;x_n^\mu,p_n^\mu) \equiv p_1^{\lambda_1}p_2^{\lambda_2}\ldots p_n^{\lambda_n} f_n(x_1^\mu,p_1^\mu;\ldots;x_n^\mu,p_n^\mu), \tag{6.38}$$

where $f_n(x_1^\mu,p_1^\mu;\ldots;x_n^\mu,p_n^\mu)$ is a scalar function. However, it should be borne in mind that the above equality is nothing but an *assumption*, although natural. Note also that the element of integration in the four-momentum-space, which we denoted generically by d^4p, is generally — with the same argument — equal to d^3p/p_0. However, this is also an assumption, since particles could well be "dressed" by the medium so as to satisfy another relation than $p^2 = m^2$, such as $p^2 = \prod(p)$ (see Chap. 13).

6.2.2. *Markovian processes*

Let us now restrict ourselves to the important case of Markovian processes and, to this end, let us define the conditional probability density that the process is in state $\{x_0^\mu, p_0^\mu\}$ knowing that it was "before" in states $\{x_1^\mu,p_1^\mu;\ldots;x_n^\mu,p_n^\mu\}$. By "before," is meant "according to the partial order of Minkowski space–time." The above sequence of states of the process is indeed partially ordered by the partial order of the x_i^μ's:

$$x_1^\mu \prec x_2^\mu \prec \cdots \prec x_n^\mu, \tag{6.39}$$

where the partial order $\prec$ is defined as $x_1^\mu \prec x_2^\mu$ if and only if x_2 is in the future of x_1, or

$$\begin{cases} (x_2 - x_1)\cdot(x_2 - x_1) \geq 0, \\ (x_2^0 - x_1^0) > 0. \end{cases} \tag{6.40}$$

The conditional density that a particle is in points $(x_1^\mu, p_1^\mu; \ldots; x_n^\mu, p_n^\mu)$ and undergoes a transition to point (x_0^μ, p_0^μ) is

$$P^\alpha\left(x_1^\mu, p_1^\mu; \ldots; x_n^\mu, p_n^\mu \left| x_0^\mu, p_0^\mu\right.\right)$$
$$= \frac{P_{n+1}^{\alpha\lambda_1\lambda_2\ldots\lambda_n}(x_0^\mu, p_0^\mu; x_1^\mu, p_1^\mu; \ldots; x_n^\mu, p_n^\mu) P_{n\lambda_1\lambda_2\ldots\lambda_n}(x_1^\mu, p_1^\mu; \ldots; x_n^\mu, p_n^\mu)}{P_n^{\lambda_1\lambda_2\ldots\lambda_n}(x_1^\mu, p_1^\mu; \ldots; x_n^\mu, p_n^\mu) P_{n\lambda_1\lambda_2\ldots\lambda_n}(x_1^\mu, p_1^\mu; \ldots; x_n^\mu, p_n^\mu)}, \tag{6.41}$$

obviously normalized by

$$\int_{\Sigma\times P^4} d\Sigma_\alpha\, d^4p\, P^\alpha\left(x_1^\mu, p_1^\mu; \ldots; x_n^\mu, p_n^\mu \left| x^\mu, p^\mu\right.\right) = 1, \tag{6.42}$$

and hence verifies the conservation relation

$$\partial_\alpha P^\alpha\left(x_1^\mu, p_1^\mu; \ldots; x_n^\mu, p_n^\mu \left| x^\mu, p^\mu\right.\right) = 0. \tag{6.43}$$

A relativistic Markov process is now defined as a stochastic process whose conditional probability P^α depends on the last state in which the process was and not on the preceding ones, i.e. as

$$\begin{cases} P^\alpha(x_1^\mu, p_1^\mu; \ldots; x_n^\mu, p_n^\mu \left| x^\mu, p^\mu\right.) = P^\alpha(x_n^\mu, p_n^\mu \left| x^\mu, p^\mu\right.) \\ x_1^\mu \prec x_2^\mu \prec \ldots \prec x_n^\mu \prec x^\mu. \end{cases} \tag{6.44}$$

The Markov property then leads to the following forms for the P_n's:

$$P_n^{\lambda_1\lambda_2\ldots\lambda_n}(x_1^\mu, p_1^\mu; \ldots; x_n^\mu, p_n^\mu) = P_1^{\lambda_n}(x_n^\mu, p_n^\mu) P^{\lambda_{n-1}}(x_n^\mu, p_n^\mu \left| x_{n-1}^\mu, p_{n-1}^\mu\right.)$$
$$\times P^{\lambda_{n-2}}(x_{n-1}^\mu, p_{n-1}^\mu \left| x_{n-2}^\mu, p_{n-2}^\mu\right.) \ldots P^{\lambda_1}(x_2^\mu, p_2^\mu \left| x_1^\mu, p_1^\mu\right.)\,. \tag{6.45}$$

Therefore, as usual, the relativistic Markovian process is completely determined by the data of both P_1 and P. From now on, the notation

$$P^\lambda(x_0^\mu, p_0^\mu \left| x^\mu, p^\mu\right.) \equiv P^\lambda\left(\{x_0^\mu, p_0^\mu\} \to \{x^\mu, p^\mu\}\right) \tag{6.46}$$

will be used, according to a current notation in physics. From the above definitions, the Chapman–Kolmogorov equation can be written as

$$P^{\lambda}\left(\{x_0^{\mu}, p_0^{\mu}\} \to \{x^{\mu}, p^{\mu}\}\right) = \int_{\Sigma \times P^4} d\Sigma_{2\nu}\, d^4 p_2\, P^{\nu}\left(\{x_0^{\mu}, p_0^{\mu}\} \to \{x_2^{\mu}, p_2^{\mu}\}\right) \times P^{\lambda}\left(\{x_2^{\mu}, p_2^{\mu}\} \to \{x^{\mu}, p^{\mu}\}\right), \tag{6.47}$$

when $p^2 = m^2$. If this condition is not verified, then one has a different normalization (see Chap. 13).

Note that if *causality* has to be satisfied, one must also have

$$P^{\lambda}\left(\{x_0^{\mu}, p_0^{\mu}\} \to \{x^{\mu}, p^{\mu}\}\right) = \begin{cases} \neq 0 & \text{for } x_0^{\mu} \prec x_1^{\mu}, \\ 0 & \text{otherwise.} \end{cases} \tag{6.48}$$

When the assumption mentioned above is valid, one can write

$$\begin{cases} P_1^{\lambda}\left(x^{\mu}, p^{\mu}\right) = p^{\mu} f_1\left(x^{\mu}, p^{\mu}\right), \\ P^{\lambda}\left(\{x_0^{\mu}, p_0^{\mu}\} \to \{x^{\mu}, p^{\mu}\}\right) = p^{\mu} f\left(\{x_0^{\mu}, p_0^{\mu}\} \to \{x^{\mu}, p^{\mu}\}\right) \end{cases} \tag{6.49}$$

and the above Chapman–Kolmogorov equation can be rewritten as

$$f\left(\{x_0^{\mu}, p_0^{\mu}\} \to \{x^{\mu}, p^{\mu}\}\right) = \int_{\Sigma \times P^4} d\Sigma_{2\nu}\, d^4 p_2\, p^{\nu} f\left(\{x_0^{\mu}, p_0^{\mu}\} \to \{x_2^{\mu}, p_2^{\mu}\}\right) \times f\left(\{x_2^{\mu}, p_2^{\mu}\} \to \{x^{\mu}, p^{\mu}\}\right). \tag{6.50}$$

When $p^2 = m^2$ and for Σ being a spacelike three-plane $t =$ const, this equation is identical to the ordinary Chapman–Kolmogorov equation, and the Fokker–Planck equation

$$p^{\mu}\partial_{\mu} f(x,p) - \frac{\partial}{\partial p^{\mu}}\left(B^{\mu}(p) f(x,p)\right) + \frac{1}{2}\frac{\partial^2}{\partial p^{\mu}\partial p^{\nu}}\left(D^{\mu\upsilon}(p) f(x,p)\right) = 0 \tag{6.51}$$

can be derived in the usual way.[4] In this last equation a stationary and homogeneous process has been assumed so that we have been able to set

$$f(x,p) \equiv f(0, p_0; x_2 - x_1, p_1). \tag{6.52}$$

6.2.3. *An alternative approach*

An alternative approach to the description of stochastic processes in μ space can be found if we think of a random mechanical system of point particles. Let us designate by X^A the coordinates in μ space:

$$X^A \equiv \{x^{\mu}, p^{\mu}\}. \tag{6.53}$$

[4]See R.L. Stratonovitch, *loc. cit.*

The stochastic process $X^A(\tau)$ is assumed, as usual, to be completely determined by the data of the distribution functions

$$W_n(X_1, \tau_1; \dots; X_n, \tau_n), \quad n = 1, 2, \dots, \tag{6.54}$$

normalized in μ space through

$$\int \prod_{\ell=1}^{\ell=n} d\mu_\ell \, W_n(X_1, \tau_1; \dots; X_n, \tau_n) = 1 \quad \text{for all } (\tau_1, \tau_2, \dots, \tau_n) \tag{6.55}$$

and satisfying the consistency relations

$$\int \prod_{\ell=k+1}^{\ell=n} d\mu_\ell \, W_n(X_1, \tau_1; \dots; X_n, \tau_n) = W_k(X_1, \tau_1; \dots; X_k, \tau_k), \quad k < n. \tag{6.56}$$

Note that $d\mu$ is the volume element in the eight-dimensional μ space. The constancy of the total weight of these distributions implies the existence of continuity equations of the general form

$$\frac{d}{d\tau_\ell} W_n \equiv \frac{\partial}{\partial \tau_\ell} W_n + \sum_\ell p_\ell \cdot \partial W_n + \sum_\ell \frac{\partial}{\partial p_{\ell\mu}} (C_\ell^\mu W_n) = 0, \tag{6.57}$$

where C_ℓ^μ can be an *operator* acting on W_n. In fact, it cannot be given a more specific form unless some statistical assumptions are provided. Such a situation occurs not only in the relativistic context but also in the Newtonian one.[5] An example is given below.

Note that notions connected with the proper time τ, such as stationarity, have no direct physical meaning; however, from the W_n's one can derive physical densities as in the case of relativistic statistical mechanics [R. Hakim (1967b)] through

$$f_n(x_1^\mu, p_1^\mu; \dots; x_n^\mu, p_n^\mu) = \int_0^\infty d\tau_1 \dots \int_0^\infty d\tau_n W_n(X_1, \tau_1; \dots; X_n, \tau_n) \tag{6.58}$$

and normalized as

$$\int \prod_{\ell=1}^{\ell=n} \left(d\Sigma_{\mu_\ell} \frac{d^3 p_\ell}{p_{\ell 0}} C_n^{\mu_\ell} \right) f_n(x_1^\mu, p_1^\mu; \dots; x_n^\mu, p_n^\mu) = 1, \tag{6.59}$$

while the consistency relations between the W_n's is accompanied by similar conditions on the P_n's:

$$\int \prod_{\ell=k+1}^{\ell=n} \left(d\Sigma_{\mu_\ell} \frac{d^3 p_\ell}{p_{\ell 0}} C_n^{\mu_\ell} \right) f_n(x_1^\mu, p_1^\mu; \dots; x_n^\mu, p_n^\mu) = f_k(x_1^\mu, p_1^\mu; \dots; x_k^\mu, p_k^\mu). \tag{6.60}$$

[5]See e.g. R.L. Stratonovich, *loc. cit.*

The $f_n(x_1^\mu, p_1^\mu; \ldots; x_n^\mu, p_n^\mu)$'s obey the continuity equation

$$\left\{\sum_\ell p_\ell.\partial_\ell + \frac{\partial}{\partial p_\ell^{\mu\ell}} C_\ell^{\mu\ell}\right\} f_n(x_1^\mu, p_1^\mu; \ldots; x_n^\mu, p_n^\mu) = 0, \tag{6.61}$$

which expresses the conservation of probabilities.

6.2.4. *Markovian processes*

Let us now restrict ourselves to the important case of Markovian processes and, to this end, let us define the conditional probability density that the process is in state $\{x_0^\mu, p_0^\mu\}$ knowing that it was "before" in states $\{x_1^\mu, p_1^\mu; \ldots; x_n^\mu, p_n^\mu\}$.

This conditional density is defined by

$$P\left(x_1^\mu, p_1^\mu; \ldots; x_n^\mu, p_n^\mu \,|x_0^\mu, p_0^\mu\right) = \frac{P_{n+1}(x_0^\mu, p_0^\mu; x_1^\mu, p_1^\mu; \ldots; x_n^\mu, p_n^\mu)}{P_n(x_1^\mu, p_1^\mu; \ldots; x_n^\mu, p_n^\mu)} \tag{6.62}$$

and verifies the conservation relation

$$\left[p \cdot \partial + \frac{\partial}{\partial p^\mu} C^\mu\right] P(x_1^\mu, p_1^\mu; \ldots; x_n^\mu, p_n^\mu \,|x^\mu, p^\mu) = 0. \tag{6.63}$$

A relativistic Markov process is now defined as a stochastic process whose conditional probability P depends on the last state of the process and not on the preceding ones, i.e. as

$$\begin{cases} P(x_1^\mu, p_1^\mu; \ldots; x_n^\mu, p_n^\mu \,|x^\mu, p^\mu) = P(x_n^\mu, p_n^\mu \,|x^\mu, p^\mu), \\ x_1^\mu \prec x_2^\mu \prec \ldots \prec x_n^\mu \prec x^\mu. \end{cases} \tag{6.64}$$

The Markov property then leads to the distributions

$$\begin{aligned} P_n(x_1^\mu, p_1^\mu; \ldots; x_n^\mu, p_n^\mu) &= P_1(x_n^\mu, p_n^\mu) P(x_n^\mu, p_n^\mu \,|x_{n-1}^\mu, p_{n-1}^\mu) \\ &\quad \times P(x_{n-1}^\mu, p_{n-1}^\mu \,|x_{n-2}^\mu, p_{n-2}^\mu) \ldots P(x_2^\mu, p_2^\mu \,|x_1^\mu, p_1^\mu). \end{aligned} \tag{6.65}$$

Therefore, as usual, the relativistic Markovian process is completely determined by the data of both P_1 and P_2.

The relativistic Chapman–Kolmogorov equation then reads

$$P(x_0^\mu, p_0^\mu \,|x_1^\mu, p_1^\mu) = \int_\Sigma d\Sigma_\mu \frac{d^3p}{p_0} C^\mu \left(P(x_0^\mu, p_0^\mu \,|x^\mu, p^\mu)\, P(x^\mu, p^\mu \,|x_1^\mu, p_1^\mu)\right), \tag{6.66}$$

from which the relativistic Fokker–Planck equation can be derived:

$$p \cdot \partial P_2 + \frac{\partial}{\partial p^\mu}\left\{-B^\mu P_2 + \frac{1}{2}\frac{\partial}{\partial p^\nu} D^{\mu\nu} P_2\right\} = 0. \tag{6.67}$$

6.2.5. *A simple illustration*

Let us consider the problem of a particle embedded in a random force field. It can be dealt with in two equivalent possible ways. Either we study the random differential system

$$\begin{cases} m\dfrac{dx^\mu(\tau)}{d\tau} = p^\mu, \\ \dfrac{dp^\mu(\tau)}{d\tau} = F^\mu, \end{cases} \tag{6.68}$$

or we directly write, as in Chap. 1, the equivalent random Liouville equation satisfied by

$$R(x,p;\tau) \equiv \delta^{(4)}\left[x - x(\tau)\right]\delta^{(4)}\left[p - p(\tau)\right], \tag{6.69}$$

namely

$$\frac{\partial}{\partial\tau}R + p\cdot\partial R + F\cdot\frac{\partial}{\partial p}R = 0, \tag{6.70}$$

and next solve it. The former method is inspired by R.L. Stratonovich,[6] while the latter is from R. Kubo.[7] The random force F is assumed to be completely specified in a statistical sense when all the moments (assumed to exist), $\langle F\otimes F\otimes\cdots\otimes F\rangle$, are known.

Let us now set

$$\begin{cases} L_0 = -p\cdot\partial, \\ L_1 = F\cdot\dfrac{\partial}{\partial p}. \end{cases} \tag{6.71}$$

With these notations, the above random Liouville equation can be written as

$$\frac{\partial}{\partial\tau}R = -\left(L_0 + L_1\right)R, \tag{6.72}$$

and, in the *interaction representation*, reads

$$\frac{\partial}{\partial\tau}\Re = -L\Re, \tag{6.73}$$

with

$$\begin{cases} \Re = \exp\left(-L_0\tau\right)R, \\ L = \exp\left(-L_0\tau\right)L_1\exp\left(+L_0\tau\right). \end{cases} \tag{6.74}$$

[6] R.L. Stratonovich, *loc. cit.*
[7] R. Kubo, *J. Math. Phys.* **4**, 174 (1963).

The random Liouville equation in interaction representation can be formally integrated and yields

$$\begin{aligned}\Re &= \mathcal{P}\exp\left(\int_0^\tau ds\, L(s)\right)\Re(0)\\ &= \mathcal{P}\exp\left(\int_0^\tau ds\, L(s)\right)\Re(0),\end{aligned} \tag{6.75}$$

where exp(...) designates symbolically the series

$$\begin{aligned}\mathcal{P}\exp\left(\int_0^\tau ds\ L(s)\right) &= \sum_{n=0}^{\infty}\int_0^\tau d\tau_1\int_0^{\tau_1} d\tau_2\ldots\int_0^{\tau_{n-1}} d\tau_n\\ &\times\mathcal{P}\left[L(\tau_1)L(\tau_2)\cdots L(\tau_n)\right],\end{aligned} \tag{6.76}$$

with $\tau_1 \leq \tau_2 \leq \cdots \leq \tau_n$, and where $\mathcal{P}$ designates the chronological operator that orders the τ_i's.

One generally wants to obtain an equation for W_1 or W_2, and hence for f_1 or f_2. For W_1 one finds that

$$\frac{\partial}{\partial\tau}\langle\Re\rangle = \frac{\partial}{\partial\tau}\left\langle\exp\left[\int_0^\tau ds\, L(s)\right]\right\rangle\times\left\langle\exp\left[\int_0^\tau ds\, L(s)\right]\right\rangle^{-1}\langle\Re\rangle, \tag{6.77}$$

so that W_1 obeys the equation

$$\begin{aligned}\frac{\partial}{\partial\tau}W_1 = L_0 W_1 + \left\{\exp\left(L_0\tau\right)\times\frac{\partial}{\partial\tau}\left\langle\exp\left[\int_0^\tau ds\, L(s)\right]\right\rangle\right\}\\ \times\left\langle\exp\left[\int_0^\tau ds\, L(s)\right]\right\rangle^{-1}\times\exp\left(-L_0\tau\right)W_1.\end{aligned} \tag{6.78}$$

The right hand side of this last equation can be explicitly written with the use of the above definitions, equations and the moments of the four-force F^μ.

If one considers that F^μ is a term of order 1 in a supposedly small coupling constant, then the various exponentials of operators involved in this last equation provide an expansion into power series in this coupling constant. This is useful in connection with the derivation of kinetic equations, for instance, and in other problems.

To the first order in this coupling constant, this equation reads

$$\begin{aligned}&\frac{\partial}{\partial\tau}W_1 + p\cdot\partial W_1 + \frac{F^\mu}{m}\frac{\partial}{\partial p^\mu}W_1\\ &= \int_0^\tau ds\, K\left\{\exp\left(-[\tau-s]p\cdot\partial\right)\times L_1(s)\times\exp\left([\tau-s]p\cdot\partial\right)\right\}W_1,\end{aligned} \tag{6.79}$$

where K{...} is the correlation function of the operator inside the brackets.

6.3. Relativistic Brownian Motion

Many articles on the relativistic Brownian motion have been published; unfortunately, none of them is fully satisfactory, either because they are formal or mathematical generalizations, lacking any physical basis, or because they contain some hidden drawback, or — more simply — because they are not covariant under the Lorentz group.

The difficulties are of several kinds. First, even though relativistic stochastic processes can be defined and studied in the covariant μ space, it is not clear whether the random process $[X^\mu(\tau), P^\mu(\tau)]$ is still Markovian. Furthermore, one also has to take account of the nonindependence of the components of $P^\mu(\tau)$ owing to the mass shell constraint, while the time coordinate $X^0(\tau)$ becomes a random function of the proper time. Secondly, when one tries to write down a covariant Langevin equation, one encounters similar difficulties. Such an equation should have the general form

$$\frac{d}{d\tau}P^\mu(\tau) + K^\mu(P) = F^\mu(\tau), \tag{6.80}$$

where K is the friction four-force and F is the random Gaussian force supposed to take account of the other part of the shocks of the particles within the medium on the massive Brownian particle. However, two kinds of difficulties do occur when one is considering such an equation. The first one deals with the specific form of the friction four-force; since the theory contains two four-vectors only — namely P^μ and u^μ, the average four-velocity of the medium — it has necessarily the general form

$$K^\mu(P) = A(P)P^\mu + B(P)u^\mu, \tag{6.81}$$

where the dependence of A and B on P must occur through the only possible invariant, $P \cdot$ u. Finally, the *assumed*[8] linearity of the relativistic Langevin equation implies the general form

$$K^\mu(P) = AP^\mu + BP \cdot u\, u^\mu \tag{6.82}$$

where A and B are now true constants, to be determined by physical considerations. On the other hand, the above form of the friction force must reduce to the ordinary one, or

$$K^i \to \beta p^i \quad (i = 1, 2, 3), \tag{6.83}$$

[8]This linearity is absolutely necessary; otherwise the usual manipulations on the Langevin equation would hardly be possible.

where β is the usual friction coefficient. This implies immediately that K^μ has the form

$$K^\mu(P) = \beta P^\mu + 2\beta P \cdot u\, u^\mu. \tag{6.84}$$

In order to obtain this last relation, use has been made of the nonrelativistic limit of the zeroth component[9] of K:

$$K^0 \to 2\beta p^0. \tag{6.85}$$

However, one has to face another problem. In order for the mass shell condition on P, i.e. $P^2 = m^2$, to be satisfied, the total force $F+K$, acting on the Brownian particle, must be orthogonal to P (in the sense of Minkowski geometry), $K^\mu(P)P_\mu - F^\mu P_\mu = 0$, or

$$\beta m^2 + 2\beta(P \cdot u)^2 = F \cdot P. \tag{6.86}$$

Accordingly, the four components of F are *not* independent. Also, if $\mathbf{F}$ is Gaussian this property is probably not satisfied by F^0. Finally, the relativistic Langevin equation cannot be dealt with as simply as in the Newtonian case.

Hence, it appears simpler to go back to the *physical* problem, namely that of a heavy particle subject to the collisions of the light particles that constitute the background medium, supposed to be in thermal equilibrium. The random motion of the Brownian particle is described by its distribution function $f(x,p)$. It obeys a Boltzmann-like kinetic equation (Chap. 2) whose collision term has the form

$$\begin{aligned} C(f) = \frac{1}{2}\int \frac{d^3p'}{p_0'}\frac{d^3p''}{p_0''}\frac{d^3\bar{p}}{\bar{p}_0} W\left(p', p'' \to p, \bar{p}\right) \\ \times\, \delta^{(4)}\left(p + p'' - p' - \bar{p}\right)\left[f_0(x,p')f(x,p'') - f_0(x,p)f(x,\bar{p})\right], \end{aligned} \tag{6.87}$$

where $f_0(x,p)$ is the thermal (i.e. Jüttner–Synge) distribution function representing the particles of the background medium. The fact that the Brownian particle is much more massive than the background particles is expressed by the fact that energy–momentum transfers are small compared to the energy–moment of the background particles; or

$$|p - p'| \ll |p|\,. \tag{6.88}$$

Expanding the above collision integral into powers of the energy–momentum transfer and keeping as usual the first two terms, one arrives

[9] See P. Mazur, *Physica*, **25**, 149 (1959).

at the covariant Fokker–Planck equation for $f(x,p)$:

$$p^\mu \partial_\mu f(x,p) - \frac{\partial}{\partial p^\mu}\left(B^\mu(p) f(x,p)\right) + \frac{1}{2}\frac{\partial^2}{\partial p^\mu \partial p^\nu}\left(D^{\mu\upsilon}(p) f(x,p)\right) = 0. \tag{6.89}$$

The Fokker–Planck coefficients $B^\mu(p)$ and $D^{\mu\upsilon}(p)$ can be evaluated from the collision integral. Taking account of the tensors available in the theory, the general form of the Fokker–Planck coefficients is given by

$$\begin{cases} B^\mu(p) &= B_1(p)p^\mu + B_2(p)u^\mu, \\ D^{\mu\upsilon}(p) &= D_1(p)\eta^{\mu\upsilon} + D_2(p)p^\mu p^\upsilon \\ & \quad +D_3(p)u^\mu u^\upsilon + D_4(p)p^{(\mu}u^{\nu)}. \end{cases} \tag{6.90}$$

Also, they must be such that an H theorem is valid; in this case, it reads

$$\partial_\mu S^\mu(x) > 0, \quad S^\mu(x) = -k_B \int d^4p \; f(x,p) \log\left[f(x,p)\right] \tag{6.91}$$

(k_B: Boltzmann's constant).
However, they must obey some general relations owing to the fact that *asymptotically* — i.e. at infinity in a timelike direction — the Jüttner–Synge equilibrium distribution must be a unique solution. The fact that the Jüttner–Synge is an asymptotic solution renders the solution to these equations easier to obtain. Asymptotically, one has

$$\left\{B^\mu - \frac{1}{2}\frac{\partial}{\partial p_\mu}\left(D_1^{\mu\nu} + D_2^{\mu\nu} + D_3^{\mu\nu} + D_4^{\mu\nu}\right)\right\} f_0 = 0, \tag{6.92}$$

which does not present an interesting equation.

Finally, let us mention the so-called stochastic quantization of G. Parisi and Y.S. Wu,[10] but it has only a *formal* resemblance to the usual Fokker–Planck equation.

6.4. Random Gravitational Fields: An Open Problem

For a variety of physical reasons, a gravitational field $g^{\mu\nu}(x)$ on a space–time manifold can often be considered as being random. For instance, the energy–momentum tensor occurring on the right hand side of Einstein's equations,

$$R_{\mu\nu}(x) - \frac{1}{2}g_{\mu\nu}(x)R(x) = 8\pi G T_{\mu\nu}(x), \tag{6.93}$$

[10] G. Parisi and Y.S. Wu, *Sci. Sinica* **24**, 483 (1981); among numerous other articles on the subject, see also H. Nakazato and Y. Yamanaka, *Phys. Rev.* **34**, 492 (1986).

can be random because the particles and/or the field's variables inside $T_{\mu\nu}(x)$ are themselves random. Another case, which can also occur when $T_{\mu\nu}(x) \equiv 0$, is that of a gravitational field whose Cauchy data are random.

Like any other classical field, the gravitational field should resort to the common methods in use in turbulence theory.[11] In particular, it should be fully characterized by the data of the quantities

$$\langle g_{\mu\nu}(x)\rangle\,,\ \langle g_{\mu\nu}(x)g_{\mu'\nu'}(x')\rangle\,,\ \langle g_{\mu\nu}(x)g_{\mu'\nu'}(x')g_{\mu''\nu''}(x'')\rangle\,,\ldots, \tag{6.94}$$

where the average values $\langle\ldots\rangle$ are to be taken on the initial gravitational field's data. Unfortunately, no explicit expression of $g_{\mu\nu}(x)$ as functions of these initial data can be explicitly calculated owing to the nonlinear character of Einstein's equations.

The random character of the metric tensor gives rise to a large variety of complications and unsolved problems. Only a few are briefly reviewed below:

(1) We first mention some new mathematical objects, the *multitensors*, which were more or less studied by several authors when dealing with propagators in general relativity.[12] The simplest example is, of course, $\langle g_{\mu\nu}(x)g_{\mu'\nu'}(x')\rangle$, although there also exist biscalars like $S(x, x')$. A particularly interesting process is the Gaussian one in which the various moments of the metric tensor are such that

$$\begin{aligned}&\langle \bar g_{\mu_1\nu_1}(x_1)\bar g_{\mu_2\nu_2}(x_2)\cdots \bar g_{\mu_n\nu_n}(x_n)\rangle\\ &\quad= \sum_{\substack{\text{permutations}\\ \text{of }\{1,2,\ldots,n\}}} \langle \bar g_{\mu_1\nu_1}(x_1)\bar g_{\mu_2\nu_2}(x_2)\rangle\langle \bar g_{\mu_3\nu_3}(x_3)\bar g_{\mu_4\nu_4}(x_4)\rangle\\ &\qquad\times\cdots\times\langle \bar g_{\mu_{n-1}\nu_{n-1}}(x_{n-1})\bar g_{\mu_n\nu_n}(x_n)\rangle,\end{aligned} \tag{6.95}$$

where we have set

$$\bar g_{\mu\nu}(x) = g_{\mu\nu}(x) - \langle g_{\mu\nu}(x)\rangle\,. \tag{6.96}$$

The Gaussian process is thus characterized by its average values and fluctuations. In the case where the stochastic process can be considered as being

---

[11] See e.g. G.K. Batchelor, *loc. cit.*

[12] A. Lichnérowicz, *Propagateurs et commutateurs en relativité générale* [Publications mathématiques de l'Institut des Hautes Etudes Scientifiques (Bures/Yvette, France), No. 10 (1961)]. They were introduced by G. de Rham [*Variétés différentiables* (Hermann, Paris, (1955)] and used by physicists trying to quantize gravitation. A more recent work is the one by N.G. Phillips and B.L. Hu, *Noise kernel and stress–energy bitensor of quantum fields in hot flat space and Gaussian approximation in the optical Schwarzschild metric* [arXiv: gr-qc/0209056 v1 (2002)].

Gaussian (or approximated by a Gaussian) and if the field correlations are "small," whatever this word means.

(2) The various physical quantities are generally tensors — say, $A^{\nu_1\nu_2\ldots\nu_q}_{\mu_1\mu_2\ldots\mu_p}$ — and in many instances it is necessary to raise or lower some indices. However, doing so renders the new tensor random, as e.g. $A^{\nu_2\ldots\nu_q}_{\nu_1\mu_1\mu_2\ldots\mu_p}$, even though the original tensor $A^{\nu_1\nu_2\ldots\nu_q}_{\mu_1\mu_2\ldots\mu_p}$ itself was not random.

A natural, albeit nonunique, idea to circumvent this problem consists in the systematic use of the *average* metric tensor $\langle g_{\mu\nu}(x)\rangle$. Let us look a bit closer at this problem and, to this end, let us split $g_{\mu\nu}(x)$ into an average and a fluctuating part $h_{\mu\nu}(x)$,

$$\begin{cases} g_{\mu\nu}(x) = \langle g_{\mu\nu}(x)\rangle + \underline{h}_{\mu\nu}(x), \\ g^{\mu\nu}(x) = \langle g^{\mu\nu}(x)\rangle + \overline{h}^{\mu\nu}(x), \end{cases} \tag{6.97}$$

where, as usual, $\|g_{\mu\nu}(x)\|$ and $\|g^{\mu\nu}(x)\|$ are mutually *inverse* matrices:

$$g_{\mu\lambda}(x)g^{\lambda\nu}(x) = \delta^{\nu}_{\mu}. \tag{6.98}$$

This property of $g_{\mu\nu}(x)$ and $g^{\mu\nu}(x)$ yields[13]

$$\begin{aligned} g_{\mu\lambda}(x)g^{\lambda\nu}(x)\langle g^{\nu}_{\lambda}(x)\rangle = {} & \langle g_{\mu\lambda}(x)\rangle \left\langle g^{\lambda\nu}(x)\right\rangle + \underline{h}_{\mu\lambda}(x)\overline{h}^{\lambda\nu}(x) \\ & + \underline{h}_{\mu\lambda}(x)\left\langle g^{\lambda\nu}(x)\right\rangle + \overline{h}^{\mu\lambda}(x)\left\langle g_{\lambda\nu}(x)\right\rangle \\ = {} & \delta^{\nu}_{\mu}, \end{aligned} \tag{6.99}$$

which after averaging leads to

$$\begin{aligned} \left\langle g_{\mu\lambda}(x)g^{\lambda\nu}(x)\right\rangle &= \langle g_{\mu\lambda}(x)\rangle \left\langle g^{\lambda\nu}(x)\right\rangle + \left\langle \underline{h}_{\mu\lambda}(x)\overline{h}^{\lambda\nu}(x)\right\rangle \\ &= \delta^{\nu}_{\mu}; \end{aligned} \tag{6.100}$$

this shows that, in general, one has *not*

$$\langle g_{\mu\lambda}(x)\rangle \left\langle g^{\lambda\nu}(x)\right\rangle = \delta^{\nu}_{\mu}. \tag{6.101}$$

One could, however, try to impose this last condition; and this would provide the following constraint on the metric fluctuations:

$$\left\langle \underline{h}_{\mu\lambda}(x)\overline{h}^{\lambda\nu}(x)\right\rangle = 0. \tag{6.102}$$

Of course, when the fluctuating part of the metric tensor is "small"[14] — and this "smallness" should be specified more precisely — this last condition

---

[13]The fluctuating part h of the metric tensor has been under(over)lined in order to emphasize that, for instance, one has $\overline{h}_{\mu\nu} \equiv \langle g_{\mu\alpha}\rangle\langle g_{\nu\beta}\rangle\overline{h}^{\alpha\beta} \neq \underline{h}_{\mu\nu}$.

[14]This requires a particular discussion on the same type as the one given for Einstein's linearized equations [see e.g. S. Weinberg, *Gravitation and Cosmology* (Wiley, New York, 1972)].

can be satisfied and the raising or lowering of indices with the average values of the metric tensor is thus acceptable.

Accordingly, the usual manipulations of tensorial quantities must involve the fluctuations of the metric tensor and require much care. On the other hand, one could also decide *a priori* that the "real" metric tensor is just $\langle g_{\mu\nu}(x)\rangle \equiv G_{\mu\nu}(x)$ and thus raise indices with $G^{\mu\nu}(x)$; this would be a quite natural possibility and certainly the simplest. Note the relationship between $G^{\mu\nu}(x)$ and $\langle g^{\mu\nu}(x)\rangle$:

$$\left\langle g^{\alpha\beta}(x)\right\rangle = G^{\alpha\beta}(x) - G^{\alpha\mu}(x)\left\langle \underline{h}_{\mu\lambda}(x)\overline{h}^{\lambda\beta}(x)\right\rangle. \tag{6.103}$$

(3) Suppose now that we take the average value of Einstein's equations. As a result of the above considerations on the average values of the metric tensor, its average does not obey Einstein's equations. Indeed, the Christoffel symbols, which are implicitly present in the curvature tensor, contain both $g_{\mu\nu}(x)$ and $g^{\mu\nu}(x)$. One can then rewrite Einstein's equations in terms of the average metric tensor $G_{\mu\nu}$ and another term, which, loosely speaking, can be called "the energy–momentum tensor of the fluctuations." It should obviously be a rather involved expression, even though we only are concerned with a Gaussian gravitational field.

(4) It should also be noted that a random metric might imply random changes of the topology of the manifold under consideration. To be more specific, let us consider a Friedmann–Lemaitre universe endowed with the Robertson–Walker metric

$$ds^2 = dt^2 - R^2(t)\frac{d\mathbf{x}^2}{\left(1+\frac{1}{4}kr^2\right)^2}. \tag{6.104}$$

Suppose now that the curvature index is random and possesses some probability distribution

$$P(k)\begin{cases} p>0, & \text{for } k=+1;\\ q>0, & \text{for } k=0; \quad \text{with } p+q+r=1;\\ r>0, & \text{for } k=-1.\end{cases} \tag{6.105}$$

Then the metric tensor is itself random, both because of the explicit presence of k in the Robertson–Walker metric and in its implicit dependence on the scale factor $R(t)$ via Friedmann's equations. Suppose now that the space–time manifold is in the state $k = 1$; then its spatial topology is closed. However, there is *a priori* a nonvanishing probability that the metric is in the state $k = 0$, where the spatial sections of the space–time manifold can be either open or closed. It *might* thus be possible that the "system" undergoes a transition in its spatial topology. The same conclusions can be drawn if, for instance, the cosmological constant is itself random.[15] It should also be noted that it seems difficult from a physical point of view to imagine how one can pass from a closed topology to an open one only via *local* fluctuations of the metric and, perhaps, the topology of the manifold under consideration (or of submanifolds) would just impose some restrictions on the stochastic nature of the metric.

6.4.1. *A simple example*

Let us now consider the example of a free particle embedded in a random gravitational field; it obeys the following (random) equation of motion (geodesic equation):

$$\frac{dp^\mu}{d\tau} + \Gamma^\mu_{\alpha\beta}(x)p^\alpha p^\beta = 0, \tag{6.106}$$

which, in terms of the average metric tensor $G_{\mu\nu}$ and the metric fluctuations, can be rewritten equivalently as

$$\frac{dp^\mu}{d\tau} + \overline{\overline{\Gamma}}^\mu_{\alpha\beta}(x)p^\alpha p^\beta = F^\mu_{\text{fluct}}, \tag{6.107}$$

where $\overline{\overline{\Gamma}}^\mu_{\alpha\beta}(x)$ denotes the Christoffel symbols constructed from the average metric tensor $G_{\mu\nu}$, and F^μ_{fluct} is the apparent force occurring because of the metric fluctuations; its explicit expression is obtained by replacing $g_{\mu\nu}$ and $g^{\mu\nu}$ with their expressions in terms of $G_{\mu\nu}$ and $h_{\mu\nu}$, and is given by

$$\begin{aligned} -F^\mu_{\text{fluct}} = \frac{1}{2}p_\alpha p^\nu G^{\alpha\rho} &\Big[\{\partial^\mu h_{\nu\rho} + \partial^\nu h^\mu{}_\rho - \partial_\rho h^{\mu\nu}\} \\ &+ \frac{1}{2}h^{\alpha\rho}\{\partial^\mu G^\nu{}_\rho + \partial^\nu G^\mu{}_\rho - \partial_\rho G^{\mu\nu}\}\Big] \\ &+ \frac{1}{2}p_\alpha p^\nu h^{\alpha\rho}\{\partial^\mu h_{\nu\rho} + \partial_\nu h^\mu{}_\rho - \partial_\rho h^{\mu\nu}\}, \end{aligned} \tag{6.108}$$

and for a Gaussian random gravitational field (with $\langle h\rangle = 0$), one gets

$$\langle F^\mu_{\text{fluct}}(x)\rangle = -\frac{1}{2}p_\alpha p^\nu \left\langle h^{\alpha\rho}\{\partial^\mu h_{\nu\rho} + \partial_\nu h^\mu{}_\rho - \partial_\rho h^\mu{}_\nu\}\right\rangle, \tag{6.109}$$

which is the simplest form one is able to obtain, the more so since we used $h \ll g$.

[15]If the cosmological constant λ is interpreted as an energy density of the vacuum, then it does actually fluctuate.

6.4.2. *The case of thermal equilibrium*

To be more specific, this can be enlarged to an assembly of identical particles whose initial data are supposed to be in thermal equilibrium and their distribution is the Jüttner–Synge function

$$f(x,p) = \frac{\beta n_{\text{eq}}}{4\pi m^2 K_2(m\beta)} \exp\left[-\beta g_{\alpha\beta}(x) u^\alpha p^\beta\right], \tag{6.110}$$

which obeys the Liouville equation

$$p \cdot \partial f(x,p) + \Gamma^\mu_{\alpha\beta}(x) p^\alpha p^\beta \frac{\partial}{\partial p^\mu} f(x,p) = 0, \tag{6.111}$$

or

$$p \cdot \partial f(x,p) + \left\{ \overline{\overline{\Gamma}}{}^\mu_{\alpha\beta}(x) p^\alpha p^\beta - F^\mu_{\text{fluct}} \right\} \frac{\partial}{\partial p^\mu} f(x,p) = 0. \tag{6.112}$$

The four-current is

$$J^\mu(x) = \sqrt{|g|} \int \frac{d^3p}{p_0} \frac{p^\mu}{m} f(x,p), \tag{6.113}$$

and by taking into account the character of the decomposition of the metric tensor as

$$g_{\mu\nu} = G_{\mu\nu} + h_{\mu\nu} \tag{6.114}$$

and from the fact that h is "small," one has

$$J^\mu(x) = \left(\sqrt{|G|}\right) \int \frac{d^3p}{p_0(G,h)|_{h=0}} \frac{p^\mu}{m} f(x,p) \tag{6.115}$$

plus a term of order h, which is exactly the same as the original term once the average has been used taking account of the Gaussian character of h. The energy–momentum tensors have thus the same form as in the nonrandom case except that $g_{\mu\nu} = G_{\mu\nu} + h_{\mu\nu}$, and their averages are just their expressions in the average gravitational field $G_{\mu\nu}$.

However, two problems have to be dealt with. The first one deals with equilibrium itself. The implicit reasoning that leads to the Jüttner–Synge function rests on the equivalence principle: thermal equilibrium is assumed to hold in a locally Lorentzian frame of reference, and hence whatever the metric fluctuations. This means that the mean free path of the particles should be much smaller than the spatial scale on which the metric fluctuations extend, and, of course, the same is true for the collision time and the timescale of the fluctuations. This requires a particular analysis of the physical case at hand.

The second problem is concerned with the Jüttner–Synge distribution itself. In spite of the fact that the main two observables $\langle J_\mu \rangle$ and $\langle T_{\mu\nu} \rangle$ can be computed from the Jüttner–Synge function in the presence of the average metric tensor $G_{\mu\nu}$, this function *is not* a solution to the Liouville equation since the Killing equation (see Chap. 4) with respect to the metric $G_{\mu\nu}$ is not satisfied. Only the (random) Killing conditions

$$\nabla_\mu (\beta u_\nu) + \nabla_\nu (\beta u_\mu) = 0 \tag{6.126}$$

are obeyed, but the average

$$\overline{\overline{\nabla}}_\mu (\beta u_\nu) + \overline{\overline{\nabla}}_\nu (\beta u_\mu) \neq 0, \tag{6.127}$$

where the double-barred covariant derivatives are defined with respect to the average metric tensor

$$\overline{\overline{\nabla}}_\mu A_\nu \equiv \partial_\mu A_\nu + \overline{\overline{\Gamma}}{}^{\alpha}_{\mu\nu} A_\alpha \tag{6.128}$$

and βu^μ does not, in general, represent a Killing four-vector.

Also these average observables cannot be found from $\langle f_{\text{eq}} \rangle$, which is given by

$$\langle f_{\text{eq}} \rangle = f_{\text{eq}}(G_{\mu\nu}) \times \exp\left[-\beta \langle h_{\mu\nu}(x) h_{\alpha\beta}(x) \rangle u^\mu u^\alpha p^\nu p^\beta\right], \tag{6.129}$$

in the case where the random metric tensor is Gaussian. Note that when the fluctuations are "small" with respect to the average metric $G_{\mu\nu}$, the equilibrium distribution function is just the Jüttner–Synge function in which the replacement $g_{\mu\nu}(x) \to G_{\mu\nu}(x)$ is made.

6.4.3. *Matter-induced fluctuations*

The expression for the four-current equilibrium fluctuations calculated in Chap. 1,

$$\delta J^{\mu\nu}(X) = \frac{m\beta n_{\text{eq}}}{4\pi K_2(m\beta)} \frac{X^\mu X^\nu}{(X \cdot X)^{5/2}} \exp\left(-m\beta \frac{u \cdot X}{(X \cdot X)^{1/2}}\right), \tag{6.130}$$

is still valid with the condition of making the change

$$\begin{aligned} &\eta_{\mu\nu} \to g_{\mu\nu} = G_{\mu\nu} + h_{\mu\nu}, \\ &|h_{\mu\nu}| \ll |G_{\mu\nu}|. \end{aligned} \tag{6.131}$$

The observables' fluctuations are thus $\langle \delta J^{\mu\nu}(x, x') \rangle$, where the average is taken over the random gravitational field

$$\begin{aligned} \delta J^{\mu\nu}(X) = {} & \frac{m\beta n_{\text{eq}}}{4\pi K_2(m\beta)} \frac{X^\mu X^\nu}{((G_{\sigma\eta} + h_{\sigma\eta}) X^\sigma X^\eta)^{5/2}} \\ & \times \exp\left(-m\beta \frac{(G_{\mu\nu} + h_{\mu\nu}) u^\mu X^\nu}{((G_{\lambda\rho} + h_{\lambda\rho}) X^\rho X^\lambda)^{1/2}}\right). \end{aligned} \tag{6.133}$$

Still separating the metric tensor into its average value and its fluctuating part and retaining only the lowest order terms, one finds that

$$\langle \delta J^{\mu\nu} \rangle = \langle \delta J^{\mu\nu} \rangle|_{h=0} + h_{\lambda\rho} B^{\lambda\rho\mu\nu} + \cdots, \tag{6.134}$$

where $B^{\cdots}$ can easily be obtained. Note that in these expressions X^μ is an involved expression of x and x', and not of the difference between the coordinates. Furthermore, the expression for $\delta J^{\mu\nu}(X)$ is valid only for values of x' not very different from x.

Accordingly, and as expected, the metric fluctuations do induce matter fluctuations: numerical four-current and energy–momentum tensor.

6.4.4. *Random Einstein equations*

In this subsection, the treatment of random Einstein equations is only outlined, owing to the complexity of the results. Such a treatment is indeed needed, in order to obtain an infinite set of equations for the moments of the gravitational field. Of course, this set is much simplified when the stochastic process is Gaussian, since only $G_{\mu\nu}$ and $h_{\mu\nu}$ are to be calculated; the equations are nevertheless not quite simple.

It should first be noted that the Riemann curvature tensor $R_{\mu\nu\alpha\beta}$ is constructed from the derivatives of the Christoffel symbols and from the metric tensor, and thus contains generic terms of the general form

$$\begin{aligned} \partial\Gamma &= g\partial\{gg\partial g\} \\ &= ggg\partial^{(2)}g + gg\partial g\partial g, \end{aligned} \tag{6.135}$$

and from the decomposition

$$g = G + h; \tag{6.136}$$

hence it has the generic form

$$R = \overline{\overline{R}} + \cdots + hhh\partial^{(2)}h + hh\partial h\partial h, \tag{6.137}$$

where the last term represents in a symbolic way the contribution of the gravitational fluctuation to the curvature tensor. There is, however, one particular case where these equations do simplify considerably; this occurs whenever the random part of the gravitational field can be considered as "small" with respect to its average part, with of course all the usual reservations about the invariance of such "smallness" in coordinates' changes.

Chapter 7

The Density Operator

In this chapter, the basis of relativistic quantum statistical mechanics is presented: its basic tool — the density operator[1] — is briefly reviewed and, as in the nonrelativistic case, it is given by

$$\rho_{\text{stat}} = \sum_n |n\rangle \varpi_n \langle n|, \quad \sum_n \varpi_n = 1, \varpi_n \geq 0, \tag{7.1}$$

where the ϖ_n's are the statistical weights of the nth state, the $\{|n\rangle\}$'s being supposed to form a complete set. From the density operator ρ, one calculates various quantities — such as Green's functions, introduced into statistical mechanics in 1959 by E.S. Fradkin, or Wigner functions on a "semiclassical" phase space [E.P. Wigner (1932)] — from which the physics of the system under study can be extracted. Average values are then obtained as

$$\langle A \rangle = \text{Tr}[\rho_{\text{stat}} A] = \sum_n \varpi_n \langle n|A|n\rangle, \tag{7.2}$$

where A is a given observable. These relations are still valid in special relativity, although some care is needed in their manipulation.

The density operator ρ_{stat} obeys a "Liouville equation" derived from the Tomonaga–Schwinger equation[2]; it reads [A.V. Prosorkevich and S.A. Smolyanskii (1976)]

$$i\frac{\delta}{\delta\Sigma}\rho_{\text{stat}} = [H, \rho_{\text{stat}}], \tag{7.3}$$

where $\sum$ is a spacelike three-surface. This approach, and some applications, have been developed by the same authors in other articles (1976, 1978).

[1] In the subsequent chapters, the energy density is called ρ, as usual. Therefore, in order to avoid some confusion with the density operator traditionally denoted by ρ, this latter is called ρ_{stat}.

[2] S. Tomonaga, *Prog. Theor. Phys.* **1**, 27 (1946).

However, it is not very easy to handle and a simpler one is dealt with in the subsequent sections.

7.1. The Density Operator for Thermal Equilibrium

In thermal equilibrium the density operator ρ_{stat} possesses the general form[3]

$$\rho_{\text{eq}} = \text{const} \times \exp\left(\sum_i \alpha_i A_i\right), \tag{7.4}$$

where A_i are *additive* first integrals of the system and where α_i are corresponding Lagrange multipliers. Whether in Newtonian physics or in relativity, there exist only seven time-independent such additive integrals: energy, impulsion and kinetic momentum. To these constants of the motion other *additive observables* must be added, such as the particle number (i.e. the difference between the particle and antiparticle numbers), the baryon number and the charge. As is usual in relativity, energy and impulsion are treated on the same footing, while rotational symmetry will not be dealt with. Therefore, the basic equilibrium density operator ρ_{eq} has the form

$$\rho_{\text{eq}} = \frac{1}{Z}\exp(-\beta_\mu P^\mu + \beta\mu N), \tag{7.5}$$

where P^μ is the total four-energy–momentum of the system; N is the particle number operator (or the charge operator, or the baryon number operator, etc.), present when the number of particles is not constant and $\langle N\rangle = \text{const}$; $\beta_\mu = \beta u_\mu$, with[4] $\beta = T^{-1}$; and μ is the chemical potential, while Z is the partition function. It is obtained from the normalization of the statistical operator as

$$\text{Tr}(\rho_{\text{stat}}) = 1 \tag{7.6}$$

or

$$Z \equiv \text{Tr}[\exp(-\beta_\mu P^\mu + \beta\mu N)]. \tag{7.7}$$

P^μ is obtained from the energy–momentum tensor $T^{\mu\nu}$ of the system as

$$P^\mu = \int_\Sigma d\Sigma_\mu T^{\mu\nu}, \tag{7.8}$$

[3]See e.g. K. Huang, *Statistical Mechanics* (Wiley, New York, (1963). Note that possibly there are other additive operators to be taken into account, like the charge of the system, or its total particle number.

[4]In what follows, a system of units where the Boltzmann constant k_B is taken to be unity; accordingly, temperatures are measured in energy units.

where $\sum$ is an arbitrary spacelike three-surface, which can always be chosen as being the spacelike three-plane $t = \text{const}$, since $T^{\mu\nu}$ is conservative. This energy–momentum tensor contains the contributions of all the fields and their interactions — both quantum and possibly external — within the system. Note also that when the system involves several particle species i with conserved "charges," one has to perform the following substitution in the equilibrium density operator:

$$\mu N \to \sum_i \mu_i N_i. \tag{7.9}$$

Therefore, the grand canonical density operator is used in the context of quantum field theory where particles are created and/or destroyed.

For later use, this chapter is mainly devoted to the free field case, either fermionic or bosonic — a case first considered by A.E. Scheidegger[5] and C.D. McKay (1951) and by A.O. Barut (1958), although the first study of the relativistic and Fermi–Dirac distributions was made by F. Jüttner (1928). The main thermodynamic properties are, however, briefly recalled below.

7.1.1. *Thermodynamic properties*

In this subsection, the main thermodynamic relations occurring in the grand canonical approach are given without proof and readers are referred to their preferred textbooks.[6] They are consequences of the exponential form of the statistical operator ρ_{stat} and of the occurrence of the physical observables therein in a linear way. They are provided in Table 7.1.

Let us now turn to the entropy; we have, successively,

$$S \equiv -\langle \log \rho_{\text{stat}} \rangle = -\langle -\beta_\mu P^\mu + \beta\mu N - \log Z \rangle, \tag{7.10}$$

or, more explicitly,

$$TS = -\mu N + U + \frac{\log Z}{\beta}, \tag{7.11}$$

from which we identify the free energy (Table 7.1) as

$$F = -\frac{\ln Z}{\beta}. \tag{7.12}$$

[5]See also A.E. Scheidegger and R.V. Krotkov (1951); A.O. Barut (1958).

[6]See e.g. K. Huang, *op.cit.*; A. Isihara, *Statistical Physics* [Academic Press; New York (1971)]; etc.

Table 7.1

Particle number (charge, etc.)	$N = \frac{1}{\beta} \frac{\partial \ln Z}{\partial \mu}\Big\|_{\beta}$
Internal energy	$U = -\frac{\partial \ln Z}{\partial \beta}\Big\|_{N,V} + \frac{\mu}{\beta} \frac{\partial \log Z}{\partial \mu}\Big\|_{\beta}$
Pressure	$P = \frac{1}{\beta} \frac{\partial \ln Z}{\partial V}\Big\|_{N,\beta}$
Chemical potential	$\mu = -\frac{1}{\beta} \frac{\partial \ln Z}{\partial N}\Big\|_{\beta,N,V}$
Particle number	$N = \frac{1}{\beta} \frac{\partial \log Z}{\partial \mu}\Big\|_{\beta}$
Gibbs free energy	$F = -\frac{\ln Z}{\beta}$
Helmholtz free energy	$G = F + \mu N$
Entropy	$S = \ln Z + \beta U - \beta \mu N$
Heat capacities at constant volume, pressure	$C_V = \frac{\partial U}{\partial T}\Big\|_{T,V}, \quad C_P = \frac{\partial U}{\partial T}\Big\|_{T,P}$

The functional derivative of the entropy with respect to the "volume"[7] of the system yields the entropy four-current $S^\nu(x)$,

$$\frac{\delta}{\delta \Sigma_\nu(x)} S = S^\nu(x), \tag{7.13}$$

and since

$$\frac{\delta}{\delta \Sigma_\nu(x)} P^\mu = T^{\mu\nu}(x), \quad \frac{\delta}{\delta \Sigma_\nu(x)} N = J^\nu(x) = n_{\rm eq} u^\nu, \tag{7.14}$$

we finally obtain

$$S^\nu(x) = \beta[u_\mu T^{\mu\nu}(x) - \mu n_{\rm eq}(x) u^\nu] - \frac{\delta}{\delta \Sigma_\nu(x)} \ln Z. \tag{7.15}$$

The last term of this expression is the Helmholtz free energy four-current. Multiplying both sides of this last equation by u^ν, we obtain

$$s(x) = \beta[\rho(x) - \mu n_{\rm eq}(x)] + f(x) \tag{7.16}$$

[7]The reader should be reminded that, in relativity, only *local* quantities have a real meaning and that the notion of a finite volume can hardly acquire a specific invariant definition. It is therefore preferable to use local quantities which are obtained by introducing into the various data some weight factors imitating the existence of a finite volume, and hence the functional derivative is actually taken with respect to Σ and this arbitrary function.

[$f(x)$ is the free energy density], which, upon multiplication by an arbitrary volume V, gives the usual thermodynamic expression:

$$U - TS - \mu N = F. \tag{7.17}$$

In terms of densities, the equation of state is obtained as

$$\begin{cases} P = -\frac{1}{\beta} n_{\text{eq}}^2 \dfrac{\partial}{\partial n_{\text{eq}}} \log Z(\beta, n_{\text{eq}}), \\ \rho = -n_{\text{eq}} \dfrac{\partial}{\partial \beta} \log Z(\beta, n_{\text{eq}}), \end{cases} \tag{7.18}$$

as a function of the invariant particle density n_{eq}; P is the pressure and ρ the energy density.

It is thus necessary to evaluate the partition function in order to obtain all physical relevant quantities, and this is done below for the case of the simple ideal gas.

7.1.2. *The partition function of the relativistic ideal gas*

There is no difference between the relativistic and the Newtonian calculations of the partition function: the general structure of the two expressions is similar except that the expression of the energy differs. Accordingly, the calculation is quite general and does apply for slightly more general systems than those composed of free particles, for instance to charged systems embedded in magnetic fields. Thus, the usual calculations are briefly repeated here. One begins with the definition of the partition function and let $\{\ell\}$ be the set of all quantum numbers that characterize an energy states $E_{\{\ell\}}$ of a generic particle in the system, and considers a representation where the particle number N (or charge, baryon number, etc.) is diagonal and has the eigenvalues $\{n\}$; one has

$$\begin{aligned} Z &= \text{Tr}\{\exp(-\beta[H - \mu N])\} \\ &= \text{Tr}\left\{\exp\left(-\beta\left[\sum_{\ell}(n_{\{\ell\}}E_{\{\ell\}} - \mu n_{\{\ell\}})\right]\right)\right\} \\ &= \sum_{n}\prod_{\ell}\exp(-\beta[n_{\{\ell\}}E_{\{\ell\}} - \mu n_{\{\ell\}}]), \end{aligned} \tag{7.19}$$

where $n_{\{\ell\}}$ is the number of particles in the state $\{\ell\}$, with of course

$$n = \sum_{\ell} n_{\{\ell\}} \tag{7.20}$$

and also

$$E = \sum_{\ell} E_{(\ell)}) n_{(\ell)} \tag{7.21}$$

More explicitly, Z can be rewritten as

$$\begin{aligned} Z &= \sum_{n} \prod_{n_{\{\ell\}}} \exp\{-\beta[n_{\{\ell\}}(E_{\{\ell\}} - \mu)]\} \\ &= \sum_{n} \prod_{n_{\{\ell\}}} (\exp\{-\beta[(E_{\{\ell\}} - \mu)]\})^{n_{\{\ell\}}} \end{aligned} \tag{7.22}$$

which is easily calculated for bosons $[n_{\{\ell\}} = 0, 1, 2, \ldots]$ or fermions $[n_{\{\ell\}} = 0, 1]$ as

$$\log Z = \begin{cases} +\sum_{\ell} \log\{1 + \exp[-\beta(E_{\{\ell\}} - \mu)]\} & \text{(fermions)}, \\ -\sum_{\ell} \log\{1 - \exp[-\beta(E_{\{\ell\}} - \mu)]\} & \text{(bosons)}. \end{cases} \tag{7.23}$$

It should be emphasized, once more, that these last expressions are valid whatever the noninteracting system at hand and, in particular, for (free) quasiparticles of whatever spectrum. They are also valid whether the system is Newtonian or relativistic.

For free particles in the absence of external fields, one has

$$\{\ell\} \equiv \mathbf{k}, \quad \frac{1}{V} \sum_{\mathbf{k}} \to \int \frac{d^3k}{(2\pi)^3}, \tag{7.24}$$

so that

$$[\log Z]^{\mu} = \begin{cases} +\dfrac{1}{(2\pi)^3} \displaystyle\int \frac{d^3k}{k_0} k^{\mu} \log(1 + \exp\{-\beta[E(\mathbf{k}) - \mu]\}) & \text{(fermions)}, \\ -\dfrac{1}{(2\pi)^3} \displaystyle\int \frac{d^3k}{k_0} k^{\mu} \log(1 - \exp\{-\beta[E(\mathbf{k}) - \mu]\}) & \text{(bosons)}. \end{cases} \tag{7.25}$$

for the four-current[8] of the "quantity $\log Z$." In other words, in the relativistic case, one calculates densities rather than global quantities and the various thermodynamic relations are intended to hold between densities or four-currents.

A last remark: the fermions '$\log Z$' contains also a term where $+\mu$ occurs in the expression; this is due to the contribution of the antiparticles. If we

[8] In this expression, the division by k^0 makes the integration element invariant, while the multiplication by k^{μ} shows the four-vector character of $\log Z$, which thus appears as a density.

look at a complex scalar field, we should recover a similar term except for the $\pm$ signs.

7.1.3. *The average occupation number*

Let us now evaluate the average occupation number. It is given by

$$\langle a_\ell^+ a_\ell\rangle = \frac{1}{Z}\mathrm{Tr}\{a_\ell^+ a_\ell \exp[-\beta(H-\mu N)]\}, \tag{7.26}$$

and since

$$H = \sum a_\ell^+ a_\ell E_\ell \tag{7.27}$$

it turns out that

$$\langle a_\ell^+ a_\ell\rangle = -\frac{1}{\beta}\frac{\delta}{\delta E_\ell}\log Z. \tag{7.28}$$

We are now in a position to calculate the average occupation number. Let us consider the expression of $(\delta/\delta E_\ell)\log Z$, with $E_\ell = E(\mathbf{k})$, in order to be more specific. It reads

$$\begin{aligned}
-\frac{1}{\beta}\frac{\delta \log Z}{\delta E(\mathbf{k})} &= +\frac{1}{(2\pi)^3}\frac{\delta}{\delta E(\mathbf{k})}\int \frac{d^3k'}{k'_0}k'^0 \log(1-\exp\{-\beta[E(\mathbf{k}')-\mu]\}) \\
&= -\frac{1}{(2\pi)^3}\int \frac{d^3k'}{k'_0}k'^0\frac{\delta E(\mathbf{k}')}{\delta E(\mathbf{k})}\frac{\exp\{-\beta[E(\mathbf{k}')]\}}{1-\exp\{-\beta[E(\mathbf{k}')-\mu]\}}
\end{aligned} \tag{7.29}$$

but, since (see App. C)

$$\frac{\delta E(\mathbf{k}')}{\delta E(\mathbf{k})} = \delta^{(3)}(\mathbf{k}-\mathbf{k}'), \tag{7.30}$$

we have

$$-\frac{1}{\beta}\frac{\delta \log Z}{\delta E(\mathbf{k})} = \frac{1}{\exp\{\beta[E(\mathbf{k})-\mu]\}-1}, \tag{7.31}$$

so that the average occupation number is

$$\langle a^+(\mathbf{k})a(\mathbf{k})\rangle = \frac{1}{\exp\{\beta[E(\mathbf{k})-\mu]\}-1}. \tag{7.32}$$

Similarly, the average occupation number of the state ℓ reads

$$n_{\{\ell\}} = -\frac{1}{\beta}\frac{\delta}{\delta E_{\{\ell\}}}\log Z = \frac{1}{\exp[\beta(E_{\{\ell\}}-\mu)]-1}. \tag{7.33}$$

The case of fermions is quite similar and we obtain

$$n_{(\ell)} = \frac{1}{\exp[\beta(E_{\{\ell\}} - \mu)] + 1}. \tag{7.34}$$

7.2. Relativistic Bosons in Thermal Equilibrium

We first look for the properties of the ideal gas of free bosons obeying the Lagrangian

$$\mathcal{L} = \frac{1}{2}(\partial\varphi \cdot \partial\varphi - m^2\varphi^2) \tag{7.35}$$

(where φ is real) and hence the equation of motion

$$\Box\varphi + m^2\varphi = 0. \tag{7.36}$$

Such a system admits the first integral

$$\begin{aligned} P^\mu &= \int_\Sigma d\Sigma_\nu T^{\mu\nu} \\ &= \int_\Sigma d\Sigma_\nu \left[\partial^\mu\varphi \cdot \partial^\nu\varphi - \eta^{\mu\nu}\left(\frac{1}{2}\partial\varphi \cdot \partial\varphi - m^2\varphi^2\right)\right] \end{aligned} \tag{7.37}$$

as an additive constant of the motion and the conservation relation

$$\partial_\mu T^{\mu\nu} = 0. \tag{7.38}$$

As usual, the field φ is decomposed into creation and annihilation operators:

$$\varphi(x) = \frac{1}{(2\pi)^{3/2}} \int \frac{d^3k}{\sqrt{2k_0}} \{a_{\mathbf{k}} \exp(+ik \cdot x) + a^+_{\mathbf{k}} \exp(-ik \cdot x)\}. \tag{7.39}$$

Choosing[9] the arbitrary spacelike three-surface, involved in the definition of P^μ, as being a three-plane $t = \text{const}$, and in the local frame of reference in which $\beta^\mu = (\beta, \mathbf{0})$, the Hamiltonian $P^0 \equiv H$ reads

$$H = \frac{1}{2}\sum_{\{\mathbf{k}\}} \omega_{\{\mathbf{k}\}} \left[a^+_{\{\mathbf{k}\}} a_{\{\mathbf{k}\}} + a_{\{\mathbf{k}\}} a^+_{\{\mathbf{k}\}}\right] = \sum_{\{\mathbf{k}\}} \omega_{\{\mathbf{k}\}} \left[a^+_{\{\mathbf{k}\}} a_{\{\mathbf{k}\}} + \frac{1}{2}\right], \tag{7.40}$$

where $\{\mathbf{k}\}$ indicates the set of those quantum numbers (i.e. the three-momentum) that determine the state of the system whose energy is $k_0 = \omega_{\{\mathbf{k}\}}$ with $w_{\{\mathbf{k}\}} = (k^2 + m^2)^{1/2}$. This Hamiltonian corresponds to a boson

[9] This can always be done without loss of generality, owing to the fact that the energy–momentum tensor is conserved.

field and the "vacuum" factor 1/2 can be absorbed into Z, so that it will be omitted. Also, the operator "number of particles" is given by

$$N = \sum_{\{\mathbf{k}\}} a^{+}_{\{\mathbf{k}\}} a_{\{\mathbf{k}\}}, \tag{7.41}$$

which is not conserved in the case of a real scalar field, since no conserved four-current is available. Finally, the equilibrium density operator is of the general form

$$\rho_{\text{eq}} = \frac{1}{Z} \exp\left(-\beta \sum_{\{\mathbf{k}\}} \omega_{\{\mathbf{k}\}} a^{+}_{\{\mathbf{k}\}} a_{\{\mathbf{k}\}} \right), \tag{7.42}$$

which is formally identical to the usual one.[10] This has the consequence that the same calculations do apply in this case and that the statistical distribution of the states $\{n\}$ is still given by the ordinary Bose–Einstein factor,

$$\text{Prob}[\{\mathbf{k}\}] = \frac{1}{\exp(\beta\omega_{\{\mathbf{k}\}}) - 1}, \tag{7.43}$$

while the partition function reads[11]

$$\begin{aligned} \log Z &= -\sum_{\{\mathbf{k}\}} \log\{1 - \exp[-\beta\omega(\mathbf{k})]\} \\ &= -\frac{d}{(2\pi)^3} \int \frac{d^3k}{\omega(\mathbf{k})} k^0 \log\{1 - \exp[-\beta\omega(\mathbf{k})]\}, \end{aligned} \tag{7.44}$$

where d is the degeneracy of the system.

For a free particle[12] $\{n\} \equiv \{\mathbf{k}\}$ and

$$\omega_{\{n\}} \equiv \sqrt{\mathbf{k}^2 + m^2} \tag{7.45}$$

so that, for such a free particle, the relativistic Bose–Einstein distribution reads

$$f_{\text{eq}}(k) = \frac{d}{(2\pi)^3} \frac{1}{\exp(\beta u \cdot k) - 1}, \tag{7.46}$$

---

[10]See e.g. K. Huang, *op. cit.*

[11]We have omitted the [(β, μ)-independent] zero-point energy since in this analysis it plays no role in the thermodynamic properties of the system.

[12]An example where the Bose–Einstein distribution is found with another set of quantum numbers $\{n\}$ can be found with the case of charged bosons embedded in an external magnetic field [Ph. Adam and R. Hakim (1982)].

where the four-momenta k^μ are connected through the mass shell relation $k^2 = m^2$. It is "normalized" via

$$n_{\rm eq} \equiv u_\alpha J^\alpha = \frac{d}{(2\pi)^3}\int \frac{d^3k}{k_0} u_\alpha k^\alpha \frac{1}{\exp(\beta u\cdot k)-1}, \tag{7.47}$$

where J^α is the particle four-current, a *nonconserved* quantity in this case. In this relation, d is a degeneracy factor depending on the internal quantum numbers (spin, etc.) and as usual, k^0 is the relativistic energy. From the expression of the energy–momentum tensor

$$T^{\alpha\beta} = \frac{d}{(2\pi)^3}\int \frac{d^3k}{k_0} k^\alpha k^\beta \frac{1}{\exp(\beta u\cdot k)-1}, \tag{7.48}$$

which has necessarily the perfect fluid form since u^μ and $\eta^{\mu\nu}$ are the only available tensors, one obtains the energy density as

$$\rho \equiv \rho(\beta m) = T^{\alpha\beta}u_\alpha u_\beta = \frac{d}{(2\pi)^3}\int \frac{d^3k}{k_0}(k\cdot u)^2 \frac{1}{\exp(u\cdot k\beta)-1} \tag{7.49}$$

and the pressure

$$\begin{aligned} P \equiv P(\beta m) &= -\frac{1}{3}\Delta_{\alpha\beta}(u)T^{\alpha\beta} \\ &= -\frac{d}{3(2\pi)^3}\int \frac{d^3k}{k_0}\Delta_{\alpha\beta}(u)k^\alpha k^\beta \frac{1}{\exp(\beta u\cdot k)-1} \\ &= \frac{d}{3(2\pi)^3}\int \frac{d^3k}{k_0}\frac{\mathbf{k}^2}{\exp(\beta u\cdot k)-1}. \end{aligned} \tag{7.50}$$

These last two expressions then provide the equation of state of the ideal Bose gas in a parametric form: $\rho = \rho(\beta m)$, $P = P(\beta m)$. Note that, for the photon field, the degeneracy is $d = 2$, and the equation of state is

$$P = \frac{1}{3}\rho, \tag{7.51}$$

resulting from the elimination of the temperature in P and ρ.

7.2.1. *The complex scalar field*

The complex boson free field, whose Lagrangian and equations of motion are

$$\begin{cases} \mathcal{L} = \frac{1}{2}(\partial\varphi^*\cdot\partial\varphi - m^2\varphi^*\varphi), \\ \Box\varphi + m^2\varphi = 0, \end{cases} \tag{7.52}$$

possesses one more additive first integral, its total charge

$$Q = \int_\Sigma d\Sigma_\nu\, J^\nu(x) = \int_\Sigma d\Sigma_\nu\, \varphi^*(x)\frac{i}{2}\overleftrightarrow{\partial}^\nu\varphi(x), \tag{7.53}$$

since the four-current is conserved,

$$\partial_\mu j^\mu(x) \equiv \partial_\mu \left(\frac{i}{2} \varphi^*(x) \overleftrightarrow{\partial}^\mu \varphi(x) \right) = 0, \tag{7.54}$$

as is easily shown from the Klein–Gordon equation obeyed by φ and φ^*. It follows that, in thermal equilibrium, the density operator reads

$$\rho_{\text{stat}} = \frac{1}{Z} \exp(-\beta u \cdot P + \beta\mu Q), \tag{7.55}$$

where μ is the chemical potential, and Q and P can be written in terms of the creation/annihilation operators

$$\begin{cases} \varphi(x) = \dfrac{1}{(2\pi)^{3/2}} \displaystyle\int \dfrac{d^3k}{\sqrt{2\omega(\mathbf{k})}} [a_{(\mathbf{k})} \exp(+ik \cdot x) + b^+_{(\mathbf{k})} \exp(-ik \cdot x)], \\ \varphi^*(x) = \dfrac{1}{(2\pi)^{3/2}} \displaystyle\int \dfrac{d^3k}{\sqrt{2\omega(\mathbf{k})}} [a^+_{(\mathbf{k})} \exp(-ik \cdot x) + b_{(\mathbf{k})} \exp(+ik \cdot x)], \end{cases} \tag{7.56}$$

as

$$\rho_{\text{stat}} = \frac{1}{Z} \exp\left(-\beta \sum_{\mathbf{k}} \left[\left\{ \sqrt{\mathbf{k}^2 + m^2} - \mu \right\} a^+_{(\mathbf{k})} a_{(\mathbf{k})} \right.\right. \\ \left.\left. + \left\{ \sqrt{\mathbf{k}^2 + m^2} + \mu \right\} b_{(\mathbf{k})} b^+_{(\mathbf{k})} \right] \right),$$

so that the average occupation number is now

$$f_{\text{eq}}(k) = \frac{d}{(2\pi)^3} \frac{\text{sgn}(u \cdot k)}{\exp[\beta(u \cdot k - \mu)] - 1}. \tag{7.57}$$

Of course, one has

$$Z(\beta, \mu) = \text{Tr}\left\{ \exp\left(-\beta \sum_{\mathbf{k}} \left[\left\{ \sqrt{\mathbf{k}^2 + m^2} - \mu \right\} a^+_{\mathbf{k}} a_{\mathbf{k}} \right.\right.\right. \\ \left.\left.\left. + \left\{ \sqrt{\mathbf{k}^2 + m^2} + \mu \right\} b_{\mathbf{k}} b^+_{\mathbf{k}} \right] \right) \right\}, \tag{7.58}$$

and, more precisely,

$$[\log Z]^\mu = -\frac{1}{(2\pi)^3} \int \frac{d^3k}{k_0} k^\mu [\log(1 - \exp\{-\beta[\omega(\mathbf{k}) - \mu]\}) \\ + \log(1 - \exp\{-\beta[\omega(\mathbf{k}) + \mu]\})], \tag{7.59}$$

as written in a manifestly covariant form and where $\omega(\mathbf{k}) \equiv |u \cdot k|$. The various physical quantities are then given either through the use of the partition function or from the expression of $f_{\text{eq}}(k)$. Accordingly, one finds that

$$n_{\text{eq}} \equiv u_\alpha J^\alpha = \frac{d}{(2\pi)^3}\int \frac{d^3k}{k_0} u_\alpha k^\alpha \times \left(\frac{1}{\exp\{\beta[\omega(\mathbf{k})-\mu]\}-1} - \frac{1}{\exp\{\beta[\omega(\mathbf{k})+\mu]\}-1}\right), \tag{7.60}$$

which connects the invariant number density n_{eq} to the chemical potential μ and the temperature β^{-1}; this relation thus normalizes the Bose–Einstein distribution. The pressure is obtained from the energy–momentum tensor

$$T^{\alpha\beta} = \frac{d}{(2\pi)^3}\int \frac{d^3k}{k_0} k^\alpha k^\beta \times \left(\frac{1}{\exp\{\beta[\omega(\mathbf{k})-\mu]\}-1} + \frac{1}{\exp\{\beta[\omega(\mathbf{k})+\mu]\}-1}\right) \tag{7.61}$$

as

$$\begin{aligned} P &= -\frac{1}{3}\Delta_{\alpha\beta}(u)T^{\alpha\beta} = -\frac{d}{3(2\pi)^3}\int \frac{d^3k}{k_0}\Delta_{\alpha\beta}(u)k^\alpha k^\beta \\ &\quad \times \left(\frac{1}{\exp\{\beta[\omega(\mathbf{k})-\mu]\}-1} + \frac{1}{\exp\{\beta[\omega(\mathbf{k})+\mu]\}-1}\right) \\ &= \frac{d}{3(2\pi)^3}\int \frac{d^3k}{k_0}\mathbf{k}^2 \left(\frac{1}{\exp\{\beta[\omega(\mathbf{k})-\mu]\}-1} + \frac{1}{\exp\{\beta[\omega(\mathbf{k})+\mu]\}-1}\right) \end{aligned} \tag{7.62}$$

and the energy density is

$$\begin{aligned} \rho &= \frac{d}{(2\pi)^3}\int \frac{d^3k}{k_0} u_\alpha u_\beta k^\alpha k^\beta \\ &\quad \times \left(\frac{1}{\exp\{\beta[\omega(\mathbf{k})-\mu]\}-1} + \frac{1}{\exp\{\beta[\omega(\mathbf{k})+\mu]\}-1}\right) \\ &= \frac{d}{(2\pi)^3}\int d^3k k^0 \left(\frac{1}{\exp\{\beta[\omega(\mathbf{k})-\mu]\}-1} + \frac{1}{\exp\{\beta[\omega(\mathbf{k})+\mu]\}-1}\right). \end{aligned} \tag{7.63}$$

Note the minus sign in the expression of n_{eq}: it corresponds to the contribution of the antiparticles. On the other hand, particles and antiparticles contribute in a similar way to the pressure and the energy density of the system; this explains the plus sign in front of the antiparticles' Bose–Einstein factor.

The various thermodynamic quantities — in particular, the three adiabatic indices — have been expressed by P.T. Landsberg and J. Dunning-Davies (1964) in terms of a number of integrals to be calculated numerically.

7.2.2. *Charge fluctuations*

The charge fluctuations of the free Bose gas is defined as

$$\delta Q^2 = \langle Q^2\rangle - \langle Q\rangle^2, \tag{7.64}$$

which can easily be calculated from the partition function since

$$\begin{cases} \langle Q\rangle = e\beta^{-1}\dfrac{\partial}{\partial\mu}\log Z, \\ \langle Q^2\rangle = e^2\beta^{-2}\dfrac{\partial^2}{\partial\mu^2}\log Z, \end{cases} \tag{7.65}$$

or

$$\langle \delta Q^2\rangle = e^2\beta^{-2}\left\{\frac{\partial^2}{\partial\mu^2}\log Z - \left(\frac{\partial}{\partial\mu}\log Z\right)^2\right\}, \tag{7.66}$$

and one finds that

$$\delta Q^2 = e^2\beta^{-4}\left[\beta\frac{\partial F}{\partial\mu} - \left(\frac{\partial F}{\partial\mu}\right)^2\frac{1}{\beta^2}\right]; \tag{7.67}$$

this is of course the same element as in Newtonian theory.

7.2.3. *A few remarks on the calculation of various integrals*

Let us add a few words about the numerical calculation of various integrals that involve the Bose–Einstein factor. They are of the general form

$$\begin{cases} I(\beta,\mu) = \displaystyle\int_0^\infty d\xi\frac{f(\xi)}{\exp[\beta(\xi-\{\mu+m\})]-1}, \\ \text{with } \xi \equiv E - m. \end{cases} \tag{7.68}$$

When $f(\xi)$ is sufficiently regular (or even not too irregular!), such integrals can be calculated via a Gauss–Laguerre method, i.e. as

$$\begin{aligned} I(\beta,\mu) &= \int_0^\infty d\xi\frac{f(\xi)}{\exp[(\xi-\{\mu+m\})]-1} \\ &= \int d\xi\exp(-\xi)\frac{f(\xi)\exp(+\xi)}{\exp[(\xi-\{\mu+m\})]-1} \\ &\approx \sum_{\ell=1}^{\ell=n} A_\ell\frac{f(\xi_\ell)\exp(+\xi_\ell)}{\exp[(\xi_\ell-\{\mu+m\})]-1}, \end{aligned} \tag{7.69}$$

where ξ_ℓ are the zeros of the Laguerre polynomial of order n and the constants A_ℓ are also connected with these polynomials and are found as well in all computing programs.

Another simple remark, which is often useful, is that the Bose–Einstein factor can be rewritten as

$$\frac{1}{\exp(x)-1} = \coth\left(\frac{x}{2}\right), \tag{7.70}$$

and that most computers have enormous precision in the calculation of a number of functions, such as $\coth(x)$.

Finally, when the simple methods do not work, one is compelled to use more sophisticated ones [see e.g. H.E. Haber and H.A. Weldon (1982a, b)].

7.2.4. *Bose–Einstein condensation*

In this subsection, the usual[13] path to Bose–Einstein condensation is followed and it is first noted that the replacement

$$\frac{1}{V}\sum_{\mathbf{k}} \to \frac{d}{(2\pi)^3}\int \frac{d^3k}{(2\pi)^3}, \tag{7.71}$$

used to calculate the partition function (and hence the Bose–Einstein distribution), is valid only when the lowest energy state, $\mathbf{k} = 0$ and $E(\mathbf{k}) = m$, is not macroscopically occupied. When this is not the case, only the other energy levels can be treated as in the above subsections (their distribution

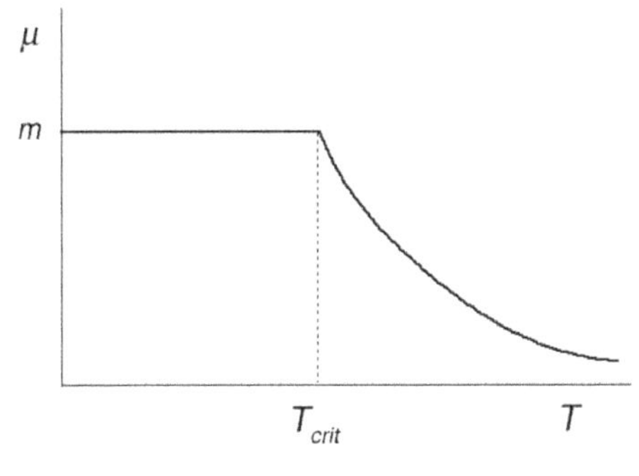

Fig. 7.1 The chemical potential as a function of the temperature. For the antiparticles — they have not been considered here — one has to change μ into $-\mu$. One always has $m^2 \leq \mu^2$.

[13]See e.g. K. Huang (1963).

function is, accordingly, the Bose–Einstein function) and the fundamental level corresponding to $\mathbf{k} = 0$ must be dealt with separately. Then the noncondensed particles have density

$$n_{\text{normal}} = \frac{d}{(2\pi)^3} \int \frac{d^3k}{k_0} u_\alpha k^\alpha \frac{1}{\exp\{\beta[\omega(\mathbf{k}) - \mu]\} - 1}, \tag{7.72}$$

where the temperature has been considered to be low enough for the contribution of the antiparticles to be neglected. The density of the condensed particles, i.e. those that lie on the lowest energy level, is then given by

$$n_{\text{cond}} = n_{\text{eq}} - n_{\text{normal}}, \tag{7.73}$$

and, as usual, the onset of the Bose–Einstein condensation phenomenon occurs when the integrand in the expression of n_{normal} diverges. This can be seen by noting that, at fixed n_{eq}, when the temperature decreases the chemical potential increases, and finally reaches its maximum value m, at a critical temperature given by

$$n_{\text{normal}} = \frac{d}{(2\pi)^3} \int \frac{d^3k}{k_0} u_\alpha k^\alpha \times \left(\frac{1}{\exp\{\beta_{\text{crit}}[\omega(\mathbf{k}) - m]\} - 1} - \frac{1}{\exp\{\beta_{\text{crit}}[\omega(\mathbf{k}) + m]\} - 1} \right). \tag{7.74}$$

This relation provides the critical temperature T_{crit} for the occurrence of condensation. In the ultrarelativistic limit, when $\beta m \to 0$, it is given by [H.E. Haber and H.A. Weldon (1981)]

$$T_{\text{crit}} = \left(\frac{3|n_{\text{eq}}|}{dm} \right)^{1/2}, \tag{7.75}$$

in contrast with the earlier result,

$$T_{\text{crit}} = \left(\frac{n_{\text{eq}}}{8\pi d\zeta(3)} \right)^{1/3} \tag{7.76}$$

[P.T. Landsberg and J. Dunning-Davies (1965)], where $\zeta(s)$ is the Riemann function, because of the neglect of antiparticles in their calculation. The above expression for T_{crit} is to be compared with the nonrelativistic value

$$T_{\text{crit}} = \frac{1}{2\pi m} \left(\frac{n_{\text{eq}}}{d\zeta(3/2)} \right)^{2/3}. \tag{7.77}$$

Below the critical temperature, the partition function can be written as

$$\log Z = -\frac{d}{(2\pi)^3}[1 - \exp(-\beta m)] \\ - \frac{d}{(2\pi)^3}\int \frac{d^3k}{\omega(\mathbf{k})} k^0 \log(1 - \exp\{-\beta[\omega(\mathbf{k})]\}), \tag{7.78}$$

where the first term corresponds to the condensed particles for which $\mathbf{k} = 0$. It can also be written as

$$\log Z = -\frac{d}{(2\pi)^3}\int \frac{d^3k}{\omega(\mathbf{k})} k^0 [(1 - \exp\{-\beta[\omega(\mathbf{k}) - m]\})\delta^{(3)}(\Delta_{\alpha\beta}(u)k^\alpha k^\beta) \\ + \log(1 - \exp\{-\beta[\omega(\mathbf{k})]\})], \tag{7.79}$$

which, upon functional differentiation with respect to $\omega(\mathbf{k})$, yields

$$f_{\text{eq}}(k) = \frac{d}{(2\pi)^3}\left\{\frac{\delta^{(3)}(\Delta_{\alpha\beta}(u)k^\alpha k^\beta)}{\exp[\beta(m-\mu)]} + \frac{\text{sgn}(u \cdot k)}{\exp[\beta(u \cdot k - \mu)] - 1}\right\}, \tag{7.80}$$

where the contribution of the antiparticles and of the vacuum have been re-established. In an interacting system both terms may contribute significantly to the final results [see e.g. H.E. Haber and H.A. Weldon (1982)]. This is the full distribution function of the bosons below the critical temperature. We come back to a deeper view of Bose–Einstein condensation in the next subsection.

7.2.5. *Interactions*

As an indication of the treatment of interactions we shall briefly consider the example of a self-interacting complex scalar field, whose Lagrangian is

$$\mathcal{L} = \frac{1}{2}(\partial\varphi^*\partial\varphi - m^2\varphi^2) - \frac{\lambda}{4!}(\varphi^*\varphi)^4. \tag{7.81}$$

This system presents the following advantage over the similar one with only a real scalar field [F. Grassi, R. Hakim and H. Sivak (1991)]: it is still relatively simple and, furthermore, it possesses a U(1) symmetry; moreover, it sheds some light on the Bose–Eintein condensation of relativistic gases.

Its free energy is thus

$$\mathcal{F} = -P + \mu j_\nu u^\nu = \frac{1}{3}T^{\alpha\beta}\Delta_{\alpha\beta}(u) + n\mu, \tag{7.82}$$

where n is the invariant charge density. Using the above Lagrangian, one obtains

$$\mathcal{F} = \frac{1}{3}\left\langle \frac{1}{2}[\partial^\alpha \varphi^* \partial^\beta \varphi + \partial^\beta \varphi^* \partial^\alpha \varphi] \right.$$
$$\left. - \eta^{\alpha\beta}\left\{\frac{1}{2}(\partial\varphi^*\partial\varphi - m^2\varphi^*\varphi) - \frac{\lambda}{4!}(\varphi^*\varphi)^4\right\}\right\rangle$$
$$\times \Delta_{\alpha\beta}(\mathrm{u}) + \mu\frac{i}{8}\langle\varphi^* \overleftrightarrow{\partial}^\nu \varphi\rangle \mathrm{u}^\nu, \tag{7.83}$$

where the brackets designate an average value calculated with the grand-canonical density operator, and it is clear that, without any approximation scheme, one cannot go very far (q is the absolute value of the charge of a typical particle of the field).

Let us separate the fluctuating part ϕ of the field from its average value σ as

$$\varphi = \phi + \sigma \tag{7.84}$$

and, rather than performing a perturbative expansion, the system will be dealt with by using a Gaussian approximation[14] for the interaction term[15]:

$$\begin{cases} \langle|\varphi^*\varphi|^2\rangle \approx 2\langle\phi^*\phi\rangle^2, \\ \langle\phi^{2\ell+1}\rangle = \langle\phi^{*(2\ell+1)}\rangle\rangle = 0. \end{cases} \tag{7.85}$$

Thence the above free energy can be rewritten as

$$\mathcal{F} = \frac{1}{3}\left\langle \frac{1}{2}[\partial^\alpha\phi^*\partial^\beta\phi + \partial^\alpha\phi\partial^\beta\phi^*] - \eta^{\alpha\beta}\left\{\frac{1}{2}\partial\phi^*\partial\phi - \frac{1}{2}m^2(\phi^*\phi + \sigma^*\sigma)\right.\right.$$
$$\left.\left. - \frac{\lambda}{4!}[2\phi^*\phi\langle\phi^*\phi\rangle + (\sigma^*\sigma)^2] + \mu\frac{i}{8}\langle\phi^* \overleftrightarrow{\partial}^\nu \phi\rangle \mathrm{u}^\nu\right\}\right\rangle \Delta_{\alpha\beta}(\mathrm{u}). \tag{7.86}$$

Let us now examine a little further this last (approximate) expression for F. The coefficient of $\eta^{\alpha\beta}$ constitutes an *effective* Lagrangian:

$$\mathcal{L}_{\text{eff}} = \frac{1}{2}\partial\phi^*\partial\phi - \frac{1}{2}m^2(\phi^*\phi + \sigma^*\sigma) - \frac{\lambda}{4!}[2\phi^*\phi\langle\phi^*\phi\rangle + (\sigma^*\sigma)^2]$$
$$+ \mu\frac{i}{8}\langle\phi^* \overleftrightarrow{\partial}^\nu \phi\rangle \mathrm{u}_\nu. \tag{7.87}$$

---

[14]The Gaussian approximation was introduced into quantum field theory by L.I. Schiff [*Phys. Rev.* **130**, 458 (1963)] and used subsequently by numerous authors [see F. Grassi *et al.* (1991), for a bibliography].

[15]R. Hakim, N. Verdon and H. Sivak, unpublished (1992).

The constant terms quadratic and quartic in σ can be omitted from *this* Lagrangian, and the corresponding *effective* Hamiltonian finally reads

$$H_{\text{eff}} = \frac{1}{(2\pi)^3}\int d^3k\{a^+(\mathbf{k})a(\mathbf{k})[\omega_{\text{eff}}(\mathbf{k}) - \mu] + b^+(\mathbf{k})b(\mathbf{k})[\omega_{\text{eff}}(\mathbf{k}) + \mu]\}, \tag{7.88}$$

where a vacuum term — irrelevant for what follows — has been omitted and where

$$\omega_{\text{eff}}(\mathbf{k}) = \sqrt{M^2 + \mathbf{k}^2}, \tag{7.89}$$

with

$$M^2 = m^2 + \frac{\lambda}{12}\langle\phi^*\phi\rangle. \tag{7.90}$$

Such an effective Hamiltonian represents the Hamiltonian of *free quasiparticles* endowed with the effective mass M. This last equation constitutes, in fact, an implicit equation for M since the term $\langle\phi^*\phi\rangle$ is to be calculated with the Bose–Einstein (M-dependent) function of these free quasiparticles, and one finds that

$$M^2 = m^2 + \frac{\lambda}{12}\frac{1}{(2\pi)^3}\int \frac{d^3k}{\sqrt{M^2 + \mathbf{k}^2}} \times \left\{\frac{1}{\exp[\beta(\sqrt{M^2 + \mathbf{k}^2} - \mu)] - 1} + \frac{1}{\exp[\beta(\sqrt{M^2 + \mathbf{k}^2} + \mu)] - 1}\right\}, \tag{7.91}$$

which has to be solved together with the equation for μ, i.e. the normalization of $f_{\text{eq}}(k)$. Also, the omitted vacuum term must be reinstated and the resulting equation then contains an infinite term and should be renormalized [see F. Grassi, R. Hakim and H. Sivak (1991) for details].

Let us now briefly examine the thermodynamics of the system. Since the quasiparticles are free, its thermodynamic properties are those of free bosons endowed with the *effective* mass M. There are, however, some differences to take into account. Apart from the renormalization of the equation for M, the energy–momentum tensor must also be renormalized since it contains an infinite vacuum term. Such a term cannot be omitted with the usual normal product, as it is (T, n)-dependent. As important is the fact that the whole thermodynamics of the system is entirely governed by the equation for M. A last feature is that these properties *a priori* depend on the average value σ of the field φ.

It remains for us to determine σ and this can be done by noting that the free energy has to be a *stable* minimum, and this provides the conditions

$$\begin{cases} \dfrac{dF}{d\sigma} = 0, \\ \dfrac{d^2F}{d\sigma^2} > 0. \end{cases} \tag{7.92}$$

The minimum condition yields

$$\frac{1}{2}(\mu^2 - M^2)\sigma = 0, \tag{7.93}$$

while the stability condition is obviously obeyed.

When there is no Bose–Einstein condensation, i.e. when $\mu^2 < M^2$, the only possible solution is $\sigma = 0$. However, when a Bose–Einstein condensation occurs in the system, i.e. when $\mu^2 = M^2$, a coherent state $\sigma \neq 0$ can develop. Note that this remark is valid for the free case as well. The fact that a condensate develops for $m^2 = M^2$ gives rise to a spontaneously broken symmetry[16] in the plane of the complex field ϕ; σ introduces a particular direction in the plane $\{\mathrm{Re}\phi, \mathrm{Im}\phi\}$ and thus breaks the U(1) symmetry existing in the absence of Bose–Einstein condensate.

It remains for us, however, to determine the precise value of the condensate σ. This can be done by looking at the charge within the condensate. Since the charge density is given by

$$n = \beta^{-1}\frac{\partial}{\partial\mu}\log Z, \tag{7.94}$$

it turns out that one has

$$n = \beta^{-1}\left\{2m\sigma^2 + \frac{d}{(2\pi)^3}\int\frac{d^3k}{k_0}u_\alpha k^\alpha \right.$$
$$\left. \times\left[\frac{1}{\exp\{\beta[\omega(\mathbf{k}) - m]\} - 1} - \frac{1}{\exp\{\beta[\omega(\mathbf{k}) + m]\} - 1}\right]\right\}, \tag{7.95}$$

which provides σ as a function of the total charge density and of the density of noncondensed particles (and antiparticles). Note that we have used $\mu = m$ (for the free case, otherwise one would have $\mu = M$ in the above interacting case[17]) but the case $\mu = -m$ (resp. M) is quite similar.

---

[16] A. Casher and M. Revzden, *Am. J. Phys.* **35**, 1154 (1967); J.M. Robinson and S.L. Trubatch, *Am. J. Phys.* **39**, 886 (1971); *ibid.* **39**, 893 (1971); *ibid.* **39**, 1190 (1971); Y. Kano, *J. Phys. Soc. Jpn.* **36**, 649 (1974); *ibid.* **37**, 310 (1974). See also J. Kapusta (1981).

[17] In this last case, σ satisfies an implicit equation since the effective mass depends on σ itself.

7.3. Free Fermions in Thermal Equilibrium

In all applications dealing with relativistic dense matter, the Fermi–Dirac function plays a very important role with various different forms and has thus been studied by many authors. Its first relativistic study was made by F. Jüttner (1928), while the first significant application was performed by S. Chandrasekhar (1939), who realized that electrons in white dwarfs were relativistic (1930), a fact that gave rise to the discovery of a limiting mass for this kind of astrophysical objects, white dwarfs — the so-called *Chandrasekhar mass.* Consequently, numerous studies were performed as to the various integrals which come into play in the calculations involving the relativistic Fermi–Dirac function, in particular treating of numerical approximations in several physical regimes. Owing to the better performances of modern computers and algorithms, many are now outdated, but nevertheless they can be useful on some occasions. We mention the articles[18] by A.W. Guess (1966), R.F. Tooper (1969), P.P. Eggleton, J. Faulkner and B.P. Flanner (1973), S.A. Bludman and K.A. van Ripper (1977), A. Wandel and A. Yahill (1979), S.I. Blinnikov (1987), and B. Pichon (1989). On the following, only elementary approximations are given (i) for temperature corrections to the completely degenerate case and (ii) for the ultrarelativistic nondegenerate regime.

In this section, only the noninteracting relativistic Fermi gas is studied. For electrons, as they occur in white dwarfs, Coulombian corrections are of the order of $e^2 n^{1/3}$, to be compared to their Fermi energy ε_F; it appears that, in white dwarfs, the Coulombian corrections constitute only a few percent or the Fermi energy.

For fermions there exist the following differences: (i) both particles and antiparticles are dealt with at the same time and (ii) the "charge" of the system is conserved, not the particle number. The fermions are chosen to be Dirac's free fields obeying

$$\begin{cases} \{i\gamma \cdot \partial - m\}\psi(x) = 0, \\ \bar{\psi}(x)\{i\gamma \cdot \overleftarrow{\partial} + m\} = 0, \end{cases} \tag{7.96}$$

so that the equilibrium density operator can be written as

$$\rho_{\text{eq}} = \frac{1}{Z}\exp(-\beta[u \cdot P - \mu Q]), \tag{7.97}$$

[18]Note that these articles often differ in their definition of energy whether they substract the rest mass energy or not.

with

$$\begin{cases} P^\mu = \int_\Sigma d\Sigma_\nu \, T^{\mu\nu}(x), \\ T^{\mu\nu}(x) = \frac{i}{2}\bar{\psi}(x)\,\overleftrightarrow{\partial}^\mu \gamma^\nu \psi(x), \end{cases} \tag{7.98}$$

$$\begin{cases} Q = \int_\Sigma d\Sigma_\nu \, J^\nu(x), \\ J^\nu(x) = e\bar{\psi}(x)\gamma^\nu\psi(x). \end{cases} \tag{7.99}$$

The decomposition of the fields $\{\psi(x), \bar{\psi}(x)\}$ into creation/annihilation operators, as

$$\begin{cases} \psi(x) = \sum_s \int d^3p\{a(\mathbf{p}, s)u(\mathbf{p}, s)\exp(-ip\cdot x) \\ \qquad + d^+(\mathbf{p}, s)v(\mathbf{p}, s)\exp(+ip\cdot x)\}, \\ \bar{\psi}(x) = \sum_s \int d^3p\{a^+(\mathbf{p}, s)\bar{u}(\mathbf{p}, s)\exp(+ip\cdot x) \\ \qquad + d(\mathbf{p}, s)\bar{v}(\mathbf{p}, s)\exp(-ip\cdot x)\}, \end{cases} \tag{7.100}$$

with

$$\begin{cases} \{a(\mathbf{p}, s), a^+(\mathbf{p}', s')\} = \delta_{ss'}\delta^{(3)}(\mathbf{p}-\mathbf{p}'), \\ \{d(\mathbf{p}, s), d^+(\mathbf{p}', s')\} = \delta_{ss'}\delta^{(3)}(\mathbf{p}-\mathbf{p}') \end{cases} \tag{7.101}$$

and the normalization for the spinors $u(\mathbf{p}, s)$ and $v(\mathbf{p}, s)$

$$\begin{cases} \sum_s u(\mathbf{p}, s)\bar{u}(\mathbf{p}, s) = \frac{|\mathbf{p}| + m}{2E_\mathbf{p}}, \\ \sum_s v(\mathbf{p}, s)\bar{v}(\mathbf{p}, s) = \frac{|\mathbf{p}| - m}{2E_\mathbf{p}} \end{cases} \tag{7.102}$$

provides

$$\rho_{\text{eq}} = \frac{1}{Z}\exp\left(-\beta\sum_\mathbf{p}[E_\mathbf{p} - \mu]a_\mathbf{p}^+ a_\mathbf{p} + [E_\mathbf{p} + \mu]d_\mathbf{p}^+ d_\mathbf{p}\right), \tag{7.103}$$

which immediately leads to the particle and antiparticle occupation numbers, in the same way as is the case for bosons:

$$n_{\text{part}}(p) \equiv \langle a_\mathbf{p}^+ a_\mathbf{p}\rangle = \frac{1}{\exp(\beta[u\cdot p - \mu]) + 1}, \tag{7.104}$$

$$n_{\text{antipart}}(p) \equiv \langle d_\mathbf{p}^+ d_\mathbf{p}\rangle = \frac{1}{\exp(\beta[u\cdot p + \mu]) + 1}, \tag{7.105}$$

with

$$u \cdot p \equiv E_{\mathbf{p}} = \sqrt{\mathbf{p}^2 + m^2}. \qquad (7.106)$$

Then one finds the expressions

$$J^\alpha = \frac{ed}{(2\pi)^3} \int \frac{d^3p}{p_0} p^\alpha \times \left(\frac{1}{\exp[\beta(u \cdot p - \mu)] + 1} - \frac{1}{\exp[\beta(u \cdot p + \mu)] + 1} \right), \qquad (7.107)$$

$$P = -\frac{d}{3(2\pi)^3} \int \frac{d^3p}{p_0} \Delta_{\alpha\beta}(p) p^\alpha p^\beta \times \left(\frac{1}{\exp[\beta(u \cdot p - \mu)] + 1} + \frac{1}{\exp[\beta(u \cdot p + \mu)] + 1} \right), \qquad (7.108)$$

$$\rho = \frac{d}{(2\pi)^3} \int \frac{d^3p}{p_0} (u \cdot p)^2 \times \left(\frac{1}{\exp[\beta(u \cdot p - \mu)] + 1} + \frac{1}{\exp[\beta(u \cdot p + \mu)] + 1} \right). \qquad (7.109)$$

In terms of n_{eq}, P and ρ, these integrals can be rewritten as

$$n_{\mathrm{eq}} = \frac{d}{2\pi^2} \int dE\, E \sqrt{E^2 - m^2} \times \left(\frac{1}{\exp[\beta(E - \mu)] + 1} - \frac{1}{\exp[\beta(E + \mu)] + 1} \right), \qquad (7.110)$$

$$P = \frac{d}{6\pi^2} \int dE (E^2 - m^2)^{3/2} \times \left(\frac{1}{\exp[\beta(E - \mu)] + 1} + \frac{1}{\exp[\beta(E + \mu)] + 1} \right), \qquad (7.111)$$

$$\rho = \frac{d}{2\pi^2} \int dE\, E^2 \sqrt{E^2 - m^2} \times \left(\frac{1}{\exp[\beta(E - \mu)] + 1} + \frac{1}{\exp[\beta(E + \mu)] + 1} \right). \qquad (7.112)$$

Note that the density of states can be obtained by a simple change of variables, $|\mathbf{p}| \to E$, since the integration element transforms as

$$\frac{d}{(2\pi)^3} \frac{d^3p}{p_0} \to \frac{d}{2\pi^2} \sqrt{E^2 - m^2} dE \qquad (7.113)$$

and is thus given by

$$g(E) = \frac{d}{2\pi^2}\sqrt{E^2 - m^2}. \tag{7.114}$$

7.4. Thermodynamic Properties of the Relativistic Ideal Fermi–Dirac Gas

The various possible regimes that govern the thermodynamic properties of the Fermi gas are essentially linked to the values of the following parameters:

- ε_F, the Fermi energy (or possibly the thermodynamic potential[19] μ),
- T, the thermal energy,
- $e^2 n^{1/3}$, the Coulomb energy (when the system is an electromagnetic plasma),

and they have to be compared to the rest mass energy m. For the system to be relativistic, it is necessary that at least one of these parameters is larger than or equal to m. However, the significance of the word "relativistic" varies according to which parameter is at stake. For instance, when $\varepsilon_F \geq m$ this means that the system is extremely dense; when $T \geq m$, the relativistic character of the system comes from its thermal agitation; finally, when $e^2 n^{1/3} \approx m$, the interaction energy is so high that pair creations are important and one has to be careful with their treatment.

As to the interplay of the basic parameters, it is clear that when $\varepsilon_F \gg T$, the system can be considered as being cold while it can be dealt with as a noninteracting system whenever

$$\varepsilon_F \gg e^2 n^{1/3} \quad \text{and/or} \quad T \gg e^2 n^{1/3}.$$

Nevertheless, the ratio

$$\Gamma = \frac{e^2 n^{1/3}}{T}$$

plays an important role in the onset of a possible crystallization (or melting) of the system; for $\Gamma \gg 1$, there is crystallization of the plasma,[20] so as to minimize the Coulomb energy.

---

[19] At $T = 0$, the chemical potential is called the "Fermi energy" (ε_F), as in the nonrelativistic case.

[20] It should be noted that the electron plasma is generally embedded in a neutralizing positive background constituted by ions of charges Ze (Z should then be added to the above formulae). Note also that the parameter Γ and its signification are the same whether the system is relativistic or not.

7.4.1. *Remarks on the numerical calculations of various physical quantities*

In the theory of white dwarfs and in many other applications, various data are needed in a numerical manner. Many calculations have been made, beginning with tables of the Fermi integrals or with more or less sophisticated Fortran codes. For instance, A.W. Guess (1966) expressed most quantities of interest in terms of the integrals

$$Q_n(\mu, m\beta) = \frac{1}{2}\int_{-\infty}^{+\infty} \frac{\cosh(n\theta)d\theta}{\exp(m\beta\cosh\theta - \beta[m-\mu]) + 1}, \tag{7.115}$$

as

$$\begin{cases} n_{\text{eq}} = \dfrac{6A}{m}[Q_3(\mu, \beta m) - Q_1(\mu, \beta m)], \\ \rho = 3A[Q_4(\mu, \beta m) - Q_3(\mu, \beta m) + Q_1(\mu, \beta m) - Q_0(\mu, \beta m)], \\ P = A[Q_4(\mu, \beta m) - 4Q_2(\mu, \beta m) + 3Q_0(\mu, \beta m)], \\ S = \dfrac{A}{m}[4\beta m\{Q_4(\mu, \beta m) - Q_2(\mu, \beta m)\} - 6q\{Q_3(\mu, \beta m) - Q_1(\mu, \beta m)\}], \\ A = \dfrac{d}{48\pi^2},\ q = \beta m - \exp(\beta\mu). \end{cases} \tag{7.116}$$

Then he proceeded to a Mellin transform of the Q_n's, particularly suitable for their detailed study, obtaining thereby recursion relations and approximations for various regimes. A similar method was used by R.F. Tooper (1969), who improved Guess' results. Finally, a large number of articles dealing with particular regimes and improving the preceding results appeared [G. Beaudet and M. Tassoul (1971); P.P. Eggleton, J. Faulkner and B.P. Flannery (1973); S.R. Hore and N.E. Frankel (1975); S.A. Bludman and K.A. van Riper (1977); B. Paczinski (1983); F.J. Fernandez Velicia (1984); A.T. Service (1986); S.I. Blinnikov (1987)]. They have been reviewed, compared and improved by B. Pichon (1989), to whom we refer.

7.4.2. *The degenerate Fermi gas*

The completely degenerate case, $T = 0\,\text{K}$, can be obtained by taking the limit $\beta \to \infty$ in the preceding expressions. For an ideal gas composed of relativistic fermions of mass m, the various energy levels are uniformly occupied untill the Fermi level ε_F is reached; the Fermi level is connected

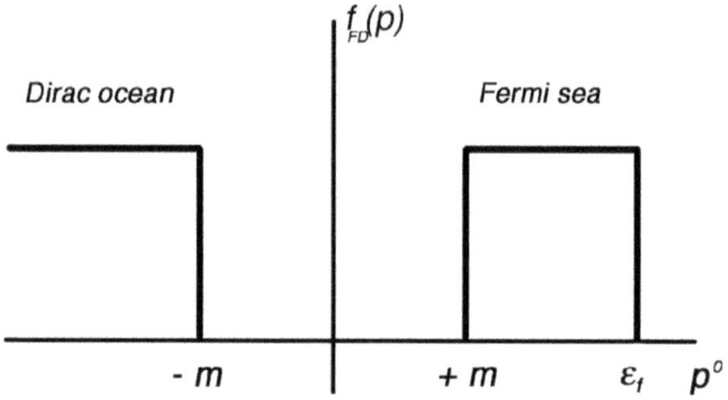

Fig. 7.2 The completely degenerate relativistic Fermi–Dirac distribution exhibits both the Fermi sea and the Dirac ocean.

with the Fermi momentum through the definition

$$p_F \equiv \sqrt{\varepsilon_F^2 - m^2}. \tag{7.117}$$

The Fermi–Dirac distribution then appears as it is depicted in Fig. 7.2 and reads

$$\begin{aligned} f_{\rm FD}(p) = {} & \frac{d}{(2\pi)^3}\delta(p^2 - m^2) \\ & \times \{\theta(u\cdot p)\theta(\varepsilon_f - u\cdot p) + \theta(-u\cdot p)\theta(u\cdot p - m)\}, \end{aligned} \tag{7.118}$$

where the first term represents the Fermi sea and the second refers to Dirac's ocean. In what follows, this "vacuum" term will be discarded although it plays an important role in renormalization problems (see Chap. 9).

The particle density is then, as in the nonrelativistic case,

$$n_{\rm eq} = \frac{d}{6\pi^2}p_F^3 \tag{7.119}$$

or

$$p_F = \left(\frac{6\pi^2}{d}\right)^{2/3} n^{2/3}, \tag{7.120}$$

whereas the energy density and the pressure assume a different form owing to the different form of relativistic energy; they read

$$\begin{aligned} \rho &= \frac{d}{(2\pi)^3}\int \frac{d^3p}{p_0}(u\cdot p)^2\theta(\varepsilon_f - u\cdot p) \\ &= \frac{d}{2\pi^2}\int_0^{p_F} d|\mathbf{p}|\mathbf{p}^2\sqrt{\mathbf{p}^2 + m^2}, \end{aligned} \tag{7.121}$$

$$P = \frac{d}{3(2\pi)^3}\int d^3p \frac{\mathbf{p}^2}{\sqrt{\mathbf{p}^2+m^2}}\theta(\varepsilon_F - u\cdot p)$$
$$= \frac{d}{6\pi^2}\int_0^{p_F} d|\mathbf{p}|\frac{\mathbf{p}^4}{\sqrt{\mathbf{p}^2+m^2}}. \tag{7.122}$$

These integrals can be calculated explicitly and the equation of state can be written in parametric form:

$$\begin{cases} \rho = \dfrac{dm^4}{16\pi^2}\left\{x(1+2x^2)\sqrt{1+x^2} - \ln\left[x+\sqrt{1+x^2}\right]\right\}, \\ P = \dfrac{dm^4}{16\pi^2}\left\{x(1+x^2)\left(\dfrac{2}{3}x^2-1\right)\sqrt{1+x^2} + \ln\left[x+\sqrt{1+x^2}\right]\right\}, \\ x = \dfrac{p_F}{m}. \end{cases} \tag{7.123}$$

7.4.3. *Thermal corrections: Sommerfeld expansion*

In many physical problems the temperature is so low that only corrections to the zero temperature Fermi–Dirac distribution are necessary. For instance, in white dwarfs, the temperature of the electrons is of the order of 10^6 K, while the Fermi energy is of the order of a few MeV; this means that (i) one should deal with relativistic electrons (their Fermi energy is larger than their rest mass) and that (ii) $k_BT \ll \varepsilon_f$. A. Sommerfeld (1928) gave an asymptotic expansion for these thermal corrections which is quite useful.[21]

When one is using the Fermi–Dirac function in practical problems, integrals of the form

$$I = \int_m^\infty dE f_{\mathrm{FD}}(E) q(E) \tag{7.124}$$

do appear, where $q(E)$ is a given function[22] supposed to have all desired regularity properties. We also assume that the function $q(E)$ possesses a known integral:

$$Q(E) = \int^E dE\, q(E). \tag{7.125}$$

[21] See also S. Chandrasekhar, *An Introduction to the Study of Stellar Structure* [University of Chicago Press, (1939); reprinted by Dover (1967)].
[22] It includes the density of states $g(E)$.

In the Sommerfeld expansion, one takes advantage of the fact that the Fermi–Dirac function is almost steplike. Integrating the integral I by parts, one finds that

$$I = f_{\mathrm{FD}}(E)Q(E)|_m^\infty - \int_m^\infty dE \frac{\partial}{\partial E} f_{\mathrm{FD}}(E) \cdot Q(E), \tag{7.126}$$

where the first term vanishes: for $E = m$, $Q(m) = 0$, by construction; and for $E \to \infty$, $f_{\mathrm{FD}}(E) \to 0$. One finally has

$$I = -\int_m^\infty dE \frac{\partial}{\partial E} f_{\mathrm{FD}}(E) \cdot Q(E). \tag{7.127}$$

Since $f_{\mathrm{FD}}(E)$ is almost a step function, its derivative is sharply peaked and hence is almost a δ function. When one expands the function $Q(E)$ into a series about $E = \mu$, the chemical potential, as

$$Q(E) = \sum_{n=0}^{\infty} \frac{1}{n!} \left. \frac{d^n Q}{dE^n} \right|_{E=\mu} (E-\mu)^n, \tag{7.128}$$

one gets

$$I = -\sum_{n=0}^{\infty} \frac{1}{n!} \left. \frac{d^n Q}{dE^n} \right|_{E=\mu} \int_m^\infty dE \frac{\partial}{\partial E} f_{\mathrm{FD}}(E)(E-\mu)^n \tag{7.129}$$

or

$$I = \sum_{n=0}^{\infty} \frac{d}{(2\pi)^3} I_n \frac{\beta^{-n}}{n!} \left. \frac{d^n Q}{dE^n} \right|_{E=\mu}, \tag{7.130}$$

with

$$I_n = \int_m^\infty dx \frac{2x^n}{\cosh\left(\frac{1}{2}x\right)} - \int_{-\infty}^\infty dx \frac{2x^n}{\cosh\left(\frac{1}{2}x\right)}. \tag{7.131}$$

In the Sommerfeld expansion, the lower bound m is replaced by $-\infty$ because of the peaked shape of the derivative of the Fermi–Dirac function; this amounts to neglecting exponentially small terms. I_n can be evaluated more precisely as

$$\begin{cases} I_{2n} = (-1)^n (2 - 2^{2n}) \pi^{2n} B_{2n}, \\ I_{2n+1} = 0, \end{cases} \tag{7.132}$$

where B_{2n} are the Bernoulli numbers of order $2n$. Finally, I turns out to be

$$I = \sum_{n=0}^{\infty} \frac{d}{(2\pi)^3} (-1)^n (2 - 2^{2n}) \pi^{2n} B_{2n} \frac{\beta^{-2n}}{n!} \left. \frac{d^{2n} Q}{dE^{2n}} \right|_{E=\mu}. \tag{7.133}$$

The first few Bernoulli numbers are

$$B_0 = 1,\ B_1 = -\frac{1}{2},\ B_2 = \frac{1}{6},\ B_4 = -\frac{1}{30},\ B_6 = \frac{1}{42},$$
$$B_8 = -\frac{1}{30},\ B_{10} = \frac{5}{66}.$$

The integral I thus appears to be a formal expansion in powers of $(k_BT)^n \equiv \beta^{-n}$; but, as in the nonrelativistic case, this is a bit fallacious since the chemical potential does depend on T itself. It is therefore necessary to evaluate first the T dependence of μ.

7.4.4. *Corrections for various thermodynamic quantities*

Let us now evaluate some thermal corrections for the main thermodynamic quantities by using the Sommerfeld expansion method.

Chemical potential. The chemical potential is obtained via the normalization of the Fermi–Dirac function

$$\begin{aligned} n_{\text{eq}} &= \frac{d}{2\pi^2}\int dE E\sqrt{E^2-m^2}\,\frac{1}{\exp[\beta(E-\mu)]+1} \\ &= \frac{d}{2\pi^2}\int_0^{\varepsilon_F} dE E\sqrt{E^2-m^2} \\ &\quad + \left[(\mu-\varepsilon_F)\varepsilon_F p_F + \frac{\pi^2T^2}{6}\frac{d}{dE}\left(E\sqrt{E^2-m^2}\right)\bigg|_{E=\mu}\right] + O(T^4). \end{aligned} \tag{7.134}$$

The first term is the zeroth order expression on n_{eq} and hence the term between the brackets is essentially vanishing, so that finally one obtains

$$\begin{aligned} \mu &= \varepsilon_F\left\{1 - \frac{\pi^2}{6}\frac{T^2}{\varepsilon_F}\frac{1}{\varepsilon_F p_F}\frac{d}{dE}\left(E\sqrt{E^2-m^2}\right)\bigg|_{E=\varepsilon_F}\right\} + O(T^4) \\ &= \varepsilon_F\left\{1 - \frac{\pi^2}{6}\frac{T^2}{\varepsilon_F^2}\left(1+\frac{\varepsilon_F^2}{p_F^2}\right)\right\} + O(T^4). \end{aligned} \tag{7.135}$$

Energy density. One has

$$q(E) = \frac{d}{2\pi^2}E^2\sqrt{E^2-m^2} \tag{7.136}$$

and therefore the energy density turns out to be

$$\rho(T) = \rho(T=0) + 2\pi^2 B_2 T^2 \varepsilon_F^2 p_F + \cdots = \frac{\pi^2}{3}T^2\varepsilon_F p_F + O(T^4). \tag{7.137}$$

Pressure. In this case the function $q(E)$ is given by

$$q(E) = \frac{d}{6\pi^2}(E^2 - m^2)^{3/2}, \tag{7.138}$$

so that one obtains

$$P(T) = P(T=0) + \frac{1}{36\pi^2}T^2 \frac{d}{dE}(E^2 - m^2)^{3/2}\bigg|_{E=\varepsilon_F} + O(T^4). \tag{7.139}$$

The quantities which appear in the above formulae, i.e. $\rho(t = 0)$ and $P(T = 0)$, were calculated previously.

Other quantities can also be obtained, such as the (volume or pressure) heat capacity.

The heat capacity (per unit volume and particle) at constant volume of the degenerate electron gas plays an important role in many astrophysical situations, such as the cooling of white dwarfs. It is given by

$$C_V = \frac{\partial \rho(n_{\rm eq}, T)}{\partial T}\bigg|_{n_{\rm eq}} = \frac{2d\pi^2}{3} T\varepsilon_F p_F, \tag{7.140}$$

which is obtained from the above approximate expression for the energy density. As in the nonrelativistic case, C_V tends to zero when $T \to 0$, and hence the third law of thermodynamics is still obeyed.

This free electron gas model has been corrected by taking into account both the electron exchange and its interaction energies to order $O(e^4 \log e^2)$ [T. Hamada and Y. Nakamura (1966)]. In Chap. 14, a more general expression is given for C_V.

7.4.5. *High temperature expansion (nondegenerate)*

The high temperature corrections may be obtained from the remark that they correspond to small β's, namely when $\beta m \ll 1$. Consequently, it is sufficient to use the following expansion of the Fermi–Dirac factor whenever it occurs in the various integrals:

$$\frac{1}{\exp[\beta(E_{\mathbf{p}} - \mu)] + 1} = \frac{\exp[-\beta(\sqrt{p^2 + m^2} - \mu)]}{1 + \exp[-\beta(\sqrt{p^2 + m^2} - \mu)]}$$

$$= \sum_{\ell=1}^{\infty} (-1)^{\ell+1} \exp[-\beta\ell(\sqrt{p^2 + m^2} - \mu)]. \tag{7.141}$$

For instance, the fermions' density is

$$\begin{aligned}
n_{\text{part}} &= \frac{d}{(2\pi)^3}\int \frac{d^3p}{p_0}p^0\frac{1}{\exp[\beta(u\cdot p-\mu)]+1}\\
&= \sum_{r=1}^{\infty}(-1)^{r+1}\frac{d}{(2\pi)^3}\int d^3p \exp\left[-\beta r\left(\sqrt{p^2+m^2}-\mu\right)\right]\\
&= \frac{dm^2}{2\pi^2\beta}\sum_{r=1}^{\infty}(-1)^{r+1}r\exp(\beta r\mu)K_2(\beta m r),
\end{aligned} \tag{7.142}$$

where K_2 is still a Kelvin function of order 2. For large r's, the Kelvin function K_2 (see App. A) is such that

$$K_2(\beta m r) \approx \left(\frac{\pi}{2\beta m r}\right)^{1/2}\exp(-\beta m r), \tag{7.143}$$

so that the above series converges since $\mu^2 < m^2$.

7.5. White Dwarfs: The Degenerate Electron Gas

White dwarf stars have a long and interesting story, beginning about 150 years ago, when F.W. Bessel discovered that Sirius seemed to orbit around a fixed point in the sky. He then assumed that in this fixed point was a star, invisible at this time: Sirius B. Owing to its weak luminosity, this star was detected only in 1862 by A.-G. Clark. The period of this binary system was of the order of 50 years and this allowed, in 1910, determination of the mass of the two components A and B as

$$\begin{cases} M_A \approx 2.3\,M_{\text{Sun}}, \; M_B \approx 1.0\,M_{\text{Sun}}\\ M_{\text{Sun}} \approx 1.98\times 10^{33}\,\text{g}, \end{cases}$$

while the luminosities were

$$\begin{cases} L_A \approx 40\,L_{\text{Sun}}, \; L_B \approx 3\times 10^{-3}\,L_{\text{Sun}},\\ L_{\text{Sun}} \approx 3.9\times 10^{33}\,\text{erg/s}, \end{cases}$$

which did not present anything special except the disproportion between the luminosity ratio (: 10^4) and the mass ratio (: 2). In 1915, J.C. Adams measured the surface temperature of Sirius B (: 8000 K), which is much more important than that of the Sun (: 5800 K) and moreover much more important than what might be expected (: 1300 K) from the star luminosity and Stefan's law.

The problems and questions then began at this stage. The radius R_B of Sirius B can indeed be estimated from the luminosity of the star and Stefan's law,

$$L_B = 4\pi\sigma T^4$$

(the energy emitted by the white dwarf being $\propto R^2 L_B$), where σ is the Stefan–Boltzmann constant,

$$\sigma = \frac{\pi^2 k_B^4}{15(hc)^3} = 5.7 \times 10^{-5}\text{erg} \times (\text{cm}^2 \times \text{s} \times \text{K}^4)^{-1},$$

and one finds that

$$\begin{cases} R_B \approx 0.26\, R_{\text{Sun}}, \\ R_{\text{Sun}} \approx 6.96 \times 10^{10}\,\text{cm}. \end{cases}$$

These figures led A. Eddington (1922) to say that Sirius B possesses an "absurd" density, of the order of $10^5\,\text{g/cm}^3$. Later, other measures led to higher surface temperatures (e.g. 32000 K) and hence to still smaller radii and, consequently, to still more "absurd" densities. The smallness of the radii of white dwarfs was confirmed in 1925 by J.C. Adams, who used the recently confirmed general relativity. This theory predicts a redshift of light rays of the order

$$\frac{\Delta\lambda}{\lambda} \approx G\frac{M}{Rc^2}, \tag{7.144}$$

which was indeed measured and confirmed the smallness of Sirius B's radius.[23]

Sirius B thus appeared as a compact (i.e. massive, small and dense) object and the question raised by this fact was: What kind of physical phenomenon would be able to provide a sufficient pressure ($P \approx 10^5\,\text{dyn}$) to balance the gravity of the star ($1.0 M_{\text{Sun}}$) so as to maintain the star in hydrostatic equilibrium.[24]

After the discovery in 1925 of the exclusion principle (W. Pauli) and of the subsequent Fermi–Dirac statistics (1926), R. Fowler realized that the pressure necessary for maintaining the white dwarf hydrostatic equilibrium was simply the pressure occurring between electrons obeying the exclusion principle.

---

[23]Similar measures have been performed for numerous other white dwarfs; see e.g. J.L. Greenstein, J.B. Okes and H.L. Shipman, *Astrophys. J.* **169**, 563 (1971).

[24]In the Sun, the pressure is only 10^{14} dyn for a mass density of the order of that of water.

A few years later, E.C. Stoner (1929) and W. Anderson (1930) remarked that, for sufficiently high densities ($\geq 10^6\,\mathrm{g/cm^3}$), the electrons were relativistic and hence the expression of their energy had to be changed in the Fermi–Dirac statistics [F. Jüttner (1928)]. This was done in 1934 by S. Chandrasekhar, who integrated numerically the equations of hydrostatic equilibrium with the first relativistic equation of state (Fermi–Dirac). He showed that beyond a certain *limiting mass* — the Chandrasekhar mass[25] — it was impossible to find any white dwarf in hydrostatic equilibrium.

After these beginnings of white dwarfs' history, these stars became a field of intensive theoretical and observational studies. In particular, their internal relativistic electron plasma was the object of important researches.

The original models of white dwarfs start from the usual hydrostatic equations for Newtonian equilibrium:

$$\begin{cases} \dfrac{d}{dr}M(r) = 4\pi r^2 \rho(r), \\[2mm] \dfrac{d}{dr}P(r) = -G\dfrac{M(r)}{r^2}\rho(r), \end{cases} \tag{7.145}$$

where $M(r)$ is the mass contained in a sphere of radius r, centered at the origin of the star, supposed to possess the spherical symmetry; $P(r)$ is the pressure at distance r from the center; and G is the gravitational constant. To this system, one must also add the boundary conditions

$$\begin{cases} M(0) = 0, \;\; P(R) = 0, \;\; \rho(0) = \rho_c, \\[2mm] M = \displaystyle\int_0^R 4\pi r^2 dr\, \rho(r). \end{cases} \tag{7.146}$$

The last condition defines the mass of the star, while the first one, i.e. $P(R) = 0$, specifies its radius; ρ_c, the central density, is a free parameter that specifies the star. It remains for one also to specify the equation of state obeyed by matter in the star.

And they show that the star is in hydrostatic equilibrium only below a critical mass, the *Chandrasekhar limit*, which corresponds to a critical number of nucleons N_{cr}, given by

$$N_{\mathrm{cr}} \approx 5.5 \times 10^{57} \left(\frac{Z}{A}\right)^2,$$

where the numerical factor comes from the various constants occurring in $E(R)$ (τ is the number of electron and A is the number of nucleons of a

[25] In the simplest case, this mass is of the order of 1.44 M_{Sun}.

typical star). It is obtained by taking the limit $R \to 0$ of $E(R_e)$ and gives rise to a total mass of the star of 1.44 M_{Sun}, for a He^4 composition. For other chemical compositions the Chandrasekhar limit is of course lower. Another fact that tends to diminish this critical mass is the β capture of electrons in the star

$$e + P \to N + \nu,$$

which is easy to understand: since this process reduces the number of electrons participating in the Fermi pressure, the star can sustain less mass. In Fig. 7.3, a plot of the radius versus the mass of a typical white dwarf has been given, with and without the effect of the neutronization process after T. Hamada and E. Salpeter (1961). Of course, this process tends to be important at high densities.

The above model is actually extremely simple and we have to mention several possible corrections to be made; in particular, the effects of temperature and of electrostatic interactions have to be examined. Also, two important physical effects in connection with the cooling of the star and the screening of ions are to be considered: they deal with the evaluation of the

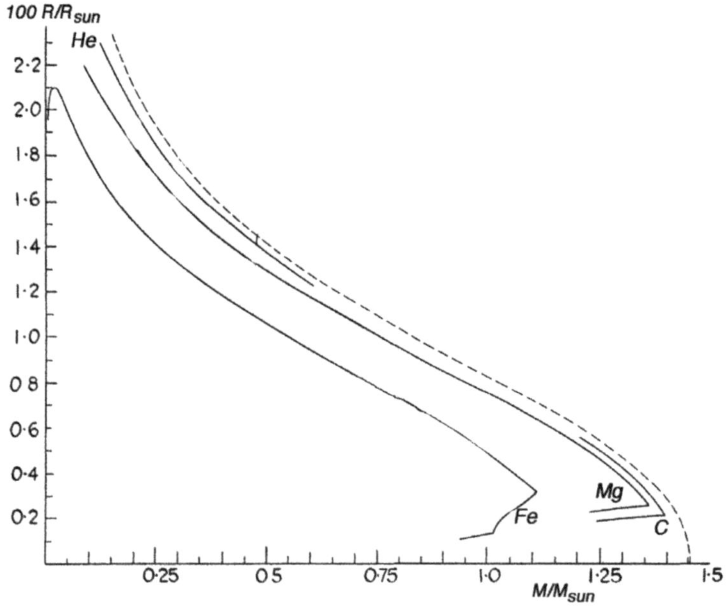

Fig. 7.3 The typical radius–mass curve of white dwarfs (a) without β capture (dashed line: Chandrasekhar curve) and (b) with neutronization for several compositions of the star (He, Mg, C, Fe) [after T. Hamada and E. Salpeter (1961)].

specific heat (at constant volume) of matter and the so-called *pycnonuclear* reactions.

7.5.1. *Cooling of white dwarfs*

The state of matter within the star depends on the various scales of energies within the system:

$$\begin{cases} \text{Fermi energy:} \varepsilon_F \\ \text{Thermal energy: } T \\ \text{Electrostatic interaction energy: } e^2/\langle r \rangle \approx e^2 n^{1/3} \end{cases}$$

and its cooling properties do depend strongly on the state of matter within the star. The luminosity — an important observable property — depends on the cooling of white dwarfs.

One of the most important physical aspects of white dwarfs is their luminosity, essentially because of the fact that — up to some physical assumptions — it leads to the age of the stars; and the question of their energy source[26] has been resolved by Mestel (1962). He showed that a satisfactory interpretation could be obtained by the assumption that the heat gathered in the core of the star could be filtered slowly through the nondegenerate envelope.

Let us briefly examine this point. The luminosity of the star is given by

$$L = 4\pi \int_0^R dr\, r^2 \rho(r) \left[\varepsilon_{\text{nucl}} - T\frac{\partial S}{\partial t}\right], \tag{7.147}$$

where $\varepsilon_{\text{nucl}}$ is the rate of energy produced by nuclear reactions (by second and by mass unit) and S is the entropy of the stellar material by unit of mass. Let us take a glance at the integrand of the latter integral; the last term can be put in the form

$$T\frac{\partial S}{\partial t} = T\left[\left.\frac{\partial S}{\partial T}\right|_\rho \frac{\partial T}{\partial t} + \frac{\partial S}{\partial \rho}\frac{\partial \rho}{\partial t}\right] \simeq C_V \frac{\partial T}{\partial t}, \tag{7.148}$$

where C_V is the specific heat at constant volume of the stellar material, and the last term has been neglected since the freezing of the star induces generally very small gravitational contraction.

[26]See, e.g. H.M. van Horn, *Cooling of White Dwarfs*, IAU Symposium No. 42, ed. W.J. Luyten (Reidel, Dordrecht, 1971).

The term $\varepsilon_{\rm nucl}$ will be neglected for other reasons (secular instability, etc.).

From the degeneracy of the electrons in the core of the white dwarf, one finds that their thermal conductivity is rather high. This can be realized by noting that the thermal conductivity λ can be estimated by

$$\lambda \approx \text{const} \times \ell \times n_{el} \times \frac{T}{\varepsilon_F} \times \frac{p_F}{m}, \tag{7.149}$$

with

$$\ell \approx \frac{1}{n_{\rm ions}} \sigma, \tag{7.150}$$

where ℓ is the average free path for collision ions–electrons and where σ, the effective cross-section of collision electrons–ions, can roughly be illustrated as

$$\sigma \approx \left(\frac{Ze^2}{\varepsilon_F} \right)^2 . \tag{7.151}$$

It follows that the average free path of the electrons effectively diffused is larger than that of nondegenerated electrons — the electrons whose energies are in the neighborhood of the Fermi energy; the diffusion of the others is inhibited by the Pauli principle. Consequently, the thermal conductivity is higher for the electrons close to the Fermi energy. Therefore, the core of the star can be considered as isothermal, at least roughly.

Finally, the relation which gives rise to the luminosity reduces to

$$L \approx -C_V M \frac{dT_C}{dt}, \tag{7.152}$$

where T_C is the temperature of the isothermal core.

Of course, the main problem is now to evaluate C_V, and one can roughly neglect the electron contribution since it contributes as T_C/ε_F. For low density and hot white dwarfs, such an approximation would not be completely correct; the thermal capacity of the electrons has then a substantial contribution.

In fact, we should have

$$\tau \approx C_V \frac{M}{L} T_C, \tag{7.153}$$

which, once included in radiative equilibrium, contains both observational data (L, M) and C_V, whose result does depend on the precise state of matter within the star.

7.5.2. *Pycnonuclear reactions*

In the presence of electrons and for high-enough densities, the heavy positively charged ions are screened and hence thermonuclear reactions will occur provided that the Coulomb barrier is lowered enough. Thus, beyond a critical temperature, which depends on the nature of the nucleus under study, some thermonuclear reactions can occur. In fact, inside white dwarfs, when the density is high enough, the electric field is screened, in particular by the electrons. It follows that, even at zero temperature, nuclear reactions can occur; such reactions are called pycnonuclear reactions.

Here are a few densities for the onset of some nuclear reactions to occur between elements:

H	He^4	C^{12}
$5 \cdot 10^4\,g/cm^3$	$8 \cdot 10^8\,g/cm^3$	$6 \cdot 10^9\,g/m^3$

The first figure shows that an average white dwarf cannot contain hydrogen in its interior since it has already been burnt out. Let us also note that all these critical densities are lower than the one calculated for the β capture: pycnonuclear reactions are produced before neutronization.

7.6. Functional Representation of the Partition Function

Although we do not use functional methods[27] in the following chapters, it is impossible to ignore the functional representation of the partition function for thermal equilibrium, especially owing to its general use and its necessity when one is dealing with gauge theories. Accordingly, a few simple elements[28] are now given still in the relativistic case and, in this subsection, only the case of a photon gas is dealt with. Furthermore, we shall not enter into the various subtleties and shall remain at a purely formal

[27] Apart from the references below, see I.M. Gelfand and A.M. Yaglom, *J. Math. Phys.* **1**, 48 (1960).

[28] See the excellent book by R.P. Feynman, *Statistical Mechanics: A Set of Lectures* (Benjamin, New York, 1972). See also R.P. Feynman and A.R. Hibbs, *Quantum Mechanics and Path Integrals* (McGraw-Hill, New York, 1965); J. Kapusta, *Finite Temperature Field Theory* (Cambridge University Press, 1989); M. LeBellac, *Thermal Field Theory* (Cambridge University Press, 2000); Ch. G. van Weert, *Statistical Field Theory: An Introduction to Real- and Imaginary-Time Thermal Field Theory* (lecture notes, Amsterdam, 2001); E. Alvarez, *Relativistic Many-Body Physics* (1984).

level. For more details see J. Kapusta (1989) and M. Le Bellac (2000). In a first reading this subsection can therefore be omitted and it is useful only when one is dealing with gauge invariance, where a number of techniques should be admitted.

7.6.1. *The partition function for gauge particles (photons)*

When one is dealing with the electromagnetic field, one has to face the problem of gauge invariance of the physical results obtained from calculations and approximations. The electromagnetic field $F^{\mu\nu}$, expressed in terms of the electromagnetic four-potential A^μ as

$$F^{\mu\nu}(x) = \partial^\mu A^\nu(x) - \partial^\nu A^\mu(x), \tag{7.154}$$

is obviously invariant under the *gauge transformations*

$$A^\mu(x) \to A^\mu(x) - \partial^\mu \Lambda(x). \tag{7.155}$$

Such an invariance must be obeyed by all physical quantities, including the density operator. On the other hand, the photon possesses two degrees of freedom — its two degrees of polarization — and hence can be described by the electromagnetic four-potential $A^\mu(x)$ only if there are imposed two conditions that suppress two of its four degrees of freedom. A well-known example is the Coulomb — or radiation — gauge conditions, which written in a manifestly covariant way read

$$\begin{cases} u_\mu A^\mu(x) = 0, \\ \Delta^\mu_\nu(u)\partial_\mu A^\nu(x) = 0, \end{cases} \tag{7.156}$$

where u^μ, in the case where matter is present, is the average four-velocity of the system under consideration. Such gauge conditions that leave only two *transverse* degrees of freedom for the electromagnetic field are generally called *physical gauges.* The Coulomb gauge breaks the Lorentz invariance since it *a priori* involves a timelike four-vector, u^μ. There exist, however, covariant gauges like

$$\partial_\mu A^\mu(x) = \chi(x) \quad \text{or} \quad A^\mu(x)A_\mu(x) = \text{const, etc.} \tag{7.157}$$

The first one, the Lorentz gauge, is among the most popular ones, while it would be quite awkward to use the second one, which is nonlinear and implies the manipulation of *ghosts.*[29]

---

[29] C. Nash, *Relativistic Quantum Fields* (Academic, New York, 1978).

A connected problem is the question of solving the equation obeyed by A^μ,

$$\Box A^\mu(x) - \partial^\mu(\partial_\nu A^\nu(x)) = J^\mu(x), \tag{7.158}$$

which results from Maxwell's equations. In Fourier space, it can indeed be rewritten as

$$\Delta^\mu_\nu(k) A^\nu(k) = -J^\mu(k), \tag{7.159}$$

and it cannot be solved for A^μ since $\Delta_{\mu\nu}(k)$ is a projector, except of course if one imposes some gauge condition and this amounts to solving the equation in the three-space orthogonal to k^μ. With the Lorentz gauge $\chi(x) \equiv 0$, the equations obeyed by A^μ read

$$\Box A^\mu(x) = J^\mu(x) \tag{7.160}$$

and can be solved in a straightforward way. Note that the Lorentz gauge imposes the condition $\Box\Lambda = 0$ on the gauge functions and, furthermore, still lets the electromagnetic field be endowed with three degrees of freedom, and we have to resort to other arguments to completely fix the gauge.

7.6.2. *The photons' partition function*

From conventional methods explained at the beginning of this chapter, the partition function of the blackbody radiation is easily found to be

$$\log Z = 2\int \frac{d^3k}{(2\pi)^3}\left\{-\frac{1}{2}\beta\omega(\mathbf{k}) - \log(1-\exp[-\beta\omega(\mathbf{k})])\right\}, \tag{7.161}$$

and a wrong result would be obtained. This is due to the fact that the functional integral used in the calculation, even though gauge-invariant, also involves an integration over the nonphysical degrees of freedom of the field A^μ:

$$Z = \int \mathcal{D}\{A^\mu\} \exp\left[-\int_0^\beta d^4x \mathcal{L}_E(A, \partial A)\right]. \tag{7.162}$$

A possible way out of this difficulty is to introduce in this ill-defined integral a constant factor, which does not break its invariance properties and allows the reduction of the unwanted degrees of freedom of A^μ. This can be done by inserting the factor

$$1 = \int \mathcal{D}\Lambda \delta[G(A_{(\Lambda)})] \det\left[\frac{\delta A_{(\Lambda)}}{\delta\Lambda}\right], \tag{7.163}$$

where $A^\mu_{(\Lambda)} \equiv A^\mu - \partial^\mu \Lambda$ is the transform of A^μ under a gauge transformation, and where $G(A_{(\Lambda)}) = 0$ is the chosen gauge condition. Note that this relation is the functional analog of the usual expression

$$1 = \int dx \delta[G(x)] \left| \frac{\partial G(x)}{\partial x} \right| = \int dG(x) \delta(G(x)). \tag{7.164}$$

The partition function then becomes

$$Z = \int \mathcal{D}A\mathcal{D}\Lambda \det \left[\frac{\delta A_{(\Lambda)}}{\delta \Lambda} \right] \delta[G(A_{(\Lambda)})] \exp \left\{ - \left[\int_0^\beta d^4x \mathcal{L}_E(A, \partial A) \right] \right\} \tag{7.165}$$

and is still gauge-invariant. It must also be noted that the insertion of the constant factor

$$1 = \int \mathcal{D}\Lambda \delta[G(A_{(\Lambda)}) - \gamma(x)] \det \left[\frac{\delta A_{(\Lambda)}}{\delta \Lambda} \right], \tag{7.166}$$

where $\gamma(x)$ is an arbitrary function independent of Λ, is quite equivalent to the one above. Accordingly, one also has

$$\begin{aligned} 1 &= \int \mathcal{D}\gamma \exp \left\{ -\frac{1}{2\xi} \int d^4x [G(A_{(\Lambda)})]^2 \right\} \\ &\quad \times \int \mathcal{D}\Lambda \delta[G(A_{(\Lambda)}) - \gamma(x)] \det \left[\frac{\delta A_{(\Lambda)}}{\delta \Lambda} \right] \\ &= \int \mathcal{D}\Lambda \exp \left\{ -\frac{1}{2\xi} \int d^4x [G(A_{(\Lambda)})]^2 \right\} \det \left[\frac{\delta A_{(\Lambda)}}{\delta \Lambda} \right], \end{aligned} \tag{7.167}$$

so that the partition function now reads

$$\begin{aligned} Z &= \int \mathcal{D}A\mathcal{D}\Lambda \det \left[\frac{\delta A_{(\Lambda)}}{\delta \Lambda} \right] \\ &\quad \times \exp \left\{ - \left[\int_0^\beta d^4x \mathcal{L}_E(A, \partial A) \right] + \frac{1}{2\xi} \int d^4x [G(A_{(\Lambda)})]^2 \right\}, \end{aligned} \tag{7.168}$$

in which the change of "variables"

$$A \to A' = A - \partial \Lambda \tag{7.169}$$

is performed so that, after using the gauge invariance of the electromagnetic Lagrangian $\mathcal{L}_E$ and relabeling A' as A, and integrating over the gauge function Λ, one obtains

$$Z = \int \mathcal{D}A \det\left[\frac{\delta A}{\delta\Lambda}\right] \exp\left\{-\left[\int_0^\beta d^4x \mathcal{L}_E(A, \partial A)\right] + \frac{1}{2\xi}\int d^4x [G(A)]^2\right\}, \tag{7.170}$$

which can be rewritten as a functional of the *effective* Lagrangian

$$\mathcal{L}_{\text{eff}} = \mathcal{L}_{\text{em}}(A, \partial A) + \frac{1}{2\xi}[G(A)]^2. \tag{7.171}$$

The parameter ξ is called the *gauge-fixing* parameter. As to the functional determinant occurring in the expression of Z, one can show that it can be cast into the form

$$\det\left[\frac{\delta A}{\delta\Lambda}\right] = \int \mathcal{D}\eta\mathcal{D}\bar{\eta} \exp[\bar{\eta}(x)M\eta(x)], \tag{7.172}$$

where M is an operator depending on the particular chosen gauge. For instance, in the Lorentz gauge, it has the form

$$M = \partial_\mu\partial_\mu. \tag{7.173}$$

The fields $\{\eta, \bar{\eta}\}$, the so-called Fadeev–Popov *ghosts*, have quite special properties[30]; in particular, they lead to a term that has the wrong sign in the effective Lagrangian:

$$\mathcal{L}_{\text{eff}} = \mathcal{L}_{\text{em}}(A, \partial A) + \frac{1}{2\xi}[G(A)]^2 + \partial\bar{\eta}\cdot\partial\eta. \tag{7.174}$$

7.6.3. *Illustration in the case of the Lorentz gauge*

In order to get further insights into what has been effected above, the case of Lorentz gauges is now considered:

$$G(A) \equiv \partial_\mu A^\mu(x) = \gamma(x). \tag{7.175}$$

In this case, the operator M reduces to ∂^2 and the effective Lagrangian is given by

$$\begin{aligned}\mathcal{L}_{\text{eff}} &= +\frac{1}{4}(\partial_\mu A_\nu - \partial_\nu A_\mu)(\partial_\mu A_\nu - \partial_\nu A_\mu)\\ &\quad - \frac{1}{\xi}(\delta_{\alpha\beta}\partial_\alpha A_\beta)^2 + \bar{\eta}\delta_{\alpha\beta}\partial_\beta\partial_\alpha\eta\end{aligned} \tag{7.176}$$

[30]Here, they are zero-spin-field but nevertheless they anticommute, thus not obeying the spin–statistics theorem. For details, see one the following books: C. Itzykson and J.B. Zuber, *Quantum Field Theory* (McGraw-Hill, New York, 1980); C. Nash, *loc. cit.*

(where repeated indices denote a summation and these equations still refer to the four-dimensional Euclidean space), so that the equations of motion for the field A read

$$\partial_\alpha \partial_\alpha A^\mu(x) - \frac{1}{\xi}\partial^\mu \partial_\alpha A_\alpha = 0, \tag{7.177}$$

which in Minkowski space read

$$\Box A^\mu(x) - \frac{1}{\xi}\partial^\mu \partial_\alpha A^\alpha = 0. \tag{7.178}$$

Several remarks are now in order. First, one sees that, in the case considered so far, the ghost field decouples from the photon field; this is not a general feature but it appears in this particular gauge. Next, the gauge invariance of subsequent averaged physical observables will result from their independence from the gauge-fixing parameter. Also, the operator

$$\eta_{\mu\nu} - \frac{1}{\xi}\partial_\mu \partial_\nu \tag{7.179}$$

is no longer a projector, as can be seen in Fourier space, and hence can be inverted.

From the preceding considerations on gauge problems, the method to be used consists in (i) choosing a gauge convenient for the calculations and (ii) checking the ξ independence of the results.

Let us now calculate the partition function of the blackbody radiation. The partition function reads

$$Z = \int \mathcal{D}A\mathcal{D}\bar{\eta}\mathcal{D}\eta \exp\left\{ - \int_0^\beta d^4x \left[\left(\frac{1}{2}\partial_\mu A_\nu \partial^\mu A^\nu - \partial_\mu A_\nu \partial^\nu A^\mu \right) \right.\right.$$
$$\left.\left. + \frac{1}{2\xi}(\partial_\alpha A_\alpha)^2 + \partial_\alpha \bar{\eta} \cdot \partial_\alpha \eta \right] \right\} \tag{7.180}$$

or, in Fourier space,

$$Z = \int \mathcal{D}A\mathcal{D}\bar{\eta}\mathcal{D}\eta \exp\left[- \frac{1}{2}\sum_n \sum_{\mathbf{k}} (\omega_n^2 + \mathbf{k}^2) A^2(\omega_n, \mathbf{k}) \right.$$
$$\left. + \left(1 - \frac{1}{2\xi}\right)(k \cdot A)^2 - 2\bar{\eta}(\omega_n^2 + \mathbf{k}^2)\eta \right]. \tag{7.181}$$

The first term of this expression corresponds to the partition function of free photons with an incorrect factor 4, which occurs because of the four degrees of freedom of A^μ. The ghost term, with the "wrong" sign, is also a free particle term and it contributes to the total partition function with a factor of -2, eliminating thereby the two spurious degrees of freedom of A^μ. The last term — the gauge-fixing term — can be chosen so as to vanish, since Z is gauge-invariant. Finally, one recovers the correct black body partition function.

Chapter 8

The Covariant Wigner Function

The Wigner function (1932) is the quantum analog of the usual distribution function on phase space with the difference that it is not always positive, particularly in domains of the size of h^3. Furthermore, because of the fact that it does not constitute a true probability density but rather a theoretical way for the computation of statistical data on quantum systems, it is not unique and many other definitions are possible.[1] In the relativistic context, the first uses of such a tool seem to be the ones by D. Biskamp (1967) and by R. Balescu (1968, 1969). However, not only are these relativistic Wigner functions not manifestly covariant but they also contain an unnecessary normal product in their definition: this normal product eliminates *ipso facto* all vacuum terms and hence some finite pertinent effects, after renormalization. Covariant, albeit not general, relativistic Wigner functions were introduced later by P.A. Carruthers and F. Zachariasen (1974, 1976, 1983), R. Hakim and R. Dominguez-Tenreiro (1976), and R. Hakim and J. Heyvaerts (1976). A covariant definition for spin 1/2 particles was provided by Ch. G. Van Weert and W.P.H. de Boer (1975). Finally, the full covariant definition occurred at about the same time [R. Hakim (1976, 1978)]. This was then given a gauge-invariant form [E.A. Remler (1977); V.V. Klimov (1982); J. Winter (1984); U. Heinz (1983, 1985); H.-Th. Elze, M. Gyulassy and D. Vasak (1986)]. An interesting attempt (see below) of a relativistic albeit nonmanifestly covariant, gauge-covariant Wigner function has also been made by I. Bialynicki-Birula, P. Gornicki and J. Rafelski (1991), with the aim of calculating the pair

[1] See L. Cohen, *J. Math. Phys.* **7**, 781 (1967), for the nonrelativistic possible Wigner-like functions.

production rate in a homogeneous electric field. Let us also add that the covariant Wigner function technique has been generalized to curved spaces [E. Calzetta and B.L. Hu (1987, 1989); E. Calzetta, S. Habib and B.L. Hu (1988)] and applied, for example, to cosmology [E. Calzetta and B.L. Hu (1989)].

8.1. The Covariant Wigner Function for Spin 1/2 Particles

As in the nonquantum case, the starting point of the construction of a covariant Wigner function is the data of the basic observables of spin 1/2 particles, namely the four-current

$$J^{\mu}(x) = \bar{\psi}(x)\gamma^{\mu}\psi(x) \tag{8.1}$$

and the energy–momentum tensor

$$T^{\mu\nu} = \frac{i}{2}\bar{\psi}(x)\gamma^{\mu}\overleftrightarrow{\partial}^{\nu}\psi(x), \tag{8.2}$$

where ψ is the fermion field and the γ's are the usual Dirac matrices. More generally, the four-current of an observable $A^{\cdots}$ is given by

$$J_A^{\mu\cdots}(x) = \bar{\psi}(x)\gamma^{\mu}A^{\cdots}\psi(x) \tag{8.3}$$

and their quantum-statistical average is given by

$$\langle J_A^{\mu\cdots}(x)\rangle = \mathrm{Tr}\,\{\rho J_A^{\mu\cdots}(x)\}, \text{ etc.}, \tag{8.4}$$

where ρ is the density operator discussed in Chap. 7.

Let us now introduce the covariant Wigner function operator as

$$F_{\mathrm{op}}(x,p) = \frac{1}{(2\pi)^4}\int d^4R\exp(-ip\cdot R)\bar{\psi}\left(x+\frac{1}{2}R\right)\otimes\psi\left(x-\frac{1}{2}R\right), \tag{8.5}$$

and the covariant Wigner function as its quantum-statistical average value:

$$\begin{aligned} F(x,p) &= \langle F_{\mathrm{op}}(x,p)\rangle = \mathrm{Tr}\,\{\rho F_{\mathrm{op}}(x,p)\} \\ &= \frac{1}{(2\pi)^4}\int d^4R\exp\left(-ip\cdot R\right)\left\langle\bar{\psi}\left(x+\frac{1}{2}R\right)\otimes\psi\left(x-\frac{1}{2}R\right)\right\rangle. \end{aligned} \tag{8.6}$$

Note that $F(x,p)$ is a matrix in the spinorial (and possibly internal) indices; explicitly, one has

$$F_{\mathrm{op}\,\alpha\beta}(x,p) = \frac{1}{(2\pi)^4}\int d^4R \exp\left(-ip\cdot R\right)\bar{\psi}_\beta\left(x+\frac{1}{2}R\right)\psi_\alpha\left(x-\frac{1}{2}R\right). \tag{8.7}$$

Notice the opposite position of the indices for the ψ's and for F.

With this definition, the above average values of observables are given by

$$\langle J^\mu(x)\rangle = \mathrm{Sp}\int d^4p\,\gamma^\mu F(x,p), \tag{8.8}$$

$$\langle T^{\mu\nu}(x)\rangle = \mathrm{Sp}\int d^4p\,\gamma^\mu p^\nu F(x,p), \tag{8.9}$$

where Sp indicates a trace over spinorial (and possibly internal) indices. As to the four-current of a given general observable A, it has also the form

$$\langle J_A^{\mu\cdots}(x)\rangle = \mathrm{Sp}\int d^4p\,\gamma^\mu A^{\cdots}(x,p)F(x,p), \tag{8.10}$$

although in each particular case one has to find out the specific form of the function $A^{\cdots}(x,p)$. In actual practice, only $\langle J^\mu(x)\rangle$ and $\langle T^{\mu\nu}\rangle$ have to be calculated, and possibly their fluctuations.

Conversely, one has the useful inverse formula

$$\bar{\psi}(x)\otimes\psi(y) = \int d^4p \exp\left[ip\cdot(x-y)\right]F_{\mathrm{op}}\left(\frac{1}{2}(x+y);p\right), \tag{8.11}$$

which is repeatedly used in the various calculations involving the covariant Wigner function.

The one-particle covariant Wigner function $F(x,p)$ can be expanded on the basis of the 16 Dirac matrices

$$\{\gamma_A\}_{A=1,2,\ldots,16} = \left(I,\gamma^\mu,\sigma_{\mu\nu}\equiv\frac{i}{2}\left[\gamma^\mu,\gamma^\nu\right],\gamma_5,\gamma_5\gamma^\mu\right) \tag{8.12}$$

as

$$\begin{aligned}F(x,p) &= \frac{1}{4}\sum_A f_A(x,p)\gamma^A\\ &\equiv \frac{1}{4}\left[f(x,p)I + f^\mu(x,p)\gamma_\mu + if^{\mu\nu}(x,p)\sigma_{\mu\nu}\right.\\ &\quad\left.\times\ f_5(x,p)\gamma^5 + f_{5\mu}(x,p)\gamma^5\gamma^\mu\right],\end{aligned} \tag{8.13}$$

with

$$f_A(x,p) = \frac{1}{4}\mathrm{Tr}\left[F(x,p)\gamma_A\right]. \tag{8.14}$$

Besides the one-particle Wigner function, other quantum distributions are needed, such as the two-body covariant Wigner function

$$F_2(x,p;x',p') = \int \frac{d^4R}{(2\pi)^4}\frac{d^4R'}{(2\pi)^4} \exp(-ip\cdot R)\exp(-ip'\cdot R')$$
$$\times \left\langle \bar{\psi}\left(x' + \frac{1}{2}R'\right) \otimes \bar{\psi}\left(x + \frac{1}{2}R\right)\right.$$
$$\left. \otimes \psi\left(x - \frac{1}{2}R\right) \otimes \psi\left(x' - \frac{1}{2}R'\right)\right\rangle, \tag{8.15}$$

whose cluster decomposition can be written as

$$F_2(x,p;x',p') = F_1(x,p) \times F_1(x',p') + g_2(x,p;x',p'), \tag{8.16}$$

where the correlation function $g_2(x,p;x',p')$ contains both the correlations occurring because of ordinary interactions and the exchange correlation. For instance, in the Hartree–Fock approximation, one should retain in $g_2(x,p;x',p')$ only its exchange part. We come back below to this question. Finally, notice the useful inverse expression

$$\bar{\psi}(x)\bar{\psi}(y)\psi(z)\psi(u) = \int d^4\xi d^4\xi' \exp[i\xi\cdot(y-z) + i\xi'\cdot(x-u)]$$
$$\times F_{2\text{op}}\left(\frac{1}{2}(y+z),\xi;\frac{1}{2}(x+u),\xi'\right). \tag{8.17}$$

Of course, three-body, ..., N-body Wigner functions can be defined, but so far we have not used them.

8.1.1. *Basic equations*

Let us now examine the equations obeyed by the Wigner function $F(x,p)$ and, in order to be specific, let us consider noninteracting particles.[2] Then the fermion field ψ satisfies the Dirac equations

$$\begin{cases} [i\gamma\cdot\partial - m]\,\psi(x) = 0, \\ \bar{\psi}(x)\,[i\gamma\cdot\partial + m] = 0, \end{cases} \tag{8.18}$$

which, once considered respectively at points $x - R/2$ and $x + R/2$, after multiplying the first one by $\bar{\psi}\left(x + \frac{1}{2}R\right)\exp(-ip\cdot R)$ from the left and the

[2]See also the analysis of this case in S.R. de Groot, W.A. van Leeuwen and Ch. G. van Weert (1980).

second one by $\psi\left(x-\frac{1}{2}R\right)\exp(-ip\cdot R)$ from the right, and integrating over R, lead to

$$\begin{cases} \{i\gamma\cdot\partial+2\left[\gamma\cdot p-m\right]\}F(x,p)=0, \\ F(x,p)\left\{i\gamma\cdot\overleftarrow{\partial}-2\left[\gamma\cdot p-m\right]\right\}=0, \end{cases} \tag{8.19}$$

where the operator $\{\ldots\}$ in the second equation acts on the left. There are two equations for F, since they represent together both the mass shell on which the particles lie and the statistical state of the system. Note that, in the nonquantum case, there are two equations as well: the Liouville equation and the mass shell equation.

One can get some further insights from the expansion of $F(x,\,p)$ on the matrices γ_A of the Diracs algebra and taking the trace of these equations. One obtains the following 32 equations after they have been added and substracted:

$$mf=p_\mu f^\mu, \tag{8.20}$$

$$\partial_\nu f^{\mu\nu}+p^\mu f-mf^\mu=0, \tag{8.21}$$

$$\tfrac{1}{2}\partial^{[\mu}f^{\nu]}-2mf^{\mu\nu}-p_\lambda f_{5\rho}\varepsilon^{\rho\lambda\mu\nu}=0, \tag{8.22}$$

$$\partial_\mu f_5^\mu+2mf_5=0, \tag{8.23}$$

$$\partial^\mu f_5-2p_\lambda\varepsilon^{\lambda\rho\nu\mu}f_{\rho\nu}-2mf^{5\mu}=0, \tag{8.24}$$

$$\partial_\mu f^\mu=0, \tag{8.25}$$

$$\partial^\mu f+4p_\lambda f^{\lambda\mu}=0, \tag{8.26}$$

$$\tfrac{1}{2}\partial_\lambda f_{5\rho}\varepsilon^{\rho\lambda\mu\nu}+p^{[\mu}f^{\nu]}=0, \tag{8.27}$$

$$p_\mu f_5^\mu=0, \tag{8.28}$$

$$\partial_\lambda f_{\rho\nu}\varepsilon^{\lambda\rho\nu\mu}+2p^\mu f_5=0. \tag{8.29}$$

Let us now investigate some of these equations and let us begin with the first one. When the system possesses u^μ as its only macroscopic four-vector, f^μ has the general form

$$f^\mu=ap^\mu+bu^\mu, \tag{8.30}$$

where a and b are functions of p. Furthermore, when the system is homogeneous and the particles are such that $p^2=m^2$, $f^\mu\propto p^\mu$. As a consequence, the second equation shows that $\partial_\nu f^{\mu\nu}\propto p^\nu$. The seventh equation, after it has been contracted with p_μ, yields $p\cdot\partial f=0$, which is nothing but the one-particle Liouville equation , suggesting thereby that f is more or less a quantum analog of the relativistic distribution function. Also, the integration over p of $\partial_\mu f^\mu = 0$ provides the four-current conservation equation $\partial_\mu J^\mu = 0$.

The second of these equations is interesting: if we go back to the field form (via an inverse Fourier transform), it reads

$$\langle \bar{\psi}\gamma^\mu\psi\rangle = \left\langle \frac{i}{2m}\bar{\psi}\overleftrightarrow{\partial}^\mu\psi \right\rangle + \frac{i}{2m}\partial_\nu \left\langle \left[\bar{\psi}\sigma^{\mu\nu}\psi\right]\right\rangle, \tag{8.31}$$

which is nothing but the Gordon decomposition of the four-current into a *convective* part (the first one on the right hand side) and a *spin* part (the second one). It follows that, when the spin effects can be neglected, one has

$$\langle \bar{\psi}\gamma^\mu\psi\rangle \approx \left\langle \frac{i}{m}\bar{\psi}\overleftrightarrow{\partial}^\mu\psi \right\rangle, \tag{8.32}$$

which, in terms of the Wigner function, is expressed as

$$f^\mu(x,p) \approx p^\mu f(x,p). \tag{8.33}$$

This shows that, when the negligence of spin is permitted, Eq. (8.33) can be chosen. Let us now look at the stationary and homogeneous solutions to the basic equations satisfied by $F(p)$, which then read

$$\begin{cases} [\gamma\cdot p - m]\,F(p) = 0, \\ F(p)\,[\gamma\cdot p - m] = 0, \end{cases} \tag{8.34}$$

and multiplying, for example, the first equation by $[\gamma\cdot p + m]$, one gets

$$\left[p^2 - m^2\right] F(p) = 0, \tag{8.35}$$

which shows that

$$F(p) \propto \delta\left(p^2 - m^2\right), \tag{8.36}$$

and hence that all its components, f_A's, are on the particle mass shell. Also, the general solution to these equations has the form

$$F(p) = [\gamma\cdot p + m]\,A(p)\,[\gamma\cdot p + m], \tag{8.37}$$

where $A(p)$ is an arbitrary 4×4 matrix. Furthermore, the above analysis indicates that when u^μ is the only macroscopic four-vector present in the system, $f^\mu \propto p^\mu$ and $F(x,\, p)$ has the general form

$$F(p) = \frac{\gamma\cdot p + m}{m} f(p). \tag{8.38}$$

Below, another stationary solution of interest is given in connection with polarized media.

When u^μ is the only possible four-vector in the theory, $F(p)$ has the form

$$F(p) = \frac{1}{4}\left\{ f(p)I + \frac{p\cdot\gamma}{m} f(p) - \frac{i}{2m}\varepsilon_{\rho\lambda\mu\nu}p^\lambda\sigma^{\mu\nu} f_5^\rho(p) + f_{5\mu}(p)\gamma_5\gamma^\mu \right\}, \tag{8.39}$$

which results from simple manipulations of the equations for $F(p)$. Indeed, $f_5(p) = 0$ since there is no available pseudoscalar. Therefore, the most general stationary Wigner function of a stationary system depends on only two unknown functions, namely $f(p)$ and $f_{5\mu}(p)$, to be determined from the explicit data of the density operator.

Finally, note also the Fourier transform of the Wigner equation

$$\begin{cases} \left[\gamma\cdot\left(p - \frac{1}{2}k\right) - m\right] F_{\rm op} = 0, \\ F_{\rm op}\left[\gamma\cdot\left(p + \frac{1}{2}k\right) - m\right] = 0, \end{cases} \tag{8.40}$$

which will be of much use in the sequel.

8.1.2. *The equilibrium Wigner function for free fermions*

Because of the general form obtained above for stationary solutions, only

$$f_{\rm eq}(p) = \frac{1}{(2\pi)^4}\int d^4R \exp\left(-ip\cdot R\right)\left\langle \bar{\psi}\left(x + \frac{1}{2}R\right)\psi\left(x - \frac{1}{2}R\right)\right\rangle_{\rm eq} \tag{8.41}$$

has to be calculated [R. Hakim and J. Heyvaerts (1978)]. The field ψ is expanded into plane waves,

$$\psi(x) = \int \frac{d^3p}{\sqrt{2p_0}}\left[a(\mathbf{p})u(\mathbf{p})\exp\left(+ip\cdot x\right) + b^+(\mathbf{p})\bar{v}(\mathbf{p})\exp\left(-ip\cdot x\right)\right], \tag{8.42}$$

where $\bar{u}\cdot u = 2m = \bar{v}\cdot v$ are the free spinors, and with the usual commutation relations and the relations

$$\begin{cases} \langle a^+(\mathbf{p})a(\mathbf{p}')\rangle_{\rm eq} = \dfrac{\delta^{(3)}(\mathbf{p}-\mathbf{p}')}{\exp\left[\beta\left(p\cdot u - \mu\right)\right]+1}, \\ \langle b^+(\mathbf{p})b(\mathbf{p}')\rangle_{\rm eq} = \dfrac{\delta^{(3)}(\mathbf{p}-\mathbf{p}')}{\exp\left[\beta\left(p\cdot u + \mu\right)\right]+1} \end{cases} \tag{8.43}$$

are used; the final result is

$$f_{\mathrm{eq}}(p) = d\frac{\delta\left[p^2 - m^2\right]}{(2\pi)^3}\left(\frac{1}{\exp\left[\beta\left(p\cdot u - \mu\right)\right]+1} + \frac{1}{\exp\left[\beta\left(p\cdot u + \mu\right)\right]+1} - \theta(-p\cdot u)\right), \tag{8.44}$$

where d is a degeneracy factor that takes account of spin (then $d = 2$) and other possible internal degrees of freedom. The first term is the contribution of the particles, the second is the antiparticle term, and the third is a vacuum term. This last one is absent when the definition of the Wigner function contains a normal product; however, after a renormalization — as we shall see in a later chapter — it leads to finite effects. Finally, the equilibrium Wigner function $F_{\mathrm{eq}}(p)$ can be written in a compact form:

$$F_{\mathrm{eq}}(p) = d\left[\gamma\cdot p + m\right]\frac{\delta(p^2 - m^2)}{(2\pi)^3}\frac{\mathrm{sgn}(p\cdot u)}{\exp[\beta(p\cdot u - \mu)]+1}, \tag{8.45}$$

where the function $\mathrm{sgn}(p\cdot u)$ is the usual sign function.

8.1.3. *Polarized media*

Let us consider another case of particular interest, the one in which there exists a macroscopic pseudovector n^μ, i.e. so that

$$\begin{aligned} n^\mu n_\mu &= -1, \\ n^\mu u_\mu &= 0. \end{aligned} \tag{8.46}$$

We now want to find out the general form for $F(x,p)$, still for a stationary solution, from the general structure of the system of equations it satisfies.

From the system obeyed, one immediately finds that the following relations hold:

$$\begin{aligned} f^\mu(p) &= \frac{p^\mu}{m}f(p), \\ f^{\mu\nu}(p) &= \frac{1}{2m}\varepsilon^{\mu\nu\alpha\beta}p_\alpha f_{5\beta}(p), \\ f_5^\mu(p) &= \frac{1}{m}\varepsilon^{\mu\nu\rho\lambda}p_\nu f_{\rho\lambda}(p), \\ f_5(p) &= 0. \end{aligned} \tag{8.47}$$

Since there exists a unique pseudovector in the theory — say, $S^\mu(p)$ — one can always write

$$f_5^\mu(p) = S^\mu(p)f(p), \tag{8.48}$$

with $p_\mu S^\mu(p) = 0$ as a consequence of $p_\mu f_5^\mu(p) = 0$. It follows that the general solution to our system of 32 equations is of the form

$$F(p) = \frac{\gamma \cdot p + m}{2m} \frac{1 + \gamma_5 \gamma_\mu S^\mu(p)}{2} f(p). \tag{8.49}$$

Note that, while the first factor of this expression is the projection over the positive energy states, the second factor cannot be interpreted as a projection over the spin states since *a priori*

$$S_\mu(p) S^\mu(p) \neq -1. \tag{8.50}$$

Note also that these two matrix operators commute due to the orthogonality of p^μ and $S^\mu(p)$.

At this point, the (pseudo-)four-vector $S^\mu(p)$ is completely arbitrary, for two reasons. First, it depends on the polarization of the system whatever its definition. Next, this arbitrariness reflects that of the Dirac spinors of the free particle. Accordingly, a physical choice must be made as to $S^\mu(p)$, which is rewritten as

$$\begin{cases} S^\mu(p) \equiv S(p) N^\mu(p), \\ N^\mu(p) N_\mu(p) = -1. \end{cases} \tag{8.51}$$

$S(p)$ can easily be shown to be directly connected with the polarization of the medium through the density operator, whose spin part reads

$$\begin{aligned} \rho_{\text{spin}} &= \xi(p) \frac{1 + \gamma_5 \gamma_\mu N^\mu(p)}{2} + [1 - \xi(p)] \frac{1 - \gamma_5 \gamma_\mu N^\mu(p)}{2} \\ &= \frac{1 + [2\xi(p) - 1] \gamma_5 \gamma_\mu N^\mu(p)}{2}, \end{aligned} \tag{8.52}$$

where $\xi(p)$ is the percentage of spin-up particles with four-momentum p. As a result, one has

$$S(p) = 2\xi(p) - 1. \tag{8.53}$$

Given a unit and constant spacelike pseudo-four-vector n^μ, orthogonal to the average four-velocity u^μ, one of the simplest *choices* for $N^\mu(p)$ is

$$N^\mu(p) = \frac{u^{[\mu} n^{\nu]} p_\nu}{\left[(u \cdot p)^2 - (n \cdot p)^2 \right]^{1/2}}, \tag{8.54}$$

whose physical meaning is that it represents a *global* spin quantization axis.

This choice is, in fact, a consequence of a simple analysis of the way a system of charged particles is usually polarized. Suppose, indeed, that the system under consideration is placed in a magnetic field (see Chap. 12):

the various spins align along its direction, thus leading to a — more or less, according to the temperature — polarized system. If the magnetic field is now switched off, the system becomes metastable and depolarizes more or less rapidly.[3] The study of electrons embedded in a magnetic field (Chap. 12) precisely provides this $N^\mu(p)$: if we now look at the Wigner $f_5^\mu(p)$ *in the presence of a magnetic field*, we can see (Chap. 12) that *it is proportional* to $N^\mu(p)$ where n^μ is the space direction of the magnetic field. The *choice* (8.54) is, accordingly, quite natural.

At this point, it should be noted that although spin does not commute with the Dirac Hamiltonian, our choice is quite sensible since we do not deal with a true equilibrium but a metastable one.

In order to get some insight into $S(p)$, let us evaluate the various macroscopic quantities $S^{\mu\nu\lambda}$ (spin density tensor) and M^μ (polarization four-vector). We have

$$\begin{aligned} M^\mu &= \int d^4p f_5^\mu(p) \\ &= \int d^4p \frac{S(p)u^{[\mu}n^{\nu]}p_\nu}{\left[(u\cdot p)^2 - (n\cdot p)^2\right]^{1/2}} f_{\text{eq}}(p), \end{aligned} \tag{8.55}$$

so that with the choice suggested by the magnetic field case,

$$\begin{cases} S(p) = \mathcal{P}[(u\cdot p)^2 - (n\cdot p)^2], \\ \mathcal{P} = \text{const}, \end{cases} \tag{8.56}$$

M^μ is easily evaluated as

$$M^\mu = -\mathcal{P}n^\mu n_{\text{eq}} m, \tag{8.57}$$

so that

$$\begin{cases} \mathcal{P} = \dfrac{1}{mn_{\text{eq}}} n_\mu M^\mu, \\ M^\mu u_\mu = 0, \end{cases} \tag{8.58}$$

which shows that M^μ is always parallel to the spin quantization axis while $\mathcal{P}$ is the polarization of the medium. The choice (8.54) has been used after previous results of the magnetic field case (see Chap. 12).

From M^μ, one gets[4]

$$S^{\mu\nu\lambda} = -\frac{1}{2}\varepsilon^{\mu\nu\lambda\alpha}M_\alpha, \tag{8.59}$$

[3]For He^3, the relaxation time is of the order of a couple of days.

[4]For a study of polarization and connected questions, see e.g. R. Hagedorn, *Relativistic Kinematics* (Benjamin, New York, 1963).

so that, locally, the spin tensor reads

$$M^{\mu\nu} \equiv S^{\mu\nu\lambda} u_\lambda = -\frac{1}{2}\varepsilon^{\mu\nu\lambda\alpha} u_\lambda M_\alpha. \tag{8.60}$$

In the rest frame of the system, where $u^\mu = (1, \mathbf{0})$, and taking n^μ along the third axis, the only nonvanishing component of $M^{\mu\nu}$ is M^{12}. Finally, our different choices appear to describe correctly a polarized medium in (metastable) thermodynamic equilibrium. Other choices are of course possible but they deal with systems prepared in particular ways: particles endowed with four-momentum p contain a prescribed p-dependent percentage of spin-up particles, etc.

8.2. Equilibrium Fluctuations of Fermions

Let us consider the thermodynamic potential Ω:

$$\Omega = -\frac{1}{\beta}\ln\{\mathrm{Tr}\exp(-\beta[H_0 + H_{\mathrm{int}} - \mu B])\}, \tag{8.61}$$

where B is the baryonic number (or charge) operator, μ the chemical potential, H_0 the free Hamiltonian and H_{int} the interaction Hamiltonian,

$$H_{\mathrm{int}} = \int dx\, J_{\mathrm{op}}(x) \cdot A(x), \tag{8.62}$$

in the case of a relativistic quantum plasma (J_{op} is the four-current operator, which is proportional to e; A is the electromagnetic four-potential). One can show that[5]

$$\Omega = \Omega_{\mathrm{free}} - \frac{2}{(2\pi)^3}\int_0^e de \int \frac{d^4k}{k^2}\langle J_{\mathrm{op}} \cdot J_{\mathrm{op}}\rangle_{(k)}, \tag{8.63}$$

an expression that exhibits the role of the four-current fluctuations in the calculation of the thermodynamics of an electromagnetic plasma. In order to calculate one-particle fluctuations like those of the four-current, of the energy–momentum tensor, etc., the quantity needed is the fluctuations of the Wigner function itself,

$$\begin{aligned} F_{\alpha\beta\mu\nu}(x, x'; p, p') \equiv \langle [&F_{\mathrm{op}\,\alpha\beta}(x,p) - F_{\alpha\beta}(x,p)] \\ &\times [F_{\mathrm{op}\,\mu\nu}(x',p') - F_{\mu\nu}(x',p')]\rangle, \end{aligned} \tag{8.64}$$

which we calculate for free particles in this section. We introduce the notations

$$\tilde{A}_{\mathrm{op}} \equiv A_{\mathrm{op}} - \langle A_{\mathrm{op}}\rangle, \tag{8.65}$$

[5]A. Fetter J. Walecka, *Quantum Theory of Many Particle System* (McGraw-Hill, New York, 1971).

so that the various fluctuations of interest read

$$\begin{cases} \langle \tilde{J}^{\mu}_{\rm op}(x)\tilde{J}^{\nu}_{\rm op}(x')\rangle = e^2{\rm Sp}\iint d^4p d^4p' \gamma^{\mu}\gamma^{\nu'}F(x,x';p,p'), \\ \langle \tilde{T}^{\mu\nu}_{\rm op}(x)\tilde{T}^{\alpha\beta}_{\rm op}(x')\rangle = {\rm Sp}\iint d^4p d^4p' \gamma^{\mu}p^{\nu}\gamma^{\alpha'}p'^{\beta}F(x,x';p,p'), \\ \langle \tilde{T}^{\mu\nu}_{\rm op}(x)\tilde{J}^{\alpha}_{\rm op}(x')\rangle = e{\rm Sp}\iint d^4p d^4p' \gamma^{\mu}p^{\nu}\gamma^{\alpha'}F(x,x';p,p'). \end{cases} \tag{8.66}$$

Since we are merely interested in a normal equilibrium state, one can assume that the system is invariant under space–time translation and hence $F(x, x'; p, p')$ has the property

$$F_{\rm eq}(x,x';p,p') = F_{\rm eq}(0,x'-x;p,p'), \tag{8.67}$$

where the spectrum of the fluctuating quantities can be obtained via the Fourier transform of $F_{\rm eq}(0,x;p,p')$:

$$F_{\rm eq}(k;p,p') = \int d^4x \exp(ik\cdot x)F_{\rm eq}(0,x;p,p'). \tag{8.68}$$

Let us now calculate the fluctuations of the Wigner function of the ideal Fermi gas in thermal equilibrium. We can write

$$\begin{aligned} &F_{\rm eq}(k;p,p') \\ &\quad = \int d^4x \exp(ik\cdot x)\langle \tilde{F}(0,p)\tilde{F}(x,p')\rangle \\ &\quad = \frac{1}{(2\pi)^8}\int d^4R\, d^4R'\, d^4x \exp(ik\cdot x - ip\cdot R - ip'\cdot R') \\ &\quad\quad \times \left\langle \bar{\psi}\left(\frac{1}{2}R\right)\otimes\psi\left(-\frac{1}{2}R\right)\otimes\bar{\psi}\left(x+\frac{1}{2}R'\right)\otimes\psi\left(x-\frac{1}{2}R'\right)\right\rangle. \end{aligned} \tag{8.69}$$

Using Wick's theorem[6] written in the form

$$\langle\bar{\psi}(1)\psi(2)\bar{\psi}(3)\psi(4)\rangle = \langle\bar{\psi}(1)\psi(2)\rangle\langle\bar{\psi}(3)\psi(4)\rangle + \langle\bar{\psi}(1)\psi(4)\rangle\langle\bar{\psi}(2)\psi(3)\rangle \tag{8.70}$$

and denoting by ${}^+F(x,p)$ the quantity

$$\begin{aligned} {}^+F(x,p) &= \frac{1}{(2\pi)^4}\int d^4R \exp(-ip\cdot R)\,\psi\left(x-\frac{1}{2}R\right)\otimes\bar{\psi}\left(x+\frac{1}{2}R\right) \\ &= -F(x,p) - \frac{i}{(2\pi)^4}\bar{S}(p), \end{aligned} \tag{8.71}$$

[6] G.C. Wick, *Phys. Rev.* **80**, 268 (1950).

where $\bar{S}(p)$ is the Fourier transform of the Dirac field anticommutator[7]

$$S(R) = i\left\{\psi\left(x + \frac{1}{2}R\right), \bar{\psi}\left(x - \frac{1}{2}R\right)\right\}, \tag{8.72}$$

one finds that

$$F_{\text{eq}}(k; p, p') = (2\pi)^4 \delta'^{(4)}(p - p') F_{\text{eq}}\left(p + \frac{1}{2}k\right) {}^{+}F_{\text{eq}}\left(p - \frac{1}{2}k\right) \tag{8.73}$$

or, equivalently,

$$\begin{aligned} F_{\text{eq}}(k; p, p') = (2\pi)^4 \delta'^{(4)}(p - p') F_{\text{eq}}\left(p + \frac{1}{2}k\right) \\ \times \left[F_{\text{eq}}\left(p - \frac{1}{2}k\right) + \frac{i}{(2\pi)^4}\bar{S}\left(p - \frac{1}{2}k\right)\right]. \end{aligned} \tag{8.74}$$

From the form[8] of $\bar{S}(p)$,

$$\bar{S}(p) = 2\pi i\,[\gamma \cdot p + m]\,\varepsilon\left(p^0\right)\delta\left(p^2 - m^2\right), \tag{8.75}$$

and that of $F_{\text{eq}}(p)$,

$$F_{\text{eq}}(p) = \frac{1}{(2\pi)^3}\varepsilon\left(p^0\right)\delta\left(p^2 - m^2\right)[\gamma \cdot p + m]\frac{1}{\exp\left(\beta[p \cdot u - \mu]\right) + 1}, \tag{8.76}$$

the final result is

$$\begin{aligned} F_{\text{eq}}(k; p, p') = &-\frac{1}{(2\pi)^2}\frac{\delta^{(4)}(p - p')}{\exp(\beta\omega) - 1} \\ &\times \left[\gamma \cdot \left(p + \frac{1}{2}k\right) + m\right] \otimes \left[\gamma \cdot \left(p - \frac{1}{2}k\right) + m\right] \\ &\times \varepsilon\left(p_0^2 - \frac{1}{4}\omega^2\right)\delta\left[\left(p + \frac{1}{2}k\right)^2 - m^2\right]\delta\left[\left(p - \frac{1}{2}k\right)^2 - m^2\right] \\ &\times \left[\frac{1}{\exp\left\{\beta\left[\left(p + \frac{1}{2}k\right) - \mu\right]\right\}} - \frac{1}{\exp\left\{\beta\left[\left(p - \frac{1}{2}k\right) - \mu\right]\right\}}\right], \end{aligned} \tag{8.77}$$

where $\varepsilon(p^0)$ is the sign function of p^0.

---

[7] There should not be confusion between this anticommutator and the polarization.

[8] N. Bogoliubov and D.V. Shirkov, *Introduction to the Theory of Quantized Fields* (Interscience, New York, 1959).

The explicit calculation of $\langle \tilde{J}^{\mu}_{\rm op}(x)\tilde{J}^{\nu}_{\rm op}(x')\rangle$ and of $\langle \tilde{T}^{\mu\nu}_{\rm op}(x)\tilde{T}^{\alpha\beta}_{\rm op}(x')\rangle$ have been made by H.D. Sivak (1985), who used them in the calculation of the dispersion relation of a quantum relativistic plasma. After some elementary calculations, one finds that

$$\begin{aligned}\langle J^{\mu}J^{\nu}\rangle_{\rm eq}(k) &= \frac{e^2}{(2\pi)^2}\frac{1}{\exp(\beta\omega)-1}\\ &\quad\times\sum_{n,\ell=\pm 1} n\ell\int\frac{d^3p}{\sqrt{m^2+p^2}\sqrt{m^2+(p+nk)^2}}(\delta_{\ell 1}-n_{\rm eq})\\ &\quad\times\delta\left[\sqrt{m^2+p^2}+\ell\sqrt{m^2+(p+nk)^2}+n\omega\right]\\ &\quad\times\left[\frac{1}{2}k^2\eta^{\mu\nu}+2p^{\mu}p^{\nu}+np^{(\mu}k^{\nu)}\right],\end{aligned} \tag{8.78}$$

where

$$n_{\rm eq}=\sum_{\pm}\frac{1}{\exp[\beta(E\mp\mu)]+1}. \tag{8.79}$$

This formula calls for several remarks. Firstly, the term $\delta_{\ell 1}-n_{\rm eq}$. The δ term is the vacuum fluctuation: it is present even though the material term, i.e. $n_{\rm eq}$, is zero. When matter is present, this term is attenuated because of the Pauli principle. The term that contains $\delta\big(\sqrt{m^2+p^2}+\ell\sqrt{m^2+(p+nk)^2}+n\omega\big)$ is an energy conservation term and comes from two different processes: fluctuations are nonzero whenever

$$\omega=\pm\left[\sqrt{m^2+p^2}+\sqrt{m^2+(p+nk)^2}\right],\quad \ell=1, \tag{8.80}$$

corresponding to either pair creation or annihilation by an electromagnetic wave of frequency ω, or

$$\omega=\pm\left[\sqrt{m^2+p^2}-\sqrt{m^2+(p+nk)^2}\right],\quad \ell=-1, \tag{8.81}$$

which expresses the existence of transitions to different forms of energy states of the electrons or of the positrons.

8.3. A Simple Example

As a simple example, we treat the case of a spin 1/2 field with a quartic self-interaction,

$$\mathcal{L}=\bar{\psi}(x)\frac{i}{2}\gamma\cdot\overleftrightarrow{\partial}\psi(x)-m\bar{\psi}(x)\psi(x)-g\left|\bar{\psi}(x)\psi(x)\right|^2, \tag{8.82}$$

which is of course nonrenormalizable, but this is immaterial since our only goal is to provide an illustration of the Wigner function technique. The basic equations for the covariant Wigner operator are easily written as

$$\{i\gamma\cdot\partial + 2\,[\gamma\cdot p - m]\}\,F_{\rm op}(x,p)$$
$$= -2g\int\frac{d^4R}{(2\pi)^4}d^4\xi\,d^4\xi'\,\exp\left[-i(p-\xi')\cdot R\right]F_{\rm 2op}\left(x,\xi;x-\frac{1}{2}R,\xi'\right), \tag{8.83a}$$

$$F_{\rm op}(x,p)\left\{i\gamma\cdot\overleftarrow{\partial} - 2\,[\gamma\cdot p - m]\right\}$$
$$= +2g\int\frac{d^4R}{(2\pi)^4}d^4\xi\,d^4\xi'\exp\left[-i(p-\xi')\cdot R\right]F_{\rm 2op}\left(x+\frac{1}{2}R,\xi;x,\xi'\right), \tag{8.83b}$$

which connect the one-particle Wigner operator to the two-particle one. Such a situation — quite analogous to that of the Newtonian BBGKY hierarchy — is quite exceptional, as the various examples given below show. One can also find an equation for F_3, which involves higher order Wigner functions, etc. The hierarchy can be cut with a supplementary assumption such as the Hartree–Vlasov ansatz

$$F_2(x,p;x',p') \approx F_1(x,p)F_1(x',p') \tag{8.84}$$

or any other. F_3 is connected to F_4 and so on.

8.4. The BBGKY Relativistic Quantum Hierarchy

In order to illustrate the above techniques in a more concrete way, it will be applied here to the so-called "scalar plasma," first studied by G. Kalman (1967, 1974). This toy model can also be considered as a particular case of the phenomenological model for nuclear matter proposed by J.D. Walecka (see Chap. 9). In addition, it has been considered by T.D. Lee and others[9] in dealing with the question of "abnormal nuclear matter," and by J. Rafelski[10] in the so-called "SLAC bag" model.[11]

---

[9] T.D. Lee and G.C. Wick, *Phys. Rev.* **D9**, 229 (1974); T.D. Lee and M. Margulies, *Phys. Rev.* **D11**, 1591 (1975); T.D. Lee, *Rev. Mod. Phys.* **47**, 267 (1975); G.C. Källmann, *Phys. Letts.* **B55**, 178 (1975); E.M. Nyman and M. Rho, *Nucl. Phys.* **A268**, 408 (1976); S.A. Moszowski and C.G. Källmann, *Nucl. Phys.* **A287**, 495 (1977); M. Wakamatsu and A. Hayashi, *Prog. Theor. Phys.* **63**, 1688 (1980).

[10] J. Rafelski, *Phys. Rev.* **D9**, 2358 (1974).

[11] A. Bardeen, M. Chanowitz, R. Giles, M. Weinstein and V.F. Weisskopf, *Phys. Rev.* **D9**, 3471 (1974).

The scalar plasma consists of a baryon field $\psi(x)$ coupled to a scalar field $\phi(x)$ and is described by the Lagrangian

$$\mathcal{L} = \frac{1}{2} i\bar{\psi}\gamma^{\mu} i\overleftrightarrow{\partial}_{\mu}\psi - \bar{\psi}(m - g\varphi)\psi + \frac{1}{2}\left(\partial_{\mu}\varphi \cdot \partial^{\mu}\varphi - \mu^2\varphi^2\right), \qquad (8.85)$$

where m is the mass of the scalar particles and g their coupling constant with the baryon field.

As in the nonquantum case, the BBGKY hierarchy is of growing complexity, involving more and more new Wigner functions and/or combination with fields. However, in practice, one employs the first few equations of the hierarchy and its use appears less involved.

Using the equations of motion

$$\begin{cases} i\gamma \cdot \partial\psi(x) - [m - g\varphi(x)]\,\psi(x) = 0, \\ i\partial\bar{\psi}(x) \cdot \gamma + \bar{\psi}(x)\,[m - g\varphi(x)] = 0, \\ \left(\Box + \mu^2\right)\varphi(x) = g\bar{\psi}(x)\psi(x), \end{cases} \qquad (8.86)$$

and the definition of the one-particle Wigner operator, one is led to the system

$$\begin{cases} \{i\gamma \cdot \partial + 2\,[\gamma \cdot p - m]\}\, F_{\mathrm{op}}(x,p) \\ \quad = -2g \displaystyle\int \frac{d^4R}{(2\pi)^4} d^4\xi \exp\left[-i(p-\xi)\cdot R\right] F_{\mathrm{op}}(x,\xi)\varphi\left(x - \frac{1}{2}R\right), \\ F_{\mathrm{op}}(x,p)\left\{i\gamma \cdot \overleftarrow{\partial} - 2\,[\gamma \cdot p - m]\right\} \\ \quad = +2g \displaystyle\int \frac{d^4R}{(2\pi)^4} d^4\xi \exp\left[-i(p-\xi)\cdot R\right] \varphi\left(x + \frac{1}{2}R\right) F_{\mathrm{op}}(x,\xi), \\ \left(\Box + \mu^2\right)\varphi(x) = g\mathrm{Sp}\displaystyle\int d^4p F_{\mathrm{op}}(x,p), \end{cases} \qquad (8.87)$$

hereafter referred to as the *generating equations* of the relativistic quantum BBGKY hierarchy, which is formally rewritten as

$$\begin{cases} LF_{\mathrm{op}} = g\displaystyle\int F_{\mathrm{op}}\varphi, \\ L^{\circ}F_{\mathrm{op}} = g\displaystyle\int \varphi F_{\mathrm{op}}, \\ KG\varphi = g\mathrm{Sp}\displaystyle\int F_{\mathrm{op}}, \end{cases} \qquad (8.88)$$

in order to facilitate the subsequent developments. After taking the average value of this system, one finds that

$$\begin{cases} LF = g \int \langle F_{\text{op}}\varphi \rangle, \\ FL^{\circ} = g \int \langle \varphi F_{\text{op}} \rangle, \\ KG\langle \varphi \rangle = g\text{Sp} \int F, \end{cases} \tag{8.89}$$

which is now briefly discussed. First, it connects F to $\langle F_{\text{op}}\varphi \rangle$ and to $\langle \varphi F_{\text{op}} \rangle$, showing that the one-particle Wigner function needs the knowledge of another function. Next, the first two equations of the system must be consistent with each other and thus must be such that the following equality is always satisfied:

$$L^{-1} \int \langle F_{\text{op}}\varphi \rangle = \int \langle \varphi F_{\text{op}} \rangle L^{\circ -1}. \tag{8.90}$$

In order to calculate $\langle F_{\text{op}}\varphi \rangle$, the first equation of the generating system is multiplied by φ from the right and then averaged; this yields

$$L\langle F_{\text{op}}\varphi \rangle = g \int \langle F_{\text{op}}\varphi\varphi \rangle \tag{8.91}$$

and the knowledge of $\langle F_{\text{op}}\varphi \rangle$ demands the data of $\langle F_{\text{op}}\varphi\varphi \rangle$, etc. Similarly, from the third equation of the generating equation of the hierarchy, with an analogous manipulation, one obtains

$$KG\langle F_{\text{op}}\varphi \rangle = g\text{Sp} \int \langle F_{\text{op}}F_{\text{op}} \rangle, \tag{8.92}$$

from which one gets

$$\langle F_{\text{op}}\varphi \rangle = gKG^{-1}\text{Sp} \int \langle F_{\text{op}}F_{\text{op}} \rangle \tag{8.93}$$

and hence

$$LF = g^2 KG^{-1}\text{Sp} \int \langle F_{\text{op}}F_{\text{op}} \rangle. \tag{8.94}$$

Note that $\langle F_{\text{op}}F_{\text{op}} \rangle$, which is not the two-particle Wigner function of the baryons, is not related to the latter through $\langle F_{\text{op}}F_{\text{op}} \rangle \neq \langle F_2 \rangle$, as remarked in the case of the calculation of fluctuations. The equations for $\langle F_{\rm op}F_{\rm op}\rangle$ involve new functions, such as $\langle F_{\rm op}F_{\rm op}\varphi\rangle$.

Finally, as in the classical case, the quantum BBGKY hierarchy cannot be solved unless some approximation scheme stops its growing complexity. The simplest way out is the Hartree approximation, which is the analog of the classical Vlasov ansatz

$$\langle F_{\rm op}\varphi\rangle \approx F\langle\varphi\rangle, \tag{8.95}$$

which breaks the hierarchy at its very beginning. The next approximation, by far more involved, consists of retaining two-operator averages and neglecting three-body correlations whatsoever.

A detailed example of the use of this hierarchy is given in the next chapter, including renormalization.

8.5. Perturbation Expansion of the Wigner Function

An alternative BBGKY hierarchy can be found from the Yang–Feldman[12] form of the equations of motion for the fields, i.e. from

$$\begin{cases} \psi(x) = \psi_0(x) + g\displaystyle\int d^4x' S_{\rm ret}(x-x')\varphi(x')\psi(x'), \\ \bar\psi(x) = \bar\psi_0(x) + g\displaystyle\int d^4x' \bar\psi(x')\bar S_{\rm ret}(x-x')\varphi(x'), \\ \varphi(x) = \varphi_0(x) + g\displaystyle\int d^4x' \Delta_{\rm ret}(x-x')\bar\psi(x')\psi(x'), \end{cases} \tag{8.96}$$

which result from a formal integration of the equations of motion of the fields. In this system, $S_{\rm ret}(x)$ is the elementary solution to the free Dirac equation (the retarded propagator) and $\Delta_{\rm ret}(x)$ the elementary solution to the free Klein–Gordon equation (the retarded propagator); the index zero indicates a free solution to the field equations at $t=-\infty$. Notice that these free field solutions obey the free field commutation relations

$$\begin{aligned} \left[\psi_0(x), \overline{\psi_0}(x')\right]_+ &= S(x-x'), \\ \left[\varphi_0(x), \varphi_0(x')\right]_- &= \Delta(x-x'), \text{etc.} \end{aligned} \tag{8.97}$$

[12] C.N. Yang and D. Feldman, *Phys. Rev.* **79**, 972 (1950).

Using once more the definition of F_{op}, the Yang–Feldman equations provide the generating system

$$\begin{aligned}
F_{\text{op}}(x,p) &= F_{0\text{op}}(x,p) + g\int \frac{d^4R}{(2\pi)^3} d^4x'\, d^4\xi \\
&\quad \times \exp\left[-iR\cdot\left(p-\frac{1}{2}\xi\right) + i\xi\cdot(x-x')\right] S_{\text{ret}}\left(x-\frac{1}{2}R-x'\right) \\
&\quad \times F\left(\frac{1}{2}\left[x'+x+\frac{1}{2}R\right];\xi\right)\varphi(x') + g\int \frac{d^4R}{(2\pi)^3} d^4x'\, d^4\xi\, \varphi(x') \\
&\quad \times F\left(\frac{1}{2}\left[x'+x-\frac{1}{2}R\right];\xi\right)\bar{S}_{\text{ret}}\left(x+\frac{1}{2}R-x'\right) \\
&\quad - g^2\int \frac{d^4R}{(2\pi)^3}\, d^4x'\, d^4x''\, d^4\xi \exp\left[-iR\cdot p + i\xi\cdot(x'-x'')\right] \\
&\quad \times \varphi(x')\bar{S}_{\text{ret}}\left(x+\frac{1}{2}R-x'\right) F\left(\frac{1}{2}\left[x'+x''\right],\xi\right) \\
&\quad \times S_{\text{ret}}\left(x-\frac{1}{2}R-x''\right)\varphi(x''),
\end{aligned} \tag{8.98}$$

which can be symbolically rewritten as

$$\begin{aligned}
F_{\text{op}}(x,p) &= F_{0\text{op}}(x,p) + g^2\int \left(\bar{S}_{\text{ret}}F_{\text{op}}\varphi + \varphi F_{\text{op}}S_{\text{ret}}\right) \\
&\quad - g^2\int \varphi\bar{S}_{\text{ret}}F_{\text{op}}S_{\text{ret}}\varphi.
\end{aligned} \tag{8.99}$$

If one introduces

$$\begin{aligned}
\varphi(x) &= \varphi_0(x) + g\int d^4x'\Delta_{\text{ret}}(x-x')\bar{\psi}(x)\psi(x) \\
&= \varphi_0(x) + g\int d^4p d^4x'\Delta_{\text{ret}}(x-x')F(x,p)
\end{aligned} \tag{8.100}$$

into the above generating equation, one is left with a generating equation that contains only the F's and φ_0 and whose general structure is of the form

$$\begin{aligned}
F_{\text{op}}(x,p) &= F_{0\text{op}}(x,p) + \int F_{\text{op}}F_{\text{op}} + \int F_{\text{op}}F_{\text{op}}F_{\text{op}} \\
&\quad + \text{terms involving } \{\varphi_0, F_0\},
\end{aligned} \tag{8.101}$$

which is particularly suitable for a perturbation expansion and its graph representation: it allows the perturbative expansion of $F(x,p)$ and $\varphi(x)$:

$$\begin{cases} \varphi(x) = \sum_{n=0}^{\infty} g^n \varphi_{(n)}(x), \\ F(x,p) = \sum_{n=0}^{\infty} g^n F_{(n)}(x,p), \end{cases} \tag{8.102}$$

The natural — but not unique — choice for the initial distribution $\langle F_{\text{op}} \rangle$ and field $\langle \varphi_0 \rangle$ is

$$F_0(x,p) = F_{\text{eq}}(p), \quad \langle \varphi_0(x) \rangle \equiv 0; \tag{8.103}$$

it corresponds to a free equilibrium of the fermions at $t = \pm\infty$ without any coherent state of the bosons. Note also that similar considerations for the boson field show that the initial value of the product $\langle \varphi_0(x'')\varphi_0(x') \rangle$ is also required. If one requires that the system be in equilibrium in the past infinity, then this last average value is directly connected to the Bose–Einstein function of the (free) bosons

$$\langle \varphi_0(x')\varphi_0(x'') \rangle = \int d^4\xi \exp\left[i\xi \cdot (x' - x'')\right] f_{\text{eq}}(\xi). \tag{8.104}$$

For instance, at order g^2, this works exactly as indicated on the equations

$$\begin{cases} F_{(2)} = g^2 \int \left(\bar{S}_{\text{ret}} F_{(0)} \varphi_{(0)} + \varphi_{(0)} F_{(0)} S_{\text{ret}}\right) - g^2 \int \varphi_{(0)} \bar{S}_{\text{ret}} F_{(0)} S_{\text{ret}} \varphi_{(0)}, \\ \varphi_{(1)} = g \int \Delta_{\text{ret}} F_{(0)}, \end{cases} \tag{8.105}$$

and the same continues indefinitely.

8.6. The Wigner Function for Bosons

The covariant Wigner function for bosons is defined exactly the same as it was for fermions with only one small difference; namely, we have to substract out the contribution of the average field. We thus have the following definition for the one-boson Wigner function:

$$\begin{aligned} f(x,p) = \frac{1}{(2\pi)^4} \int d^4R \exp\left(-ip \cdot R\right) \left[\phi^*\left(x + \frac{1}{2}R\right) - \left\langle \phi^*\left(x + \frac{1}{2}R\right)\right\rangle\right] \\ \times \left[\phi\left(x - \frac{1}{2}R\right) - \left\langle \phi\left(x - \frac{1}{2}R\right)\right\rangle\right], \end{aligned} \tag{8.106}$$

where, for simplicity, we have considered a scalar field whose possible internal indices are implicit and hence products of the field are implicitly tensor products in the internal space.

For such a field, the main observables are the four-current

$$J^{\mu}_{\text{op}}(x) = \frac{i}{2}\phi^*(x)\overleftrightarrow{\partial}^{\mu}\phi(x) \tag{8.107}$$

and the energy–momentum tensor

$$T^{\mu\nu}_{\text{op}}(x) = \frac{i}{2}\partial^{(\mu}\phi^*(x)\partial^{\nu)}\phi(x) - \mathcal{L}\eta^{\mu\nu}, \tag{8.108}$$

where L is the Lagrangian and whose average values, written in terms of the bosons' Wigner function, read

$$J^{\mu}(x) = \int d^4p\, p^{\mu} f(x,p), \tag{8.109}$$

$$T^{\mu\nu} = \int d^4p\, p^{\mu}p^{\nu} f(x,p). \tag{8.110}$$

If the equation of motion of the field $\phi(x)$ is written in the form

$$\left(\Box + \mu^2\right)\phi(x) = S(x), \tag{8.111}$$

where $S(x)$ contains the effects of possible other fields and/or $\phi(x)$ itself, then the Wigner function can be shown to obey the equations [P. Carruthers and F. Zachariasen (1976, 1983)]

$$\begin{aligned} p\cdot\partial f_{\text{op}}(x,p) = \frac{1}{2}\int \frac{d^4R}{(2\pi)^4}\exp\left(-ip\cdot R\right)&\left(\left\{\phi^*\left(x+\frac{1}{2}R\right)S\left(x-\frac{1}{2}R\right)\right\}\right.\\ &\left.-\left\{S^*\left(x+\frac{1}{2}R\right)\phi\left(x-\frac{1}{2}R\right)\right\}\right), \end{aligned} \tag{8.112}$$

$$\begin{aligned} &\left(p^2-\mu^2+\frac{1}{4}\Box\right)f_{\text{op}}(x,p)\\ &\quad= \frac{1}{2}\int \frac{d^4R}{(2\pi)^4}\exp\left(-ip\cdot R\right)\times\left[\left\{\phi^*\left(x+\frac{1}{2}R\right)S\left(x-\frac{1}{2}R\right)\right\}\right.\\ &\quad\quad\left.+\left\{S^*\left(x+\frac{1}{2}R\right)\phi\left(x-\frac{1}{2}R\right)\right\}\right], \end{aligned} \tag{8.113}$$

which result immediately from the definition of the Wigner function and from the equations of motion. For a free field, where $S(x)\equiv 0$, this system reduces to

$$\begin{cases} p\cdot\partial f(x,p) = 0,\\ \left(p^2-\mu^2\right)f(x,p) = 0, \end{cases} \tag{8.114}$$

which shows — as in the spin 1/2 case — that the first equation is connected with the flow in phase space while the second one concerns the mass shell of the particles.

The equilibrium Wigner distribution for scalar bosons is given by

$$f_{\mathrm{eq}}(p) = \frac{\delta\left(p^2 - m^2\right)}{(2\pi)^3}\left\{\frac{1}{\exp(\beta u \cdot p) - 1} - \frac{1}{2}\theta\left(-u \cdot p\right)\right\}, \tag{8.115}$$

while for a complex scalar field it is given by

$$\begin{aligned} f_{\mathrm{eq}}(p) &= \frac{\delta\left(p^2 - m^2\right)}{(2\pi)^3}\left(\frac{\theta\left(p \cdot u\right)}{\exp\left[\beta\left(u \cdot p - \mu\right)\right] - 1}\right. \\ &\quad \left. + \frac{\theta\left(p \cdot u\right)}{\exp\left[\beta\left(u \cdot p + \mu\right)\right] - 1} - \theta\left(-p \cdot u\right)\right) \\ &= \frac{\delta\left(p^2 - m^2\right)}{(2\pi)^3}\frac{\mathrm{sgn}\left(p \cdot u\right)}{\exp\left[\beta\left(u \cdot p - \mu\right)\right] - 1}, \end{aligned} \tag{8.116}$$

where these equilibrium distributions can easily be obtained from the definition of the bosons' Wigner function definition, from the decomposition of the free fields into plane waves

$$\phi(x) = \frac{1}{(2\pi)^{3/2}}\int \frac{d^3k}{\sqrt{2\omega_{\mathbf{k}}}}\left\{a(\mathbf{k})\exp[-ik \cdot x] + a^+(\mathbf{k})\exp[+ik \cdot x]\right\},$$
$$\text{(real scalar field)} \tag{8.117}$$

$$\phi(x) = \frac{1}{(2\pi)^{3/2}}\int \frac{d^3k}{\sqrt{2\omega_{\mathbf{k}}}}\left\{a_{(\mathbf{k})}\exp(+ik \cdot x) + b^+_{(\mathbf{k})}\exp(-ik \cdot x)\right\},$$
$$\text{(complex scalar field)} \tag{8.118}$$

from the usual equilibrium value

$$\left\langle a^+_{(\mathbf{k})} a_{(\mathbf{k})}\right\rangle = \frac{1}{\exp\left[\beta\omega(\mathbf{k}) - 1\right]} \quad \text{(real scalar field)}, \tag{8.119}$$

$$\left\langle a^+_{(\mathbf{k})} a_{(\mathbf{k})}\right\rangle = \frac{1}{\exp\{\beta\left[\omega(\mathbf{k}) - \mu\right] - 1\}} \quad \text{(complex scalar field)} \tag{8.120}$$

and from the comutation relations

$$\left[a^+_{(\mathbf{k})}, a_{(\mathbf{k}')}\right] = \delta^{(3)}\left(\mathbf{k} - \mathbf{k}'\right), \text{etc.} \tag{8.121}$$

8.6.1. *The example of the $\lambda\varphi^4$ theory*

The case of bosons is best illustrated on the particularly simple[13] model of the $\lambda\varphi^4$ theory, whose Lagrangian is

$$\mathcal{L} = \frac{1}{2}\partial\varphi \cdot \partial\varphi - \mu^2\varphi^2 - \frac{\lambda}{4!}\varphi^4, \tag{8.122}$$

so that, in this case, one has

$$S(x) = -\frac{\lambda}{3!}\varphi^3(x), \tag{8.123}$$

and thus the above equation turns out to be

$$p \cdot \partial f_{\mathrm{op}}(x,p) = -\frac{\lambda}{2.3!}\int \frac{d^4R}{(2\pi)^4}\exp\left(-ip\cdot R\right)\left(\left\{\varphi\left(x+\frac{1}{2}R\right)\right.\right.$$
$$\left.\left.\times\,\varphi^3\left(x-\frac{1}{2}R\right)\right\} - \left\{\varphi^3\left(x+\frac{1}{2}R\right)\phi\left(x-\frac{1}{2}R\right)\right\}\right), \tag{8.124}$$

so that

$$p \cdot \partial f_{\mathrm{op}}(x,p) = -\frac{\lambda}{2.3!}\int \frac{d^4R}{(2\pi)^4}\exp\left[-i(p-\xi)\cdot R\right]$$
$$\times\left\{f_{2\mathrm{op}}\left(x,\xi;x-\frac{1}{2}R,\xi'\right) - f_{2\mathrm{op}}\left(x,\xi;x+\frac{1}{2}R,\xi'\right)\right\}, \tag{8.125}$$

where the two-particle Wigner function operator is defined as

$$f_{2\mathrm{op}}(x,p,x',p') = \int \frac{d^4R}{(2\pi)^4}\frac{d^4R'}{(2\pi)^4}\exp\left(-ip\cdot R - ip'.R'\right)\varphi\left(x+\frac{1}{2}R\right)$$
$$\times\,\varphi\left(x'+\frac{1}{2}R'\right)\varphi\left(x'-\frac{1}{2}R'\right)\varphi\left(x-\frac{1}{2}R\right), \tag{8.126}$$

while the second equation, obeyed by f_{op}, reads

$$\left(p^2 - \mu^2 + \frac{1}{4}\Box\right) f_{\mathrm{op}}(x,p)$$
$$= \frac{\lambda}{2.3!}\int \frac{d^4R}{(2\pi)^4}\exp\left[-i(p-\xi)\cdot R\right]\times\left\{f_{2\mathrm{op}}\left(x,\xi;x-\frac{1}{2}R,\xi'\right)\right.$$
$$\left. + f_{2\mathrm{op}}\left(x,\xi;x+\frac{1}{2}R,\xi'\right)\right\}. \tag{8.127}$$

Thus, taking both sides' average, we obtain an equation that connects f_1 and f_2. Then, multiplying for instance by $f_{1\mathrm{op}}$ and averaging, we obtain an

[13]For simplicity, we use $\langle\varphi\rangle = 0$.

equation connecting f_1, f_2, f_3 and others. And so on: we therefore realize that a complete hierarchy is obtained.

Of course, it cannot be solved except when it is closed by some assumption, such as the Hartree–Vlasov one:

$$f_2(x,p;x',p') \approx f_1(x,p)f_1(x',p'). \tag{8.128}$$

For instance, in the case of equilibrium the last equation of the first order equations of the hierarchy reads

$$[p^2 - \mu^2]f_1(p) = \frac{\lambda}{3!}\int \frac{d^4R}{(2\pi)^4} d^4\xi' \exp\left[-i(p-\xi)\cdot R\right] f_1(\xi)f_1(\xi') \tag{8.129}$$

or

$$M^2 - \mu^2 = \frac{\lambda}{3!}\frac{1}{(2\pi)^3}\int \frac{d^3\xi}{\sqrt{p^2+M^2}}\left\{\frac{1}{\exp\left(\beta_\mu p^\mu\right)-1} + \frac{1}{2}\right\}, \tag{8.130}$$

where M is the effective mass of the bosons (see Chap. 13). Note that the term $1/2$ is infinite and thus should be renormalized.

8.6.2. *Four-current fluctuations of the complex scalar field*

We now evaluate the fluctuations of the four-current operator in the case of the complex scalar field φ, assuming for simplicity the case $\langle\varphi\rangle = 0$. The four-current operator is given by

$$J^\mu_{\rm op}(x) = \int d^4p\, p^\mu f_{\rm op}(x,p), \tag{8.131}$$

so that the four-current fluctuation tensor $J^{\mu\nu}(x,x')$ is given by

$$J^{\mu\nu}(x,x') \equiv \langle J^{\mu\nu}_{\rm op}(x,x')\rangle = \int d^4p\, d^4p' p^\mu p'^\nu \langle f_{\rm op}(x,p) f_{\rm op}(x',p')\rangle, \tag{8.132}$$

and thus it is sufficient to calculate $\langle f_{\rm op}(x,p)f_{\rm op}(x',p')\rangle$ in order to get the result with the *bonus* of the possibility of calculating all one-particle operator fluctuations, such as the energy–momentum fluctuations. However, this path is not followed here and the calculation of the four-current fluctuations is directly performed by replacing the plane wave expansion of the complex field in the definition of $J^{\mu\nu}(x,x')$. A straightforward calculation then provides (see details in Chap. 13)

$$J^{\mu\nu}(x-x',0) = (2\pi)^4 \int\int d^4k\, d^4p \exp\left[ik\cdot(x-x')\right] p^\mu p^\nu \times \frac{\Delta(p+\frac{1}{2}k)\Delta(p-\frac{1}{2}k)}{\exp(\beta\omega)-1}\left\{n\left(p-\frac{1}{2}k\right) - n\left(p+\frac{1}{2}k\right)\right\}, \tag{8.133}$$

where $n(p \pm \frac{1}{2}k)$ is the usual Bose–Einstein factor and $\Delta(p)$ the commutator of the scalar field.

8.7. Gauge Properties of the Wigner Function

For the sake of simplicity, the gauge properties of the covariant Wigner functions are studied in the case of a QED plasma only. Under a gauge transformation of the electromagnetic four-vector A^μ,

$$A^\mu(x) \to A^\mu(x) + \partial^\mu \Lambda(x), \tag{8.134}$$

where $\Lambda(x)$ is an arbitrary function, a Dirac spinor is changed as

$$\begin{cases} \psi(x) \to \psi(x) \exp\left[+ie\Lambda(x)\right], \\ \bar{\psi}(x) \to \bar{\psi}(x) \exp\left[-ie\Lambda(x)\right], \end{cases} \tag{8.135}$$

so that the Wigner function is changed as

$$\begin{aligned} F(x,p) \to \int \frac{d^4R}{(2\pi)^4} \exp\left(-ip \cdot R\right) \exp\left\{ ie\left[\Lambda\left(x + \frac{1}{2}R\right) - \Lambda\left(x - \frac{1}{2}R\right)\right]\right\} \\ \times \left\langle \bar{\psi}\left(x + \frac{1}{2}R\right) \otimes \psi\left(x - \frac{1}{2}R\right)\right\rangle, \end{aligned} \tag{8.136}$$

which can be rewritten as

$$\begin{aligned} F(x,p) \to \int \frac{d^4R}{(2\pi)^4} d^4\xi \exp\left[-i(p-\xi)\cdot R\right] \\ \times \exp\left\{ ie\left[\Lambda\left(x + \frac{1}{2}R\right) - \Lambda\left(x - \frac{1}{2}R\right)\right]\right\} F(x,\xi). \end{aligned} \tag{8.137}$$

8.7.1. *Gauge-invariant Wigner functions*

In order to eliminate the problem of proving the gauge invariance of the results in each case, a gauge-invariant relativistic Wigner function has been defined by several authors [E.A. Remler (1977); V.V. Klimov (1982); J. Winter (1984); U. Heinz (1985); H.-T. Elze, M. Gyulassy and D. Vasak (1986a,b); see also A.V. Selikhov (1988); H.T. Elze and U. Heinz (1989)]. Such a formalism is very elegant and will probably have interesting developments; however, at the present moment, it leads to very involved equations for which approximation schemes are not easy to work out. Furthermore, if a gauge-covariant formalism has been erected for gauge bosons (photons, gluons, etc.), ghosts are not yet fully taken into account. For details, references and discussions, see the review article by H. Th. Elze and U. Heinz (1989).

In this subsection, a one-particle gauge-invariant Wigner function is defined in the case of QED only.[14] There are three reasons for this. Firstly, QED plasmas are relatively well studied and constitute a "laboratory" for gauge theories. Secondly, no problems of ghosts are to be dealt with, when using linear gauges[15] in QED. Finally, by treating the case of a QED plasma rather than the general case of a quark–gluon plasma, the use of an unnecessarily complex algebra is avoided without losing the basic ideas.

Let us start from the definition of the one-particle Wigner operator

$$F_{\mathrm{op}}(x,p) = \frac{1}{(2\pi)^4}\int d^4R \exp\left(-ip\cdot R\right)\bar{\psi}(x)\left[\frac{i}{2}\overleftrightarrow{\partial} - eA(x)\right]\psi(x) \quad (8.138)$$

it can be rewritten as

$$F_{\mathrm{op}}(x,p) = \frac{1}{(2\pi)^4}\int d^4R \exp\left(-ip\cdot R\right)\exp\left(+i\frac{1}{2}D\cdot R\right)\bar{\psi}\left(x\right)$$

$$\otimes \exp\left(-i\frac{1}{2}D\cdot R\right)\psi\left(x\right), \quad (8.139)$$

where D is the ordinary differential operator $D_\mu = -i\partial_\mu$, which is the generator of space–time translations. Those authors first rewrite the above Wigner operator as

$$F_{\mathrm{op}}(x,p) = \frac{1}{(2\pi)^4}\int d^4R \exp\left(-ip\cdot R\right)\bar{\psi}(x)\exp\left(+i\frac{1}{2}D^+\cdot R\right)$$

$$\otimes \exp\left(-i\frac{1}{2}D\cdot R\right)\psi\left(x\right), \quad (8.140)$$

where D^+ is nothing but the same operator as D acting on the left. One might think that D should be replaced by the kinetic (gauge-invariant) four-momentum

$$\begin{cases} D_\mu \to \partial_\mu + ieA_\mu, \\ D_\mu^+ \to \partial_\mu - ieA_\mu, \end{cases} \quad (8.141)$$

defining a "gauge-invariant" translation operator, suggesting thereby a way to reach gauge invariance; however, some care is needed when one is applying this operator to the spinors ψ and $\bar{\psi}$, and it must be carefully defined.

---

[14]In another subsection, the gauge-invariant Wigner function studied by E.A. Remler (1977) is briefly examined.
[15]See e.g. C. Nash, *Relativistic Quantum Fields* (Academic, New York, 1978).

Let us now examine this line of thought with some details and, to this end, we shall closely follow the articles devoted by D. Vasak, M. Gyulassy and H. Th. Elze (1986) and by P. Zhuang and U. Heinz (1996) to the Abelian plasmas. They constitute extensions and simplifications of earlier, more developed works.[16]

These authors first remarked that if one introduces a factor $f(A;x,R)$ such that

$$\begin{cases} f(A;x,R=0)=0, \\ f(A-\partial\Lambda;x,R)=f(A;x,R)+\Lambda\left(x+\frac{1}{2}R\right)-\Lambda\left(x-\frac{1}{2}R\right), \end{cases} \tag{8.142}$$

then the insertion of the term

$$U\left(A;x+\frac{1}{2}R,x-\frac{1}{2}R\right)\equiv\exp\left[ief(A;x,R)\right] \tag{8.143}$$

in the definition of the Wigner operator allows the removal of the gauge dependence, and finally they realized that the ansatz

$$\begin{aligned} f(A;x,R) &\equiv -R^\mu\int_0^1 dsA_\mu\left(x+R\left\{s-\frac{1}{2}\right\}\right) \\ &= -\int_{x-1/2R}^{x+1/2R} dz^\mu A_\mu(z), \end{aligned} \tag{8.144}$$

where the integral is a path integral along a straight line, from point $(x-R/2)$ to $(x+R/2)$, obeys the above conditions, and does lead to a gauge-invariant definition of the Wigner operator. The above expression of $U\left(A;x+\frac{1}{2}R,x-\frac{1}{2}R\right)$ should be specified more precisely. Indeed, when $A^\mu(x)$ is an operator, U must be supplemented by a path ordering of the exponential,

$$U\left(A;x+\frac{1}{2}R,x-\frac{1}{2}R\right)=P\exp\left[-R^\mu\int_0^1 ds\,A_\mu\left(x+R\left\{s-\frac{1}{2}\right\}\right)\right], \tag{8.145}$$

where P is the chronological order of $\exp(\cdots)$. A few words are now in order about the choice used for the path. Indeed, there exist an infinity of other possible paths that would lead to other gauge-invariant Wigner operators.

[16]H.-T. Elze, M. Gyulassy and D. Vasak (1986a,b).

In fact, this linear path has been chosen not only for reasons of simplicity but also for the following physical reason. Indeed, in doing so, the quantity

$$\hat{\pi}^{\mu} = \hat{p}^{\mu} - eA^{\mu}(x) = \frac{i}{2}\left(D - D^{+}\right) \tag{8.146}$$

acquires the signification of the kinetic momentum of a generic particle while $\hat{p}$ is the canonically conjugate momentum to $\hat{x}$ [see D. Vasak, M. Gyulassy and H. Th. Elze (1986a)]. The latter authors also remarked that the two definitions, i.e. the intuitive one and the more sophisticated one with a path integral, are equivalent.

Another point of importance is that the main observables, i.e. the four-current and the energy–momentum tensor, are still provided by the same relations as in the non-gauge-invariant Wigner function

$$\begin{cases} j^{\mu}(x) = \mathrm{Sp}\displaystyle\int d^4p\, \gamma^{\mu} F(x,p), \\ T^{\mu\nu}(x) = \mathrm{Sp}\displaystyle\int d^4p\, p^{\mu}\gamma^{\nu} F(x,p). \end{cases} \tag{8.147}$$

Finally, starting from the Dirac equations derived from the Lagrangian

$$L = \bar{\psi}(x)\left\{\frac{i}{2}\overleftrightarrow{\partial}\right\}\psi(x) - \frac{1}{4}F^{\mu\nu}F_{\mu\nu} \tag{8.148}$$

and using the same techniques as at the beginning of this chapter, P. Zhuang and U. Heinz (1996) arrived at the dynamical equations obeyed by the gauge-invariant Wigner function [see also D. Vasak, M. Gyulassy and H. Th. Elze (1986a)], which they wrote as

$$\begin{cases} \{\gamma^{\mu}\left[iD_{\mu}(x,p) + 2\Pi_{\mu}(x,p)\right] - 2m\}\, F_{\mathrm{op}}(x,p) = 0, \\ F_{\mathrm{op}}(x,p)\,\{\gamma^{\mu}\left[iD_{\mu}(x,p) - 2\Pi_{\mu}(x,p)\right] + 2m\} = 0, \end{cases} \tag{8.149}$$

with the definitions

$$\begin{cases} D_{\mu}(x,p) \equiv \partial_{\mu} - e\displaystyle\int_{-1/2}^{+1/2} ds\, F_{\mu\nu}\left(x - s\nabla_{(p)}\right)\nabla_p^{\nu}, \\ \Pi_{\mu}(x,p) \equiv p_{\mu} - e\displaystyle\int_{-1/2}^{+1/2} s ds\, F_{\mu\nu}\left(x - s\nabla_{(p)}\right)\nabla_p^{\nu}, \end{cases} \tag{8.150}$$

where $\nabla_{(p)} \equiv \partial/\partial p$; D_{μ} and Π_{μ} play the role of gauge-invariant operators in place of ∂_{μ} and p_{μ}. The above dynamical equations satisfied by the gauge-invariant Wigner operator can also be decomposed on the Dirac algebra in order to get further insights [P. Zhuang, and U. Heinz (1996)].

For a constant and homogeneous magnetic field, the above equations read

$$\begin{cases} \left[i\gamma\cdot\left(\partial - eF_{\mu\nu}\nabla^{\nu}_{(p)}\right) + 2\left(\gamma\cdot p - m\right)\right] F_{\rm op}(x,p) = 0, \\ F_{\rm op}(x,p)\left[i\gamma\cdot\left(\overleftarrow{\partial} - eF_{\mu\nu}\overleftarrow{\nabla}^{\nu}_{(p)}\right) - 2\left(\gamma\cdot p - m\right)\right] = 0, \end{cases} \tag{8.151}$$

which are exactly the equations obtained with the "naïve" non-gauge-invariant Wigner function. This is an illustration of a result by A.V. Selikhov (1988), according to whom there always exists a gauge where the two functions do coincide; in this case, it is the Landau gauge (see Chap. 15).

8.7.2. *A few remarks*

(1) The above gauge-invariant formalism for the Wigner functions of the electrons and photons is very elegant and has been used for the study of the classical limit by expanding both the equations and the Wigner function in powers of h, and this allowed the obtaining of deeper insights into their interpretation [D. Vasak, M. Gyulassy and H. Th. Elze (1986a); P. Zhuang and U. Heinz (1996); etc.].

(2) In order to solve the problem of pair emission in an external homogeneous electric field, I. Bialyncki-Birula, P. Gornicki and J. Rafelski (1991) used a gauge-invariant (but not manifestly Lorentz-invariant) Wigner function and the transport equation it satisfies in the Hartree–Vlasov approximation. This line was followed by several authors.[17] The equal time gauge-invariant Wigner functions used in this approach have been connected to the manifestly Lorentz-invariant ones by P. Zhuang and U. Heinz (1996). They showed that the equal time Wigner function can be obtained from the fully invariant one by an integration over the variable p^0,

$$f(\mathbf{x},t;\mathbf{p}) = \int dp^0 f(x,p), \tag{8.152}$$

and they found that the I. Bialyncki-Birula *et al.* equal time approach was only one of several possibilities. Also, P. Zhuang and U. Heinz

[17]C. Best, P. Gornicki and W. Greiner, *Ann. Phys.* **225**, 169 (1993); J.M. Eisenberg and G. Kalbermann, *Phys. Rev.* **D37**, 1197 (1988); C. Best and J.M. Eisenberg, **D47**, 4639 (1993); G.R. Shin and J. Rafelski, *Phys. Rev.* **A48**, 1869 (1993); I. Bialynicki-Birula, E.D. Davis and J. Rafelski (1993); O.T. Serimaa, J. Javainen and S. Varro, *Phys. Rev.* **A33**, 2913 (1986); J. Javainen, S. Varro and O.T. Serimaa, *Phys. Rev.* **A35**, 2791 (1987); S. Varro and J. Javainen (2003).

(1996) realized that, in this approach, the off-shell effects contained in the full covariant Wigner function are eliminated by the approximations. In addition, they discussed the advantages and disadvantages of both approaches for the problem under consideration, i.e. pair creation in an electric field. In particular, they recalled that the complete covariant Wigner function poses some difficulties when one is solving the initial value problem present in this question [I. Bialyncki-Birula, P. Gornicki and J. Rafelski (1991)]. This problem could, however, be solved with some approximation methods: a very simple example of solution is indeed provided in Chap. 12.

8.7.3. *Gauge-invariant Wigner functions for the photon field*

The case of the electromagnetic field is much simpler than that of the electrons; indeed, a definition of the Wigner function for $A^\mu(x)$ similar to that of the scalar field, i.e.

$$\begin{aligned} g_{\text{op}}^{\mu\nu}(x,k) &= \int \frac{d^4R}{(2\pi)^4} \exp(-ik\cdot R) \left\{ A^\mu\left(x+\frac{1}{2}R\right) A^\nu\left(x-\frac{1}{2}R\right) \right. \\ &\quad \left. - \left\langle A^\mu\left(x+\frac{1}{2}R\right)\right\rangle \left\langle A^\nu\left(x-\frac{1}{2}R\right)\right\rangle \right\}, \end{aligned} \tag{8.153}$$

which is obviously gauge-invariant. From the (gauge-invariant) equation of motion satisfied by the electromagnetic field

$$\Box A^\mu(x) - \partial^\mu(\partial\cdot A(x)) = 4\pi J^\mu(x) = 4\pi e\mathrm{Sp}\int d^4p\,\gamma^\mu F(x,p), \tag{8.154}$$

it is not very difficult to show that $g_{\text{op}}^{\mu\nu}(x,k)$ obeys the equations

$$\begin{cases} k\cdot\partial g_{\text{op}}^{\mu\nu}(x,k) \\ \quad = \displaystyle\int \frac{d^4R}{(2\pi)^3} \exp(-ip\cdot R) \left[A^\mu\left(x+\frac{1}{2}R\right) J^\nu\left(x-\frac{1}{2}R\right) \right. \\ \quad \left. - J^\nu\left(x+\frac{1}{2}R\right) A^\mu\left(x-\frac{1}{2}R\right) \right], \\ \left(k^2-\frac{1}{4}\Box\right) g_{\text{op}}^{\mu\nu}(x,k) \\ \quad = \displaystyle\int \frac{d^4R}{(2\pi)^3} \exp(-ip\cdot R) \left[A^\mu\left(x+\frac{1}{2}R\right) J^\nu\left(x-\frac{1}{2}R\right) \right. \\ \quad \left. -J^\nu\left(x+\frac{1}{2}R\right) A^\mu\left(x-\frac{1}{2}R\right) \right]. \end{cases} \tag{8.155}$$

It is, however, clear that in order to solve these equations, one has to choose a gauge since they are equivalent to the equations of motion, and the simplest covariant gauge is the Lorentz one, $\partial \cdot A(x) = 0$. With such a choice $g^{\mu\nu}_{\text{op}}(x,k)$ must satisfy

$$k_\mu g^{\mu\nu}_{\text{op}}(x,k) = 0, \quad k_\nu g^{\mu\nu}_{\text{op}}(x,k) = 0. \tag{8.156}$$

Indeed, Eq. (8.159) is not gauge invariant.

On the other hand, a glance at the right hand side of the equations obeyed by the Wigner operator shows that, when one is writing the BBGKY hierarchy, products of various Wigner operators and complicated expressions involving products of the electromagnetic field $F^{\mu\nu}$ occur, suggesting thereby to define a gauge-covariant operator as

$$\begin{aligned} G^{\mu\nu\alpha\beta}_{\text{op}}(x,k) = \int \frac{d^4R}{(2\pi)^4} \bigg\{ & F^{\mu\nu}\left(x + \frac{1}{2}R\right) F^{\alpha\beta}\left(x - \frac{1}{2}R\right) \\ & - \left\langle F^{\mu\nu}\left(x + \frac{1}{2}R\right)\right\rangle \left\langle F^{\alpha\beta}\left(x - \frac{1}{2}R\right)\right\rangle \bigg\}, \end{aligned} \tag{8.157}$$

and since $F^{\mu\nu}$ and A^μ are connected through

$$F^{\mu\nu}(x) = \partial^\mu A^\nu(x) - \partial^\nu A^\mu(x), \tag{8.158}$$

the above Wigner functions are interconnected through

$$\begin{aligned} & G^{\alpha\beta\mu\nu}_{\text{op}}(x,k) \\ & = -\left(k^\nu + \frac{i}{2}\partial^\nu\right)\left(k^\alpha - \frac{i}{2}\partial^\alpha\right) g^{\mu\beta}_{\text{op}} + \left(k^\nu + \frac{i}{2}\partial^\nu\right)\left(k^\beta - \frac{i}{2}\partial^\beta\right) g^{\mu\alpha}_{\text{op}} \\ & \quad - \left(k^\alpha + \frac{i}{2}\partial^\alpha\right)\left(k^\mu - \frac{i}{2}\partial^\mu\right) g^{\nu\beta}_{\text{op}} + \left(k^\mu + \frac{i}{2}\partial^\mu\right)\left(k^\beta - \frac{i}{2}\partial^\beta\right) g^{\nu\alpha}_{\text{op}} \end{aligned} \tag{8.159}$$

and it obeys the equations

$$\begin{cases} \left(k_\mu + \frac{i}{2}\partial_\mu\right) G^{\mu\nu\alpha\beta}_{\text{op}}(x,k) \\ \quad = i\text{Sp} \displaystyle\int\int d^4p \frac{d^4R}{(2\pi)^4} \exp(-ip\cdot R)\, F^{\alpha\beta}\left(x + \frac{1}{2}R\right) \gamma^\nu F_{\text{op}}\left(x - \frac{1}{2}R\right), \\ \left(k^\lambda + \frac{i}{2}\partial^\lambda\right) G^{\mu\nu\alpha\beta}_{\text{op}}(x,k) + \left(k^\mu + \frac{i}{2}\partial^\mu\right) G^{\nu\lambda\alpha\beta}_{\text{op}}(x,k) \\ \quad + \left(k^\nu + \frac{i}{2}\partial^\nu\right) G^{\lambda\mu\alpha\beta}_{\text{op}}(x,k) = 0, \end{cases} \tag{8.160}$$

which are nothing but the Wigner form of the Maxwell equations.

From the expression of $G^{\alpha\beta\mu\nu}(x,k)$, one can obtain the energy–momentum tensor as

$$T^{\mu\nu}(x) = \left\langle F^{\mu\alpha}(x)F^{\nu}_{\alpha}(x) + \frac{1}{4}\eta^{\mu\nu}F^{\alpha\beta}(x)F_{\alpha\beta}(x)\right\rangle$$
$$= \int d^4k \left\{G^{\mu\alpha\,\nu}_{\alpha}(x,k) + \frac{1}{4}\eta^{\mu\nu}G^{\alpha\beta}_{\alpha\beta}(x,k)\right\}, \qquad (8.161)$$

which is to be compared with the nonquantum results by T.W. Marshall (1963, 1965) together with the covariant Wigner function for $F^{\mu\nu}$.

8.7.4. *Another gauge-invariant Wigner function*

The above analysis of the transformation properties of the Wigner function indicates that the factors Λ which appear in the exponential of the transformed F could be compensated for with A terms. This is possible and has been achieved by E.A. Remler (1977) with the following new definition for the Wigner function:

$$F(x,p) \to F_G(x,p) = \int \frac{d^4R}{(2\pi)^4} \exp\left[-i(p-\mathcal{A})\cdot R\right]$$
$$\times \left\langle \bar{\psi}\left(x+\frac{1}{2}R\right) \otimes \psi\left(x-\frac{1}{2}R\right)\right\rangle, \qquad (8.162)$$

where $\mathcal{A}$ is given by

$$\mathcal{A}_\mu(x,R) \equiv \left(R\cdot\partial_{(x)}\right)^{-1}\left\{A_\mu\left(x+\frac{1}{2}R\right) - A_\mu\left(x-\frac{1}{2}R\right)\right\}. \qquad (8.163)$$

This operator can be formally defined through its Fourier transform as being

$$\mathcal{A}_\mu(x,R) = \int d^4k\, \exp(ik\cdot x)\left\{\frac{A_\mu(k)}{ik\cdot R}\left[\exp(ik\cdot R) - \exp(-ik\cdot R)\right]\right\}; \qquad (8.164)$$

and when one expands the exponentials within the brackets, one obtains

$$\mathcal{A}_\mu(x) = \int d^4k\, \exp(ik\cdot x)\left\{\frac{A_\mu(k)}{ik\cdot R}\left(\sum_{n=0}^{\infty}\left[\frac{1-(-1)^n}{n!}\left(\frac{1}{2}ik\cdot R^n\right)\right]\right)\right\}$$
$$= A_\mu(x) + \frac{1}{24}\left(R\cdot\partial\right)^2 A_\mu(x) + \frac{1}{16\times 5!}\left(R\cdot\partial\right)^4 A_\mu(x) + \cdots. \qquad (8.165)$$

It is clear that, in this definition of F_G, the exponential should be ordered since A_μ is *a priori* not a c number. However, in view of another version of the Hartree–Vlasov equation, A_μ will be considered as a mean field and hence as a c number.

With this definition of the new Wigner function, the exact form of the four-current is the same as to the older one and the energy–momentum tensor is

$$T^{\mu\nu} = \int d^4p \left\{ \left(p^\mu p^\nu - \frac{1}{4}\eta^{\mu\nu} \right) (p^2 - m^2) + \left(\partial^\mu \partial^\nu - \frac{1}{4}\eta^{\mu\nu}\Box \right) \right\} f_G(x,p). \tag{8.166}$$

Let us now turn to an equation obeyed by the Wigner function and let us consider the simpler case of scalar particles — hence, the Wigner function reduces to a scalar function f_G — embedded in a mean electromagnetic field,

$$\begin{cases} \Box\langle A^\mu(x)\rangle = 4\pi e\mathrm{Sp} \displaystyle\int d^4p\, p^\mu f_G(x,p) \\ \partial_\mu \langle A^\mu(x)\rangle = 0, \end{cases} \tag{8.167}$$

so as to avoid problems of ordering the exponential of field operators. After some lengthy but straightforward algebra, E.A. Remler (1977) arrive at the equation for f_G

$$\left(\Gamma_\mu \Gamma^\mu + m^2\right) f_G(x,p) = 0, \tag{8.168}$$

with the operator Γ_μ given by

$$\Gamma_\mu \equiv \frac{1}{2}\partial_\mu - ip_\mu + ieF_\mu\left(x, -i\nabla_{(p)}\right), \tag{8.169}$$

where F_μ has been defined as

$$F_\mu = \left\{ \frac{1}{2}\frac{\partial}{\partial x^\mu} + \frac{\partial}{\partial y^\mu} \right\} A_\mu(x,y)\cdot y - A_\mu\left(x + \frac{1}{2}y\right). \tag{8.170}$$

The resulting Hartree–Vlasov equation is highly nonlinear, quite intricate, and very difficult to use. The same is true of the case of spin 1/2 particles.

Unfortunately, this line of research has not yet been fully developed.

8.7.5. *Gauge invariance and approximations*

The main problem, however, is raised by the question of the gauge transformation properties of the approximations performed during an actual calculation. It is clear that one cannot make assumptions of an almost arbitrary nature even though they are apparently "physical." Some do preserve the transformation properties of the equations which are approximated but this is not the case for others. To be specific, the case of the Hartree–Vlasov approximation,

$$\langle F(x,p)A^\mu(x)\rangle \approx F(x,p)\langle A^\mu(x)\rangle, \tag{8.171}$$

for the non-gauge-invariant Wigner function is first dealt with. For a QED plasma (see Chap. 15) the equations obeyed by the Wigner function read

$$\begin{aligned}&\{i\gamma\cdot\partial + 2\,(\gamma\cdot p - m)\}\,F(x,p)\\&\quad = 2e\int\frac{d^4x'}{(2\pi)^4}d^4p'\exp\left[-ip'\cdot(x-x')\right]\gamma_\mu\left\langle F_{\rm op}(x,p-\frac{1}{2}p')A^\mu(x')\right\rangle\end{aligned} \tag{8.172}$$

$$\begin{aligned}&F(x,p)\left\{i\gamma\cdot\overleftarrow{\partial} - 2(\gamma\cdot p - m)\right\}\\&\quad = -2e\int\frac{d^4x'}{(2\pi)^4}d^4p'\,\exp\left[-ip'\cdot(x-x')\right]\left\langle A^\mu(x')F_{\rm op}(x,p+\frac{1}{2}p')\right\rangle\gamma_\mu,\end{aligned} \tag{8.173}$$

$$\Box\,\langle A^\mu(x)\rangle - \partial^\mu\left[\partial_\nu\,\langle A^\nu(x)\rangle\right] = 4\pi e{\rm Sp}\int d^4p\,\gamma^\mu F(x,p). \tag{8.174}$$

Under a gauge transformation the third equation remains unchanged while the first two have the same transformation properties whether they are exact or are approximated with the above ansatz. The same is true of other approximations for various correlations, for instance

$$\begin{aligned}\langle F(x,p)A(x')A(x'')\rangle &\approx \langle F(x,p)\rangle\,\langle A(x')A(x'')\rangle\\&+\langle F(x,p)A(x')\rangle\,\langle A(x'')\rangle + \langle F(x,p)A(x'')\rangle\,\langle A(x')\rangle\,.\end{aligned} \tag{8.175}$$

This is one of the disadvantages of the nonmanifestly gauge-invariant Wigner function: in each case one has to check the gauge invariance of what is effectively performed.

Of course, the full covariant Wigner function has not this inconvenience. For instance, the Hartree–Vlasov approximation (absence of correlations between the electromagnetic field and the electrons) reads

$$\langle F_{\rm op}(x,p)F^{\mu\nu}(x')\rangle \approx \langle F_{\rm op}(x,p)\rangle\,\langle F^{\mu\nu}(x')\rangle\,, \tag{8.176}$$

which is certainly a gauge-invariant approximation; it behaves in the same way on both sides of the approximation. Suppose, for instance, that $\langle F_{\rm op}(x,p)F^{\mu\nu}(x')\rangle$ has to be approximated by something like

$$\langle F_{\rm op}(x,p)F^{\mu\nu}(x')\rangle \approx \langle F_{\rm op}(x,p)\rangle^{1/2}\,\langle F_{\rm op}(x,p)\rangle^{1/2}\,\langle F^{\mu\nu}(x')\rangle\,;$$

it is clear that the two sides of the approximation would not behave in the same way and the gauge symmetry would be destroyed.

Chapter 9

Fermions Interacting via a Scalar Field: A Simple Example

The case of fermions interacting through a scalar field occurs many times: in the model of J.D. Walecka (1974), in the abnormal nuclear matter of T.D. Lee (1978), in one form of the phenomenological model for quark interactions of J. Rafelski (1974), etc. However, we study this model not so much for the applications it could have, but for the ease it can have in operating the covariant Wigner function, and particularly the renormalization. An interesting article devoted to the Walecka model has been written at $T = 0^\circ$ K by S.A. Chin (1977).

In this chapter, we study the properties of the system characterized by the Lagrangian

$$\mathcal{L} = \bar{\psi}(i\gamma \cdot \partial - m)\psi + g\bar{\psi}\psi\phi + \frac{1}{8\pi}[(\partial\phi)^2 - \mu_R^2\phi^2] - \mathcal{L}_c, \qquad (9.1)$$

where $(\bar{\psi}, \psi)$ refers to the fermion field, ϕ is a scalar field mediating the interaction and $\mathcal{L}_c$ is the Lagrangian's counterterms:

$$4\pi\mathcal{L}_c = \frac{\alpha}{1!}\phi + \frac{\delta\mu^2}{2!}\phi^2 + \frac{\gamma}{3!}\phi^3 + \frac{\lambda}{4!}\phi^4. \qquad (9.2)$$

However, we shall forget this term for some time and consider it in the second part of this chapter. Here we consider this model as a simple tool to show the use of the Wigner function. However, it may be looked at as a part of the D.A. Walecka (1974) model for nuclear matter, the abnormal nuclear matter model of T.D. Lee (1975), etc. Also, the Higgs mechanism enters partly in the context of this model.

Such a system gives one the simplest quantum-field-theoretical models for the description of relativistic dense matter or, more precisely, it represents a good "laboratory" for its study. Therefore, we shall be concerned, in this chapter, with the thermodynamic properties of the system, its renormalization and the scalar field spectrum.

It has sometimes been called the "scalar plasma" [G. Kalman (1974)], both for brevity and owing to the fact that plasma methods were used to deal with it.

We now come back to the case of the scalar plasma at finite density and temperature [J. Diaz Alonso and R. Hakim (1978, 1984, 1988)], and the Hartree–Vlasov approximation will mainly be dealt with. In this approximation $\langle\varphi\rangle$ is only the collective field whose source is the fermion field and correlations of whatever nature are neglected:

$$\begin{cases} \langle F_{\rm op}\varphi\rangle \approx F\langle\varphi\rangle, \\ \langle F_{\rm op}\otimes F_{\rm op}\rangle \approx \langle F_{\rm op}\rangle\otimes\langle F_{\rm op}\rangle, \\ \langle\varphi\varphi\rangle \approx \langle\varphi\rangle\langle\varphi\rangle, \\ \text{etc.} \end{cases} \tag{9.3}$$

The basic equations of the hierarchy then reduce to the set

$$\begin{cases} \{i\gamma\cdot\partial + 2[\gamma\cdot p - m]\}F(x,p) = -2g_S\displaystyle\int\frac{d^4R}{(2\pi)^4}d^4\xi \\ \quad\times\exp[-i(p-\xi)\cdot R]F(x,\xi)\left\langle\varphi\left(x-\frac{1}{2}R\right)\right\rangle, \\ F(x,p)\{i\gamma\cdot\overleftarrow{\partial} - 2[\gamma\cdot p - m]\} = +2g_S\displaystyle\int\frac{d^4R}{(2\pi)^4}d^4\xi \\ \quad\times\exp[-i(p-\xi)\cdot R]F(x,\xi)\left\langle\varphi\left(x+\frac{1}{2}R\right)\right\rangle, \\ (\Box + m_S^2)\langle\varphi(x)\rangle = g_S\mathrm{Sp}\displaystyle\int d^4p\,F(x,p). \end{cases} \tag{9.4}$$

9.1. Thermal Equilibrium

In equilibrium, the system is stationary and homogeneous, and therefore one has

$$\begin{cases} F(x,p) = F_{\rm eq}(p), \\ \langle\varphi(x)\rangle = \text{const} \equiv \varphi_{\rm eq}, \end{cases} \tag{9.5}$$

and the above equations reduce to

$$\begin{cases} [\gamma\cdot p - m]F_{\rm eq}(p) = -g_S F_{\rm eq}(p)\varphi_{\rm eq}, \\ F_{\rm eq}(p)[\gamma\cdot p - m] = -g_S F_{\rm eq}(p)\varphi_{\rm eq}, \\ m_S^2\varphi_{\rm eq} = g_S Sp\int d^4p\,F_{\rm eq}(p). \end{cases} \tag{9.6}$$

The first two equations are rewritten as

$$\begin{cases} [\gamma \cdot p - M]F_{\text{eq}}(p) = 0, \\ F_{\text{eq}}(p)[\gamma \cdot p - M] = 0, \\ M = m - g_S \varphi_{\text{eq}}. \end{cases} \tag{9.7}$$

This means that the fermions of the system are endowed with the *effective mass* M and in equilibrium, as shown in a preceding chapter, F_{eq} is identical with the equilibrium Wigner function for particles with such an effective mass. Then F_{eq} is replaced in the third equation (Klein–Gordon), and one gets

$$M = m - \Gamma \frac{M^3}{m^2} \sum_{\pm} \int_0^\infty \frac{\xi^2 d\xi}{\sqrt{\xi^2+1}} \frac{1}{\exp\left(\beta M\left[\sqrt{\xi^2+1} \mp \mu\right]\right) + 1}, \tag{9.8}$$

where the vacuum term has been dropped until a later subsection, where it will be discussed. G. Kalman (1974) first obtained this equation at $T = 0\,\text{K}$, to which it reduces at this temperature. Like Kalman, we have set

$$\Gamma = \frac{4}{\pi} g_S^2 \left(\frac{m}{m_S}\right)^2, \tag{9.9}$$

which characterizes the strength of the interaction. This self-consistent transcendental equation for M is strongly reminiscent of the usual gap equation of superconductivity; hence, in the sequel, it is referred to as the *gap equation.* This equation controls all the thermal equilibrium properties of the system, since M occurs in the equilibrium Wigner function and hence in the thermodynamic quantities (Fig. 9.1) [J. Diaz-Alonso and R. Hakim (1978)].

The basic data — pressure, energy density, baryon number density — are obtained through the standard formulae given at the beginning of Chaps. 7 and 8, with the replacement $m \to M(\mu, \beta)$, and by adding to the energy–momentum tensor the contribution of the scalar field, namely (Fig. 9.2)

$$\begin{aligned} T_\varphi^{\mu\nu} &= -\mathcal{L}\eta^{\mu\nu} \\ &= \left(\frac{1}{2} m_S^2 \langle\varphi\rangle_{\text{eq}}^2\right)\eta^{\mu\nu} \\ &= \frac{m_S^2}{8\pi g_S^2}(M-m)^2\,\eta^{\mu\nu} \end{aligned} \tag{9.10}$$

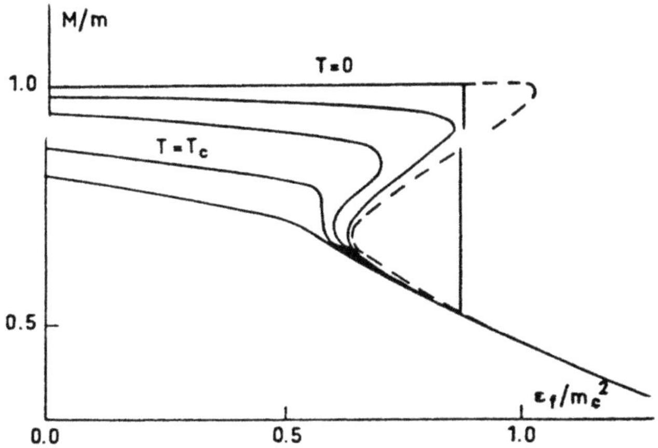

Fig. 9.1 The effective fermion mass is displayed as a function of the chemical potential (in units of the fermion mass m). Note that below a critical temperature, the effective mass presents a discontinuity expressing a first order phase transition. The vertical line corresponds to a Maxwell construction.

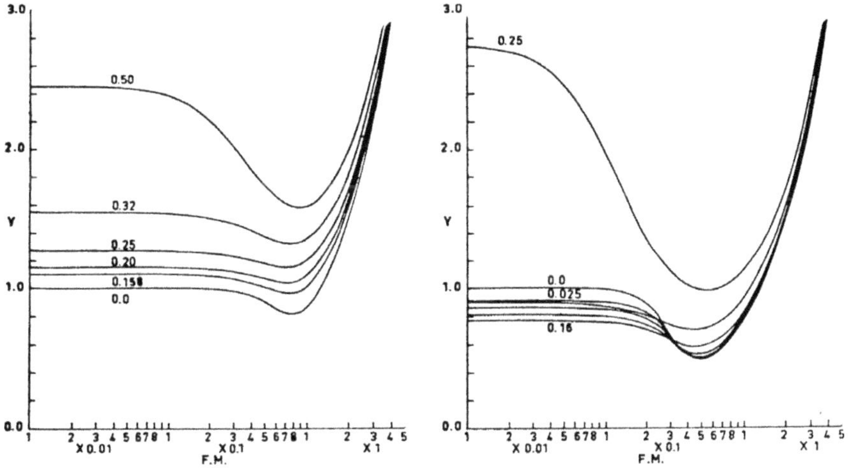

Fig. 9.2 The energy per fermion is plotted against the parameter k_F for several values of the temperature. In the left hand side diagram Γ has been chosen to be 10, while its value is 100 in the right hand side one. Note that there is a critical temperature below which there exists a collective bound state of the fermions: it occurs whenever the energy per fermion is smaller than 1 (in units of m).

(remember that, in equilibrium, $\partial_\mu \langle \varphi \rangle_{\text{eq}} = 0$), and the four-current

$$J^\mu = \text{Sp} \int d^4p \, \gamma^\mu F_{\text{eq}}(p)$$
$$= \int d^4p \, f^\mu_{\text{eq}}(p). \tag{9.11}$$

Figure 9.3 shows the behavior of the pressure as a function of the chemical potential for several temperatures. The coupling constant Γ has been chosen to be equal to 4. The double point is characteristic of a first order phase transition and the Maxwell construction amounts to suppressing the "loop" of the curve, represented by a dotted line. As usual, the observables (n, ρ, P) are obtained without difficulty from J^μ and $T^{\mu\nu}$.

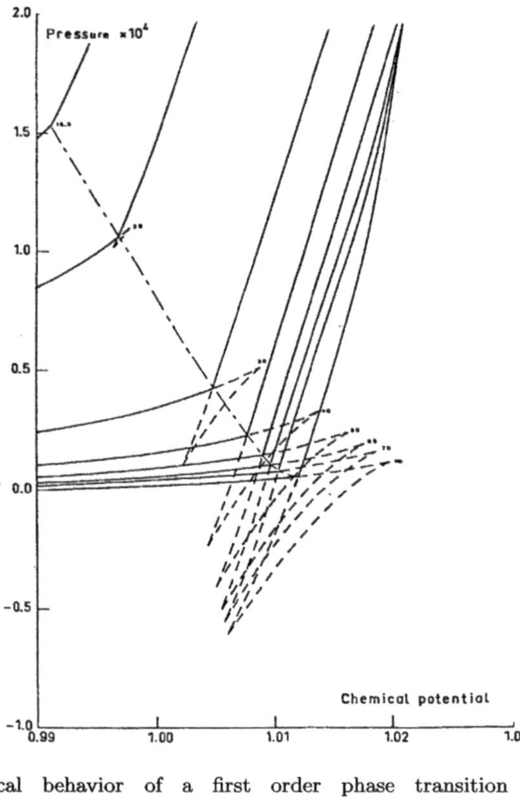

Fig. 9.3 Typical behavior of a first order phase transition shown for the pressure/chemical potential, at several temperatures. The cut points correspond to the Maxwell construction for every temperature, giving the transition pressures. The dashed–dotted line connects the branch to a point between T and the critical temperature T_C.

9.2. Collective Modes

The collective oscillation modes of the relativistic scalar plasma [J. Diaz-Alonso and R. Hakim (1988)] are obtained in the random phase approximation, i.e. from the linearized Hartree–Vlasov system obtained by replacing the perturbed functions:

$$\begin{cases} F(x,p) \approx F_{\text{eq}}(p) + F_1(x,p), \\ \langle \varphi(x) \rangle \approx \langle \varphi \rangle_{\text{eq}} + \varphi_1, \end{cases} \tag{9.12}$$

where the higher order terms are neglected. With this linearization, the above Hartree–Vlasov system, once Fourier-transformed, can be written as

$$\begin{cases} \left[\gamma \cdot \left(p - \frac{1}{2}k\right) - M\right] F_1(k,p) = -g_S F_{\text{eq}}\left(p + \frac{1}{2}k\right) \varphi_1(k), \\ F_1(k,p)\left[\gamma \cdot \left(p + \frac{1}{2}k\right) - M\right] = -g_S F_{\text{eq}}\left(p - \frac{1}{2}k\right) \varphi_1(k), \\ (k^2 - m_S^2)\varphi_1(k) = -4\pi g_S \mathrm{Sp} \int d^4p\, F_1(k,p), \end{cases} \tag{9.13}$$

whose solution is

$$\begin{aligned} F_1(k,p) = -g_S &\left[\gamma \cdot \left(p - \frac{1}{2}k\right) - M\right]^{-1} F_{\text{eq}}\left(p + \frac{1}{2}k\right) \\ &- F_{\text{eq}}\left(p - \frac{1}{2}k\right)\left[\gamma \cdot \left(p + \frac{1}{2}k\right) - M\right]^{-1} \end{aligned} \tag{9.14}$$

or, more explicitly,

$$\begin{aligned} F_1(k,p) = -g_S &\left\{ \frac{\gamma \cdot (p - \frac{1}{2}k) + M}{\left(p - \frac{1}{2}k\right)^2 - M^2 + i\varepsilon} F_{\text{eq}}\left(p + \frac{1}{2}k\right) \right. \\ &\left. + g_S\, F_{\text{eq}}\left(p - \frac{1}{2}k\right) \frac{\gamma \cdot (p + \frac{1}{2}k) + M}{\left(p + \frac{1}{2}k\right)^2 - M^2 - i\varepsilon} \right\} \varphi_1(k), \end{aligned} \tag{9.15}$$

where the $i\varepsilon$ terms reflect the boundary conditions (compare with the Yang–Feldman form of the equation of motion for F; see Chap. 8). Note also that the first term of the solution (resp. the second) is a solution to the first equation of the linearized system while the remaining term is a homogeneous solution to the other. Inserting now this solution into the equation for φ_1, one obtains an equation of the form

$$[(k^2 - m_S^2) + \Pi(k)]\varphi_1(k) = 0, \tag{9.16}$$

with

$$\Pi(k) = -2\pi g_S^2 \,\mathrm{Sp} \int d^4p \left\{ \frac{\gamma\cdot(p - \frac{1}{2}k) + M}{k\cdot p - i\varepsilon} F_{\mathrm{eq}}\left(p + \frac{1}{2}k\right) \right.$$
$$\left. - F_{\mathrm{eq}}\left(p - \frac{1}{2}k\right) \frac{\gamma\cdot(p + \frac{1}{2}k) + M}{k\cdot p - i\varepsilon} \right\}, \tag{9.17}$$

where use has been made of the fact that $F_{\mathrm{eq}}(p \pm \frac{1}{2}k)$ are on the *mass shells*:

$$\left(p \pm \frac{1}{2}k\right)^2 = M^2 = \pm 2k\cdot p. \tag{9.18}$$

The dispersion equation obeyed by the collective modes is thus

$$[(k^2 - m_S^2) + \Pi(k)] = 0, \tag{9.19}$$

and once the traces have been performed one obtains an explicit form for the polarization tensor $\Pi(k)$,

$$\Pi(k) = \frac{2\pi g_S^2}{M} \int d^4p \frac{p^2 - \frac{1}{4}k^2 + M^2}{k\cdot p - i\varepsilon} \left[f_{\mathrm{eq}}\left(p + \frac{1}{2}k\right) - f_{\mathrm{eq}}\left(p - \frac{1}{2}k\right) \right], \tag{9.20}$$

where divergent vacuum contributions are included. The modes are discussed in a later section after the theory has been renormalized.

9.3. Two-Body Correlations

(1) First, an alternative treatment of the collective modes of the scalar plasma is briefly outlined; it is based on a particular treatment of correlation, showing thereby how fluctuations are connected with such modes: see the fluctuation–dissipation theorem.[1]

[1] A.G. Sitenko, *Electromagnetic Fluctuation in Plasmas* (Academic, New York, 1967).

Let us rewrite the first two equations of the BBGKY hiererarchy in symbolic form:

$$\begin{cases} LF = \int \langle F_{\text{op}}\varphi \rangle, \\ KG\langle \varphi \rangle = \int \langle F_{\text{op}} \rangle, \\ L\langle F_{\text{op}}F_{\text{op}} \rangle = \int \langle F_{\text{op}}F_{\text{op}}\varphi \rangle, \\ L\langle F_{\text{op}}\varphi \rangle = \int \langle F_{\text{op}}\varphi\varphi \rangle, \\ KG\langle \varphi\varphi \rangle = \int \langle F_{\text{op}}\varphi \rangle, \\ KG\langle F_{\text{op}}\varphi \rangle = \int \langle F_{\text{op}}F_{\text{op}} \rangle, \\ \dots\ \dots\ \dots \end{cases} \tag{9.21}$$

Neglecting three-body correlations, while retaining only two-body ones, the latter are contained in the ansatz

$$\langle F_{\text{op}}\varphi\varphi \rangle \approx F\langle \varphi\varphi \rangle + \sum \langle F_{\text{op}}\varphi \rangle \langle \varphi \rangle - 2F\langle \varphi \rangle \langle \varphi \rangle, \tag{9.22}$$

where the sum is over a permutation of the variable occurring in the scalar field operator. In fact, $\langle \varphi\varphi \rangle$ can be known, in equilibrium, in an approximate way once the excitation spectrum $\omega = \omega(k)$ is given through a Bose–Einstein factor, so that the hierarchy is finally closed.

Let us now briefly and symbolically rederive the excitation spectrum of the scalar plasma. From the fourth equation of the above system, the use of the other equations and the ansatz on the vanishing of three-body correlations, we get

$$\langle F_{\text{op}}\varphi \rangle = L^{-1} \int F\langle \varphi\varphi \rangle, \tag{9.23}$$

which, once inserted in the fifth one, can be written as a homogeneous equation for $\langle \varphi\varphi \rangle$,

$$KG\langle \varphi\varphi \rangle = \int L^{-1} \int F\langle \varphi\varphi \rangle, \tag{9.24}$$

showing that the polarization operator is essentially

$$\Pi = \int L^{-1} \int F. \tag{9.25}$$

A detailed calculation indeed indicates that we actually find anew, in thermodynamic equilibrium, the same Π as in our preceding Hartree–Vlasov calculation (see below).

(2) Let us first briefly implement this derivation of Π for the quasibosons of the scalar plasma model via the use of the second order equations of the BBGKY. With the notations

$$\begin{cases} C_{\varphi\varphi}(x) \equiv \langle\varphi(x+x')\varphi(x')\rangle_{\rm eq} - \langle\varphi\rangle_{\rm eq}^2, \\ C_{F\varphi}(x,p) \equiv \langle F_{\rm op}(x',p)\varphi(x-x')\rangle - F_{\rm eq}(p)\langle\varphi\rangle_{\rm eq}, \end{cases} \tag{9.26}$$

the first equations read[2]

$$\begin{cases} (\gamma\cdot p - M)F_{\rm eq}(p) = -g_S \displaystyle\int d^4k\, C_{F\varphi}\left(k, p+\frac{1}{2}k\right), \\ F_{\rm eq}(p)(\gamma\cdot p - M) = -g_S \displaystyle\int d^4k\, C_{F\varphi}\left(k, p-\frac{1}{2}k\right), \end{cases} \tag{9.27}$$

while the next order is written as

$$\begin{cases} \left\{\gamma\cdot\left(p-\frac{1}{2}k\right) - M\right\} C_{F\varphi}(k,p) = -g_S C_{\varphi\varphi}(k) F_{\rm eq}\left(p+\frac{1}{2}k\right), \\ C_{F\varphi}(k,p)\left\{\gamma\cdot\left(p+\frac{1}{2}k\right) - M\right\} = -g_S C_{\varphi\varphi}(k) F_{\rm eq}\left(p-\frac{1}{2}k\right), \end{cases} \tag{9.28}$$

where several assumptions discussed below have been made, like

$$C_{F\varphi}(k,p) \approx C_{\varphi F}(k,p), \tag{9.29}$$

or the space–time homogeneity of the system in equilibrium. Also, the above ansatz,

$$\langle F_{\rm op}\varphi\varphi\rangle \approx F_{\rm eq}C_{\varphi\varphi} + \sum C_{F\varphi}\langle\varphi\rangle_{\rm eq} - 2F_{\rm eq}\langle\varphi\rangle_{\rm eq}\langle\varphi\rangle_{\rm eq}, \tag{9.30}$$

has been used. Similarly, from the Klein–Gordon equation obeyed by the scalar field φ, one gets

$$\begin{cases} m_S^2\langle\varphi\rangle_{\rm eq} = g_S {\rm Sp}\displaystyle\int d^4p\, F_{\rm eq}(p) \quad \text{(gap equation)}, \\ (k^2 - m_S^2)C_{\varphi\varphi}(k) = -g_S{\rm Sp}\displaystyle\int d^4p\, C_{F\varphi}(k,p). \end{cases} \tag{9.31}$$

When $F_{\rm eq}(p)$ is on the mass shell $p^2 = m^2$ — this is the case when $F_{\rm eq}(p)$ is approximated by its Hartree–Vlasov form — the system can be solved in

[2] Remember that $a(k)$ is the Fourier transform of $a(x)$.

the same manner as above and one gets

$$C_{F\varphi}(k,p) = -g_S C_{\varphi\varphi}(k)\left[\frac{\gamma\cdot\left(p-\frac{1}{2}k\right)+M}{\left(p-\frac{1}{2}k\right)^2-M^2}F_{\text{eq}}\left(p+\frac{1}{2}k\right)\right.$$

$$\left. + F_{\text{eq}}\left(p-\frac{1}{2}k\right)\frac{\gamma\cdot\left(p+\frac{1}{2}k\right)+M}{\left(p+\frac{1}{2}k\right)^2-M^2}\right]. \qquad (9.32)$$

Once introduced into the "second order" Klein–Gordon equation, one gets exactly the same excitation spectrum as above. Note that $C_{\varphi\varphi}(k)$ is essentially the quasibosons' Wigner function.

9.3.1. *A brief discussion*

A few remarks are now in order and, in particular, some assumptions made have to be explained.

(1) Let us first come back to the assumption $C_{F\varphi}(k,p) \approx C_{\varphi F}(k,p)$. It can be considered either as an assumption or as the result of a symmetrization of the basic Lagrangian. As a matter of fact, the quantization of the field should be done *after* it is symmetrized: instead of $C_{F\varphi}$ and $C_{\varphi F}$, one must quantize

$$\frac{1}{2}(C_{\varphi F} + C_{\varphi F}).$$

It follows that whether or not the equality is satisfied is immaterial since the resulting equations are those given above. Another implicit assumption is

$$\operatorname{Im} C_{\varphi\varphi}(k) = \operatorname{Im} C_{\varphi\varphi}(-k).$$

Indeed, in the second equation for $C_{F\varphi}$, it is actually $C_{\varphi\varphi}(-k)$ that occurs on its right hand side. However, the compatibility of two these equations (i.e. when one changes k into $-k$ in one of these equations, one should recover the other) joined to the (implicit) symmetrization actually leads to $C_{\varphi\varphi}(-k)$. This relation may be interpreted as expressing the symmetry beween the particles and their antiparticles: $C_{\varphi\varphi}(\pm k)$ is directly related to the Wigner function of the particles (antiparticles).

(2) It may be somewhat strange that our two derivations of the excitation spectrum of scalar quasiparticles yield exactly the same results: in the first derivation, correlations were neglected, while two-body correlations

were retained in the second one; moreover, the first derivation dealt with an (slightly) off-equilibrium state whereas this was not so in the second case.

In fact, there is no paradox at all. We have already mentioned that a small perturbation of the Hartree–Vlasov equilibrium state can also be considered as a quasiparticle and hence is linked to a small two-body correlation. Next, it should be recalled that, in our second derivation, $F_{\text{eq}}(p)$ was *approximated* by the Hartree–Vlasov equilibrium Wigner function: correlations of spin 1/2 particles were still neglected. On the other hand, when this approximation is not performed, one goes beyond the random phase approximation. For instance, $F_{\text{eq}}(p)$ can be expanded as successive "powers" of two-body correlations (the zeroth order term being the Hartree–Vlasov one) of ϕ; then terms involving for example $\langle\phi\phi\rangle\langle\phi\phi\rangle$ occur and one obtains the nonlinear effects of the usual Hartree–Vlasov approximation, i.e. the interactions of quasiparticles.

9.3.2. *Exchange correlations*

It is well known that the higher the density the better the Hartree approximation. However, at moderate densities exchange effects cannot be neglected and contain a great deal of physically interesting features. Accordingly, the modifications introduced by the Hartree–Fock approximation have to be considered. We still use the example of the "scalar plasma."

To gain some feeling for what should be a Hartree–Fock ansatz, the Klein–Gordon equation for the scalar field φ is formally solved as

$$\varphi = -g_S \int d^4x' \Delta(x - x')\overline{\psi}(x')\psi(x'), \tag{9.33}$$

where $\Delta(x)$ is an elementary solution to the free Klein–Gordon equation

$$\Box\Delta(x) + \mu^2\Delta(x) = 4\pi\delta^{(4)}(x). \tag{9.34}$$

Although we choose the retarded solution,[3] this is of no importance for what we are seeking: a Hartree–Fock ansatz. Next, this formal solution is inserted into the right hand side of the first equation of the BBGKY

[3]See e.g. S. Schweber, *An Introduction to Relativistic Quantum Theory of Fields* (Harper and Row, New York, 1961).

hierarchy and provides

$$\begin{aligned}&\{i\gamma\cdot\partial+2[\gamma\cdot p-m]\}F(x,p)\\&\quad=-32g_S^2\int d^4x'\,d^4x''\,d^4p'\exp[-2ip'\cdot(x-x')]\Delta(x-x')\\&\quad\times\langle F_{\rm op}(x,p-p')F_{\rm op}(x'',p'')\rangle\end{aligned}\tag{9.35}$$

and, of course, another similar equation. The last term of this equation involves a term of the form

$$\left\langle\overline{\psi}\left(x+\frac{1}{2}R\right)\otimes\psi\left(x-\frac{1}{2}R\right)\otimes\overline{\psi}\left(x''+\frac{1}{2}R''\right)\otimes\psi\left(x''-\frac{1}{2}R''\right)\right\rangle,$$

whose Hartree approximation simply corresponds to

$$\begin{aligned}&\left\langle\overline{\psi}\left(x+\frac{1}{2}R\right)\otimes\psi\left(x-\frac{1}{2}R\right)\otimes\overline{\psi}\left(x''+\frac{1}{2}R''\right)\otimes\psi\left(x''-\frac{1}{2}R''\right)\right\rangle\\&\approx\left\langle\overline{\psi}\left(x+\frac{1}{2}R\right)\otimes\psi\left(x-\frac{1}{2}R\right)\right\rangle\left\langle\overline{\psi}\left(x''+\frac{1}{2}R''\right)\otimes\psi\left(x''-\frac{1}{2}R''\right)\right\rangle,\end{aligned}\tag{9.36}$$

which is equivalent to $\langle F_{\rm op}\varphi\rangle \approx F\langle\varphi\rangle$. The Hartree–Fock term is now obtained by taking the average value of all possible pairs (with the correct signs) among the four fermion operators occurring above. Note that, in general, one has *not*

$$\langle\psi\otimes\psi\rangle=0=\langle\overline{\psi}\otimes\overline{\psi}\rangle,\tag{9.37}$$

but this property is true in equilibrium, provided that the total Fermion number (baryonic or charge number) is conserved. Finally, the Hartree–Fock term arises from the pairing

$$\begin{aligned}(\alpha\beta)\text{ component}=-\sum_\lambda&\left\langle\overline{\psi_\beta}\left(x+\frac{1}{2}R\right)\psi_\lambda\left(x''-\frac{1}{2}R''\right)\right\rangle\\&\times\left\langle\overline{\psi}_\lambda\left(x''-\frac{1}{2}R''\right)\psi_\alpha\left(x-\frac{1}{2}R\right)\right\rangle.\end{aligned}\tag{9.38}$$

As a consequence, the Hartree–Fock term to be added to the Hartree term has the form

$$32g_S^2\int d^4X\,d^4\xi\,d^4\xi'K(X,\xi,\xi';x,p)F(X,\xi)F(X,\xi'),\tag{9.39}$$

where the kernel K is given by

$$\begin{aligned}
K(X,\xi,\xi';x,p) = \int d^4x' d^4x'' d^4p\, d^4p \frac{d^4R' d^4R''}{(2\pi)^8} \Delta(x'-x'') \exp[-i(p-p')\cdot R'] \\
\times \exp(-ip''\cdot R'') \exp[-2ip'\cdot(x-x')] \\
\times \exp\left\{ i\xi \cdot \left[x - x'' + \frac{1}{2}(R'+R'') \right] \right\} \\
\times \exp\left\{ i\xi' \cdot \left[x'' - x + \frac{1}{2}(R'+R'') \right] \right\} \\
\times \delta^{(4)} \left\{ X - \left[x'' + x + \frac{1}{2}(R'-R'') \right] \right\}.
\end{aligned} \tag{9.40}$$

In thermal equilibrium this term can be somewhat simplified, owing to the fact that (i) the Hartree term can be included in the left hand side of the equation for F_{eq} via the effective mass term, and that (ii) the space–time invariance of the system can be used. Finally, the equilibrium Hartree–Fock term can be written as

$$32g_S^2 \int d^4k d^4k' K(k,k';p) F_{\text{eq}}(k') F_{\text{eq}}(k), \tag{9.41}$$

with

$$K(k,k';p) = 8g_S^2 \pi \delta^{(4)}(p-k)\Delta(k'-k), \tag{9.42}$$

so that the first equations of the hierarchy read

$$\begin{cases} [\gamma\cdot p - M_{\text{op}}(p)] F_{\text{eq}}(p) = 0, \\ F_{\text{eq}}(p)[\gamma\cdot p - M_{\text{op}}(p)] = 0, \end{cases} \tag{9.43}$$

where M_{op} is a mass operator (acting on the right in the first equation and on the left in the second) which is a functional of F_{eq} itself.

9.4. Renormalization — An Illustration of the Procedure

So far, in our scalar plasma model, the vacuum Wigner function has been neglected. It represents essentially a modified Dirac ocean: the modification comes from the change $m \to M$. The fermions' vacuum is thus modified by the temperature and matter density. As has long been advocated by T.D. Lee,[4] the vacuum appears to be a physical medium *per se*. However,

[4]See, for instance, T.D. Lee, *Particle Physics and Introduction to Field Theory* (Harwood, 1988).

and as expected, the vacuum Wigner function gives rise to infinities (i) in the gap equation, (ii) in the energy–momentum tensor and (iii) in the excitation spectrum. So these quantities have to be renormalized. As usual, after counterterms are introduced into the Lagrangian, first the physical quantities are regularized and next the infinities are removed while arbitrary constants that occur in these processes are determined through the values of the physical constants of the problem. As indicated above, the vacuum Wigner function is given by

$$F_{\text{vac}}(p) = -\frac{\gamma \cdot p + m}{(2\pi)^3}\delta(p^2 - m^2)\theta(-p \cdot u). \tag{9.44}$$

Therefore, the following counterterms are introduced into the Lagrangian and one has

$$\mathcal{L} \to \mathcal{L} + \mathcal{L}_c, \tag{9.45}$$

with

$$\mathcal{L}_c = \frac{\alpha}{1!}\varphi + \frac{\delta\mu^2}{2!}\varphi^2 + \frac{\gamma}{3!}\varphi^3 + \frac{\lambda}{4!}\varphi^4 - \frac{Z}{2}(\partial\varphi)^2. \tag{9.46}$$

9.4.1. *Regularization of the gap equation*

With the latter counterterms the gap equation is easily found to have the form

$$\begin{aligned}(m_R^2 + \delta m_S^2)(m - M) + \alpha g_S + \frac{\gamma}{2g_S}(m - M)^2 + \frac{\lambda}{6g_S^2}(m - M)^3 \\ = 4\pi g_S^2 \text{Sp}\int d^4p\{F_{\text{mat}}(p) + F_{\text{vac}}(p)\},\end{aligned} \tag{9.47}$$

where m_R is the renormalized (finite) mass of the scalar field and $F_{\text{mat}}(p)$ is nothing but the "matter part" (i.e. the Fermi–Dirac terms). The vacuum term is explicitly given by

$$-\frac{16\pi g_S^2}{(2\pi)^3}M\int d^4p\,\theta(-p^0)\delta(p^2 - M^2) = -\frac{8\pi g_S^2}{(2\pi)^3}M\int\frac{d^3p}{\sqrt{\mathbf{p}^2 + M^2}}, \tag{9.48}$$

which is obviously infinite and is now regularized via the *dimensional regularization* procedure,[5] a method possessing the advantages of manifest covariance, elegance and, very often, simplicity. The final result, proven

[5]C. Itzykson and J.B. Zuber, *Quantum Field Theory* (MacGraw-Hill, New York, 1981).

hereafter, is

$$-\frac{8\pi g_S^2}{(2\pi)^3}M\int\frac{d^3p}{\sqrt{\mathbf{p}^2+M^2}}=\frac{2m^3}{\pi\varepsilon}g_S^2\left(\frac{M}{m}\right)^3\left[1-\varepsilon\ln\left(\frac{M}{m}\right)\Lambda\right]+O(\varepsilon^2), \tag{9.49}$$

where the integral is calculated in $4-\varepsilon$ dimensions. It exhibits clearly a pole term in $1/\varepsilon$, while the term Λ is to be determined below.

In order to write down the vacuum Wigner function in $4-\varepsilon$ dimensions, the following expressions are used[6]:

$$\mathrm{Tr}(\gamma^\mu\gamma^\nu)=\phi(4-\varepsilon)\eta^{\mu\nu} \tag{9.50}$$

and[7]

$$\int\frac{d^nk}{(k^2+b^2)^p}=\pi^{n/2}\frac{\Gamma\left(p-\frac{1}{2}n\right)}{\Gamma(p)}\frac{1}{(b^2)^{p-n/2}}, \tag{9.51}$$

where p and n are integers and $\phi(4-\varepsilon)$ is an arbitrary function with a continuous derivative such that $\phi(0)=4$. The vacuum Wigner function written in $4-\varepsilon$ dimensions reads

$$F_{\text{vac}}(p)=-\frac{\gamma\cdot p+m}{(2\pi)^3\phi(4-\varepsilon)}\chi(4-\varepsilon)\delta(p^2-m^2)\theta(-p\cdot u), \tag{9.52}$$

where $\chi(4-\varepsilon)$ is another arbitrary function of the same kind as $\phi(4-\varepsilon)$. As a matter of fact, the function $\phi(4-\varepsilon)$ does not play any role, since it is eliminated while one is taking the trace of F_{vac}. Furthermore, as will appear below, $\chi(4-\varepsilon)$ can always be taken to have the form

$$\chi(4-\varepsilon)=(\text{const})^\varepsilon. \tag{9.53}$$

Accordingly, the vacuum part of the integral appearing in the gap equation can be rewritten in $4-\varepsilon$ dimensions as

$$\begin{aligned}I\equiv\frac{1}{2}\int\frac{d^{3-\varepsilon}p}{\sqrt{p^2+m^2}}&=\frac{1}{2}\pi^{1-\varepsilon/2}\Gamma\left(\frac{1}{2}\varepsilon-1\right)M^{2-\varepsilon}\\&=\frac{1}{2}\pi M^2\exp\left(-\frac{1}{2}\varepsilon\ln\pi\right)\exp(-\varepsilon\ln M)\Gamma\left(\frac{1}{2}\varepsilon-1\right).\end{aligned} \tag{9.54}$$

[6]C. Itzykson and J.B. Zuber, *op. cit.*

[7]C. Nash, *Relativistic Quantum Fields* (Academic, New York, 1978).

Using the well-known functional relation $\Gamma(x+1) = x\Gamma(x)$ for defining $\Gamma\left(\frac{1}{2}\varepsilon - 1\right)$ and expanding the various terms of this last equation in powers of ε, one gets

$$I = -\frac{\pi M^2}{\varepsilon} + \frac{1}{2}\pi M^2(\ln(\pi M^2) - \ln\{\exp[1+\Gamma'(1)]\}) + O(\varepsilon). \tag{9.55}$$

It remains for one to expand the quantity $\chi(4-\varepsilon)I$, and one obtains

$$\chi(4-\varepsilon)I = -\frac{\pi M^2}{\varepsilon} + \pi M^2 \ln\left(\frac{M}{m}\Lambda\right) + O(\varepsilon), \tag{9.56}$$

with[8]

$$\frac{\Lambda}{m} \equiv \sqrt{\pi}\exp\left\{-\left[\frac{1}{2} + \frac{1}{2}\Gamma'(1) - \chi'(4)\right]\right\}. \tag{9.57}$$

Introducing now these results into the gap equation, one obtains

$$\begin{aligned}(m_R^2 + \delta m_S^2)(m-M) + \alpha g_S + \frac{\gamma}{2g_S}(m-M)^2 + \frac{\lambda}{6g_S^2}(m-M)^3 \\ = 4\pi g_S^2\left[\left\{\int d^4p\, f_{\text{eq}}(p)\right\} + \frac{M^3}{2\pi^2}\frac{1}{\varepsilon} - \frac{M^3}{2\pi^2}\ln\left(\frac{M}{m}\Lambda\right)\right] + O(\varepsilon).\end{aligned} \tag{9.58}$$

The pole term is now absorbed into the counterterms, leading to a redefinition of the various constants that appear on the left hand side. This is of course possible since the pole term is in M^3, the same power with which it appears on the left hand side. Then one is led to the following redefinition of the various constants:

$$\begin{cases} \delta\mu^2 = -\dfrac{6m^2 g_S^2}{\pi\varepsilon} + B_F, \\ \alpha = \dfrac{2m^3 g_S}{\pi\varepsilon} + A_f, \\ \gamma = \dfrac{12 m g_S^3}{\pi\varepsilon} + C_F, \\ \lambda = -\dfrac{12 g_S^4}{\pi\varepsilon} + D_F, \end{cases} \tag{9.59}$$

[8] χ being arbitrary, so is χ' and hence Λ is an arbitrary constant; moreover, since only the first derivative of χ is involved in this calculation, it can always be chosen as $\chi(4-\varepsilon) = (\text{const})^{\varepsilon}$.

where the constants (A_F, B_F, C_F and D_F) are finite constants to be related to the renormalized experimental values of (α_R, m^2_{SR}, γ_R, λ_R). We come back to this question in a later section.

9.4.2. *Regularization of the energy–momentum tensor*

The vacuum Wigner function gives rise to other infinities which occur in the energy–momentum tensor of the fermions

$$T^{\mu\nu}_{\text{fermions}} = \text{Sp} \int d^4p\, p^\mu p^\nu [F_{\text{mat}}(p) + F_{\text{vac}}(p)] \tag{9.60}$$

and it is *a priori* not obvious that the infinities due to the vacuum term can also be absorbed in the *same* counterterms as above.

Let us show these properties. The total energy–momentum tensor is

$$T^{\mu\nu} = T^{\mu\nu}_{\text{matter}} + T^{\mu\nu}_{\text{vac}} + T^{\mu\nu}_{\text{scal}}, \tag{9.61}$$

where $T^{\mu\nu}_{\text{matter}}$ is the finite temperature and density–dependent part of the energy–momentum tensor of the fermions and $T^{\mu\nu}_{\text{scal}}$ is the energy–momentum tensor of the scalar particles,

$$T^{\mu\nu}_{\text{scal}} = \frac{\eta^{\mu\nu}}{4\pi}\left\{\alpha(m-M) + \frac{1}{2}(m_S^2 + \delta\mu^2)(m-M)^2 \right.$$
$$\left. + \frac{\gamma}{3!}(m-M)^3 + \frac{\lambda}{4!}(m-M)^4\right\}; \tag{9.62}$$

it includes the various counterterms. Only $T^{\mu\nu}_{\text{vac}}$ is infinite and, owing to Lorentz invariance, it is necessarily proportional to $\eta^{\mu\nu}$,

$$T^{\mu\nu}_{\text{vac}} = X\eta^{\mu\nu}, \tag{9.63}$$

and X is given by

$$X = \frac{1}{4-\varepsilon}\,\text{Tr} \int d^{4-\varepsilon}p\, p\cdot\gamma\, F_{\text{vac}}(p), \tag{9.64}$$

where use has been made of[9]

$$\eta_{\mu\nu}\eta^{\mu\nu} = 4 - \varepsilon = \varphi(\varepsilon). \tag{9.65}$$

Using the above results, one is led to

$$X = -\frac{4M^2}{(2\pi)^3(4-\varepsilon)} I\chi(4-\varepsilon), \tag{9.66}$$

[9]See e.g. C. Itzykson and J.B. Zuber, *op. cit.*

where I is the same integral as the one occurring in the regularization of the gap equation. Therefore, one has

$$X = -\frac{M^2}{(2\pi)^3}\left\{-\frac{\pi M^2}{\varepsilon} + \pi M^2 \ln\left(\frac{M}{m}\Lambda\right) - \frac{\pi M^2}{4}\right\}, \tag{9.67}$$

where the constant pole term $m^4/8\pi^2\varepsilon$ has been eliminated on the ground that the energy–momentum tensor is defined up to a constant; in a vacuum $T^{\mu\upsilon} = 0$.

Finally, the regularized energy–momentum tensor can be written as

$$\begin{aligned} T^{\mu\nu}_{\text{finite}} &= T^{\mu\nu}_{\text{matter}} + \frac{\eta^{\mu\nu}}{4\pi}\left\{\frac{A_F}{g_S}(m-M) + \frac{B_F}{2g_S^2}(m-M)^2 + \frac{m_S^2}{2g_S^2}(m-M)^2\right. \\ &\quad \times \left.\frac{C_F}{3!g_S^3}(m-M)^3 + \frac{D_F}{4!g_S^4}(m-M)^4 - \frac{M^4}{2\pi}\ln\left(\frac{M}{m}\Lambda\exp\left(-\frac{1}{4}\right)\right)\right\}. \end{aligned} \tag{9.68}$$

Note that the constant Λ is redundant and could have been absorbed into the various counterterms.

9.4.3. *Determination of the constants* (A_F, B_F, C_F, D_F)

It remains for one to connect the constants (A_F, B_F, C_F, D_F and Λ) to the renormalized (i.e. physical) data $(\alpha_R, m_{SR}^2, \gamma_R, \lambda_R)$. First, Λ is determined in such a way that, in a normal vacuum where $\langle\varphi\rangle = 0$, and thus when $m = M$, the renormalized energy–momentum tensor $T_R^{\mu\nu} \equiv 0$. This is achieved by choosing $\Lambda = \exp(1/4)$.

Next, by definition, the "experimental" constants $(\alpha_R, m_{SR}^2, \gamma_R, \lambda_R)$ are respectively the coefficients of $(\varphi/1!, \varphi^2/2!, \varphi^3/3!, \varphi^4/4!)$, in a renormalized Lagrangian. Note also that the coefficient of $\eta^{\mu\nu}$ is nothing but an effective Lagrangian that accounts for the quantum fluctuations in the Hartree–Vlasov approximation.[10] In this effective Lagrangian the coefficients of $\langle\varphi\rangle$, $\langle\varphi^2\rangle$, $\langle\varphi^3\rangle$ and $\langle\varphi^4\rangle$, are not A_F, B_F, C_F and D_F, since the term $M \ln M$ also contains such terms when expanded in a power series of $\langle\varphi\rangle$. The Taylor formula up to fourth order in $\langle\varphi\rangle$ for the function

$$\eta(\langle\varphi\rangle) = -\frac{M^4}{8\pi^2}\ln\left(\frac{M}{m}\right) \tag{9.69}$$

[10] It is rather an effective potential, since there is no kinetic energy term in this coefficient, owing to the space–time translational invariance of the equilibrium state.

is

$$\eta(\langle\varphi\rangle) = \langle\varphi\rangle\frac{g_S m^3}{8\pi^2} - \langle\varphi\rangle^2\frac{7g_S^2 m^2}{16\pi^2} + \langle\varphi\rangle^3\frac{13g_S^3 m}{24\pi^2} - \langle\varphi\rangle^4\frac{25g_S^4}{96\pi^2} + \frac{\langle\varphi\rangle^5}{\pi^2 m[1-(g_S/m)\theta(\langle\varphi\rangle)]}\frac{1}{5!}, \tag{9.70}$$

where θ is an unknown constant such that $0 < \theta < 1$ where θ comes from the Taylor–McLaurin formula. Identifying now the various renormalized constants with the corresponding coefficients of $\langle\varphi\rangle, \ldots, \langle\varphi\rangle^4/4!$, one obtains

$$\begin{aligned} \frac{\alpha_R}{1!} &= \frac{A_F}{1!} + \frac{g_S m^3}{2\pi}, \\ 0 &= \frac{B_F}{2!} + \frac{7g_S^2 m^2}{4\pi}, \\ \frac{\gamma_R}{3!} &= \frac{C_F}{3!} + \frac{13g_S^3 m}{6\pi}, \\ \frac{\lambda_R}{4!} &= \frac{D_F}{4!} - \frac{25g_S^4}{24\pi}. \end{aligned} \tag{9.71}$$

The various unknown constants are thus completely determined in the function of the renormalized coupling constants and the mass. The above expressions are identical to the ones first obtained by S.A. Chin (1977) in an interesting article, with the use of conventional diagramatic techniques.

9.5. Qualitative Discussion of the Effects of Renormalization

In this section we discuss the qualitative behavior of the role of the vacuum on thermodynamics, whether in a normal or abnormal state. To this end we consider the energy density T_R^{00} of the system; it reads

$$T_R^{00} = T_{\text{mat}}^{00} + \left\{\frac{\mu_R^2}{2!}\langle\phi\rangle^2 + \frac{\gamma_R}{3!}\langle\phi\rangle^3 + \frac{\lambda_R}{4!}\langle\phi\rangle^4 + \frac{3g^5\langle\phi\rangle^5}{\pi^2 m[1-(g/m)\langle\phi\rangle\theta 5!]}\right\}, \tag{9.72}$$

where θ is, as above, a parameter such that $0 < \theta < 1$. This expression is obtained by using Eqs. (9.69)–(9.72) and (9.68). The effect of renormalization is twofold: (i) it leads to the observable values of the various constants μ_R^2, γ_R and λ_R, and (ii) it adds to the total energy density T_R^{00} a vacuum contribution, i.e. the last term on the right hand side of Eq. (9.68). It is not difficult to realize that this contribution is always positive, since $m - g\langle\phi\rangle > 0$ and $0 < \theta < 1$.

First, we note that the minimum (or minima) of the energy density T_R^{00} with respect to $\langle\phi\rangle$, i.e.

$$\frac{\partial}{\partial\langle\phi\rangle} T_R^{00}(\langle\phi\rangle) = 0, \tag{9.73}$$

leads to the renormalized gap equation, as it should.

The qualitative features of $T_R^{00}(\langle\phi\rangle)$ are shown on Figs. 9.4 and 9.5. When the various constants μ_R^2, γ_R and λ_R are such that matter is in a normal state — only one minimum in $T_R^{00}(\langle\phi\rangle)$ — the vacuum energy term does not change the general shape of the energy curve: it only displaces the position of the minimum and also the numerical value of the corresponding energy density, which is increased since we add a positive quantity.

The situation is, however, not so simple in the case of abnormal matter. In such a case, depending on the values of the constants μ_R^2, γ_R and λ_R

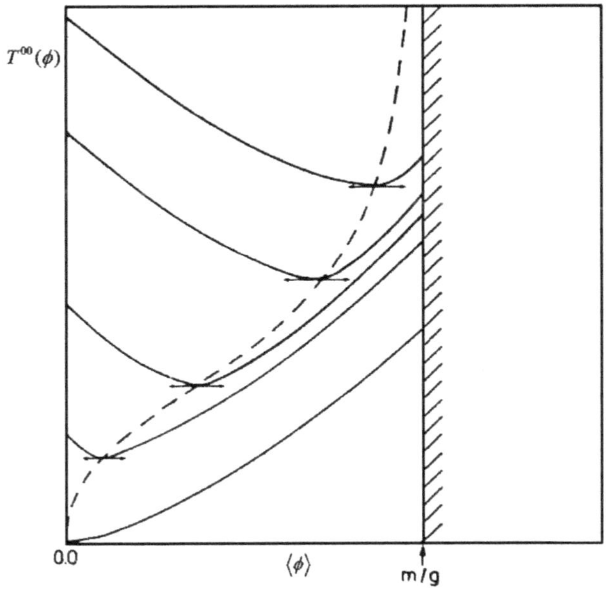

Fig. 9.4 The energy density is plotted as a function of $\langle\phi\rangle$. The various continuous curves correspond to various Fermi energies. On the dashed line lie the minima of the continuous lines. At $T = 0\,\mathrm{K}$ the equation providing these minima is the gap equation. At $T \neq 0\,\mathrm{K}$ the free energy should be plotted.

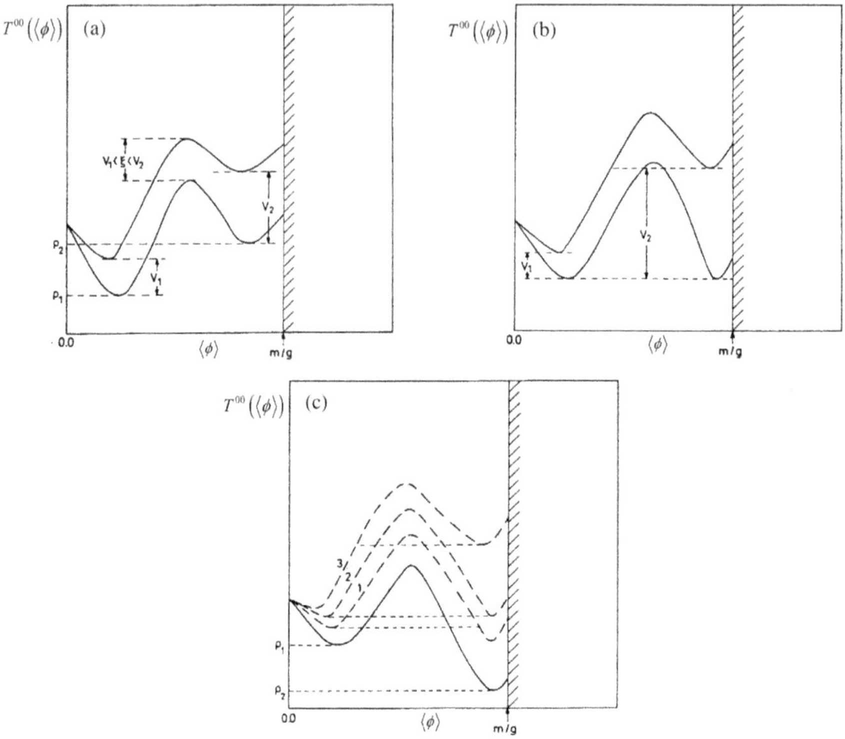

Fig. 9.5 The effect of quantum fluctuation on the energy density $T^{00}(\langle\phi\rangle)$ at $T = 0\,\mathrm{K}$ and for a given value of p_f is represented. The quantum term in $T^{00}(\langle\phi\rangle)$ has two effects: (i) it increases the energy density (the larger $\langle\phi\rangle$ and the larger the increase) and (ii) it shifts the minima in the direction of decreasing $\langle\phi\rangle$. Various cases are shown in diagrams (a)–(c). (a) Normal matter remains normal. (b) The degenerate ground state becomes nondegenerate and the first minimum is the lowest: matter becomes normal. (c) In the semiclassical case (continuous curve), matter is in an abnormal state; the effect of quantum fluctuations may *a priori* be represented by one of the dashed lines labeled 1, 2 or 3: (1) matter remains abnormal, (2) the ground state becomes degenerate, and (3) matter becomes normal. The existence of these various cases depends on the possible value of m/g.

[a detailed discussion on the various cases has been given by T.D. Lee and M. Margulies (1975)], there are two minima (separated by a maximum) corresponding to two possibilities: either the lowest minimum is obtained for the smallest values of $\langle\phi\rangle$ (in such a case it corresponds to a stable normal state while the other minimum is metastable and the maximum is unstable) or it is obtained at a larger value, $\langle\phi\rangle$ [in this case, the first minimum is metastable — see Fig. 9.5(c) — while the second is stable and represents abnormal matter]. Moreover, whether these two cases [shown on Fig. 9.5(c)] exist or not does depend on the possible limiting values of $\langle\phi\rangle$, i.e. of m/g. What is now the effect of quantum fluctuations on this situation? In fact, it is easy to realize that the term brought by the vacuum fluctuations is not only positive but also monotically increasing with $\langle\phi\rangle$. This means that the value of the energy density corresponding to the first minimum to the right of $\langle\phi\rangle = 0$ is *less* increased than for the second one, corresponding to a larger value of $\langle\phi\rangle$. It follows that several cases have to be considered. For the sake of the discussion we label ρ_1 and ρ_2 the minima corresponding to $\langle\phi\rangle_1$ and $\langle\phi\rangle_2$, respectively, and take $0 < \langle\phi\rangle_1 \leq \langle\phi\rangle_2$; let us also call V_1 and V_2 the corresponding vacuum contribution (ρ_1 refers to the quasiclassical case). Therefore, we examine the following different cases:

(i) $\rho_1 < \rho_2$ (metastable abnormal state — ρ_2; stable normal state — ρ_1). In this case $\rho_1 + V_1 < \rho_2 + V_2$ and hence the quantum fluctuations enhance the stability of the normal state; also, they can possibly suppress the second (metastable) minimum, depending on its depth relative to the maximum (the energy density of the maximum is less increased than that of the metastable minimum).

(ii) $\rho_1 = \rho_2$ (and $\langle\phi\rangle_1 < \langle\phi\rangle_2$). In this case, the quasiclassical degenerate normal state is split into a normal state (state 1) and, possibly, a metastable state 2 (or no state 2 at all).

(iii) $\rho_1 > \rho_2$ (normal metastable state and abnormal stable state). This is the most complicated case, since there are several possibilities: (a) $\langle\phi\rangle_2 > m/g$, (b) $\rho_1 + V_1 = \rho_2 + V_2$, (c) $\rho_1 + V_1 < \rho_2 + V_2$; of course, these various cases depend on the values of the constants at hand.

Finally, it should be mentioned that all these cases have to be considered according to the value of m/g, since for example when $\langle\phi\rangle_2 > m/g$ there is no physical second minimum.

9.6. Thermodynamics of the System

Before looking at the gap equation and its properties, we first indicate how it can be obtained from the free energy of the system.

9.6.1. *The gap equation as a minimum of the free energy*

We start from the usual thermodynamical relations

$$\begin{cases} a = \rho - Ts, \\ \mu = \dfrac{\rho - Ts + P}{n_{\rm eq}}, \end{cases} \tag{9.74}$$

where a is the free energy density, s the entropy density, $n_{\rm eq}$ the fermion density, and ε_f the chemical potential. Note that ε_f is the Fermi energy and μ reduces to the latter only at $T = 0\,\mathrm{K}$. Explicitly, one has

$$\begin{aligned} \rho = {} & \frac{M^4}{\pi^2}\int_0^\infty \varepsilon^2\sqrt{\varepsilon^2+1}(n_+ + n_-)d\varepsilon \\ & + \frac{\mu_R^2}{8\pi}\langle\phi\rangle^2 + \frac{\gamma_R}{24\pi}\langle\phi\rangle^3 + \frac{\lambda_R}{96\pi}\langle\phi\rangle^4 - \frac{M^4}{8\pi^2}\ln\left(\frac{M}{m}\right) \\ & - \frac{gm^3}{8\pi^2}\langle\phi\rangle + \frac{7m^2g^2}{16\pi^2}\langle\phi\rangle^2 - \frac{13mg^2}{24\pi^2}\langle\phi\rangle^3 + \frac{25g^4}{96\pi^2}, \end{aligned} \tag{9.75}$$

$$\begin{aligned} P = {} & \frac{M^4}{3\pi^2}\int_0^\infty \varepsilon^4\frac{1}{\sqrt{\varepsilon^2+1}}(n_+ + n_-)d\varepsilon \\ & - \frac{\mu_R^2}{8\pi}\langle\phi\rangle^2 - \frac{\gamma_R}{24\pi}\langle\phi\rangle^3 - \frac{\lambda_R}{96\pi}\langle\phi\rangle^4 + \frac{M^4}{8\pi^2}\ln\left(\frac{M}{m}\right) \\ & + \frac{gm^3}{8\pi^2}\langle\phi\rangle - \frac{7m^2g^2}{16\pi^2}\langle\phi\rangle^2 + \frac{13mg^2}{24\pi^2}\langle\phi\rangle^3 - \frac{25g^4}{96\pi^2}, \end{aligned} \tag{9.76}$$

$$n_{\rm eq} = \frac{M^4}{\pi^2}\int_0^\infty \varepsilon^2(n_+ - n_-)d\varepsilon. \tag{9.77}$$

In Eqs. (9.75)–(9.77), $n_\pm$ are the usual Fermi–Dirac factors:

$$n_\pm = \frac{1}{\exp[\beta(E \mp \mu)] + 1}. \tag{9.78}$$

From Eqs. (9.74)–(9.76) one gets

$$a = n_{\rm eq}\mu - P \tag{9.79}$$

and, in thermal equilibrium, one has

$$\left.\frac{\partial a}{\partial\langle\phi\rangle}\right|_{T,n_{\rm eq}} = 0 = \left.\frac{\partial a}{\partial M}\right|_{T,n_{\rm eq}}. \tag{9.80}$$

On the other hand, it is not difficult to check that

$$\begin{aligned}\frac{\partial P_{\text{mat}}}{\partial M} &= -\frac{M^3}{\pi^2}\int_0^\infty \frac{d\varepsilon}{\sqrt{\varepsilon^2+1}}\varepsilon^2(n_+ + n_-) + \frac{M^3}{\pi^2}\frac{\partial \varepsilon_f}{\partial M}\int_0^\infty d\varepsilon\,\varepsilon^2(n_+ - n_-)\\ &= -\frac{M^3}{\pi^2}\int_0^\infty \frac{d\varepsilon}{\sqrt{\varepsilon^2+1}}\varepsilon^2(n_+ + n_-) + n_{\text{eq}}\frac{\partial \varepsilon_f}{\partial M},\end{aligned} \tag{9.81}$$

which has been obtained with various integrations by parts and also by using the properties of the Fermi–Dirac factors n_{eq}. In Eq. (9.81) P_{mat} is the pressure of the fermions, i.e. the first term of Eq. (9.81). Finally, Eq. (9.79) yields

$$\begin{aligned}0 = \left.\frac{\partial a}{\partial M}\right|_{T,n_{\text{eq}}} &= n_{\text{eq}}\frac{\partial \varepsilon_f}{\partial\langle\phi\rangle} - \frac{\partial P}{\partial\langle\phi\rangle}\\ &= n_{\text{eq}}\frac{\partial \varepsilon_f}{\partial\langle\phi\rangle} - g\frac{M^3}{\pi^2}\int_0^\infty d\varepsilon\frac{\varepsilon^2}{\sqrt{\varepsilon^2+1}}(n_+ + n_-)\\ &\quad + \frac{1}{4\pi}\left[\mu_R^2\langle\phi\rangle + \frac{\gamma_R}{2}\langle\phi\rangle^2 + \frac{\lambda_R}{6}\langle\phi\rangle^3 + \frac{2gM^3}{\pi}\ln\left(\frac{M}{m}\right)\right.\\ &\quad \left. + \frac{gM^3}{2\pi} - \frac{gm^3}{2\pi} + \frac{7m^2g^2}{2\pi}\langle\phi\rangle - \frac{13mg^3}{2\pi}\langle\phi\rangle^2 + \frac{25g^4}{6\pi}\langle\phi\rangle^3\right],\end{aligned} \tag{9.82}$$

which is nothing but the renormalized gap equation.

9.6.2. *Thermodynamics*

With this full renormalization of the theory in Hartree–Vlasov approximation, one can finally write the basic equations that determine the thermodynamic state, i.e. the renormalized gap equation

$$\begin{aligned}&g_S\left(\alpha_R - \frac{g_S m^3}{2\pi}\right) + \left(\frac{m_S^2}{2!} + \frac{7g_S^2m^2}{2\pi}\right)(m-M)\\ &\quad + \left(\gamma_R - \frac{13g_S^3m}{\pi}\right)\frac{(m-M)^2}{2g_S} + \left(\lambda_R - \frac{25g_S^4}{\pi}\right)\frac{(m-M)^3}{6g_S^2}\\ &= 4\pi g_S^2\,\text{Sp}\int d^4p\,F_{\text{mat}}(p) - \frac{2m^3g_S^2}{\pi}\left(\frac{M}{m}\right)^3\ln\left(\frac{M}{m}e^{1/4}\right)\end{aligned} \tag{9.83}$$

and the renormalized energy–momentum tensor

$$T_R^{\mu\nu} = T_{\text{mat}}^{\mu\nu} + \frac{1}{4\pi}\eta^{\mu\nu}\left\{\left(\alpha_R - \frac{g_S m^3}{2\pi}\right)\langle\varphi\rangle + \left(\frac{m_S^2}{2!} + \frac{7g_S^2 m^2}{2\pi}\right)\langle\varphi\rangle^2 + \left(\gamma_R - \frac{13g_S^3 m}{\pi}\right)\langle\varphi\rangle^3 + \left(\lambda_R - \frac{25g_S^4}{\pi}\right)\langle\varphi\rangle^4 - \frac{(m - g_S\langle\varphi\rangle)^4}{2\pi}\right\}\ln\left(1 - \frac{g_S\langle\varphi\rangle}{m}\right). \tag{9.84}$$

Note that, owing to the values obtained earlier for A_F and its relation to α_R, one has $\alpha_R = 0$.

Since the thermodynamic state of the system depends on the effective mass of the fermions, it is necessary to study briefly the gap equation it satisfies.

As an example of the role of the vacuum fluctuations in the thermodynamics of the system, several figures are drawn below which allow a comparison of the renormalized and semiclassical data.

Although these figures refer to the zero temperature case, results at finite temperatures were also obtained elsewhere [J. Diaz Alonso and R. Hakim (1984)]. The conclusions are quite similar: the effective mass is higher in the renormalized case (Fig. 9.6) than in the semiclassical one and the pressure

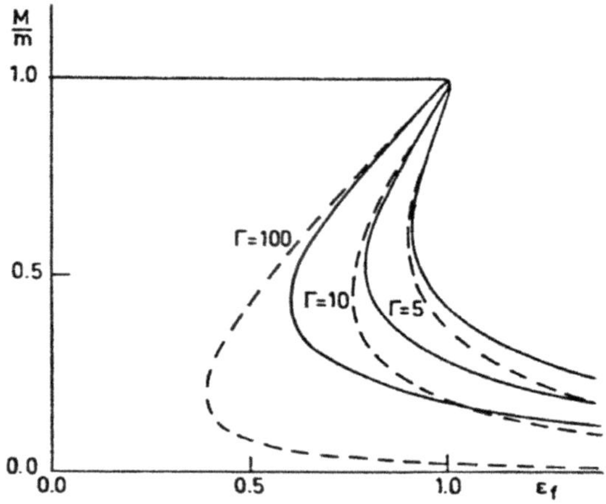

Fig. 9.6 Plot of the effective mass (in units of the baryon mass) at 0 K as a function of the Fermi energy of the system for several values of the coupling constant. The dotted line represents the semiclassical case (Hartree–Vlasov), while the continuous line corresponds to the renormalized case [after J. Diaz Alonso (1985); J. Diaz Alonso and R. Hakim (1984)].

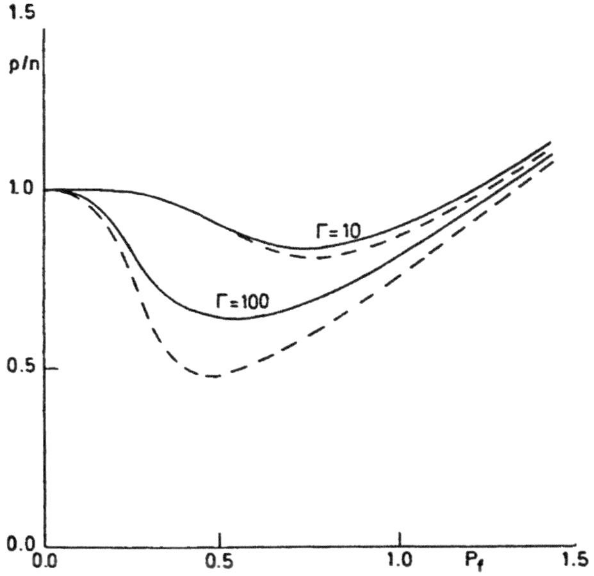

Fig. 9.7 The energy per baryon as a function of the Fermi momentum, at 0 K (same notations as in the preceding figure [after J. Diaz Alonso (1985); J. Diaz Alonso and R. Hakim (1984)].

(Fig. 9.8) is generally higher so that the first order phase transition is somewhat softened, depending on the value adopted for the coupling constant Γ. Accordingly, the phase diagram is also slightly modified (Fig. 9.9). The energy per fermion is also depicted in Fig. 9.7.

9.7. Renormalization of the Excitation Spectrum

When the vacuum terms are no longer omitted, the dispersion equation (or the Fourier transform of the Klein–Gordon equation) — to which the contribution of the counterterms has been added — can be written as

$$- k^2(1+Z)\varphi_1 + \left[m_S^2 + \delta\mu^2 + \gamma\langle\varphi_{\rm eq}\rangle + \frac{\lambda}{2}\langle\varphi_{\rm eq}\rangle^2\right]\varphi_1$$
$$= [\Pi_{\rm mat}(k) + \Pi_{\rm vac}(k^2)], \tag{9.85}$$

where $\Pi_{\rm mat}(k)$ has been calculated above and $\Pi_{\rm vac}(k^2)$ has a similar form with the difference that $f_{\rm eq}(p)$ has to be replaced by $f_{\rm vac}(p)$. Explicitly,

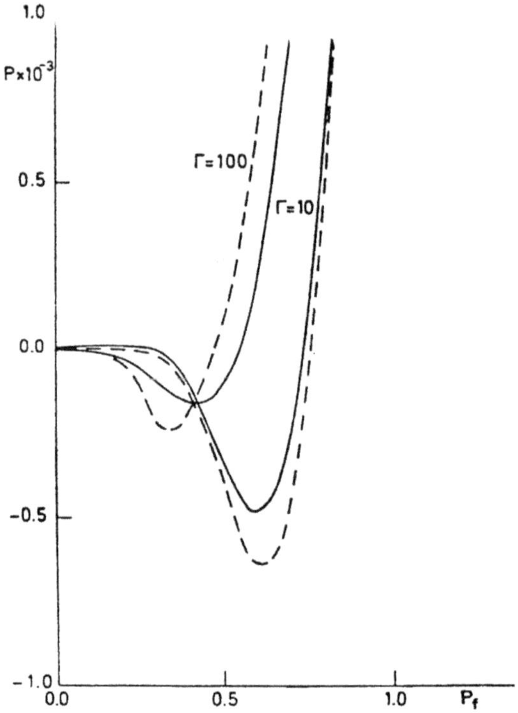

Fig. 9.8 The pressure as a function of the Fermi momentum, at 0 K (same notations as in the preceding figure [after J. Diaz Alonso (1985); J. Diaz Alonso and R. Hakim (1984)].

one has

$$\Pi_{\text{vac}}(k^2) = -\frac{2g_S^2}{2\pi^2}(4M^2 - k^2)k^2 \int d^4\xi \frac{\theta(-\xi^0)\delta(\xi^2 - M^2)}{(\xi \cdot k)^2 - \frac{1}{4}k^2} + \frac{2g_S^2}{\pi^2} \int d^4\xi\, \theta(-\xi^0)\, \delta(\xi^2 - M^2). \tag{9.86}$$

The two integrals occurring in $\Pi_{\text{vac}}(k^2)$ are clearly divergent and must be regularized as integrals in the gap equation and the energy–momentum tensor. These integrals are respectively called I_0 and I_1, so that

$$\Pi_{\text{vac}}(k^2) = -\frac{2g_S^2}{2\pi^2}(4M^2 - k^2)k^2 I_0 + \frac{2g_S^2}{\pi^2} I_1. \tag{9.87}$$

I_1 has already been calculated in $4 - \varepsilon$ dimensions in the regularization of the gap equation, while I_0 is given by [J. Diaz Alonso and

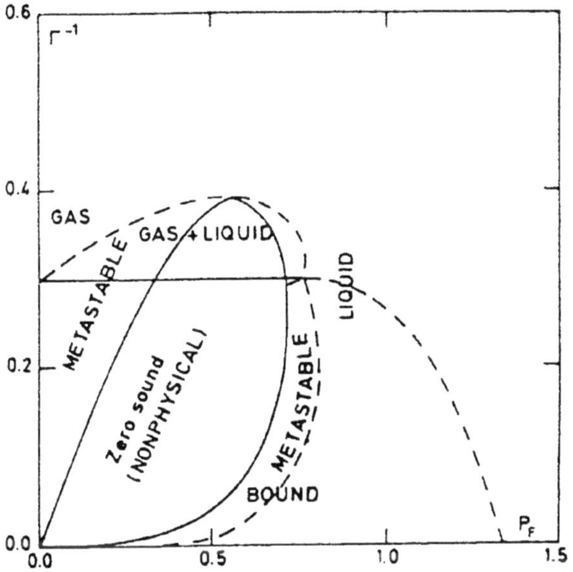

Fig. 9.9 The phase diagram — Γ^{-1} as a function of the Fermi momentum — at 0 K (same notations as in the preceding figure [after G. Kalman (1974); J. Diaz Alonso (1985); J. Diaz Alonso and R. Hakim (1984)].

R. Hakim (1988)]

$$I_0 = \frac{1}{k^2}\frac{2\pi}{\varepsilon} - \frac{\pi}{k^2}\left\{2\ln\left(\frac{\Lambda M}{m}\right) + 1 + \frac{1}{2}\int_0^1 \frac{d\alpha}{\sqrt{1-\alpha}}\ln\left(1 - \frac{\alpha k^2}{4M^2}\right)\right\}. \tag{9.88}$$

Finally, the vacuum polarization "tensor" reads

$$\Pi_{\text{vac}}(k^2) = -\frac{6M^2 g_S^2}{\pi\varepsilon} + \frac{k^2 g_S^2}{\pi\varepsilon} + \Pi_{\text{vac}}^f(k^2), \tag{9.89}$$

where the finite part, $\Pi_{\text{vac}}^f(k^2)$, is given by

$$\begin{aligned}\Pi_{\text{vac}}^f(k^2) = \frac{2g_S^2 M^2}{\pi}&\left\{3\ln\left(\frac{\Lambda M}{m}\right) + 1 + \frac{1}{2}\int_0^1 \frac{d\alpha}{\sqrt{1-\alpha}}\ln\left(1 - \frac{\alpha k^2}{4M^2}\right)\right\}\\ - \frac{g_S^2 k^2}{\pi}&\left\{2\ln\left(\frac{\Lambda M}{m}\right) + 1 + \frac{1}{2}\int_0^1 \frac{d\alpha}{\sqrt{1-\alpha}}\ln\left(1 - \frac{\alpha k^2}{4M^2}\right)\right\}.\end{aligned} \tag{9.90}$$

The pole terms are now canceled by choosing the various counterterms $\delta\mu^2$, λ and Z as

$$\begin{cases} \delta\mu^2 = -\dfrac{6m^2 g_S^2}{\pi\varepsilon} + B_F, \\ \gamma = \dfrac{12 m g_S^3}{\pi\varepsilon} + C_F, \\ \lambda = -\dfrac{12 g_S^4}{\pi\varepsilon} + D_F, \\ Z = -\dfrac{g_S^2}{\pi\varepsilon} + Z_F, \end{cases} \tag{9.91}$$

where the constants indexed by F are as yet arbitrary unknown constants to be determined as above.

It should be noted that the poles occurring in these last expressions are exactly the same as those obtained previously in the renormalization of the gap equation. However, the infinite constants are *a priori* different from the ones already obtained. It should indeed be borne in mind that while the Hartree approximation is also used — and this explains the same poles as before — the calculation of the excitation modes implies a slightly off-equilibrium state: as a matter of fact, this off-equilibrium situation is connected with a *higher order* equilibrium approximation. It follows that the above finite constants are not necessarily the same as those obtained in the renormalization of the gap equation.

In order to determine these constants, we first demand that, in a normal vacuum (i.e. at $T = 0$ and $p_f = 0$), $\langle\varphi\rangle_{\text{vac}} = 0$, the dispersion equation reduces to

$$k^2 - m_S^2 = O((k^2 - m_S^2)^2), \tag{9.92}$$

when $k^2 \approx m_S^2$; i.e. the dispersion curve possesses the usual meson branch at least at low three-momenta. This condition leads, as usual, to the two equations

$$\begin{cases} \tilde{\Pi}(k)\Big|_{\substack{k^2 = m_S^2 \\ \langle\varphi\rangle_{\text{eq}} = 0}} = 0, \\ \dfrac{d}{dk^2}\tilde{\Pi}(k)\Big|_{\substack{k^2 = m_S^2 \\ \langle\varphi\rangle_{\text{eq}} = 0}} = 0, \end{cases} \tag{9.93}$$

where $\tilde{\Pi}(k)$ is the vacuum polarization tensor plus the finite part of the counterterm contributions:

$$\tilde{\Pi}(k) = \left\{ -k^2 Z_F + B_F + C_F \langle\varphi\rangle_{\text{eq}} + \frac{1}{2} D_F \langle\varphi\rangle_{\text{eq}}^2 - \Pi_{\text{vac}}^f(k^2) \right\}. \tag{9.94}$$

The second condition provides Z_F as a function of g_S^2, m_S^2 and m, while the first one allows the determination of B_F as a linear function of Z_F. C_F and D_F are now determined from the definitions of γ_R and λ_R,

$$\begin{cases} \dfrac{d}{d\langle\varphi\rangle_{\text{eq}}} \tilde{\Pi}(k) \bigg|_{\substack{k^2 = m_S^2 \\ \langle\varphi\rangle_{\text{eq}} = 0,}} = \gamma_R, \\ \dfrac{d}{d\langle\varphi\rangle_{\text{eq}}^2} \tilde{\Pi}(k) \bigg|_{\substack{k^2 = m_S^2 \\ \langle\varphi\rangle_{\text{eq}} = 0}} = \lambda_R, \end{cases} \tag{9.95}$$

and immediately lead to an explicit form of C_F and D_F as functions of g_S^2, m_S^2, and of the "experimental" values of γ_R and λ_R.

Finally, the arbitrary constant Λ is given the same value as above and is thus chosen in such a way that the vacuum energy density is zero.

9.7.1. *Comparison with the semiclassical case*

We now compare briefly the renormalized and the second order cases.

First, we compare the dispersion curves $\omega = \omega(\mathbf{k})$ in the case of $\gamma_R = \lambda_R = 0$, still in the case $\Gamma = 100$ and a temperature of 0 K. Furthermore, several values of p_F have been taken: $p_F = 0.41, 0.50, 0.60$, in units of the fermion mass m. Also, we take the value $m/\mu = 3$.

In both cases there exist four branches: (i) the usual meson branch whose asymptote is the straight line $k^0 = |\mathbf{k}|$, (ii) a zero sound branch and (iii) two other meson branches at high frequencies and large wave numbers related to the above vacuum polarization term.

Let us comment a little bit on these curves in the two approximations under consideration. In both cases the meson branch is quite similar to the usual one. In fact, when the fermion density (or the Fermi momentum p_F) tends to zero, this curve tends to the usual one, $k^{02} - \mathrm{k}^2 = \mu_R^2$. As to the zero sound branch in the Hartree approximation, it exists only in the unphysical region of the renormalized phase diagram while, in the perturbative case, it still occurs in the physical region and furthermore it appears to be larger and larger as the density is increased. Besides this property, the Hartree curve is much smaller than the perturbative one.

The second boson-like branch, at high frequency, stems from the vacuum contribution. It also occurs in the perturbative context at order 2 in the coupling constant. In a vacuum the dispersion relation has the form

$$(k^2 - \mu_R^2)A(k^2) = 0, \tag{9.96}$$

where $A(k^2)$ is a known function; it follows that the new branch obeys the dispersion relation $A(k^2) = 0$. Such a branch does occur in QED, but it is usually considered as being unphysical, since it exists at ultrahigh energies [$\sim 2m \exp(137)$]. However, it also exists in the $\pi - N$ system at much lower density ($\sim$4 GeV). Morever, even in QED, it has been shown that, in the presence of a magnetic field $\sim 1.9 \cdot 10^{16}$ G, it does correspond to the possible existence of a massive longitudinal photon and has a finite lifetime. In our case also, a preliminary study of the stability of this branch indicates that it is probably stable (i.e. it has a finite lifetime).

Consequently, it is no surprise if such a branch is also present in the case of the system nucleon-scalar particle itself. In the case of a 4 GeV high energy "pion," this should lead to immediate experimental consequences; in the scalar case, the situation is less clear because of the fact that an actual physical system should first be associated with the case of the "scalar plasma."

9.8. A Short Digression on Bosons

Before closing this chapter, it should be worth having a few words on the case of bosons. In other words, let us consider (Fig. 9.10) the case of a Lagrangian,

$$L = \frac{1}{2}(\partial\phi)^2 - \frac{\mu_0^2}{2}\phi^2 - \frac{\lambda_0}{4!}\phi^4, \tag{9.97}$$

where ϕ is a real field, in order to simplify the discussion.

In such a case, and since the chemical potential is zero,

$$F = -P, \tag{9.98}$$

where the free energy F is only the opposite of the total pressure P, and the pressure is obtained from the energy–momentum tensor

$$T_{\mu\nu} = \langle \partial_\mu\phi \cdot \partial_\nu\phi - \eta_{\mu\nu}L \rangle \tag{9.99}$$

via its spatial components,

$$P = -\frac{1}{3}\Delta_{\mu\nu}(u)T^{\mu\nu}, \tag{9.100}$$

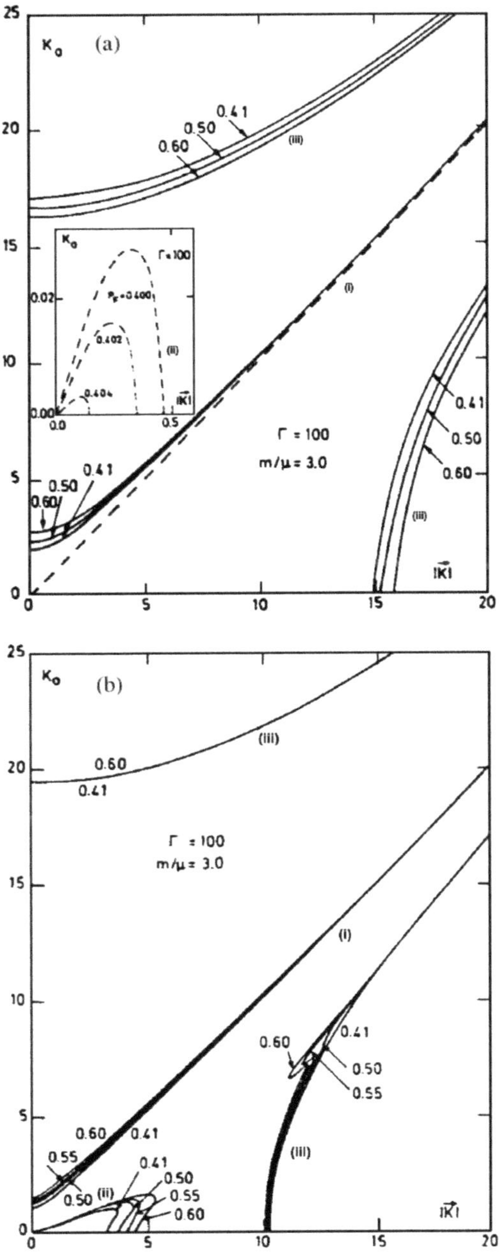

Fig. 9.10 The Hartree approximation for the dispersion curves as compared to the g^2 order; right — second order (renormalized) curve; left — renormalized Hartree–Vlasov case. The curves indexed (iii) represent a renormalized contribution.

$\Delta_{\mu\nu}(u)$ being the usual projector over the spatial components of the energy–momentum tensor. The result is that

$$F = +\frac{1}{3}\Delta_{\mu\nu}(u)\left\langle \partial^\mu\phi\,\partial^\nu\phi\right\rangle + \frac{1}{2}\mu_0^2\left\langle\phi^2\right\rangle + \frac{\lambda}{4!}\left\langle\phi^4\right\rangle - \frac{1}{2}\left\langle(\partial\phi)^2\right\rangle. \qquad (9.101)$$

Let us now be more specific by assuming that $\langle\phi\rangle = 0$, and let us also assume that a Gaussian approximation,

$$\begin{cases} \langle\phi^{(2n+1)}\rangle \approx 0, \\ \quad \langle\phi^{2n}\rangle \approx (2n-1)!!\langle\phi^2\rangle^n, \end{cases} \qquad (9.102)$$

does hold. This corresponds to the Hartree–Vlasov approximation of fermions.

We also set

$$K = \langle\phi^2\rangle \qquad (9.103)$$

and since

$$\phi(x) = \frac{1}{(2\pi)^{3/2}}\int\frac{d^3k}{\sqrt{2\omega(\mathbf{k})}}\{a_{\mathbf{k}}e^{-ik\cdot x} + a_{\mathbf{k}}^+e^{ik\cdot x}\} \qquad (9.104)$$

results from the freelike equations obeyed by ϕ, K has the value

$$K = \frac{1}{(2\pi)^3}\int\frac{d^3k}{\omega(\mathbf{k})}\left\{\frac{1}{e^{\beta\omega(\mathbf{k})}-1} + \frac{1}{2}\right\}. \qquad (9.105)$$

With this approximation, the free energy has the form

$$F = \frac{1}{(2\pi)^3}\int\frac{d^3k}{\omega(\mathbf{k})}\left\{\frac{1}{e^{\beta\omega(\mathbf{k})}-1} + \frac{1}{2}\right\} \times \left\{\frac{1}{2}\left[M^2 - \mu_0^2 - \frac{\lambda_0}{4}K - \frac{1}{3}\mathbf{k}^2\right]\right\}, \qquad (9.106)$$

the minimization of which, with respect to M, yields

$$M^2 = \mu_0^2 + \frac{\lambda_0}{2}K, \qquad (9.107)$$

which is infinite due to the term $1/2$ inside K.

A few comments are now in order. First, this last equation, the so-called gap equation, is an implicit equation for the parameter M, and second, it should be renormalized in much the same way as above, since it contains the bare parameters μ_0^2 and λ_0 and also the already-known infinite integral

$$\frac{1}{2(2\pi)^3}\int\frac{d^3k}{\sqrt{\mathbf{k}^2 + M^2}}, \qquad (9.108)$$

which we have encountered in the case of fermions. As to the gap equation itself, it can also be obtained from the equation

$$\Box\langle\phi(x)\phi(0)\rangle + \mu_0^2\langle\phi(x)\phi(0)\rangle + \frac{\lambda_0}{2}\langle\phi(x)\phi(0)\rangle K = 0, \tag{9.109}$$

which results from the equation of motion for ϕ, its multiplication by another ϕ, and also taking the approximation into consideration.

These considerations are fully considered in an article by F. Grassi *et al.* (1991).

Chapter 10

Covariant Kinetic Equations in the Quantum Domain

In the last few years, many attempts have been made to calculate the transport coefficients of relativistic dense matter, whether for nuclear (symmetric or pure neutrons) or quark matter. Two main motivations are behind these attempts. Firstly, experiments involving heavy ion collisions demand a better understanding of an assumed hydrodynamical stage[1] of the initial fireball and an eventual return to a state of local equilibrium. Secondly, the description of dense objects that occur in relativistic astrophysics (white dwarfs, neutron stars, strange stars, etc.) necessitates the knowledge of transport properties in many important problems, such as their cooling, or the energy and momentum transfer from inner to outer parts of the star.

Although the various transport coefficients can be evaluated via the use of Kubos relations,[2] or connected models, the transport coefficients are generally calculated on the basis of kinetic theory, i.e. through the use of a Boltzmann equation, or of its quantum version, or of any other kinetic equation. Such an equation involves the dynamics of dense matter under consideration through a cross-section. Unfortunately, the various cross-sections are generally not completely reliable, because they are calculated in a domain where little is known (deconfining transition for hadron/quark matter) or where the perturbative expansion does not converge (nuclear matter); also, the equation of state, from an experimental point of view, is not at all satisfactory.

Below, a systematic study of a generalization of the relaxation time approximation is performed, but let us briefly mention a few kinetic equations found in the literature. One of the most studied and used, particularly

[1]D. Bjorken, *Phys. Rev.* **D27**, 140 (1983).
[2]R. Kubo, *Lectures in Theoretical Physics*, Vol. 1 (Wiley, Interscience, New York, 1959).

in heavy ion collision, is the relativistic version of the Boltzmann equation, modified by Uhlenbeck and Uehling[3] in such a way that the relativistic Fermi–Dirac or Bose–Einstein distributions are stationary solutions of the collision term, which is not the case for the usual relativistic Boltzmann equation. This relativistic version, merely phenomenological, is often called the *BUU equation.* It reads

$$p \cdot \partial f(x,p) = \frac{1}{2} \int \frac{d^3p'}{p'_0} \frac{d^3p''}{p''_0} \frac{d^3\bar{p}}{\bar{p}_0} W(p',p'' \to p,\bar{p}) \delta^{(4)}(p+p''-p'-\bar{p})$$
$$\times \{f(x,p')f(x,p'')[1 \mp f(x,p)][1 \mp f(x,\bar{p})]$$
$$- f(x,p)f(x,\bar{p})[1 \mp f(x,p')] \times [1 \mp f(x,p'')]\}, \qquad (10.1)$$

where the plus sign refers to bosons and the minus one to fermions, and $W(p',p'' \to p,\bar{p})$ is (up to the δ term) the transition probability per unit of time for two colliding particles, which is assumed to obey the detailed balance property:

$$W(p',p'' \to p,\bar{p}) = W(p,\bar{p} \to p',p''). \qquad (10.2)$$

Note that this equation is not an equation for $F(x,\ p)$ but only for $f(x,p)$; it follows that spin effects are generally neglected in this merely phenomenological approach.

For the reasons mentioned above, a more phenomenological approach is preferred, based on the use of a relaxation time model of the collision term, where all the dynamics is supposed to be contained in a single parameter, via the relaxation time which should be estimated with another model. Possibly, the relaxation time can be replaced by a momentum-dependent "relaxation time function," $\tau(p)$. Next, the transport coefficients are calculated via the usual approximation methods, such as the Chapman–Enskog and 14-moment ones.

Such an approach is, of course, more modest than a general one but, besides its pedagogical value — by avoiding the complexity of involved equations — it presents the advantage of giving the general structure of the transport coefficients as functions of the temperature T, the particle density n, etc., *and* the relaxation time τ. Furthermore, when this last quantity is roughly evaluated as

$$\tau = \frac{1}{\sigma n v_{\rm th}}, \qquad (10.3)$$

[3]E.A. Uehling, G.E. Uhlenbeck, Phys. Rev., **43**, 552 (1933); E.A. Uehling, *ibid.*, **46**, 197 (1934).

where n is the particle — or possibly quasiparticle — density, σ is the total cross-section of the process under study and v_{th} is the average relative thermal velocity of the two colliding particles, reasonable orders of magnitude can be expected. Following this line, many authors have used a relativistic relaxation time model of the general form

$$p \cdot \partial f(x,p) = -\frac{f(x,p) - f_{\text{eq}}(x,p)}{\tau(p)}, \tag{10.4}$$

where $f(x,p)$ is a semiclassical distribution function for the particles (or the quasiparticles), $f(x,p)$ a local equilibrium distribution and $\tau(p)$ a given "relaxation time function" given *a priori*. For instance, Ch. Marle (1969) used $\tau(p) = \text{const}$, while the choice

$$\tau(p) = \frac{\tau}{u \cdot p} \tag{10.5}$$

is the one used by J.L. Anderson and H.R. Witting (1974) for reasons given previously (see Chap. 2). This last choice for $\tau(p)$ leads, as already indicated, to the Landau–Lifschitz form of relativistic hydrodynamics, a form that leads to more sensible results at low densities, as discussed by P. Danielewicz and M. Gyulassy (1985).

For the various reasons given in a subsequent section, J.L. Anderson's and H.R. Witting's approach is generalized in this chapter.[4]

10.1. General Form of the Kinetic Equation

In the context of the Wigner function approach used throughout this book, relativistic quantum kinetic equations — whether for fermions or for bosons — are obtained by replacing the interaction term of the equations obeyed by the one-particle Wigner function with a suitable collision term; for fermions, one can write

$$\begin{cases} \{i\gamma \cdot \partial + 2[\gamma \cdot p - m]\}F(x,p) = C_1[F(x,p)], \\ F(x,p)\{i\gamma \cdot \partial - 2[\gamma \cdot p - m]\} = C_2[F(x,p)], \end{cases} \tag{10.6}$$

where $C_1[F]$ and $C_2[F]$ represent the collision term. However, they are not independent and must satisfy the consistency relation (see Chap. 8)

$$\{i\gamma \cdot \partial + 2[\gamma \cdot p - m]\}^{-1} C_1[F(x,p)] = C_2[F(x,p)]\{i\gamma \cdot \partial - 2[\gamma \cdot p - m]\}^{-1}, \tag{10.7}$$

[4] R. Hakim, L. Mornas, P. Peter and H. Sivak (1992).

which is given a simpler form in the sequel. Furthermore, they must be connected to the mass shell relation, as shown in a subsequent section. Next, they must be such that the equilibrium Wigner function is a solution of the collision terms

$$C_1[F_{\text{eq}}(p)] = C_2[F_{\text{eq}}(p)] = 0. \tag{10.8}$$

As to a possible H theorem, the question is quite delicate and a possible way out is indicated below.

Let us briefly outline the boson form of the relativistic kinetic equation (see Chap. 13) and let us write

$$[k^2 - \Pi(k)]\phi(k) = 0 \tag{10.9}$$

as the dynamics of the boson field ϕ. In terms of the Wigner operator $f(k,p)$, the kinetic equations (without an *ad hoc* collision term) read

$$\left\{p \cdot k - \frac{1}{2}\left[\Pi\left(p + \frac{1}{2}k\right) - \Pi\left(p - \frac{1}{2}k\right)\right]\right\} f(p,k) = C_1\{f(k,p)\}, \tag{10.10}$$

$$\left\{p^2 + \frac{1}{4}k^2 - \frac{1}{2}\left[\Pi\left(p + \frac{1}{2}k\right) + \Pi\left(p - \frac{1}{2}k\right)\right]\right\} f(k,p) = C_2\{f(k,p)\}. \tag{10.11}$$

We write them in a Fourier transform and the relativistic kinetic equation contains obviously a collision term in the first equation while the second equation, which represents the mass shell, has no collision. In the case where Π is equal to m^2, exactly the Anderson–Whitting term is recovered.

10.2. An Introductory Example

Before studying the form of a general relaxation time kinetic equation, let us begin with a very simple example: a version of the BGK equation for spin 1/2 particles. This equation is written as the system

$$\begin{cases} \{i\gamma \cdot \partial + 2[\gamma \cdot p - m]\}F(x,p) = -i\gamma \cdot u \dfrac{F(x,p) - F_{\text{eq}}(p)}{\tau}, \\ F(x,p)\{i\gamma \cdot \partial - 2[\gamma \cdot p - m]\} = -\dfrac{F(x,p) - F_{\text{eq}}(p)}{\tau} i\gamma \cdot u, \end{cases} \tag{10.11}$$

where the brackets in the second equation act leftward. $F_{\text{eq}}(p)$ is the equilibrium Wigner function of a relativistic ideal Fermi gas. This system possesses all the characteristics of the BGK equation. It is linear and the equilibrium solution is a stationary solution. Four-current and energy–momentum conservations are satisfied provided that the Landau–Lifschitz matching conditions are verified. The two "collision terms" are of course mutually compatible. Finally, when one neglects the spin, i.e. when one neglects the spin part of $F_{\text{eq}}(p)$, one recovers the formal[5] BGK equation

$$p \cdot \partial f(x,p) = -p \cdot u \frac{f(x,p) - f_{\text{eq}}(p)}{\tau}, \tag{10.12}$$

given by J.L. Anderson and H.R. Witting (1976). Besides its simplicity, this kinetic equation possesses the interest of indicating a number of problems to be found in other relativistic and quantum kinetic equations.

Let us first try to solve this system by a *naive* Chapman–Enskog expansion in the parameter τ and let us limit ourselves to order 1:

$$F(x,p) = F_{\text{eq}}(x,p) + \tau F_1(x,p) + \tau^2 F_2(x,p) + \cdots, \tag{10.13}$$

where $F_{\text{eq}}(x,p)$ is a local equilibrium Wigner distribution. Solving the first equation, one obtains

$$F_1(x,p) = -\gamma \cdot u \gamma^\lambda \frac{\gamma \cdot p + m}{4m} \partial_\lambda f_{\text{eq}}(x,p), \tag{10.14}$$

while the solution to the second one is

$$F_1(x,p) = -\frac{\gamma \cdot p + m}{4m} \gamma^\lambda \gamma \cdot u \, \partial_\lambda f_{\text{eq}}(x,p), \tag{10.15}$$

which is different from, and in contradiction with, the first one; in the same time, the two equations are consistent.

In order to see exactly the reason why this naive Chapman–Enskog expansion provides wrong results, let us rewrite the above system as

$$\begin{cases} \gamma \cdot \partial F(x,p) = -\left\{\frac{\gamma \cdot u}{\tau} - 2i(\gamma \cdot p - m)\right\}[F(x,p) - F_{\text{eq}}(x,p)] \\ \partial F(x,p)\gamma = -[F(x,p) - F_{\text{eq}}(x,p)]\left\{\frac{\gamma \cdot u}{\tau} + 2i(\gamma \cdot p - m)\right\}, \end{cases} \tag{10.16}$$

which leads to

$$\begin{aligned} F(x,p) - F_{\text{eq}}(x,p) &= -\frac{\gamma \cdot \left(\frac{u}{\tau} - 2ip\right) - 2im}{\left(\frac{u}{\tau} - 2ip\right)^2 + 4m^2} \gamma \cdot \partial F(x,p) \\ F(x,p) - F_{\text{eq}}(x,p) &= -\partial F(x,p) \cdot \gamma \frac{\gamma \cdot \left(\frac{u}{\tau} + 2ip\right) + 2im}{\left(\frac{u}{\tau} + 2ip\right)^2 + 4m^2}. \end{aligned} \tag{10.17}$$

[5]Note that this equation is fully quantal despite its "classical" appearance: f_{eq} is indeed a quantum distribution.

After some rearrangements and setting $p^\mu = m\xi^\mu$, this system can be rewritten as

$$\begin{cases} F(x,p) - F_{\rm eq}(x,p) = -\dfrac{\gamma \cdot \left(\dfrac{u}{m\tau} - 2i\xi\right) - 2i}{\left(\dfrac{u}{m\tau} - 2ip\right)^2 + 4} \dfrac{\gamma \cdot \partial F(x,p)}{m}, \\ F(x,p) - F_{\rm eq}(x,p) = -\dfrac{\partial F(x,p) \cdot \gamma}{m} \dfrac{\gamma \cdot \left(\dfrac{u}{m\tau} + 2i\xi\right) + 2i}{\left(\dfrac{u}{m\tau} + 2i\xi\right)^2 + 4}, \end{cases} \tag{10.18}$$

which shows clearly that there are two expansion parameters in the system. The first one is the normal parameter occurring in the Chapman–Enskog expansion, i.e. $\varepsilon \equiv \tau/L$; and the second one is $\eta \equiv 1/mL$, which is the ratio of the Compton wavelength of the particles (remember that $\mathsf{h} = c = 1$) to a typical hydrodynamical length, L. Accordingly, we expand the solution of the BGK system in powers of ε and keeping the first order only; we get

$$F_1(x,p) = -\frac{1}{2} \frac{\gamma \cdot \xi + 1}{\eta \xi \cdot u} \frac{\gamma \cdot \partial F_{\rm eq}(x,p)}{m}, \tag{10.19}$$

or, equivalently,

$$F_1(x,p) = -\frac{\tau}{p \cdot u} p \cdot \partial F_{\rm eq}(x,p), \tag{10.20}$$

and the *same* expression is obtained from the second BGK equation, achieving thereby a consistent approximate solution. The source of the previous problem was, of course, the incoherent mixing of the two expansion parameters.

It might be argued that the particle Compton wavelength is always extremely small compared to the other lengths present in a physical system. This is generally exact except when one deals with quasiparticles whose mass may approach zero. In Chaps. 11, 13 and 14, we shall study such a case.

Another remark is that this solution for the relativistic BGK equation leads exactly to the Anderson–Witting results. Let us now briefly show this. What is needed is to get the transport coefficients in $T^{\mu\nu}$ and J^μ, which can be obtained from $F_{(1)}$. These last equations can be rewritten as

$$J^\mu = \int d^4p f^\mu, \quad T^{\mu\nu} = \int d^4p\, p^\mu f^\nu, \tag{10.21}$$

indicating that only the function $f^\mu_{(1)}(x,p)$ is of interest in view of our goal. It is given by

$$
\begin{aligned}
f^\mu_{(1)}(x,p) &\equiv \mathrm{Tr}(\gamma^\mu F_{(1)}) \\
&= -\frac{\tau}{p\cdot u}\frac{p^\mu}{m}p\cdot\partial f_{\mathrm{eq}} \\
&= \frac{p^\nu}{m}f_{(1)},
\end{aligned}
\tag{10.22}
$$

where the last equation is derived from the former with the use of the Anderson–Whitting first order form.

In other words, it has exactly the same form as (and is identical with) the Anderson–Witting results (1974b). This, of course, does not mean that the physical content of the kinetic equation for F is identical with that of Anderson and Witting; for instance, higher approximations do differ; or while it is appropriate for polarized media, this is not the case for the Anderson–Witting equation. However, it has been found, for unpolarized media, that most collision terms do possess the property that the first order Chapman–Enskog solution coincides with the Anderson–Witting solution.

Let us now take a glance at polarized media, i.e. those whose equilibrium Wigner function is given: $f^\mu_{5(1)}$.

It is not difficult to see that Eq. (10.20) is still valid and, more important, so is the case for Eq. (10.22). It follows that the transport equations so obtained are identical with those given by Anderson and Witting. However, from Eq. (10.20), one can obtain the relaxation of the average polarization four-vector $f^\mu_{5(1)}(x,p)$,

$$
f^\mu_{5(1)}(x,p) \equiv \frac{1}{4}\mathrm{Tr}[\gamma_5\gamma^\mu F_{(1)}(x,p)], \tag{10.23}
$$

as

$$
\begin{aligned}
M^\mu_{(1)}(x,p) &= \int d^4p f^\mu_{(1)5}(x,p) \\
&= -\tau\int d^4p\frac{p\cdot\partial}{p\cdot u}f_{5\mathrm{eq}}(x,p),
\end{aligned}
\tag{10.24}
$$

and the corresponding transport coefficients

$$
\begin{aligned}
M^\mu_{(1)}(x) = u^\mu &\left\{-\tau\lambda\mathcal{P}\,n\cdot X + \frac{\tau}{3}i_{4-1}(\pi^{\lambda\nu}\partial_\lambda n_\nu\mathcal{P} + n\cdot\partial\mathcal{P})\right\} \\
&+ n^\mu\left\{\tau\left(i_{21} - \frac{1}{3}i_{4-1}\right)\dot{\mathcal{P}}\right\} \\
&+ \pi^{\mu\alpha}\left\{\frac{\tau}{3}\mathcal{P}i_{4-1}\dot{n}\cdot\partial u_\alpha + \tau\mathcal{P}i_{21}\dot{n}_\alpha\right\},
\end{aligned}
\tag{10.25}
$$

where λ is the thermal conductivity, X^α is

$$X^\alpha \equiv \Delta^{\alpha\lambda}(u)[\partial_\lambda \beta + \beta \mathsf{u}_\lambda], \tag{10.26}$$

and i_{nm} and I_{mn} are integrals given by

$$\begin{cases} i_{nm} = \displaystyle\int_0^\infty \sinh^n x \cosh^m x \frac{1}{e^{\gamma \cosh x + \alpha} + 1}, \\ I_{nm} = \displaystyle\int_0^\infty \sinh^n x \cosh^m x \frac{e^{\gamma \cosh x - \alpha}}{e^{\gamma \cosh x + \alpha} + 1}. \end{cases} \tag{10.27}$$

It is finally clear that the system (10.11) does not allow — at least at first order in the Chapman–Enskog expansion — a coupling between polarization and four-current.

10.3. A General Relaxation Time Approximation

Although quite natural and valid more or less for dilute unpolarized systems, the Anderson–Witting equation possesses some obvious limitations, which we first briefly review. First, the concept of a distribution function does possess a well-known domain of validity and, instead, one should use a covariant Wigner function. Next, nucleons or quarks are fermions obeying some Dirac equations and spin is taken via a 4×4 covariant Wigner matrix, or a larger one when internal degrees of freedom are taken into account, while it is not so in the Anderson–Witting equation. On the other hand, while this latter equation must be supplemented by a mass shell constraint on p^μ, it is not so in the Wigner function approach. Finally, polarized matter can be dealt with more completely using our Wigner function approach. Moreover, when used in its original form, the Anderson–Witting equation does not allow the existence of spin waves (or internal quantum numbers waves).

Unfortunately, neither the obtaining of a relaxation time term for a relativistic kinetic equation obeyed by the Wigner function nor its Chapman–Enskog expansion is a trivial problem. As to the collision term, it is indeed difficult to infer its general form due to the matrix character of the Wigner function. As to the Chapman–Enskog expansion, quantum theory *and* relativity do introduce, as indicated by the previous example, a supplementary length and, accordingly, one more expansion parameter,

$$\varepsilon \equiv \frac{\tau}{L}, \quad \eta \equiv \frac{1}{mL}, \tag{10.28}$$

or the following one:

$$\bar{\eta} \equiv \frac{1}{m\tau} = \frac{\text{Compton wavelength}}{\text{relaxation time}}. \tag{10.29}$$

In ordinary cases, however, η is generally much smaller than ε, or $\bar{\eta} \ll 1$, and its contribution is perfectly negligible. Nevertheless, when one thinks of systems of quasifermions, their effective mass depends on T, ε_f, etc. and can in principle be arbitrarily small, leading thereby to arbitrarily large values of $\bar{\eta}$. A well-known example is the J.D. Walecka model (1974) or G. Kalman's *scalar plasma* (1976), where the effective mass of the quasifermions is given by

$$M = m - g\langle\varphi\rangle \tag{10.30}$$

(see Chaps. 9 and 13) and tends to zero at high densities and/or temperatures.[6]

10.3.1. *Properties of the kinetic system*

The basic system of the relativistic quantum kinetic equations reads

$$\begin{cases} \{i\gamma\cdot\partial + 2[\gamma\cdot p - m]\}F(x,p) = C[F(x,p)], \\ F(x,p)\{i\gamma\cdot\partial - 2[\gamma\cdot p - m]\} = \bar{C}[F(x,p)], \end{cases}$$

where $\bar{C}$ is chosen in such a way that this system is consistent, i.e. so that

$$\bar{C}[F] = -\gamma^0 C^\dagger[F]\gamma^0. \tag{10.31}$$

This property results from the following one

$$\gamma^0 F^\dagger \gamma^0 = F, \tag{10.32}$$

and from the requirement of consistency.[7]

In what follows, the collision terms are decomposed on the basis of the γ matrices as

$$\begin{cases} C = \frac{1}{4}\{cI + c_\mu\gamma^\mu + c_{\mu\nu}\sigma^{\mu\nu} + c_{5\mu}\gamma_5\gamma^\mu + c_5\gamma_5\}, \\ \bar{C} = \frac{1}{4}\{\tilde{c}I + \tilde{c}_\mu\gamma^\mu + \tilde{c}_{\mu\nu}\sigma^{\mu\nu} + \tilde{c}_{5\mu}\gamma_5\gamma^\mu + \tilde{c}_5\gamma_5\}. \end{cases} \tag{10.33}$$

[6] For instance, this property would not be true for nucleons interacting via a pseudoscalar meson field [see e.g. J. Diaz Alonso (1985)]; in such a case, one has indeed $M^2 = m^2 + g^2\langle\phi\rangle^2 > m^2$. It is, however, unstable.

[7] Note that this condition should be valid whatever the second member, be it a collision term or an interaction. Only the subsequent results specialized to a linear collision term are not general.

In our relaxation time approximation, F appears linearly in $C[F]$ and, since $C[F_{\rm eq}]=0$, $C[F]$ must depend on F through the combination

$$\delta F(x,p)\equiv F(x,p)-F_{\rm eq}(x,p). \tag{10.34}$$

The general BGK system does not look like a relativistic kinetic equation and this is due to the fact that it includes both transport properties and mass shell constraints. They can be disentangled in several ways and after decomposing the system on the basis of the Dirac algebra, after some algebra (after adding and subtracting these equations) one gets the "transport" system

$$\begin{aligned}
&p\cdot\partial f=\frac{1}{2i}p^{\mu}(c_{\mu}+\tilde{c}_{\mu}),\\
&p\cdot\partial f_{\nu}-p_{\mu}\partial f_{\nu}+m\partial_{\nu}f=\frac{1}{i}p^{\mu}(c_{\mu\nu}-\tilde{c}_{\mu\nu})+\frac{m}{2i}(c_{\nu}+\tilde{c_{\nu}}),\\
&p\cdot\partial f^{\mu\nu}+\partial^{[\mu}f^{\nu]\lambda}p_{\lambda}=\frac{1}{4i}\varepsilon^{\mu\nu\alpha\beta}p_{\alpha}(c_{5\beta}+\tilde{c}_{5\beta}),
\end{aligned} \tag{10.35a}$$

$$\begin{aligned}
&p\cdot\partial f_{5}=\frac{1}{2}p^{\mu}(c_{5\mu}-\tilde{c}_{5\mu})-\frac{m}{2}(c_{5}+\tilde{c}_{5}),\\
&p\cdot\partial f_{5}^{\lambda}-p_{\mu}\partial^{\lambda}f_{5}^{\mu}=\frac{1}{2}\varepsilon^{\mu\nu\alpha\lambda}p_{\alpha}(c_{\mu\nu}+\tilde{c}_{\mu\nu}),
\end{aligned} \tag{10.35b}$$

and the "mass shell" system

$$(p^{2}-m^{2})f=\frac{1}{4}p^{\mu}(c_{\mu}-\tilde{c}_{\mu})+\frac{m}{4}(c-\tilde{c})+\partial^{\mu}p^{\nu}f_{\mu\nu}, \tag{10.36}$$

$$\begin{aligned}
(p^{2}-m^{2})f_{\nu}=\;&\frac{1}{2}p^{\mu}(c_{\mu\nu}-\tilde{c}_{\mu\nu})+\frac{m}{4}(c_{\nu}-\tilde{c}_{\nu})+\frac{p_{\nu}}{4}(c-\tilde{c})\\
&+m\partial^{\mu}f_{\mu\nu}-\frac{1}{2}\varepsilon_{\rho\lambda\mu\nu}p^{\mu}\partial^{\lambda}f_{5}^{\rho},
\end{aligned} \tag{10.37}$$

$$\begin{aligned}
(p^{2}-m^{2})f_{\mu\nu}=&-\frac{1}{8}\varepsilon_{\rho\lambda\mu\nu}(c_{5}^{\rho}-\tilde{c_{5}^{\rho}})p^{\lambda}+\frac{m}{4i}(c_{\mu\nu}-\tilde{c}_{\mu\nu})\\
&+\frac{1}{8i}p_{[\mu}(c_{\nu]}+\tilde{c}_{\nu]})+\frac{1}{4}\varepsilon_{\rho\lambda\mu\nu}p^{\lambda}\partial^{\rho}f_{5}\\
&-\frac{1}{4}p_{[\mu}\partial_{\nu]}f-\frac{m}{4}\partial_{[\mu}f_{\nu]},
\end{aligned} \tag{10.38}$$

$$\begin{aligned}
(p^{2}-m^{2})f_{5}^{\lambda}=&-\frac{1}{4i}\varepsilon^{\mu\nu\alpha\lambda}p_{\alpha}(c_{\mu\nu}-\tilde{c}_{\mu\nu})+\frac{m}{4}(c_{5}^{\lambda}-\tilde{c_{5}^{\lambda}})\\
&-\frac{1}{4}p^{\lambda}(c_{5}+\tilde{c_{5}})+\frac{1}{2}\varepsilon^{\mu\nu\alpha\lambda}p_{\alpha}\partial_{\mu}f_{\nu}-\frac{m}{2}\partial^{\lambda}f_{5},
\end{aligned} \tag{10.39}$$

$$
\begin{aligned}
(p^2 - m^2)f_5 = &-\frac{1}{4i}p^\mu(c_{5\mu} + \tilde{c_{5\mu}}) + \frac{m}{4i}(c_5 - \tilde{c_5}) \\
&+ \frac{m}{2}\partial_\mu f_5^\mu - \frac{1}{2}\varepsilon_{\lambda\rho\nu\mu}p^\mu\partial^\lambda f^{\rho\nu}.
\end{aligned} \tag{10.40}
$$

These last equations — the "mass shell" system — need some discussion. In the absence of any external field or condensate whatsoever, in the kinetic regime we are considering, collisions are pointlike and hence colliding particles lie on the mass shell $p^2 = m^2$. This property can be seen in another way: when the solution to the transport equations is expanded into a convergent approximation whose zeroth order is such that $p^2 = m^2$, as is the case for an equilibrium Wigner function, then owing to the linearity of the collision term, each order is itself on the mass shell and, consequently, this is the case for the complete solution.

10.3.2. *The collision term*

The collision term, which must be *linear* in $(F - F_{\text{eq}})$, might be chosen to have the form

$$
C(F) = M \cdot (F - F_{\text{eq}}) \cdot N, \tag{10.41}
$$

where M and N are 4×4 complex matrices. *A priori* they depend on 2×16 complex parameters, while the most general relaxation term assumes the form

$$
[C(F)]^{ab} = \chi^{abcd}(F - F_{\text{eq}})_{cd}, \tag{10.42}
$$

where the indices $\{a, b, c, d\}$ are spinor indices running from 1 to 4 and, accordingly, it depends on 4^4 parameters, and of course much less when symmetries are taken into account. However, despite this lack of generality, such a collision term possesses a sufficient degree of complexity to accommodate most useful physical cases. It will be briefly studied in what follows, and for the moment let us make a few comments on the general collision term $C[F]$.

A first remark is that $C[F]$ is necessarily built up from the most general scalars that can be constructed from what is available, i.e. from

$$
\begin{cases}
\delta f, \delta f^\mu, \delta f_{\mu\nu}, \delta f_5^\mu, \delta f_5, \\
p^\mu, u^\mu, n^\mu, \varepsilon^{\mu\nu\alpha\beta},
\end{cases} \tag{10.43}
$$

where n^μ is the unit (spacelike) four-vector in the direction of a possible polarization of the system. They obey exactly the same relations similar to the equilibrium relations:

$$\begin{aligned} f^\mu &= \frac{p^\mu}{m} f, \\ f_5 &= 0, \end{aligned} \tag{10.44a}$$

$$\begin{aligned} f^{\mu\nu} &= \varepsilon^{\mu\nu\alpha\beta} p_\alpha f_{5\beta}, \\ f_5^\mu &= \frac{1}{m} \varepsilon^{\mu\nu\alpha\beta} p_\nu f_{\mu\nu}. \end{aligned} \tag{10.44b}$$

Note that we have suppressed the indices 1 and ε of the f's; they will be re-established whenever necessary. It follows that, for instance, a scalar such as $u_\mu f^\mu$ is proportional to f and hence should not be considered as essentially different from f. Finally, with all these constraints in mind, we can write

$$p \cdot \partial f_{\mathrm{eq}} = \frac{m}{2i}(c + \tilde{c}) = a f_{(1)\varepsilon} + b_\lambda f^\lambda_{5(1)\varepsilon}, \tag{10.45}$$

$$\begin{aligned} p \cdot \partial f^\mu_{5\mathrm{eq}} &= -\frac{im}{2} \Delta^{\mu\nu}(p)(c_{5\nu} + \tilde{c}_{5\nu}) \\ &= c_5^\mu f_{(1)\varepsilon} + d^{\mu\nu} f_{5\nu(1)\varepsilon}. \end{aligned} \tag{10.46}$$

Moreover, because of the relation $p_\mu f^\mu_{5\mathrm{eq}} = 0 = p_\mu f^\mu_{5(1)\varepsilon}$, the tensors $a^\mu, b^\mu, c^\mu, d^{\mu\lambda}$ should be chosen as being orthogonal to p^μ, so that these last equations contain $1 + 3 + 3 + 3 \times 3 = 16$ independent parameters. The various components of a, b^μ, c_5^μ and $d^{\mu\lambda}$ are functions of the various components of p^μ and of constant relaxation times.

Let us now discuss the above system:

(1) The fact that $f^\mu \propto p^\mu f$ has the interesting consequence that the energy–momentum tensor is now symmetric. As a result, the local polarization tensor is conserved at order ε:

$$\partial_\lambda S^{\mu\nu\lambda} = -\frac{1}{2}\varepsilon^{\mu\nu\lambda\rho} \int d^4p f_{5\rho} = O(\varepsilon). \tag{10.47}$$

Unlike the results of the preceding subsection, where the first order solution — in the parameter ε — implied a complete decoupling between f and f_5^λ, here there exists a possible coupling via the functions b^μ and c^μ. As a consequence, spin diffusion and other transport phenomena connected with polarization do appear at this order and not at order η.

(2) When the medium is not polarized, $n^\mu \equiv 0$ and the system (10.45), (10.46) decouples still at order ε.
(3) Equations (10.45) and (10.46) for f^μ and $f^{\mu\nu}$ do not contain any new information, but these functions can be obtained from the relations (10.44) once the system (10.45), (10.46) has been solved.
(4) One could be tempted to take as the most general form of the relaxation quantum relaxation time approximation Eqs. (10.45) and (10.46), with the left hand side now containing the f^A's instead of the f^A_{eq}'s and on the right hand side $f^A - f^A_{\text{eq}}$ instead of $f^A_{(1)\varepsilon}$. In fact, this would be equivalent to an equation (or, rather, a system) in which $F_{(1)\eta} \equiv 0$; indeed, this would imply that $F_{(1)\eta} \equiv 0$ obeys both

$$(\gamma \cdot p - m)F_{(1)\varepsilon} = 0$$

and $C(F_{(1)\eta}) = 0$, and it is not difficult to realize that their only solution is precisely $F_{(1)\eta} \equiv 0$. This choice would also mean that the nontrivial part of the solution would be at least $O(\eta^2)$.
(5) The system (10.45), (10.46) is linear in the unknown functions $f_{(1)\varepsilon}$ and $f^\lambda_{5(1)\varepsilon}$ and hence can be solved without any particular difficulty and, consequently, allows the calculation of the transport coefficient at $O(\varepsilon)$ and $O(\eta)$. On the other hand, the explicit calculation of $F_{(1)\eta}$ — although straightforward since all our equations are linear — is much more involved.

10.3.3. *General form of $F_{(1)}$*

First, the BGK system is expanded into a series of ε and η, on the basis of

$$\begin{cases} \partial = \partial_{(0)} + \varepsilon\partial_{(1)\varepsilon} + \eta\partial_{(1)\eta} + \cdots, \\ F = F_{\text{eq}} + \varepsilon F_{(1)\varepsilon} + \eta F_{(1)\eta} + \cdots. \end{cases} \tag{10.49}$$

After some algebra, one finds the first order correction in ε as

$$F_{(1)\varepsilon} = -\frac{p\cdot\partial F_{\text{eq}}}{mA(p)}, \tag{10.50}$$

where $A(p)$ is a *known* function when the matrices M and N are given. Note that despite its apparent generality, it is essentially similar to the form given in the example discussed at the end of the last subsection, with $\tau = \tau(p)$. Also, it is valid whether the system is polarized or not. However, if we require that, at order ε, the Landau–Lifschitz conditions be true, then there exist simplified relations among the M and N relations.

The calculation of $F_{(1)\eta}$ is much more involved and finally leads to a quite complex expression which is given elsewhere [R. Hakim, L. Mornas, P. Peter and H. Sivak (1992)]. We only give the result as

$$\begin{aligned}
F_{(1)\eta} = I \cdot \Bigg[& -\frac{i}{4}(\beta-\beta^*)\Delta^{\alpha\beta}(p)u_\alpha r_\beta + i(\delta-\delta^*)\frac{p\cdot r}{4m} \\
& - \frac{1}{4m}(\beta+\beta^*)\varepsilon^{\lambda\rho\alpha\beta}p_\lambda u_\rho t_{\alpha\beta}\Bigg] + \gamma_\mu\Bigg[-\frac{i}{4m}(\beta-\beta^*)\Delta^{\alpha\beta}(p)u_\alpha r_\beta p^\mu \\
& + i(\delta-\delta^*)\frac{p\cdot r}{4m}\frac{p^\mu}{m} - i(\alpha-\alpha^*)\frac{p\cdot r}{4m}\Delta^{\mu\nu}(p)u_\nu \\
& + (\alpha+\alpha^*)\frac{p^\lambda t_{\lambda\rho}}{4m}\varepsilon^{\rho\alpha\nu\mu}u_\alpha p_\nu + \frac{1}{4}(\beta+\beta^*)\Delta^\alpha{}_\lambda(p)\Delta^\beta{}_\nu(p)u_\alpha t_{\beta\rho}\varepsilon^{\rho\lambda\nu\mu} \\
& - \varepsilon^{\rho\lambda\nu\mu}\frac{p_\nu}{m}t_{\lambda\rho}\Bigg] + \sigma^{\mu\nu}\Bigg[\frac{i}{4m}r_{[\mu}p_{\nu]} - \frac{i}{4m}(\alpha+\alpha^*)\frac{p\cdot r}{2m}u_{[\mu}p_{\nu]} \\
& - \frac{i}{8}(\beta+\beta^*)\Delta^\alpha{}_\mu(p)\Delta^\beta{}_\nu(p)u_\alpha r_{\beta]} + \frac{(\alpha-\alpha^*)}{8m}p^\lambda t^\rho{}_\lambda\Delta^{\alpha\beta}(p)u_\alpha\varepsilon_{\beta\rho\mu\nu} \\
& - \frac{1}{8m}(\beta-\beta^*)\varepsilon_{\lambda\alpha\beta[\mu}p^\lambda u^\alpha t^\beta{}_{\nu]} + \frac{(\beta-\beta^*)}{8m}\Delta^{\alpha\beta}(p)u_\alpha t^\rho_\beta p^\lambda\varepsilon_{\lambda\rho\mu\nu} \\
& - \frac{\delta-\delta^*}{8m}\varepsilon_{\alpha\beta\mu\nu}\frac{p^\alpha p_\lambda}{m}t^{\lambda\beta}\Bigg] + \gamma_5^\mu\Bigg[\frac{1}{4}(\beta+\beta^*)z_\mu + \frac{i}{4m}(\alpha-\alpha^*)\frac{p^\mu p^\lambda}{m}u^\rho t_{\lambda\rho} \\
& - \frac{i}{4}(\beta-\beta^*)\Delta^{\alpha\beta}(p)u_\alpha t_{\beta\mu} + \frac{i}{4m}(\delta-\delta^*)p^\lambda t_{\lambda\mu} \\
& - \frac{i}{4}(\beta-\beta^*)\Delta_{\mu\alpha}(p)\Delta_{\nu\beta}(p)u^{[\alpha}t^{\beta]\nu}\Bigg] + \gamma_5\left[\frac{i}{4m}(\alpha+\alpha^*)p^\lambda t_{\lambda\rho}u^\rho\right],
\end{aligned} \tag{10.51}$$

where we have set

$$r_\mu = \frac{1}{4m}\partial_\mu f_{\text{eq}}, \tag{10.52}$$

$$t_{\mu\nu} = \frac{1}{4m}\partial_\mu(S_\nu f_{\text{eq}}), \tag{10.53}$$

$$z_\alpha = \varepsilon_{\lambda\mu\nu\alpha}\frac{p^\lambda}{m}u^\mu r^\nu, \tag{10.54}$$

$$\alpha = \frac{\mu_3 + m\mu_4}{\mu_1 + m\mu_2 + \frac{p\cdot u}{m}\mu_3}, \tag{10.55}$$

$$\beta = \frac{\mu_3 - m\mu_4}{\mu_1 + m\mu_2 + \frac{p\cdot u}{m}\mu_3}, \tag{10.56}$$

$$\delta = \alpha\beta\Delta^{\mu\nu}(p)u_\mu u_\nu. \tag{10.57}$$

The above solution for $F_{(1)\eta}$ finally depends on four real parameters only, namely the real and imaginary parts of α and β. In the case of an unpolarized system, the above expression for $F_{(1)\eta}$ has a slightly simpler form, obtained by setting $S^{\mu}(p) \equiv 0, t^{\alpha\beta}(p) \equiv 0$.

We must add that a number of subtleties have been avoided and that the actual calculations are somewhat more complex [see R. Hakim, L. Mornas, P. Peter and H. Sivak (1992)]. We have only outlined an involved calculation and the interested reader should refer to the given article for more details, where he can see, for instance, the expression of the transport coefficients.

Chapter 11

Application to Nuclear Matter

An interesting phenomenological model[1] for nuclear matter has been proposed by J.D. Walecka (1974) and S.A. Chin and J.D. Walecka (1974, 1979), which we use hereafter in order to illustrate the covariant Wigner function techniques. This model is based on the remark that the nucleon–nucleon potential is repulsive at short distances and attractive at longer ones (see Fig. 11.1). In order to mimic such a behavior, Walecka proposed using a short range vector field $A^\mu(x)$, repulsive in the static limit, and a long range scalar field $\varphi(x)$, attractive in the same limit; these fields are sometimes identified with the ω and σ meson fields, respectively. The parameters of the model are then fitted with the nuclear saturation density and binding energy per nucleon.

In spite of some known inadequacies,[2] such a model is certainly worth studying, because it is the archetype of numerous other models for nuclear matter, and as such has become a reference on the basis of which many improvements have been made by taking into account other mesons, chiral symmetry, the low energy nucleon–nucleon scattering data and nonlinear couplings.

[1] See the general review by B.D. Serot and J.D. Walecka (1986).

[2] One of the problems of the model is the too-high value obtained for the nuclear compressibility.

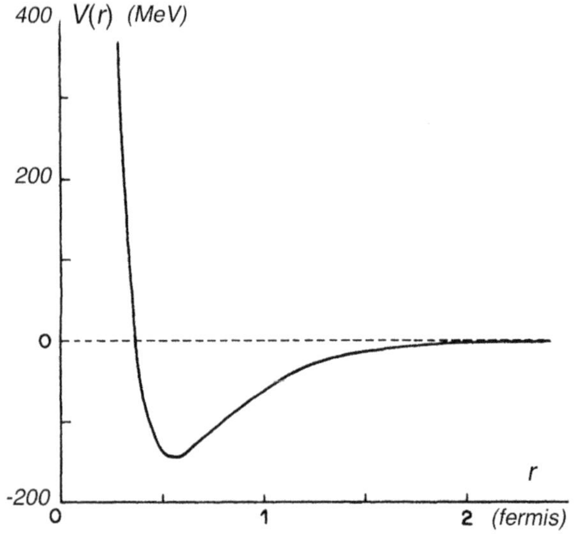

Fig. 11.1 General shape of the nuclear potential.

The Lagrangian considered is given by

$$\begin{aligned} L = &\frac{1}{2}\bar{\psi}[\gamma^\mu(i\partial_\mu - g_V A_\mu) - (m - g_S\varphi]\psi \\ &- \frac{1}{2}\bar{\psi}[\gamma^\mu(i\overleftarrow{\partial}_\mu + g_V A_\mu) + (m - g_S\varphi]\psi \\ &+ \frac{1}{2}(\partial_\mu\varphi \cdot \partial^\mu\varphi - m_S^2\varphi^2) - \frac{1}{4}F_{\mu\nu}F^{\mu\nu} + \frac{1}{2}m_V^2 A_\mu A^\mu, \end{aligned} \tag{11.1}$$

with

$$F_{\mu\nu} \equiv \partial_\mu A_\nu - \partial_\nu A_\mu, \tag{11.2}$$

from which one obtains the equations of motion for the nucleons

$$\begin{cases} [\gamma^\mu(i\partial_\mu - g_V A^\mu) - (m - g_S\varphi)]\psi = 0, \\ \bar{\psi}[\gamma^\mu(i\overleftarrow{\partial}_\mu + g_V A^\mu) + (m - g_S\varphi)] = 0, \end{cases} \tag{11.3}$$

and for the mesons

$$\begin{cases} (\Box + m_S^2)\varphi = g_S\bar{\psi}\psi, \\ (\Box + m_V^2)A^\mu = g_V\bar{\psi}\gamma^\mu\psi, \\ \partial_\mu A^\mu = 0. \end{cases} \tag{11.4}$$

Note that the last equation is a constraint which imposes the fields A^μ to have only three degrees of freedom.

The generating equation of the BBGKY hierarchy then reads

$$\begin{cases} \{i\gamma \cdot \partial + 2[\gamma \cdot p - m]\}F(x,p) \\ \quad = -2g_S \int \frac{d^4R}{(2\pi)^4} d^4\xi \exp[-i(p-\xi)\cdot R] \left\langle F(x,\xi)\varphi\left(x - \frac{1}{2}R\right)\right\rangle \\ \quad\quad + 2g_V \int \frac{d^4R}{(2\pi)^4} d^4\xi \exp[-i(p-\xi)\cdot R] \left\langle F(x,\xi)\gamma_\mu A^\mu\left(x - \frac{1}{2}R\right)\right\rangle, \\ F(x,p)\{i\gamma \cdot \vec{\partial} - 2[\gamma \cdot p - m]\} \\ \quad = +2g_S \int \frac{d^4R}{(2\pi)^4} d^4\xi \exp[-i(p-\xi)\cdot R] \left\langle F(x,\xi)\varphi\left(x + \frac{1}{2}R\right)\right\rangle \\ \quad\quad - 2g_V \int \frac{d^4R}{(2\pi)^4} d^4\xi \exp[-i(p-\xi)\cdot R] \left\langle \gamma_\mu A^\mu\left(x + \frac{1}{2}R\right)F(x,\xi)\right\rangle, \\ (\Box + m_S^2)\langle\varphi(x)\rangle = g_S \mathrm{Sp}\int d^4p\, F(x,p), \\ (\Box + m_V^2)\langle A^\mu(x)\rangle = g_V \mathrm{Sp}\int d^4p\, \gamma^\mu F(x,p), \end{cases} \tag{11.5}$$

which leads to the Hartree–Vlasov system by neglecting two-body correlations, including exchange ones:

$$\langle F_{\mathrm{op}}\varphi\rangle \approx F\langle\varphi\rangle, \langle F_{\mathrm{op}}A^\mu\rangle \approx F\langle A^\mu\rangle, \langle F_{\mathrm{op}} \otimes F_{\mathrm{op}}\rangle \approx F \otimes F. \tag{11.6}$$

11.1. Thermodynamic Properties at Finite Temperature

In equilibrium the system is assumed to be homogeneous and stationary. Hence, the various derivative terms disappear and one is left with

$$\begin{cases} [\gamma \cdot (p - g_V A_{\mathrm{eq}}) - M]F_{\mathrm{eq}}(p) = 0, \\ F_{\mathrm{eq}}(p)[\gamma \cdot (p - g_V A_{\mathrm{eq}}) - M] = 0, \\ m_S^2 \varphi_{\mathrm{eq}} = g_S \mathrm{Sp}\int d^4p\, F_{\mathrm{eq}}(p), \\ m_V^2 A^\mu_{\mathrm{eq}} = g_V \mathrm{Sp}\int d^4p\, \gamma^\mu F_{\mathrm{eq}}(p) = g_V n_{\mathrm{eq}} u^\mu, \end{cases} \tag{11.7}$$

where the index eq refers to the equilibrium values; it can also be written as

$$\begin{cases} [\gamma \cdot \left(p - \frac{g_V^2 n_{\text{eq}}}{m_V^2}\right) - M] F_{\text{eq}}(p) = 0, \\ F_{\text{eq}}(p) \left[\gamma \cdot \left(p - \frac{g_V^2 n_{\text{eq}}}{m_V^2}\right) - M\right] = 0, \\ m_S^2 \varphi_{\text{eq}} = g_S \text{Sp} \int d^4 p \, F_{\text{eq}}(p), \end{cases} \tag{11.8}$$

where the effective mass M is given by

$$\begin{cases} M \equiv m - g_S \varphi_{\text{eq}}, \\ \varphi_{\text{eq}} \equiv \langle \varphi \rangle. \end{cases} \tag{11.9}$$

This system is formally the same as the one obtained in the case of the relativistic scalar plasma with the following small differences:

(i) p is to be replaced by

$$p \to \left(p - \frac{g_V^2}{m_V^2} n_{\text{eq}} A_{\text{eq}}\right); \tag{11.10}$$

(ii) the mass shell condition is now

$$\left(p - \frac{g_V^2}{m_V^2} n_{\text{eq}}\right)^2 = M^2; \tag{11.11}$$

(iii) the equilibrium distribution becomes

$$F_{\text{eq}}(p) = \frac{\gamma \cdot (p - g_V^2 n_{\text{eq}}/m_V^2) + M}{4M} f_{\text{eq}}(p), \tag{11.12}$$

where

$$f_{\text{eq}}(p) = \frac{d}{(2\pi)^3} \delta[p^{*2} - M^2] \left\{ \frac{\theta(p^* \cdot u)}{\exp(\beta p^* \cdot u - \mu^*) + 1} \right.$$
$$\left. + \frac{\theta(-p^* \cdot u)}{\exp(-\beta p^* \cdot u - \mu^*) + 1} - \theta(-p^* \cdot u) \right\}, \tag{11.13}$$

with

$$\begin{cases} p^{*\lambda} = p^\lambda - \frac{g_V^2 n_{\text{eq}}}{m_V^2} u^\lambda, \\ \mu^* = \mu - \frac{g_V^2}{m_V^2} n_{\text{eq}}, \end{cases} \tag{11.14}$$

and d is the spin–isospin degeneracy factor [in symmetric nuclear matter $d = 4$ (two spin states and two isospin states) and in pure neutron matter $d = 2$ (two spin states only).

The φ field equation of the equilibrium Hartree–Vlasov system is formally the same as the gap equation studied in Chap. 9 with the proviso that f_{eq} possesses now the form given above. Note also that besides the gap equation, another transcendental equation must be solved (except at 0 K, where it is trivial) — the one that connects the chemical potential μ and the equilibrium density n_{eq},

$$\frac{g_V^2 n_{\text{eq}}}{m_V^2} = \int d^3p f_{\text{eq}}(p; \mu, n_{\text{eq}}) \tag{11.15}$$

where the dependence of f_{eq} on μ and n_{eq} has been made apparent in the notations. Note that the degeneracy factor d allows the distinction between symmetric nuclear matter and pure neutron matter ($d = 2$; two spin states).

In this chapter, the vacuum term of f_{eq} will be dropped: it is the only term that gives rise to infinities — the field A^μ does not give rise to divergences in this approximation of thermal equilibrium — and the renormalization procedure is exactly identical to the one performed for the scalar plasma. Moreover, a renormalization process in a merely phenomenological model does not make much sense, except for esthetical reasons.

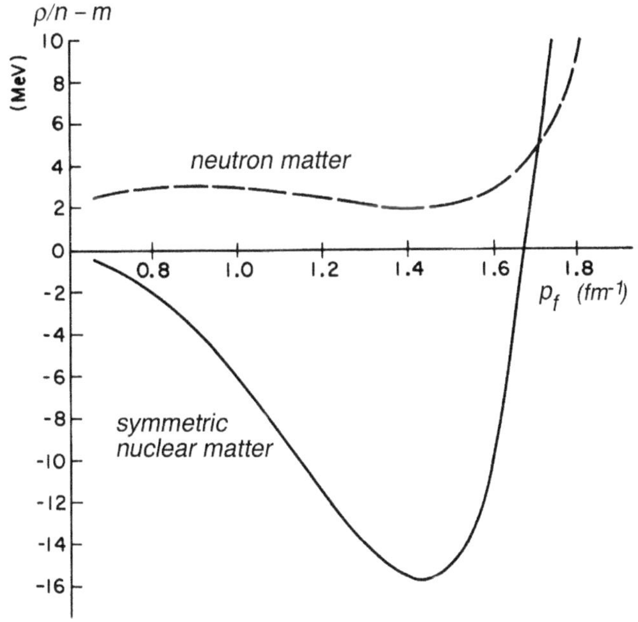

Fig. 11.2 The energy per nucleon (in MeV) as a function of the Fermi momentum (in fm^{-1}), in the Walecka model for pure neutron matter and symmetric nuclear matter. The binding energy is adjusted to the experimental value -15.75 MeV at Fermi momentum $1.42\,\text{fm}^{-1}$ [after B.D. Serot and J.D. Walecka (1986)].

As in the case of the scalar plasma, one can obtain the other thermodynamic quantities through the calculation of the energy–momentum tensor:

$$T^{\mu\nu} = T^{\mu\nu}_{\rm mat} + T^{\mu\nu}_{\varphi} + T^{\mu\nu}_{A} = T^{\mu\nu}_{\rm mat} + m_V^2 \left(\frac{g_V^2 n_{\rm eq}}{m_V^2}\right)^2 u^\mu u^\nu$$
$$+ \left\{ \frac{m_S^2}{8\pi g_S^2}(M-m)^2 + m_V^2 \left(\frac{g_V^2 n_{\rm eq}}{m_V^2}\right)^2 \right\} \eta^{\mu\nu}. \tag{11.16}$$

There remains for one to determine the as yet arbitrary constants $\{m_S^2, g_S^2; m_V^2, g_V^2\}$. Let us first note that in this mean field approximation these last quantities always appear through the combination

$$C_V^2 \equiv g_V^2 \left(\frac{m^2}{m_V^2}\right), \quad C_S^2 \equiv g_S^2 \left(\frac{m^2}{m_S^2}\right), \tag{11.17}$$

for which B.D. Serot and J.W. Walecka gave the values

$$C_V^2 = 195.9, \quad C_S^2 = 267.1.$$

This can be achieved by identifying the binding energy per nucleon (Fig. 11.2) in nuclear matter calculated in the model and the experimentally known value ($-15.75\,$MeV), fitted at the saturation density ($2.24 \times 10^{14}\,$g/cm^3 or $p_f = 1.42\,$fm^{-1}) of nuclear matter.

11.1.1. *Thermodynamics in some important cases*

Note that one can expect a first order phase transition, of the gas–liquid type, because of the role played by the (attractive) scalar field, and indeed this is what appears after the calculations are performed. Qualitatively, the thermodynamic properties obtained in this model are similar to those of the relativistic scalar plasma studied in Chap. 9.

Therefore, only some important limiting cases are briefly given below: (i) the degenerate case at low temperatures, (ii) the high temperature and nondegenerate case, and (iii) the low temperature and nondegenerate case. Use is made of the notations

$$\begin{cases} \vartheta \equiv \dfrac{\mu^*}{M}, \\ x \equiv \sqrt{\vartheta^2 - 1}. \end{cases} \tag{11.18}$$

In what follows we shall set

$$\gamma^* = m^*\beta, \quad \alpha^* = \frac{\gamma^*\mu^*}{m^*}, \quad \mu^* = \mu - g_V^2 u_\lambda \langle A^\lambda \rangle. \tag{11.19}$$

Degenerate matter and low temperature ($\gamma^* \gg 1$ and $\alpha^* \gg \gamma^*$). A Sommerfeld expansion (see Chap. 7) of the integrals for the various thermodynamic quantities yields

$$n_{\text{nucleon}} \approx d\frac{4\pi M^3}{3(2\pi)^3}\left\{\vartheta^3 + \frac{\pi^2}{6\gamma^{*2}}\frac{6x^2-3}{\vartheta}\right\}, \tag{11.20}$$

$$\rho \approx d\frac{4\pi M^4}{(2\pi)^3}\left\{\frac{x\vartheta}{8}(2x^2-1) - \frac{1}{8}\ln(x+\vartheta) + \frac{m_S^2}{2g_S^2}(m-M)^2 + \frac{g_V^2}{2m_V^2}n_{\text{nucleon}}^2\right\}, \tag{11.21}$$

$$P \approx d\frac{4\pi M^4}{3(2\pi)^3}\left\{\frac{x\vartheta}{8}(2x^2-5) + \frac{3}{8}\ln(x+\vartheta) + \frac{\pi^2}{2\gamma^{*2}}x\vartheta - \frac{g_S^2}{2m_S^2}(m-M)^2 + \frac{g_V^2}{2m_V^2}n_{\text{nucleon}}^2\right\}, \tag{11.22}$$

$$S \approx d\frac{4\pi M^3}{3(2\pi)^3}\left\{\frac{\pi^2}{\gamma^*}x\vartheta + \frac{7\pi^4}{120\gamma^{*3}}\frac{8x^5-23x^3+18x}{\vartheta^7}\right\}, \tag{11.23}$$

while the gap equation reads

$$M \approx m - \frac{g_S^2}{m_S^2}d\frac{4\pi M^4}{(2\pi)^3}\left\{\frac{1}{2}x\vartheta - \frac{1}{2}\ln(x+\vartheta) + \frac{\pi^2}{6\gamma^{*2}}\frac{x}{\vartheta}\right\}. \tag{11.24}$$

High temperature and nondegenerate matter. In this case, the various Fermi–Dirac factors occurring in the thermodynamic expressions can be expanded into geometric series,

$$\frac{1}{\exp(-\gamma^*[\cosh(x) - \mu^*/m^*]) + 1} = \sum_{k=0}^{\infty}(-1)^k \exp(-k\gamma^*[\cosh(x) - \mu^*/m^*]), \tag{11.25}$$

$$\frac{1}{\{\exp(-\gamma^*[\cosh(x) - \mu^*/m^*]) + 1\}^2} = \sum_{k=0}^{\infty}(-1)^k(k+1)\exp(-k\gamma^*[\cosh(x) - \mu^*/m^*]), \tag{11.26}$$

of the exponentials, leading thereby to rapidly converging series of Kelvin functions (see App. A). Replacing each Kelvin function by its high temperature approximation (small arguments of K_n), one gets

$$n_{\text{nucleon}} \approx d\frac{M^2}{6}\frac{\mu^*}{\gamma^{*2}}, \tag{11.27}$$

$$\rho \approx \left\{ d\frac{7M^4\pi^2}{120} + d^2\frac{g_S^2}{m_S^2}\frac{M^6}{288} + d^2\mu^{*2}\frac{g_V^2}{m_V^2}\frac{M^4}{72} \right\} \frac{1}{\gamma^{*4}}, \tag{11.28}$$

$$P \approx \left\{ d\frac{7M^4\pi^2}{360} - d^2\frac{g_S^2}{m_S^2}\frac{M^6}{288} + d^2\mu^{*2}\frac{g_V^2}{m_V^2}\frac{M^4}{72} \right\} \frac{1}{\gamma^{*4}}, \tag{11.29}$$

$$S \approx d\frac{7\pi^2}{90}\left(\frac{M}{\gamma^*}\right)^3, \tag{11.30}$$

and the gap equation is written as

$$M \approx m + d\frac{g_S^2}{m_S^2}\frac{M^3}{12\gamma^{*2}}. \tag{11.31}$$

Low temperature and nondegenerate matter. In this case, one has $\gamma^* \gg 1$ (low temperatures) and $\alpha^* \ll \gamma^*$, with

$$\alpha^* \equiv \frac{\gamma^*\mu^*}{M} = \beta\left(\mu - \frac{g_V^2}{m_V^2} n_{\text{nucleon}}\right). \tag{11.32}$$

The various integrals are still expanded into a geometric series, giving rise to a series of Kelvin functions which are replaced by their asymptotic expressions (large values of the argument of K_n), and one finds that

$$n_{\text{nucleon}} \approx d\frac{M^2}{2\pi^2}\left(\frac{\pi}{2}\right)^{1/2}\frac{1}{\gamma^{*3/2}}\exp\left(\gamma^*\left[\frac{\mu^*}{M} - 1\right]\right), \tag{11.33}$$

$$\begin{aligned} \rho &\approx d\frac{M^4}{2\pi^2}\left(\frac{\pi}{2}\right)^{1/2}\frac{1}{\gamma^{*3/2}}\exp\left(\gamma^*\left[\frac{\mu^*}{M} - 1\right]\right) \\ &\quad + d^2\frac{g_S^2}{m_S^2}\frac{M^6}{16\pi^3}\frac{1}{\gamma^{*3}}\exp\left(2\gamma^*\left[\frac{\mu^*}{M} - 1\right]\right) \\ &\quad + d^2\frac{g_V^2}{m_V^2}\frac{M^6}{8\pi^3}\frac{1}{\gamma^{*3}}\exp\left(2\gamma^*\left[\frac{\mu^*}{M} - 1\right]\right), \end{aligned} \tag{11.34}$$

$$\begin{aligned} P &\approx d\frac{M^4}{2\pi^2}\left(\frac{\pi}{2}\right)^{1/2}\frac{1}{\gamma^{*3/2}}\exp\left(\gamma^*\left[\frac{\mu^*}{M} - 1\right]\right) \\ &\quad - d^2\frac{g_S^2}{m_S^2}\frac{M^6}{16\pi^3}\frac{1}{\gamma^{*3}}\exp\left(2\gamma^*\left[\frac{\mu^*}{M} - 1\right]\right) \\ &\quad + d^2\frac{g_V^2}{m_V^2}\frac{M^6}{8\pi^3}\frac{1}{\gamma^{*3}}\exp\left(2\gamma^*\left[\frac{\mu^*}{M} - 1\right]\right), \end{aligned} \tag{11.35}$$

$$S \approx \rho \approx d\frac{M^3}{2\pi^2}\left(\frac{\pi}{2}\right)^{1/2}\left\{\frac{1-\frac{\mu^*}{M}}{\gamma^{*1/2}}+\frac{1}{\gamma^{*3/2}}\right\}\exp\left(\gamma^*\left[\frac{\mu^*}{M}-1\right]\right), \quad (11.36)$$

and the gap equation has the approximation

$$M \approx m + d\frac{g_S^2}{m_S^2}\frac{M^3}{2\pi^2}\left(\frac{\pi}{2}\right)^{1/2}\frac{1}{\gamma^{*3/2}}\exp\left(\gamma^*\left[\frac{\mu^*}{M}-1\right]\right). \quad (11.37)$$

11.2. Remarks on the Oscillation Spectra of Mesons

We give a few words on the spectra of the oscillations of φ and A^λ. Therefore, we shall use the equations of the system only in a symbolic way. It is thus written as

$$\begin{aligned}\mathcal{L}F_{\text{op}} &= g_V \text{Sp}\int \gamma\cdot F_{\text{op}}A - g_S\text{Sp}\int F_{\text{op}}\varphi,\\ F_{\text{op}}\bar{\mathcal{L}} &= -g_V\text{Sp}\int A\cdot\gamma F_{\text{op}} + g_S\text{Sp}\int \varphi F_{\text{op}},\end{aligned} \quad (11.38)$$

$$\begin{aligned}\text{KG}\varphi &= g_S\text{Sp}\int F_{\text{op}},\\ \text{KG}A &= g_V\text{Sp}\int \gamma F_{\text{op}}.\end{aligned} \quad (11.39)$$

Now the same approximation as that in Chap. 9 is made; it is

$$\begin{cases}\langle\varphi\rangle \approx \varphi_{\text{eq}} + \varphi_{(1)},\\ \langle A\rangle \approx A_{\text{eq}} + A_{(1)},\\ \langle F_{\text{op}}\rangle \approx F_{\text{eq}} + F_{(1)},\end{cases} \quad (11.40)$$

which are introduced into the first two equations of the hierarchy,

$$\mathcal{L}^*F_{(1)} = g_V\text{Sp}\int \gamma F_{\text{eq}}A_{(1)} - g_S\text{Sp}\int F_{\text{eq}}\varphi_{(1)}, \quad (11.41)$$

and another one that is similar, where we have introduced terms like φ_{eq}, in the star of $\mathcal{L}^*$. We now "solve" these last equations as

$$F_{(1)} = \mathcal{L}^{*-1}\left\{g_V\text{Sp}\int \gamma F_{\text{eq}}A_{(1)} - g_S\text{Sp}\int F_{\text{eq}}\varphi_{(1)}\right\} \quad (11.42)$$

and introduce this into the Klein–Gordon equations

$$\begin{cases}\text{KG}\varphi_{(1)} = \mathcal{L}^{*-1}\left\{g_V\text{Sp}\int \gamma F_{\text{eq}}A_{(1)} - g_S\text{Sp}\int F_{\text{eq}}\varphi_{(1)}\right\},\\ \text{KG}A_{(1)} = \mathcal{L}^{*-1}\left\{g_V\text{Sp}\int \gamma\otimes\gamma F_{\text{eq}}A_{(1)} - g_S\text{Sp}\int \gamma F_{\text{eq}}\varphi_{(1)}\right\};\end{cases} \quad (11.43)$$

these equations can be rewritten in the form

$$\begin{cases} \mathrm{KG}\varphi_{(1)} = \Pi_{\varphi A} A_{(1)} + \Pi_{\varphi}\varphi_{(1)}, \\ \mathrm{KG}A_{(1)} = \Pi_A A_{(1)} + \Pi_{A\varphi}\varphi_{(1)}. \end{cases} \tag{11.44}$$

Once Fourier-transformed they lead immediately to the excitation spectra of the (φ, A^λ).

When the coefficients $\Pi_{\varphi A}$ and $\Pi_{A\varphi}$ are negligible, one recovers the excitation spectrum of the scalar particle and that of the vector particle. Note that in the latter case the excitation spectrum is quite similar to the one obtained in QED (Chap. 15), with necessarily a Lorentz "gauge" and with the change

$$\frac{1}{p^2 \pm i\varepsilon} \to \frac{1}{p^2 - m^2 \pm i\varepsilon}. \tag{11.45}$$

Note that in general the two spectra are coupled and that an interesting discussion can be found in the article of S.A. Chin (1977).

11.3. Transport Coefficients of Nuclear Matter

The various transport coefficients of nuclear matter have usually been calculated via the Boltzmann–Uhlenbeck–Uehling equation or Kubo's formula. However, as was repeatedly emphasized, it is at least as reasonable to perform such a calculation in the relaxation time approximation studied in Chap. 10. This has been done by L. Mornas (1992) and by R. Hakim and L. Mornas (1993), and this approach is followed below, still in the case of the Walecka model.

We shall use the collision term studied in the example given at the beginning of Chap. 10, i.e.

$$\begin{cases} C[F] = -i\gamma \cdot u \dfrac{F - F_{\mathrm{eq}}}{\tau}, \\ \bar{C}[F] = -i\dfrac{F - F_{\mathrm{eq}}}{\tau}\gamma \cdot u, \end{cases} \tag{11.46}$$

so that our relaxation time kinetic equation reads (we have not included the equations for the average scalar and vector fields, which remain identical to the ones given above)

$$\{i\gamma \cdot \partial + 2[\gamma \cdot p - m]\}F(x,p)$$
$$+ 2g_S \int \frac{d^4R}{(2\pi)^4} d^4\xi \exp[-i(p-\xi)\cdot R]F(x,\xi)\left\langle \varphi\left(x - \frac{1}{2}R\right)\right\rangle$$

$$- 2g_V \int \frac{d^4R}{(2\pi)^4} d^4\xi \exp[-i(p-\xi)\cdot R] F(x,\xi)\gamma_\mu \left\langle A^\mu\left(x-\frac{1}{2}R\right)\right\rangle$$
$$= C[F(x,p)], \tag{11.47a}$$
$$F(x,p)\{i\gamma\cdot\overleftarrow{\partial} - 2[\gamma\cdot p - m]\}$$
$$- 2g_S \int \frac{d^4R}{(2\pi)^4} d^4\xi \exp[-i(p-\xi)\cdot R] F(x,\xi) \left\langle \varphi\left(x+\frac{1}{2}R\right)\right\rangle$$
$$+ 2g_V \int \frac{d^4R}{(2\pi)^4} d^4\xi \exp[-i(p-\xi)\cdot R] \left\langle \gamma_\mu A^\mu\left(x+\frac{1}{2}R\right)\right\rangle F(x,\xi)$$
$$= \bar{C}[F(x,p)]. \tag{11.47b}$$

In these equations, the only correlations retained are those due to collisions between particles; this system thus describes nucleons interacting via mean fields and pointlike collisions.

Before solving this system, a remark is in order. Since a nonpolarized medium is to be considered, the 16 components of the Wigner function F are not all needed in the calculation of the transport coefficients: only f and f^μ are required to calculate the off-equilibrium part of the four-current and of the energy–momentum tensor. Accordingly, it would be desirable to obtain a kinetic equation which would involve these two functions only.

To this end, one must realize that the Chapman–Enskog approximation — or almost any others — does imply a weak gradient assumption for the average fields $\langle\varphi\rangle$ and $\langle A^\mu\rangle$. Practically, this means that these quantities are expanded into a Taylor series which we limit to its first order:

$$\begin{cases} \left\langle \varphi\left(x\pm\frac{1}{2}R\right)\right\rangle \approx \langle\varphi(x)\rangle \pm \frac{1}{2}R\cdot\partial\,\langle\varphi(x)\rangle, \\ \left\langle A^\mu\left(x\pm\frac{1}{2}R\right)\right\rangle \approx \langle A^\mu(x)\rangle \pm \frac{1}{2}R\cdot\partial\,\langle A^\mu(x)\rangle. \end{cases} \tag{11.48}$$

Once introduced into our kinetic system for F and after performing lengthy but straightforward algebraic manipulations, it is possible to eliminate the components f_5^μ and $f^{\mu\nu}$ and arrive at the system

$$\begin{cases} \tilde{p}\cdot\partial f - \partial\tilde{m} f^\mu - \frac{1}{2}\partial_\lambda(\tilde{p}^2-\tilde{m}^2)\frac{\partial}{\partial p_\lambda} f = -\tilde{p}\cdot u\frac{f-f_{\rm eq}}{\tau}, \\ \tilde{p}\cdot\partial f^\beta - \partial^\beta\tilde{m} f - \frac{1}{2}\partial_\lambda(\tilde{p}^2-\tilde{m}^2)\frac{\partial}{\partial p_\lambda} f^\beta + \partial^{[\beta}\tilde{p}^{\alpha]} f_\alpha \\ \quad = -\tilde{p}\cdot u\frac{f^\beta - f^\beta_{\rm eq}}{\tau}, \end{cases} \tag{11.49}$$

which is somewhat simpler than the original system. Note also that the same kind of approximation — first order Chapman–Enskog — for the effective mass shell provides

$$(\tilde{p}^2 - \tilde{m}^2)\begin{bmatrix} f \\ f^\mu \end{bmatrix} = 0. \tag{11.50}$$

As a matter of fact, a consistent Chapman–Enskog expansion would have led to the same final result, although in a much more involved way. In this last system we have set[3]

$$\begin{cases} \tilde{m} \equiv m - g_S\langle\varphi(x)\rangle, \\ \tilde{p}^\mu \equiv p^\mu - g_V\langle A^\mu(x)\rangle. \end{cases} \tag{11.51}$$

11.3.1. *Chapman–Enskog expansion*

As in Chap. 10, the various physical quantities F, $\langle\varphi\rangle$ and $\langle A^\mu\rangle$ are expanded in the parameter τ as

$$\begin{cases} F = F_{\text{eq}} + \tau F_1 + \cdots, \\ \langle\varphi\rangle = \langle\varphi\rangle_{\text{eq}} + \langle\varphi\rangle_1 + \cdots, \\ \langle A^\mu\rangle = \langle A^\mu\rangle_{\text{eq}} + \langle A^\mu\rangle_1 + \cdots, \end{cases} \tag{11.52}$$

and, instead of F's, it is more appropriate to use the expansion

$$\begin{cases} f = f_{\text{eq}} + \tau f_1 + \cdots, \\ f^\lambda = f^\lambda_{\text{eq}} + \tau f^\lambda_1 + \cdots. \end{cases} \tag{11.53}$$

These expansions, introduced into the system for f and f^μ, then yield

$$\begin{cases} f_1 = -\dfrac{1}{p^* \cdot u}\left(p^* \cdot \partial - \dfrac{1}{2}\partial_\lambda[p^{*2} - m^{*2}]\dfrac{\partial}{\partial p_\lambda}\right) f_{\text{eq}}, \\ f^\mu_1 = \dfrac{p^{*\mu}}{m^*} f_{\text{eq}}, \end{cases} \tag{11.54}$$

with the notations

$$\begin{cases} m^* \equiv m - g_S\langle\varphi\rangle_{\text{eq}}, \\ p^{*\mu} \equiv p^\mu - g_V\langle A^\mu\rangle_{\text{eq}}. \end{cases} \tag{11.55}$$

---

[3] Note that $M = \langle\tilde{m}\rangle_{\text{eq}}$ and that $p^* = p - \langle A(x)\rangle_{\text{eq}}$.

The off-equilibrium quantities J_1^μ and T_1^μ are then calculated and their final expression reads

$$J_1^\mu = \tau \frac{4\pi d m^{*4}}{3(2\pi)^3} \left\{ \frac{I_{4,1}^* I_{4,-1}^*}{I_{4,0}^*} - I_{4,0}^* \right\} \Delta^{\mu\lambda}(u)(\partial_\lambda \beta + \beta \dot{u}_\lambda) \tag{11.56}$$

for the baryon four-current, and

$$\begin{aligned} T_1^{\mu\nu} = {} & \tau \frac{4\pi d m^{*4}}{(2\pi)^3} \left\{ \frac{2}{15} I_{6,-1}^* \gamma^* \sigma^{\mu\nu} + \theta^* \Delta^{\mu\nu}(u) \right. \\ & \left. \times \frac{\gamma^*}{9} \left(\frac{I_{4,0}^{*2} I_{2,3}^* + I_{4,1}^{*2} I_{2,1}^* - 2 I_{2,2}^* I_{4,0}^* I_{4,1}^*}{I_{2,2}^{*2} - I_{2,1}^* I_{2,3}^*} + I_{6,-1}^* \right) \right\} \\ & + \tau \frac{4\pi d m^{*4}}{3(2\pi)^3} \frac{g_V^2 n_B}{m_V^2} [u^\mu \Delta^{\nu\lambda}(u) + u^\nu \Delta^{\mu\lambda}(u)] \\ & \times \left\{ \frac{I_{4,1}^* I_{4,-1}^*}{I_{4,0}^{*2}} - I_{4,0}^{*2} \right\} (\partial_\lambda \beta + \beta \dot{u}_\lambda) + \tau \gamma^* \frac{4\pi d m^{*3}}{3(2\pi)^3} (m - m^*) \\ & \times \left(\frac{I_{4,0}^{*2} I_{2,3}^* + I_{4,1}^{*2} I_{2,1}^* - 2 I_{2,2}^* I_{4,0}^* I_{4,1}^*}{I_{2,2}^{*2} - I_{2,1}^* I_{2,3}^*} + I_{6,-1}^* \right) \theta^* \eta^{\mu\nu} \end{aligned} \tag{11.57}$$

for the energy–momentum tensor. Let us specify the notations used in the expressions of J_1^μ and T_1^μ. The quantities $I_{m,n}$ are the integrals

$$I_{n,m}^{*\pm} = \int dy \sinh^n y \cosh^m y \frac{\exp(\gamma^* \cosh y \pm \alpha^*)}{[\exp(\gamma^* \cosh y \pm \alpha^*) + 1]^2}, \tag{11.58}$$

with

$$\begin{cases} \gamma^* \equiv m^* \beta, \\ \alpha^* \equiv \dfrac{\gamma^* \mu^*}{m^*} = \beta \left(\mu - g_V^2 \dfrac{n_B}{m_V^2} \right), \\ \mu^* \equiv \mu - g_V^2 u_\lambda \langle A^\lambda \rangle_{\text{eq}}. \end{cases} \tag{11.59}$$

In the expression of $T_1^{\mu\nu}$, $\sigma^{\mu\nu}$ is the (traceless) shear tensor (see Chap. 2), not to be mistaken for a Dirac matrix, whose main properties are

$$\begin{cases} \sigma^{\mu\nu} \equiv \left[\Delta^{\mu\alpha}(u) \Delta^{\nu\beta}(u) - \dfrac{1}{3} \Delta^{\mu\nu}(u) \Delta^{\alpha\beta}(u) \right] \partial_\alpha u_\beta, \\ u_\mu \sigma^{\mu\nu} = 0, \\ \Delta^{\mu\alpha}(u) \sigma_{\mu\nu} = 0. \end{cases} \tag{11.60}$$

In the determination of J_1^μ and $T_1^{\mu\nu}$, implicit use was made of the conservation relations which lead *in fine* to the equations:

$$\begin{cases} \dot{\alpha}^* = \dfrac{\gamma^*}{3} \dfrac{I^*_{4,0} I^*_{2,3} - I^*_{2,2} I^*_{4,1}}{(I^*_{2,2})^2 - I^*_{2,1} I^*_{2,3}} \theta^*, \\ \dot{\gamma}^* = \dfrac{\gamma^*}{3} \dfrac{I^*_{4,0} I^*_{2,2} - I^*_{2,1} I^*_{4,1}}{(I^*_{2,2})^2 - I^*_{2,1} I^*_{2,3}} \theta^*, \\ \theta^* \equiv \theta \left(1 + \dfrac{3\dot{m}^*}{m\theta}\right). \end{cases} \tag{11.61}$$

These equations are identical to those obtained by J.L. Anderson and H.R. Witting (1974) for particles moving "freely" (i.e. influenced only by the mean fields) between collisions, the only difference being that nonstarred quantities are replaced by starred ones. Note also that θ^* is a kind of effective divergence of the four-velocity which appears (see below) in the entropy production.

11.3.2. *Transport coefficients: Eckart versus Landau–Lifschitz representations*

Once the off-equilibrium quantities have been calculated, they must be cast into a form appropriate to the identification of transport coefficents, i.e. in Eckart's form (see Chap. 2). They have the general structure

$$\begin{cases} J^\mu_{\text{off}} = \kappa \Delta^{\mu\nu}(u)(\partial_\nu \beta + \beta \dot{u}_\nu), \\ T^{\mu\nu}_{\text{off}} = 2\eta\, \sigma^{\mu\nu} + \frac{1}{2}\varsigma\, \theta \Delta^{\mu\nu}(u) + A^{\mu\nu}, \end{cases} \tag{11.62}$$

where $A^{\mu\nu}$ can easily be obtained from the explicit expression of $T_1^{\mu\nu}$ as

$$A^{\mu\nu} = \tau \frac{4\pi d m^{*4}}{3(2\pi)^3} \frac{g_V^2 n_B}{m_V^2} [u^\mu \Delta^{\nu\lambda}(u) + u^\nu \Delta^{\mu\lambda}(u)] \times \left\{ \frac{I^*_{4,1} I^*_{4,-1}}{I^{*2}_{4,0}} - I^{*2}_{4,0} \right\} (\partial_\lambda \beta + \beta \dot{u}_\lambda). \tag{11.63}$$

While the equation for J^μ_{off} obeys the first Landau–Lifschitz condition, this is not the case for $T^{\mu\nu}_{\text{off}}$, which is not orthogonal to u^μ. This is due to the fact that the energy–momentum tensor includes the contribution of the collective fields $\langle\varphi\rangle$ and $\langle A^\mu\rangle$; although its matter (baryonic) part does satisfy these matching conditions, this is of little use since it is the total off-equilibrium part of the energy–momentum tensor which enters into the dissipation processes.

In order to cast the off-equilibrium quantities into Eckart's form, the pressure, the energy density and the local hydrodynamical four-velocity have to be redefined. This is achieved with the following changes, all of order $0(\tau)$:

$$\begin{cases} \rho \to \rho + \frac{1}{3}\varsigma\theta^*(m - m^*), \\ P \to P - \frac{1}{3}\varsigma\theta^*(m - m^*), \\ u^\mu \to U^\mu = u^\mu + \xi^\mu, \end{cases} \tag{11.64}$$

with

$$\xi^\mu = \frac{\kappa}{n_B}\Delta^{\mu\nu}(u)(\partial_\nu\beta + \beta\dot{u}_\nu). \tag{11.65}$$

Finally, one gets for the shear viscosity (Fig. 11.3)

$$\eta = \tau\frac{4\pi d m^{*4}}{15(2\pi)^3}I^*_{6,-1}\gamma^*, \tag{11.66}$$

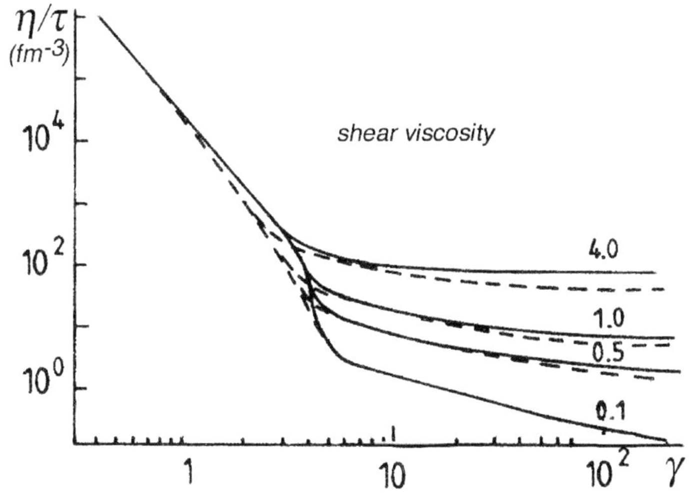

Fig. 11.3 The shear viscosity as a function of the temperature parameter $\gamma = m\beta$ for pure neutron matter. Continuous lines represent the above expression, while dashed ones refer to J.L. Anderson and H.R. Witting's result.

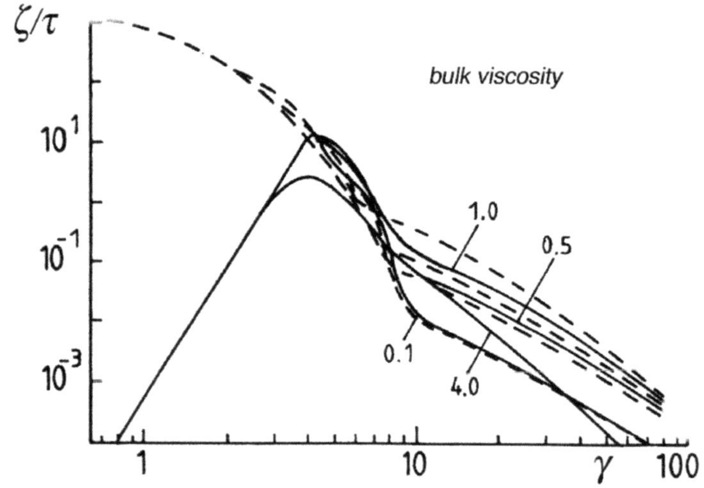

Fig. 11.4 The bulk viscosity as a function of the temperature parameter γ for pure neutron matter. Continuous lines represent the above expression, while dashed ones refer to J.L. Anderson and H.R. Witting's result.

for the bulk viscosity (Fig. 11.4)

$$\varsigma = \tau \frac{4\pi d m^{*4}}{3(2\pi)^3} I^*_{6,-1} \gamma^* \left\{ \frac{I^*_{4,0} I^*_{2,3} + I^*_{4,1} I^*_{2,1} - 2 I^*_{2,2} I^*_{4,0} I^*_{4,1}}{I^{*2}_{2,2} - I^*_{2,1} I^*_{2,3}} + I^*_{6,-1} \right\} \quad (11.67)$$

and for the thermal conductivity (Fig. 11.5)

$$\lambda = \tau \frac{4\pi d m^{*3}}{3(2\pi)^3} \gamma^{*2} \frac{I^*_{4,1}}{I^*_{4,0}} \left\{ \frac{I^*_{4,1} I^*_{4,-1}}{I^*_{4,0}} - I^*_{4,0} \right\}. \quad (11.68)$$

These expressions are *formally* identical to those obtained by J.L. Anderson and H.R. Witting (1974), and they differ only by the occurrence of starred quantities. Although this was not *a priori* obvious, it seems quite natural. However, in spite of their analogy — which gives rise to similar behaviors for the shear viscosity and the thermal conductivity — this is not the case for the bulk viscosity.

In order to assess the importance of collective effects with their absence in transport coefficients, several curves representing the bulk viscosity of pure neutron matter have been plotted in Fig. 11.6, with and without the effects of the scalar and/or vector fields.

The figures show that, at high temperatures, the attractive scalar field is responsible for the behavior of the bulk viscosity, at least for the greatest part. As to the other transport coefficients, it is the repulsive vector field

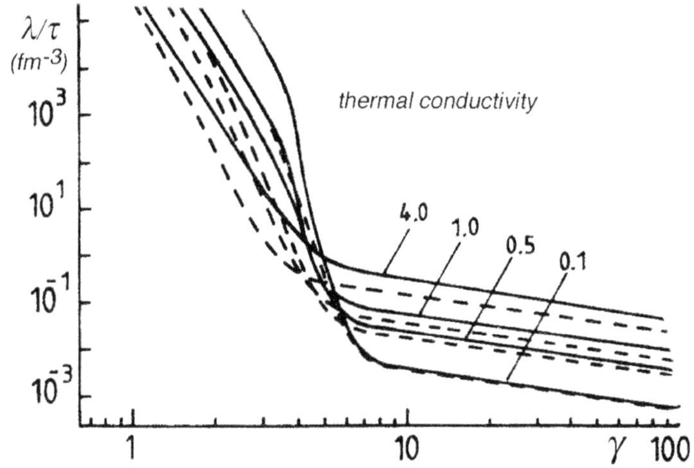

Fig. 11.5 The thermal conductivity as a function of the temperature parameter γ for pure neutron matter. Continuous lines represent the above expression, while dashed ones refer to J.L. Anderson and H.R. Witting's result.

which plays the main role in the modification of the transport coefficients. Details and analytical forms for extreme cases are given elsewhere [R. Hakim and L. Mornas (1993)].

11.3.3. *Entropy production*

It remains for one to justify the passage to the starred quantities in the above calculations. This can be shown by looking at the entropy production in these dissipative processes; further arguments justifying the formal analogy with the Anderson–Witting approach — through the use of effective quantities — can be put forward when studying the entropy production.

This can be done in several ways; here we begin with a microscopic point of view by noting that $f(x,p)$ plays the role of a distribution function on the mass shell $p^{*2} = m^{*2}$. Therefore, the entropy four-flux is defined as usual[4] for fermions as

$$S^\mu(x) = -k_B \sum_\pm \int d^4p^* \frac{p^{*\mu}}{m^*} \{f_\pm(x,p) \ln f_\pm(x,p) \\ + [1 - f_\pm(x,p)] \ln[1 - f_\pm(x,p)]\}, \tag{11.69}$$

[4] See e.g. S.R. de Groot, W. A. van Leeuwen and Ch. G. van Weert, *Relativistic Kinetic Theory* (North-Holland, Amsterdam, 1980).

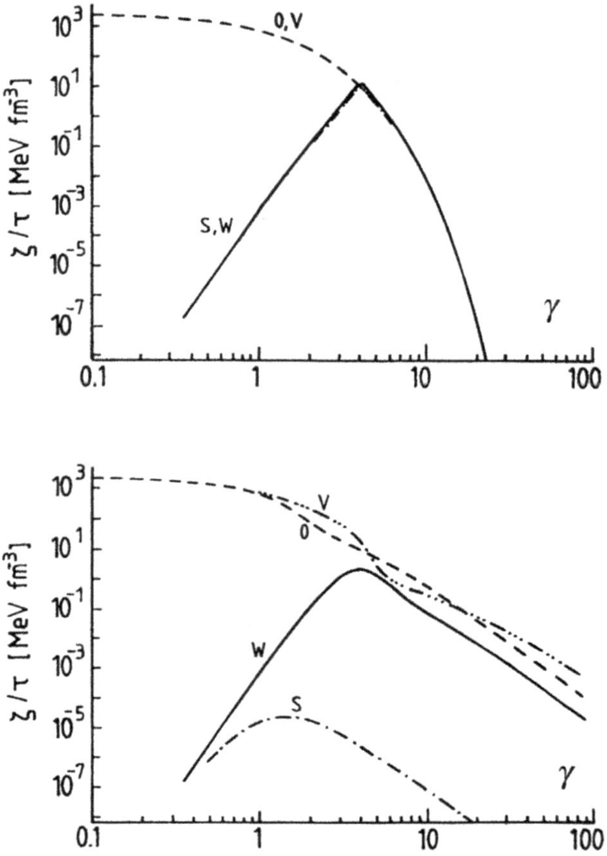

Fig. 11.6 The effect of the various collective fields is indicated in the case of bulk viscosity for pure neutron matter, for two different values of the chemical potential. W designates the result with the Walecka model, S with the scalar field only, V with the vector field only and 0 without any collective effects.

where the plus sign refers to nucleons and the minus one to antinucleons. Note that this definition of the entropy four-flux could be wrong but it is correct in our case, due to the fact that we deal with *free* quasi-particles and, in such a case, f is actually a distribution function.

(1) Here the entropy production of the off-equilibrium system is calculated on the basis of the covariant formulation of the Gibbs relation

$$S^\mu = P_{\text{Eck}}\beta^\mu - \alpha J^\mu_{\text{Eck}} + \beta u_\lambda T^{\mu\lambda}_{\text{Eck}}, \tag{11.70}$$

where $\beta \equiv T^{-1}$ and $\alpha \equiv \mu T^{-1}$, and S^μ is the entropy flux four-vector; the total entropy of the system is thus

$$S = \int_\Sigma d\Sigma_\mu S^\mu \tag{11.71}$$

at "time" Σ.

At order 0 (i.e. in local equilibrium), this last equation can be rewritten as

$$S^\mu_{\text{eq}} = \left[(\rho + P)_{\text{nucleons}} \frac{\gamma^*}{m^*} - \alpha^* n_B\right] u^\mu \tag{11.72}$$

and can be shown to obey $\partial_\mu S^\mu_{\text{eq}} = 0$, with the help of Eqs. (11.61a) and (11.61.b), as it should.

At order 1 in τ, it is difficult to show that Eq. (11.61) leads to

$$S^\mu = P_{\text{nucleons}} \frac{\gamma^*}{m^*} u^\mu - \alpha^* J^\mu + \frac{\gamma^*}{m^*} u_\lambda T^{\mu\lambda}_{\text{nucleons}}, \tag{11.73}$$

where the index "nucleons" refers to quantities calculated with the Wigner function of the nucleons only and not, for example, to quantities connected with the collective fields $\langle\phi\rangle$ and $\langle A^\lambda\rangle$.

From Eq. (11.72) the entropy production rate σ is given by

$$\sigma \equiv \partial_\mu S^\mu_{(1)} = -\partial_\mu \alpha^* \cdot J^\mu_{(1)} + \partial_\mu(\beta u_\lambda) T^{\lambda\mu}_{(1)} \tag{11.74}$$

$$\begin{aligned} &= -\partial_\mu \alpha^* \cdot J^\mu_{(1)} + T^{\mu\lambda}_{(1)\text{nucleons}} \frac{\partial_\mu(\gamma^* u_\lambda)}{m^*} \\ &\quad - \frac{\gamma^*}{m^*}(m_S^2 \dot{\phi}_{\text{eq}} \cdot \phi_{(1)} + g_V F^{\mu\lambda} u_\lambda J_{\mu(1)}). \end{aligned} \tag{11.75}$$

This expression is finally cast into sums of squares,

$$\sigma = K\left\{-[\Delta^{\mu\lambda}(u)\partial_\lambda \alpha^* + \beta g_V F^{\mu\lambda} u_\lambda]^2\right\} + \eta \frac{\gamma^*}{m^*}(\sigma^{\mu\lambda})^2 + \frac{1}{3}\zeta \frac{\gamma^*}{m^*}(\theta^*)^2, \tag{11.76}$$

where use has been made of the energy–momentum conservation relation written in the form

$$\partial_\mu T^{\mu\nu}_{\text{fields}} = g_S \partial^\nu \phi \int d^4p\, f + g_V F^{\lambda\nu} J_\lambda. \tag{11.77}$$

In Eq. (11.75) the first square, $[\cdots]^2$, is negative, owing to the space character of the tensor involved therein, and the other terms are positive. The entropy production rate is positive, as demanded by the second principle of thermodynamics. Furthermore, it can be written in the general form

$$\sigma = \sum_i \lambda_i \chi^{i2}, \tag{11.78}$$

where λ_i are the transport coefficients (11.66) and (11.67) and χ_i are the associated thermodynamic forces. The latter appear to be modified by the presence of the scalar and vectorial fields: for instance, θ is replaced by θ^* [see Eq. (11.61c)], which involves $\dot{\phi}_{\text{eq}}$ and ϕ_{eq}. It should be noted that had we not modified the pressure term in the energy–momentum tensor (equivalently, had we not decomposed $T^{\mu\nu}_{(1)}$ as was done), then we would not have obtained a decomposition of the entropy production rate of the general form (11.77) and, consequently, our transport coeffficients would have been ill-defined.

(2) We now calculate the entropy production from a microscopic point of view, and let us show that it leads exactly to the same developments.

Note that since f plays the role of a distribution function on the mass shell $p^{*2} = m^{*2}$, the entropy can be defined as usual. Then calculation of the entropy production rate yields

$$\begin{aligned} S^\mu = -m^{*3} \sum_{\pm} \int \frac{d^3\omega}{\omega_0} \omega^\mu \{ f_\pm(x,p) \ln f_\pm(x,p) \\ + [1 - f_\pm(x,p)] \ln[1 - f_\pm(x,p)] \}, \end{aligned} \tag{11.79}$$

where $\omega^\mu = p^{\mu *}/m^*$; the entropy production rate σ is given by

$$\begin{aligned} \sigma \equiv \partial S^\mu &= 3\partial_\mu \ln(m^*) S^\mu - m^{*3} S^\mu \\ &= -m^{*3} \sum_{\pm} \int \frac{d^3\omega}{\omega_0} \omega^\mu \left\{ \ln \left[\frac{f_\pm}{1 - f_\pm} \right] \right\}, \end{aligned} \tag{11.80}$$

using the expansion of f and $\bar{f}$ in powers of τ,

$$\ln \left[\frac{f}{1-f} \right] \sim \ln \left[\frac{f_{\text{eq}}}{1 - f_{\text{eq}}} \right] + \frac{f_{(1)}}{f_{\text{eq}}(1 - f_{\text{eq}})}, \tag{11.81}$$

and a similar expression for $\bar{f}$. In the calculation of σ, we must also use

$$\begin{aligned} \partial_\mu f_{\text{eq}} &= [\partial_\mu \alpha^* - \partial_\mu(\gamma^* u_\lambda) \cdot \omega^\lambda] f_{\text{eq}}(1 - f_{\text{eq}}), \\ \partial_\mu \bar{f}_{\text{eq}} &= -[\partial_\mu \alpha^* + \partial_\mu(\gamma^* u_\lambda) \cdot \omega^\lambda] \bar{f}_{\text{eq}}(1 - \bar{f}_{\text{eq}}), \end{aligned} \tag{11.82}$$

so that we finally obtain

$$\sigma = 3\partial_\mu \ln(m^*) \left[S^\mu + \alpha^* J^\mu - \frac{4}{3}\frac{\gamma^*}{m^*} u_\lambda T^{\lambda\mu}_{\text{nucleons}} \right]$$
$$- \frac{\gamma^*}{m^*} u^\lambda \left[\frac{1}{2} m_S^2 \partial_\lambda \phi^2 + g_V F_{\mu\lambda} J^\mu \right] - \partial_\mu \alpha^* \cdot J^\mu_{(1)}$$
$$+ \partial_\mu(\gamma^* u_\lambda) \left(\frac{T^{\lambda\mu}_{(1)\text{nucleons}}}{m^*} \right). \tag{11.83}$$

At order 0, as it should be, while the first order entropy four-flux reduces to

$$S^\mu_{(1)} = -\alpha^* J^\mu_{(1)} + \left(\frac{\gamma^*}{m^* u_\lambda} \right) T^{\mu\lambda}_{(1)\text{nucleons}} \tag{11.84}$$

and the total entropy production rate

$$\sigma = \partial_\mu (S^\mu_{\text{eq}} + S^\mu_{(1)}), \tag{11.85}$$

finally gives rise to Eq. (11.75),

$$\partial_\mu S^\mu = \kappa q^2 + \eta \sigma^{\mu\nu} \sigma_{\mu\nu} + \varsigma \theta^{*2}, \tag{11.86}$$

which does coincide with the expression obtained through the use of the thermodynamic relation (11.69):

$$S^\mu = P\beta^\mu + \beta u_\lambda T^{\lambda\mu} - \alpha J^\mu$$
$$= -\frac{1}{3}\beta^\mu \Delta_{\lambda\nu}(u) T^{\lambda\nu} + \beta u_\lambda T^{\lambda\mu} - \alpha J^\mu. \tag{11.87}$$

11.3.4. *A brief comparison: BGK versus BUU*

It was pointed out several times that, due to the present state of the theoretical and experimental arts, it was often useless to deal with involved kinetic equations, the relaxation time approximation providing sufficient results in a first study. A comparison has been done by L. Mornas (1994) for the transport coefficients of symmetric nuclear matter at four times the nuclear saturation density and for $T \leq 200\,\text{MeV}$. This is depicted in Fig. 11.7, and below one can see the curves representing the various transport coefficients calculated with the relaxation time model (dot–dashed lines), with P. Danielewicz (1984) (where in-medium effects are not taken into account) and from the Boltzmann–Uhlenbeck–Uehling (BUU) equation with in-medium effects.

To perform this comparison, the relaxation time inserted in the BGK equation has been taken to be the average value of the one calculated with the BUU equation

$$\tau^{-1} = \frac{\int d^3p\, \tau^{-1}(p) f_{\text{eq}}(p)}{\int d^3p\, f_{\text{eq}}(p)} = \frac{1}{n_{\text{eq}}} \int d^4p\, \tau^{-1}(p) p \cdot u\, f_{\text{eq}}(p), \tag{11.88}$$

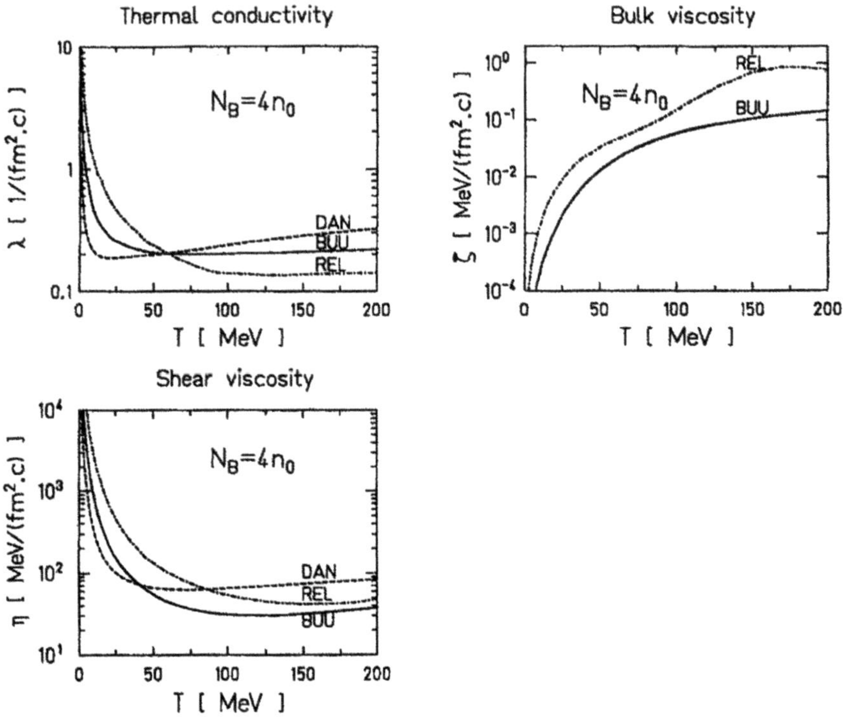

Fig. 11.7 A comparison between the results obtained for transport coefficients with and without in-medium effects and from the use of either the BGK or BUU transport equations [after L. Mornas (1994)].

with

$$\tau^{-1}(p) = \frac{1}{p^0}\int d^4p_2\, d^4p_3\, d^4p_4\, W(p,p_2 \to p_3,p_4) \times \{f_{\rm eq}(p_3)f_{\rm eq}(p_4) \times [1-f_{\rm eq}(p_2)] + [1-f_{\rm eq}(p_3)]\,[1-f_{\rm eq}(p_4)]\,f_{\rm eq}(p_2)\}\,[f_{\rm eq}(p)]\,. \tag{11.89}$$

This provides

$$\tau^{-1} = \frac{1}{n_{\rm eq}}\int d^4p_1\, d^4p_2\, d^4p_3\, d^4p_4\, W(p_1,p_2 \to p_3,p_4) \times \{f_{\rm eq}(p_3)f_{\rm eq}(p_4)\,[1-f_{\rm eq}(p_2)]\}\,[f_{\rm eq}(p)], \tag{11.90}$$

which is depicted in Fig. 11.8 and compared with previous results by K.H. Müller, and J. Randrup.[5]

[5]J. Randrup, *Nucl. Phys.* **A314**, 429 (1979); K.H. Müller, *Phys. Lett.* **B93**, 247 (1980).

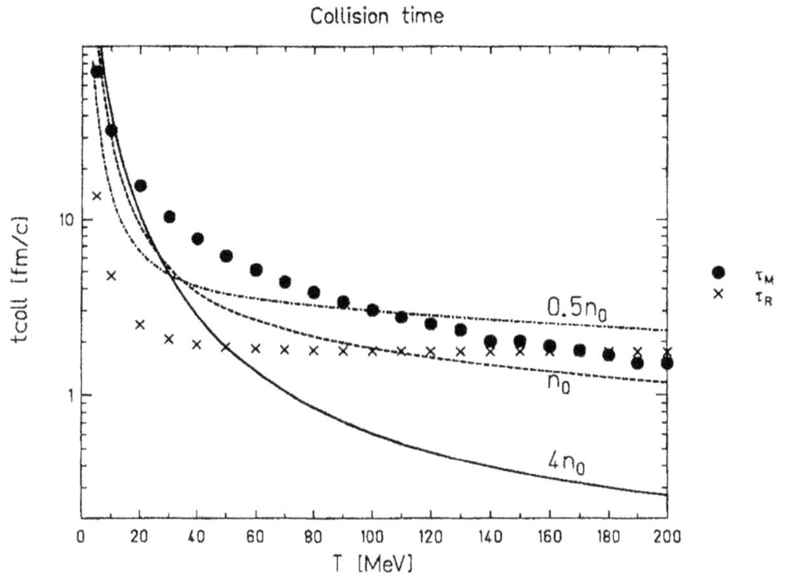

Fig. 11.8 Collision time (for three values of the baryon density; n_0 is the nuclear saturation density) according to the above relation, and comparison with the estimates by J. Randrup (1979) and K.H. Müller (1980), respectively indicated by crosses and dots [after L. Mornas (1994)].

It can be checked that the calculated τ possesses the expected properties of a relaxation time: it tends to a constant as $T \to \infty$ and is proportional to n_{eq}^{-1}; at low temperatures it behaves as T^{-2}.

11.4. Discussion

Let us now summarize and discuss the various assumptions behind our calculations.

(1) A specific model was chosen as to the description of relativistic nuclear matter and the Walecka model was the dynamical basis for our calculations. This particular model was chosen essentially because it is "canonical" in the sense that it is used as a reference for almost all other relativistic models. Of course, it contains its own problems, such as the much too large value of the compressibility coefficient[6] of nuclear matter, but this can be remedied by

[6]B.D. Serot and J.D. Walecka (1986).

adding suitable terms (or fields) to the basic Lagrangian: for instance, the addition of self-coupling of the scalar field constitutes one such possibility.[7] Therefore, our choice is not unreasonable even though it is not the only possible one: our calculations can always be improved.

(2) Since our aim was an evaluation of some collective effects, the simplest collision term was chosen: a relaxation time approximation, one that reduces to the Anderson–Witting one in the absence of collective effects. This choice was motivated both by the necessity of comparing similar results (with and without collective effects) and by the physical content of this collision term.

(3) The dynamics of the system (within the relaxation time approximation) was entireley involved in the relaxation time, itself to be evaluated with a more detailed analysis. However, the relaxation time could well be dependent on the collective fields ϕ and V^μ. In order to discuss this point, let us limit ourselves to the case of the scalar field. In the collision term, $\tau(\phi)$ appears only as τ^{-1}, so that in a first order Chapman–Enskog expansion one has

$$\begin{aligned}\frac{f_{(1)}}{\tau(\phi)} &= \frac{f_{(1)}}{\tau(\phi_{\text{eq}}) + \phi_{(1)}\tau'(\phi_{\text{eq}}) + \cdots} \\ &= \frac{f_{(1)}}{\tau(\phi_{\text{eq}})} + 0(\tau^2).\end{aligned} \tag{11.89}$$

$f - f_{\text{eq}}$ is a first order quantity and any (new) first order term like $\phi_{(1)}$ gives rise to a second order term. Therefore, our results are formally not affected by the possible ϕ dependence of the relaxation time. It is clear, however, that the numerical value of τ has changed and that the comparison of our results and those obtained by Anderson and Witting makes sense only for those τ's such that $\tau(\phi_{\text{eq}}) = \tau$. In order to get a more precise idea of the influence of ϕ on τ, we obviously need a specific calculation. Nevertheless, an estimation of τ as $\tau \approx 1/n\sigma_{\text{tot}}$ can give some clues in the absence of ϕ: in σ_{tot}, the mass of the nucleon m has to be replaced by its effective mass, $m^* = m - g_S\phi_{\text{eq}}$. A simple calculation shows that the one-boson-exchange nucleon–nucleon total cross-section is proportional to m^{-2}. This leads to $\tau \propto m^2$ and hence $\tau \propto (m - g_S\phi_{\text{eq}})^2$. It follows that ϕ makes τ smaller: $\tau(\phi_{\text{eq}}) \leq \tau(0)$. It should also be noted that in a one-boson-exchange calculation (or in higher order processes), it is sufficient to use

[7] R.M. Waldhauser, J.A. Maruhn, H. Stocker and W. Greiner, *Phys. Rev.* **C38**, 1003 (1988); C.M. Ko and Q. Li, *ibid.* **C37**, 2270 (1988).

the customary vacuum boson propagator: in the domain of temperatures and densities considered, both T and μ are much smaller than m_S.

(4) A few words have now to be said about the Chapman–Enskog expansion of the solution to the transport equation. This expansion was a series in powers of the small parameter $\varepsilon \equiv \tau/L$, where L is a (macroscopic) hydrodynamic scale. As a matter of fact, there exist several other scales, namely the ones defined by the various wavelengths occuring in the system, i.e.

$$\lambda_N = \frac{1}{m}, \quad \lambda_S = \frac{1}{m_S}, \quad \lambda_V = \frac{1}{m_V}. \tag{11.90}$$

Therefore, a complete Chapman–Enskog expansion should be an expansion in powers of several dimensionless parameters, besides ε, such as $\eta \equiv 1/mL$, $\chi \equiv 1/m_S L$ and $\xi \equiv 1/m_V$. In fact, instead of η we would rather use the parameter $\eta^* \equiv 1/m^* L$. As discussed elsewhere [J. Diaz Alonso and R. Hakim (1984)], while the parameter η^* is negligible, it is generally not so at high densities and/or temperatures for the parameter η^* and, consequently, a multiparameter expansion should be dealt with. On the other hand, the remaining parameters χ and ξ are also negligible in the approximation under study; however, when collective effects involve the consideration of quasibosons (Chap. 13), their effective mass might lead to effective parameters whose values are not negligible (compared to unity).

(5) The problem of renormalization has now to be discussed. In our calculation no infinity occurred: this was due to the fact that, systematically, the vacuum contribution to the Wigner function, i.e. terms involving

$$F_{\text{eq}}(p) = -\frac{d}{(2\pi)^3}\theta(-p^{*0})\delta\left(p^{*2} - m^{*2}\right), \tag{11.91}$$

was discarded. Does this procedure make sense? The answer to this question is twofold and it depends on the fact that the system is dominated either by collisions or by collective effects. When collective effects dominate, the thermal equilibrium state of the medium is controlled by a *renormalized* gap equation arising from the regularization of the vacuum term occurring in the gap equation [S.A. Chin (1977); J. Diaz Alonso and R. Hakim (1984)]. On the other hand, if the system is dominated by collisions (as, for instance, is the case for a dilute "gas"), the renormalization processes reduce to the usual renormalization procedure, leading to a finite cross-section and hence to a finite relaxation time. However, it should be borne in mind that we are dealing with a merely phenomenological theory and also that the length scale at hand (kinetic scale) is much larger than most scales where quantum fluctuations do show up, i.e. of the order of the Compton wavelengths. Accordingly, it is not necessary to take quantum fluctuations (via the vacuum Wigner function) into account. However, in order to be consistent with what is usually done in the case of thermodynamical equilibrium (in the Hartree approximation), one can use the renormalized gap equation studied elsewhere (see Chap. 10) instead of the nonrenormalized equation *and* various counterterms.

(6) The limiting cases (low densities and/or temperatures; high densities and/or temperatures) can easily be understood in this model; m^* is close to the nucleon mass m, while for the effective chemical potential μ^* one has $\mu^* = \mu + O(n_B)$. Accordingly, our result should be close to those already obtained by Anderson and Witting (1974): these properties can be checked through the figures where the various transport coefficients are computed as functions of the energy density expressed in units of the nuclear saturation density, for several temperatures. In the other limiting cases, the effective mass of the nucleon is almost vanishing [B.D. Serot and J.D. Walecka (1986); S.A. Chin (1977); G. Kalman (1974); J. Diaz Alonso and R. Hakim (1984)] and hence the general behavior of the transport coefficients can be obtained from the extreme relativistic limit of Anderson and Witting's results. Finally, our results mainly differ in the intermediate regime, as witnessed through the figures above. Note, however, the change in the slope of the curves λ and ξ when $\gamma \to 0$.

11.5. Dense Nuclear Matter: Neutron Stars

Relativistically dense matter occurs in quite different situations. A first instance is met on earth in large particle accelerators, where droplets of compressed and heated nuclear matter are formed during very short times in heavy ion collisions. Secondly, it is present in neutron stars. Thirdly, the physics of the primeval universe involves relativistically dense matter owing to both its densities and temperatures. All three situations represent different physical conditions which are intensively studied. The main challenge is to reconcile all the regimes within one theory. Here, we shall focus on the second situation.

Neutron stars were imagined by L. Landau, just after the discovery of the neutron by J. Chadwick in 1932, on the basis of what was known after S. Chandrasekhar (1932) about white dwarf stars. The principle was identical in the two cases: the Fermi pressure, due to the Pauli principle, balanced the gravity due to the mass of the star. Shortly after (1934), W. Baade and F. Zwicky, while observing the star then known as the Baade object, in the center of the Crab nebula,[8] published a short note concluding with these words: "With all reserve we advance the view that supernovae represent the transitions from ordinary stars into neutron stars, which in their final stages consist of extremely closely packed neutrons." In 1939, R. Oppenheimer and G. Volkov[9] proved that Landau's assumption of the existence of neutron stars was consistent with general relativity. With the assumption that neutrons form a relativistic ideal Fermi gas, they were able to show that neutron stars were objects with a radius of about 15 km and endowed with a mass of 0.7 solar mass, which represented a central density of $3.6 \cdot 10^{15}\,\text{g/cm}^3$, slightly more than 10 times the nuclear saturation density. In 1967, J. Bell and A. Hewish discovered the first pulsars, which were soon identified by T. Gold[10] (1969) with neutron stars.

Neutron stars are very interesting objects to study, essentially because they involve many domains on the frontiers of physics. In particular, relativistic statistical mechanics is involved in several aspects connected with the bulk data (mass and radius) of the star: its stability and its rotational accidents ("glitches"), which might be connected with a superfluid interior; its radiation, due to its magnetosphere; its strong magnetic field, which gives rise to interesting effects (see Chap. 12); new states of matter (such as "magnetic solids"; LOFER states; quark matter; meson condensation; superfluidity and/or superconductivity, colored or not); etc. In this section, a brief outline of some of the problems encountered will be presented.

11.5.1. *The static equilibrium of a neutron star*

The static equilibrium of a neutron star is determined by the general relativity equations of static equilibrium,[11] otherwise called the

---

[8]W. Baade and F. Zwicky, *Phys. Rev.* **45**, 138 (1934); see also *ibid.* **46**, 76 (1934). The Crab nebula was the residue of the explosion of a supernova, in 1054, observed by the Chinese, Korean and Japanese astronomers. Today, the "Baade object" is known as the Crab pulsar.

[9]R. Oppenheimer and G. Volkov, *Phys. Rev.* **55**, 374 (1939).

[10]T. Gold, *Nature* **221**, 25 (1969).

[11]See e.g. S. Weinberg, *Gravitation and Cosmology* (Wiley, New York, (1972), or R. Hakim, *Introduction to Relativistic Gravitation* Cambridge University Press, 1995).

Tolman–Oppenheimer–Volkov equations, which read

$$\begin{cases} \dfrac{dP(r)}{dr} = -G\dfrac{M(r)\rho(r)}{r^2}\left(1+\dfrac{P(r)}{\rho(r)}\right)\left(1+\dfrac{4\pi r^3 P(r)}{m(r)}\right)\left(1-2G\dfrac{M(r)}{r}\right)^{-1} \\ M(r) = \displaystyle\int_0^r 4\pi r'^2 dr' \rho(r'), \end{cases} \tag{11.92}$$

where $P(r)$ is the pressure at distance r from the center of the star; $\rho(r)$ the energy density, $M(r)$ the mass contained in a sphere of radius r, and G the gravitational constant. Note that the total mass of the star is $M(R)$, where R is its radius. Note that this system supposes that the energy–momentum tensor of matter is of the perfect fluid form.

This system, often called the Tolman[12]–Oppenheimer–Volkov (TOV) system, is based on Einstein's equations

$$R_{\mu\nu} - \frac{1}{2}g_{\mu\nu}R = 8\pi G T_{\mu\nu}, \tag{11.93}$$

and various assumptions, such as the spherical symmetry of the star, or its stationarity. As to the energy–momentum tensor $T_{\mu\nu}$, it is supposed to have the perfect fluid form.

In order to integrate this system, the first problem met is that of the knowledge of a reliable equation of state, $P = P(\rho)$. The second problem — a merely technical one — is the question of the initial conditions and when to stop the (numerical) integration of the TOV system.

The second problem is not difficult to solve: $M(0) = 0$ and the integration must be stopped when the pressure vanishes or so. This last condition defines the radius of the star: $P(R) = 0$. Finally, the only remaining parameter is the central energy density, ρ_C. Outside the star, the metric is the Schwarzschild (1916) metric, which matches smoothly the interior one.

As to the equation of state, which appears to be an essential ingredient in all possible models of neutron stars, it depends on the composition and the state of matter at different depths inside the star, and just as important, on the assumptions used to describe matter.

11.5.2. *The composition of matter in a neutron star*

In the initial R. Oppenheimer and G. Volkov model (1939), a neutron star was composed of an ideal Fermi gas of neutrons. However, one knows that

[12]R.C. Tolman, *Relativity, Thermodynamics and Cosmology* (Oxford University Press, 1934).

the neutron can desintegrate via the β-desintegration,

$$N \to P + e^- + \bar{\nu}_e$$

(into proton, electron and antineutrino), so that a pure neutron matter state is simply not possible and the neutron matter always contains a small admixture of protons and electrons, the antineutrinos instantly escaping the star due to their very weak interactions with matter. Conversely, protons and electrons tend to recombine into neutrons:

$$P + e^- \to N + \nu_e.$$

The balance of these reactions results in a chemical equilibrium. Inside the star, the high density displaces the equilibrium in such a way that only a few percent of electrons (and protons) subsist[13]:

$$\mu_e + \mu_P = \mu_N, \tag{11.94}$$

since $\mu_\nu = \mu_{\bar{\nu}} = 0$, owing to the fact that the neutrinos do not stay within the star.

As to the nuclei composition of the star, it should be noted that since the neutron–neutron force is weaker than the neutron–proton one, the excess of neutrons is more and more loosely bound to their parent nucleus. Finally, beyond a density of the order of $4 \times 10^{11}\ \mathrm{g/cm^3}$ — the so-called *drip point* — all nucleons are essentially "free" inside the neutron star.

The calculation of the chemical composition of a neutron star was initiated by B.K. Harrison and J.A. Wheeler (1958). They had a detailed calculation by G. Baym, H.A. Bethe and C.J. Pethick (1971) which rests on a semiempirical mass formula $E(A, Z)$ for the energy of a nucleus with A nucleons and Z protons, and this is valid as long as the nuclei preserve their identity. Such a formula — of the Bethe–Weizäcker type — takes account of various contributions to the energy of a nucleus calculated with some models reliable at low energies, and then extrapolated at high densities where it is supposed to be still valid. An example of such a mass formula is

$$\begin{cases} E(A, Z) = 15.68A - 18.56A^{2/3} - 0.717\dfrac{Z^2}{A^{1/3}} - 28.1\dfrac{(A-2Z)^2}{A} + E_P, \\[2ex] E_P = \dfrac{1}{2}(-1)^Z[1 + (-1)^A] \times \dfrac{12}{A^{1/2}}, \end{cases} \tag{11.95}$$

[13]Details can be found in the book by S.L. Shapiro and S.A. Teukolsky, *Black Holes, White Dwarfs and Neutron Stars* (Wiley, New York, 1983).

where one can recognize the energy per nucleon, the surface energy and the electrostatic energy, and where E_P is a pairing energy. From the empirical relation that connects the radius of a nucleus and its number of nucleons,

$$R = R_0 A^{1/3} \quad (R_0 = 1.25\,\mathrm{fm}), \tag{11.96}$$

one can infer that the centers of the nuclei all have the same density, the so-called saturation density.

$$\rho_{\mathrm{sat}} = \frac{A}{4\pi R_0^3/3}. \tag{11.97}$$

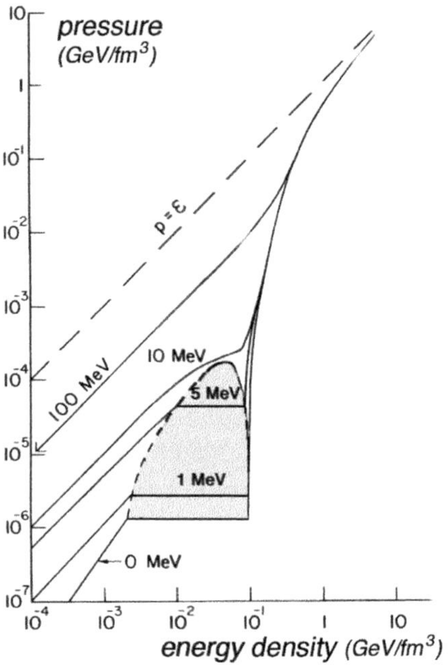

Fig. 11.9 The equation of state of pure neutron matter in the Walecka model. In our system of units the pressure and the energy density are measured in GeV/cm^3. Temperatures are measured in MeV and indicated along the various isotherms. The gray area corresponds to a region where there exists a first order phase transition (it occurs because of the scalar field). The horizontal lines are coexistence lines resulting from a Maxwell construction. The curve $p = \rho$ is the causal limit to which the Walecka equation of state is asymptotic. This characteristic "stiffness" of this equation of state is due to the (repulsive) vector field occurring in the model [after B.D. Serot and J.D. Walecka (1986)].

The total energy density of cold matter ($T = 0\,\mathrm{K}$) is thus

$$\rho = n_{\text{nuclei}} E(A,\, Z) + \rho_{\text{electrons}} + \rho_{\text{neutrons}}. \qquad (11.98)$$

The chemical composition of matter is obtained by minimizing ρ with respect to the parameters A and Z, treated as being continuous:

$$\frac{\partial}{\partial A}\rho = 0, \quad \frac{\partial}{\partial Z}\rho = 0. \qquad (11.99)$$

One then finds hypothetical nuclei — such as ${}^{982}_{32}\mathrm{Ge}$, ${}^{1350}_{50}\mathrm{Sn}$, ${}^{1800}_{50}\mathrm{Sn}$ and ${}^{1500}_{40}\mathrm{Zr}$ — that contain more and more neutrons as the density is increased.

From the expression for ρ, one obtains the equation of state (for densities below the drip point) — necessary for completing the TOV system — through

$$P = -n^2 \frac{\partial(n\rho)}{\partial n}. \qquad (11.100)$$

11.5.3. *Beyond the drip point*

Beyond the neutron drip point, the equation of state of neutron matter (Fig. 9.9) is lesser and lesser known. It can be considered as reliable until the nuclear saturation density, and also at slightly higher densities. The equation of state obtained from models such as that of Walecka described above needs to be matched smoothly to the low density results around the saturation density, where the domains of validity of the approaches overlap. At higher densities, one then extrapolates the equation of state at hand, whether it be the Walecka one or any other extension. As an example, the solution of the TOV system is given in the case of the Walecka equation of state in Fig. 11.10.

Presently, there exist dozens of plausible equations of state for relativistic nuclear matter, most of which agree with low energy data, and the possibility of observational discrimination between them is problematic. Note, however, the possibility of eliminating a few equations of state by looking at the measured masses of pulsars belonging to binary systems: if the maximum mass for a neutron star, predicted by the use of a given equation of state, is lower than the measured mass, then it has to be rejected. One of the very few equations to be eliminated is the equation of state of the ideal neutron gas used by R. Oppenheimer and G. Volkoff (1939): it predicted a maximum mass of 0.7 solar mass, whereas one observes neutron stars with 1.4 solar mass, for instance.

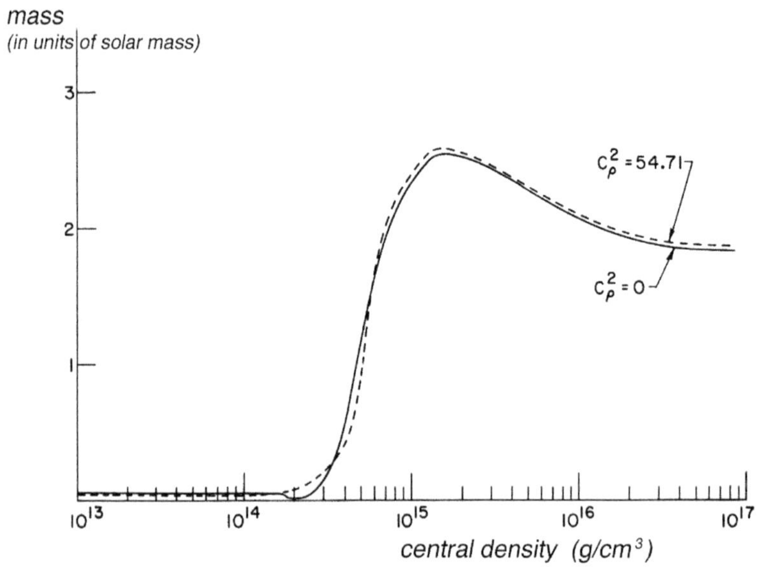

Fig. 11.10 The mass of a neutron star as a function of its central density. The continuous line corresponds to the Walecka model, while the dashed one takes account of the influence of the ρ meson. Compared with observed masses of neutron stars — often of the order of 1.4 solar mass — such an equation of state provides a too-important maximum [after B.D. Serot and J.D. Walecka (1986)].

The state of matter below the drip point is quite problematic. Indeed, it ranges from a neutron superfluid liquid to superfluid and superconducting protons, quark matter, without forgetting pion condensation, color superconducting fluids, etc.

Chapter 12

Strong Magnetic Fields

V. Canuto and H.Y. Chiu (1968*ff*) have studied in great detail the electron gas embedded in a strong magnetic field,[1] while the QED plasma modes (still in a strong magnetic field) have been studied by many authors [V. Canuto (1969, 1970); P. Bakshi, R. Cover and G. Kalman (1976); D.B. Melrose and R.J. Stoneham (1976, 1977); D.B. Melrose (1983, 1997); H. Sivak (1985); A.E. Shabad and H. Perez-Rojas (1976*ff*), etc.], and the equation of state and the equilibrium properties of such a plasma have been investigated by D.H. Constantinescu (1972*ff*), P. Rehak (1975), G.A. Schulman (1972*ff*), S. Visvanathan (1962), etc. The results obtained via the use of the covariant Wigner function techniques by several authors [R. Dominguez-Tenreiro (1977); R. Hakim and H. Sivak (1982)] are presented here. An interesting domain where the strong magnetic fields are involved deals with magnetized solids, a field initiated by M.A. Ruderman (1971, 1972) [see Dong Lai (2001)], and it was also studied by V.V. Kadomtsev and V.S. Kudriatsev (1971), M.L. Glasser and J.I. Kaplan (1975), and others. These magnetic solids play an important role in the study of phenomena connected with the external crust of a neutron star. However, they are not considered here.

By "strong magnetic field" is meant a magnetic field whose intensity is of the order of the critical value defined by the approximate equality of the Larmor radius and the Compton wavelength of a particle (4.414×10^{13} G).

Since 1968, when the systematic study of the electron gas embedded in strong magnetic fields was undertaken by V. Canuto and H.Y. Chiu, the magnetic field of neutron stars has been evaluated to be slightly subcritical ($\approx 10^{12}$ G), and more recently new types of magnetic stars, *magnetars*, have

[1]See the excellent review by V. Canuto and J. Ventura (1977).

been discovered with much more intense fields, of the order of 10^{15}–10^{16} G. On the other hand, J.P. Ostriker and F.D.A. Hartwick[2] speculated that even within some white dwarfs, magnetic fields as intense as 10^{11}–10^{13} G could be found. Such magnetic white dwarfs were studied by D. Adams (1985) and D. Adams and H. Sivak (1985).

Let us briefly mention how such intense magnetic fields are evaluated in neutron stars.[3] Assuming that pulsars are neutron stars — and there exists a *general consensus* as to this assumption — and that the magnetic field is a dipole field, the energy radiated per unit of time during the rotation is

$$\dot{E} = -\frac{1}{6}B^2R^6\omega^4\sin^2\theta, \tag{12.1}$$

where ω is the rotation velocity, B the magnetic field, R the radius of the star, and θ the angle between the dipole and the rotation axis. The radiation emitted by the rotating dipole is at the expense of the rotation kinetic energy of the star,

$$E = \frac{1}{2}I\omega^2, \tag{12.2}$$

and accordingly

$$\dot{E} = I\omega\dot{\omega}. \tag{12.3}$$

Assuming that the Crab pulsar is a uniform sphere of 1.4 solar mass and of radius 12 km, one gets

$$E = 2.5 \times 10^{49}\ \text{erg}, \quad \dot{E} = 6.4 \times 10^{38}\ \text{erg/s}. \tag{12.4}$$

The measure of the slowing down, or of $\dot{\omega}$, for the same pulsar provides

$$\dot{E} \approx 5 \times 10^{38}\ \text{erg/s}, \tag{12.5}$$

which is very close to the above evaluation. Finally, with $\theta = \pi/2$, the magnetic dipole model yields

$$B \approx 5.2 \times 10^{12}\ \text{G}.$$

Most pulsars, indeed, provide similar orders of magnitude. To explain the additional three orders of magnitude recently discovered in magnetars, one invokes the threading of magnetic field lines during a turbulent convection phase in the progenitor supernova, thus bringing into action a transient but

---

[2]J.P. Ostriker and F.D.A. Hartwick, *Astrophys. J.* **161**, 541 (1968).

[3]See details in S.L. Shapiro and S.A. Teukolski, *Black Holes, White Dwarfs and Neutron Stars* (J. Wiley and Sons, New York, 1983).

powerful dynamo. The recently discovered *magnetars* give rise to the figure given above, $B \approx 2 \times 10^{15}$ G.

In this chapter, the problem of a QED plasma embedded in a strong magnetic field is addressed and, to this end, the thermal equilibrium properties of a magnetic electron gas are first dealt with. Note that when one is speaking of an "electron gas," it is always assumed that it exists within a positive neutralizing background of ions, whether in the form of a fluid or possibly as a lattice.

The calculations with magnetic fields are particularly long and involved and require much notation. Therefore, only the main results are presented here and only a brief outline of the calculations is given.

The free electron field, embedded in a magnetic field $A^\mu(x)$, obeys the Dirac equations

$$\begin{cases} \{i\gamma \cdot (\partial - ieA) - m\}\psi(x) = 0, \\ \bar{\psi}(x)\{i\gamma \cdot (\overleftarrow{\partial} + ieA) + m\} = 0, \end{cases} \tag{12.6}$$

which has first to be solved. H.Y. Chiu and V. Canuto (1968) used the solution obtained by M.H. Johnson an B.A. Lippmann[4]; it is, however, simpler to use the covariant solutions derived later[5] [R. Hakim and H. Sivak (1982)].

The energy levels of an electron within the magnetic field, the so-called *Landau levels*, are given by

$$E_{n,\sigma,p_{||},\chi} = \chi\sqrt{m^2 + p_{||}^2 + \frac{B}{B_{\text{crit}}}(2n + \sigma + 1)}, \tag{12.7}$$

B_{crit} being the critical field where the Larmor radius of an electron is equal to its Compton wavelength,

$$B_{\text{crit}} = \frac{m^2c^3}{eh} = 4.414 \times 10^{13}\,\text{G}, \tag{12.8}$$

where

$$\chi = \pm 1;\ \sigma = \pm 1;\ n = 0, 1, 2 \ldots;\ p_{||} \in \Re. \tag{12.9}$$

χ characterizes the positive/negative energies of an electron, σ its spin state, and n is the main quantum number describing, so to speak, the size of the orbit of an electron; finally, $p_{||}$ is the momentum parallel to the direction of the magnetic field.

[4]M.H. Johnson and B.A. Lippmann, *Phys. Rev.* **76**, 828 (1949); see also H. Robl, *Acta Phys. Austriaca* **6**, 105 (1952).
[5]H. Sivak, unpublished (1979) and thesis (1985).

The spinors associated with these eigenvalues are[6]

$$\psi_{1n\chi} = c_{1n\chi} \begin{vmatrix} (\chi E_{1n} + m)\varphi_n \\ 0 \\ p_{||}\varphi_n \\ [2(n+1)eh]^{1/2}\varphi_{n+1} \end{vmatrix} \exp(-i\chi E_{1n} x \cdot u), \tag{12.10}$$

(spin up)

$$\psi_{-1n\chi} = c_{-2n\chi} \begin{vmatrix} 0 \\ (m + \chi E_{-2n})\varphi_n \\ (2neh)^{1/2}\varphi_{n-1} \\ -p_{||}\varphi_n \end{vmatrix} \exp(-i\chi E_{-1n} x \cdot u), \tag{12.11}$$

(spin down)

with

$$\begin{aligned} \varphi_n = {} & N_n H_n \left\{ (eh)^{1/2}[x_{(r)} - a] \right\} \\ & \times \exp\left[-(2eh)(x_{(r)} - a)^2 + ip_{||} \cdot x_{(n)} + (2eh)^{1/2} i x_{(s)} \cdot (x_{(r)} - a) \right], \end{aligned} \tag{12.12}$$

where H_n is a Hermite polynomial of order n and $x_{(r)}, x_{(n)}, x_{(s)}$ are defined by

$$x_{(r)} \equiv -x \cdot r, \quad x_{(s)} \equiv -x \cdot s, \quad x_{(n)} \equiv -x \cdot n, \tag{12.13}$$

and N_n is the normalization coefficient. The definitions of the quadrivectors s, r and n are given in the next section; however, they reduce to x, y and z, respectively.

12.1. Relations Obeyed by the Magnetic Field

The electromagnetic field $F^{\mu\nu}$ possesses a magnetic character that is expressed by the relations

$$\begin{cases} F^{\mu\nu} F_{\mu\nu} > 0, \\ \varepsilon_{\mu\nu\alpha\beta} F^{\mu\nu} F^{\alpha\beta} \equiv {}^* F^{\mu\nu} F_{\mu\nu} = 0. \end{cases} \tag{12.14}$$

The second relation tells us that the electric and magnetic fields in an arbitrary inertial frame are orthogonal to each other, while the first one

[6]M.H. Johnson and B.A. Lippmann, *loc. cit.*

means the existence of an inertial frame where the electromagnetic field is purely magnetic. Then it can be shown[7] that there exists a timelike four-vector u^μ and a spacelike four-vector h^μ endowed with the properties

$$\begin{cases} u_\mu u^\mu = 1, \\ h_\mu h^\mu = -h^2, \\ u_\mu h^\mu = 0, \end{cases} \tag{12.15}$$

such that

$$\begin{cases} F^{\mu\nu} = \varepsilon^{\mu\nu\alpha\beta} h_\alpha u_\beta, \\ {}^*F^{\mu\nu} = -2h^{[\mu}u^{\nu]}. \end{cases} \tag{12.16}$$

Thus, there exists a frame of reference where the electromagnetic field is of the form

$$F^{\mu\nu} = \begin{bmatrix} 0 & 0 & 0 & 0 \\ 0 & 0 & -h & 0 \\ 0 & +h & 0 & 0 \\ 0 & 0 & 0 & 0 \end{bmatrix} = F_{\mu\nu}. \tag{12.17}$$

The unit spacelike four-vector antiparallel to h^μ is denoted by n^μ:

$$n^\mu = -\frac{h^\mu}{h}.$$

Note that the four-vectors (u^μ, h^μ) are not unique and that from one pair one can find another related through

$$\begin{cases} u'^\mu = u^\mu \cosh\chi + n^\mu \sinh\chi, \\ n'^\mu = u^\mu \sinh\chi + n^\mu \cosh\chi, \end{cases} \tag{12.18}$$

which can be interpreted as a kind of "Lorentz transformation" in a two-dimensional Minkowski space, and the use of such "vectors" should be accompanied by the requirement for an invariance under these transformations.

We now introduce the following projectors which are repeatedly used in what follows[8]:

$$\begin{cases} \Pi^{\mu\nu} = \eta^{\mu\nu} - u^\mu u^\nu + n^\mu n^\nu, \\ \Omega^{\mu\nu} = u^\mu u^\nu - n^\mu n^\nu. \end{cases} \tag{12.19}$$

---

[7]A. Lichnérowicz, *Relativistic Magnetohydrodynamics* (Benjamin, New York, 1971).

[8]This projector should not be confused with the polarization operator used repeatedly in this book with the same notation. For the polarization operator, one generally indicates the variables on which it depends.

The projector $\Pi^{\mu\nu}$ is a projection over the two-plane orthogonal to the one spanned by u^μ and n^μ. It possesses u^μ as timelike and n^μ as spacelike eigenvectors; it has two other spacelike eigenvectors — say, r^μ and s^μ — which can always be chosen as being mutually orthogonal and of length -1:

$$\begin{cases} \Pi^{\mu\nu} u_\nu = \Pi^{\mu\nu} n_\nu = 0, \\ \Pi^{\mu\nu} r_\nu = r_\nu, \\ \Pi^{\mu\nu} s_\nu = s_\nu. \end{cases} \tag{12.20}$$

The four-vectors (r_ν, s_ν) are no more unique than (u_ν, n_ν) and the whole family of possible such eigenvectors are given by

$$\begin{cases} s'^\mu = s^\mu \cos\phi + r^\mu \sin\phi, \\ r'^\mu = -s^\mu \sin\phi + r^\mu \cos\phi. \end{cases} \tag{12.21}$$

With these notations, one easily finds the useful relations

$$\begin{aligned} \Pi^\mu_\nu F^{\nu\alpha} &= F^{\mu\alpha}, \\ \Pi^{\mu\nu} &= -(r^\mu r^\nu + s^\mu s^\nu) = -\frac{1}{h^2} F^{\mu\alpha} F^\mu{}_\nu, \\ r^\mu &= \frac{1}{h} F^\mu_\nu s^\nu = -\varepsilon^{\mu\nu\alpha\beta} n_\alpha u_\beta s_\nu, \\ F^{\mu\nu} &= -h(r^\mu s^\nu - r^\nu s^\mu). \end{aligned} \tag{12.22}$$

The projector $\Omega^{\mu\nu}$ is a projection over the two-plane spanned by u^μ and n^μ. Finally, the four-potential associated with the magnetic field $F^{\mu\nu}$ reads

$$A^\mu(x) = -\frac{1}{2} F^\mu{}_\nu x^\nu \tag{12.23}$$

in the Lorentz gauge. Note also the useful relations

$$\begin{cases} \Pi^{\mu\nu} = \dfrac{1}{h^2} F^{\mu\alpha} F^\nu{}_\alpha, \\ \Omega^{\mu\nu} = \dfrac{1}{4h^2} {}^*F^{\mu\alpha} {}^*F^\nu{}_\alpha. \end{cases} \tag{12.24}$$

12.2. The Partition Function

Let us now study the various thermodynamical functions of the electron gas in a strong magnetic field. The additive first integrals of the movement are essentially:[9] $\hat{Q}$, the charge of the system; $u_\mu \hat{P}^\mu$, the energy; and $n_\mu \hat{P}^\mu$,

[9]There also exists the rotation around the magnetic field; however, we do not consider it. Neither do we deal with the collective movements along the magnetic field.

the moment along the magnetic field. The partition function of the system then reads

$$\rho_{\text{stat}} = \frac{1}{Z} \exp[-\beta(u_\mu \hat{P}^\mu - \mu\hat{Q})], \tag{12.25}$$

where $H \equiv \hat{P}^0$ and the parallel component of $\hat{P}$ has been "forgotten": there are no parallel collective fields. From this equation the following formulae are obtained:

$$\begin{cases} n_{\text{eq}} = \dfrac{1}{\beta}\dfrac{\partial \ln Z}{\partial \mu}, \\ \rho = u_\mu u_\nu T^{\mu\nu}, \\ P_\perp = -\dfrac{1}{2}\Omega_{\mu\nu}T^{\mu\nu}, \\ P_{||} = n_\mu n_\nu T^{\mu\nu}, \end{cases} \tag{12.26}$$

from which several derivations of these quantities can easily be obtained.

The partition function is thus

$$Z = \{\exp[-\beta(H - \mu\hat{Q})]\}, \tag{12.27}$$

where we have retained only the main additive first integrals, H and $\hat{Q}$, leaving the remaining one. Let us now calculate this partition function. First, the Hamiltonian has the form

$$H = \sum_{n,\sigma,p_{||}} \sqrt{m^2 + p_{||}^2 + \frac{B}{B_{\text{crit}}}(2n+\sigma+1)}\,[a^+_{n,\sigma,p_{||}} a_{n,\sigma,p_{||}} + d^+_{n,\sigma,p_{||}} d_{n,\sigma,p_{||}}],$$

$$\hat{Q} = \sum_{n,\sigma,p_{||}} [a^+_{n,\sigma,p_{||}} a_{n,\sigma,p_{||}} - d^+_{n,\sigma,p_{||}} d_{n,\sigma,p_{||}}], \tag{12.28}$$

where the creation (destruction) operators of the electrons read $\{a^+_{n,\sigma,p_{||}}, a_{n,\sigma,p_{||}}\} \equiv \{a^+_{n,\sigma,p_{||}} a_{n,\sigma,p_{||}} + a_{n,\sigma,p_{||}} a^+_{n,\sigma,p_{||}}\} = I$ and similarly for the positrons (d's and d^+'s), and obey the commutation relations

$$\begin{aligned} \{a^+_{n,\sigma,p_{||}}, a_{n',\sigma',p'_{||}}\} &\equiv \{a^+_{n,\sigma,p_{||}} a_{n',\sigma',p'_{||}} + a_{n',\sigma',p'_{||}} a^+_{n,\sigma,p_{||}}\} \\ &= \delta_{nn'}\delta_{\sigma\sigma'}\delta_{p_{||}p'_{||}}, \end{aligned} \tag{12.29}$$

and zero when one of the quantum numbers is different. They are related as

$$\begin{aligned} \psi &= \sum_\ell [a_\ell u_\ell e^{-iE_\ell t} + d^+_\ell \bar{v}_\ell e^{+iE_\ell t}], \\ \bar{\psi} &= \sum_\ell [a^+_\ell \bar{u}_\ell e^{+iE_\ell t} + d_\ell v_\ell e^{-iE_\ell t}]. \end{aligned} \tag{12.30}$$

Next, since this is a system without interaction, it can be treated as in Chap. 7, or as

$$
\begin{aligned}
Z &= \sum_q \prod_{n_\ell} \exp(-\beta[n_\ell E_\ell - \mu]) \\
&= \sum_q \prod_{n_\ell} \{\exp(-\beta[E_\ell - \mu])\}^{n_\ell}, \qquad (12.31)
\end{aligned}
$$

where $n_\ell (n_\ell = 0, 1)$ is the eigenvalue of a_ℓ, where q is the total charge and $\ell = \{\chi; \sigma; n; p_{||}\}$. Finally, the partition function appears to be

$$
\begin{aligned}
\log Z &= \sum_\ell \{1 + \exp[-\beta(E_\ell - \mu)]\} \\
&= \sum_{p_{||}, n, \sigma} \left\{1 + \exp\left[-\beta\left(\sqrt{m^2 + p_{||}^2 + \frac{B}{B_{\text{crit}}}(2n + \sigma + 1)} - \mu\right)\right]\right\},
\end{aligned}
\qquad (12.32)
$$

where we have restricted this formula to electrons.

Now let us look at the sum $\sum$ which occurs in this last expression, and let us use the developments of V. Canuto and H.Y. Chiu (1970) to calculate the degeneracy level. We then consider the sum

$$
\frac{1}{2\pi} \sum_{p_{\perp 1}} \sum_{p_{\perp 2}} \to \int_{-\infty}^{+\infty} \int_{-\infty}^{+\infty} dp_{\perp 1}\, dp_{\perp 2}, \qquad (12.33)
$$

which, in cylindrical coordinates, reads

$$
\int_{-\infty}^{+\infty} \int_{-\infty}^{+\infty} dp_{\perp 1} dp_{\perp 2} = \int_0^\infty p_\perp dp_\perp \int_0^{2\pi} d\phi = \pi \int_0^\infty dp_\perp^2. \qquad (12.34)
$$

Classical quantization yields

$$
p_\perp^2 \to 2nm^2 \frac{B}{B_0} \qquad (12.35)
$$

and from the quantization of energy it is actually a harmonic oscillator. To obtain this degeneracy, it is sufficient to evaluate the equation

$$
\frac{dp_{\perp 1} dp_{\perp 2}}{2\pi}, \qquad (12.36)
$$

between two successive levels of the energy, the way one quantum of action of continuous level coalesces into

$$
d_n = \frac{1}{2\pi} \int_n^{n+1} dp_{\perp 1}\, dp_{\perp 2}, \qquad (12.37)
$$

and one gets

$$d_n = \frac{m^2}{2\pi}\frac{B}{B_0}. \tag{12.38}$$

Therefore, one has

$$\sum_f \to \frac{1}{(2\pi)^2} m^2 \frac{B}{B_0} \sum_n \sum_s \int_{-\infty}^{+\infty} \tag{12.39}$$

for the summation over the states of the system.

Finally, $\log Z$ is written as

$$\log Z = \frac{e}{(2\pi)^2} \frac{B}{|e|} \sum_{r=1}^{r=2} \sum_{n=0}^{\infty} \int_{-\infty}^{+\infty} d\xi \, \{\ln(1 + \exp -\beta[E_{r,n}(\xi) - \mu]) + \ln(1 + \exp -\beta[E_{r,n}(\xi) + \mu])\}, \tag{12.40}$$

where the first bracket in the integral corresponds to the electrons, while the other one is that of positrons, which we have restituted.

12.2.1. *Magnetization of an electron gas*

As an example where the above calculations of the partition function work, we now calculate the magnetization. From the magnetic moment operator $\hat{M}$,

$$\hat{M} = -\frac{\partial H}{\partial B}, \tag{12.41}$$

we are going to take the average value

$$M = \langle \hat{M} \rangle = \mathrm{Tr}(\rho_{\mathrm{stat}} \hat{M}); \tag{12.42}$$

however, before this is done a few explanatory words are in order. In the magnetic moment operator, it is the derivative with respect to B which plays a role and not with respect to h; B is the magnetic induction, while h is the magnetic field. They are interrelated via the formula

$$B = H + M(B) \tag{12.43}$$

and, unlike what was done previously where there was no distinction between H and B, now we have to be a little bit more serious. In particular, the Larmor radius should contain B and not H:

$$r_0 = \frac{eB}{m}. \tag{12.44}$$

It is, indeed, shown that an orbiting electron is sensitive to the magnetic induction and not to the ambient magnetic field. In particular, the eigenenergy of H is

$$E_{n,s,p_{||}} = \sqrt{m^2 + p_{||}^2 + \frac{B}{B_{\text{crit}}}(2n+s+1)}. \tag{12.45}$$

Let us now calculate the average value M. We have

$$\begin{aligned}\langle\hat{M}\rangle &= -\text{Tr}\left(\rho_{\text{stat}}\frac{\partial H}{\partial B}\right) \\ &= \beta\frac{\partial}{\partial B}\ln Z(T,B)\end{aligned} \tag{12.46}$$

(H is the Hamiltonian) or, in terms of the partition function,

$$M = \frac{1}{\beta}\frac{\partial}{\partial B}\sum_\alpha d_n \ln[1+\exp(-\beta E_\alpha + \beta\mu)]. \tag{12.47}$$

With the substitution

$$\sum_\alpha \to \sum_{p,n,s} d_n, d_n = \frac{1}{2\pi}m^2\frac{B}{B_0}, \tag{12.48}$$

introduced into Eq. (12.46), a form of the magnetization is written as

$$\begin{aligned}M = \frac{e}{\pi^2}m^2\Bigg(&-\frac{B}{B_{\text{crit}}}\sum_{n=0}^{\infty} n\int_0^\infty \frac{d\xi}{\varepsilon_n}\frac{1}{1+e^{\beta m\varepsilon_n}} \\ &+\frac{1}{2}\frac{1}{\beta m}\int_0^\infty d\xi \ln\{1+\exp[-\beta m\varepsilon_0(\xi)+\beta\mu]\} \\ &+\frac{1}{\beta m}\sum_{n=1}^{\infty}\int_0^\infty d\xi\ln\{1+\exp[-\varepsilon_n(\xi)m\beta+\beta\mu]\}\Bigg).\end{aligned} \tag{12.49}$$

There exists, however, a simpler form for the magnetization obtained after an integration by parts,

$$\int_0^\infty d\xi \ln\{1+\exp[-\beta m\varepsilon_n(\xi)+\beta\mu]\} \equiv \beta m\int_0^\infty d\xi\frac{\xi^2}{\varepsilon_n(\xi)}\frac{1}{1+e^{\beta m\varepsilon_n(\xi)-\beta\mu}}, \tag{12.50}$$

which reads

$$\begin{aligned}M = \frac{e}{\pi^2}m^2\Bigg[&-\frac{B}{B_{\text{crit}}}\sum_{n=0}^{\infty} n\int_0^\infty \frac{d\xi}{\varepsilon_n(\xi)}\frac{1}{1+e^{\beta m\varepsilon_n(\xi)}} \\ &+\frac{1}{2}\int_0^\infty \frac{d\xi}{\varepsilon_0(\xi)}\xi^2\frac{1}{1+\exp[\beta m\varepsilon_0(\xi)-\beta\mu]} \\ &+\sum_{n=1}^{\infty}\int_0^\infty\frac{d\xi}{\varepsilon_n(\xi)}\xi^2\frac{1}{1+e^{\varepsilon_n(\xi)m\beta-\beta\mu}}\Bigg].\end{aligned} \tag{12.51}$$

12.3. Relativistic Quantum Liouville Equation

In this chapter, a slight modification of the covariant Wigner function is used and it reads

$$F_{\text{new}}(x,p) = \frac{1}{(2\pi)^4}\int d^4R \exp(-i\pi\cdot R)\left\langle \bar{\psi}\left(x+\frac{1}{2}R\right)\otimes\psi\left(x-\frac{1}{2}R\right)\right\rangle, \tag{12.52}$$

where we have set

$$\pi^\mu = p^\mu + eA^\mu(x). \tag{12.53}$$

With this definition, the ideal electron gas thermodynamic quantities are gauge-invariant since the energy–momentum tensor is itself gauge-invariant,

$$^*T^{\mu\nu} = \text{Sp}\int d^4p\, p^\mu\gamma^\nu F_{\text{new}}(x,p), \tag{12.54}$$

as mentioned in Chap. 8. In what follows, the index "new" is suppressed.

From this definition for $F(x,\, p)$ and Dirac's equation, one easily finds [R. Dominguez Tenreiro and R. Hakim (1982)] the relativistic quantum Liouville equations as

$$\begin{cases} \{i\gamma\cdot\partial + 2(\gamma\cdot p - m) + ieF^\alpha{}_\mu\gamma^\mu\nabla_\alpha\}F(x,p) = 0, \\ F(x,p)\{i\gamma\cdot\overleftarrow{\partial} - 2(\gamma\cdot p - m) + ieF^\alpha{}_\mu\gamma^\mu\cdot\nabla_\alpha\} = 0. \end{cases} \tag{12.55}$$

Note that the Lorentz gauge condition, which is linear in x^μ, gives rise in these equations to the terms

$$ieF^\alpha{}_\mu\gamma^\mu\frac{\partial F}{\partial p^\alpha}. \tag{12.56}$$

This equation is now investigated a bit further by looking at its form for x-independent solutions $F(x,\, p) = F(p)$. To this end, F is decomposed on the basis of the algebra of the 16 Dirac matrices as

$$F(x,p) = \frac{1}{4}\left\{f(x,p)I + f_\mu(x,p)\gamma^\mu + \frac{1}{2}f(x,p)\sigma^{\mu\nu} + f_5(x,p)\gamma^5 + f_5^\mu(x,p)\gamma^\mu\gamma_5\right\}, \tag{12.57}$$

with

$$f_A(x,p) = \frac{1}{4}\text{Sp}[\gamma_A F(x,p)], \quad A = 1,2,\ldots,16, \tag{12.58}$$

and one finds the system

$$\begin{cases} 2ip_\mu f^\mu - 2imf - eF^{\mu\alpha}\dfrac{\partial}{\partial p^\alpha} f^\mu = 0, \\ 2ip_\mu f^\mu - 2imf + eF^{\mu\alpha}\dfrac{\partial}{\partial p^\alpha} f^\mu = 0, \end{cases} \tag{12.59}$$

$$\begin{cases} 2ip_\mu f_5^\mu + 2imf_5 - eF^\alpha{}_\mu \dfrac{\partial}{\partial p^\alpha} f_5^\mu = 0, \\ 2ip_\mu f_5^\mu - 2imf_5 + eF^\alpha{}_\mu \dfrac{\partial}{\partial p^\alpha} f_5^\mu = 0, \end{cases} \tag{12.60}$$

$$\begin{cases} 2ip_\mu f - 2imf_\mu - 2ip^\alpha f_{\mu\alpha} - eF^\alpha{}_\mu \dfrac{\partial}{\partial p^\alpha} f + eF^{\alpha\beta}\dfrac{\partial}{\partial p^\alpha} f_{\mu\beta} = 0, \\ 2ip_\mu f - 2imf_\mu + 2ip^\alpha f_{\mu\alpha} + eF^\alpha{}_\mu \dfrac{\partial}{\partial p^\alpha} f + eF^{\alpha\beta}\dfrac{\partial}{\partial p^\alpha} f_{\mu\beta} = 0, \end{cases} \tag{12.61}$$

$$\begin{cases} -2ip_{[\mu} f_{\nu]} + 2imf_{\mu\nu} + 2\varepsilon_{\alpha\lambda\mu\nu} p^\alpha f_5^\lambda + eF^\alpha{}_\mu \dfrac{\partial}{\partial p^\alpha} f_{\nu]} \\ \qquad + ie\varepsilon_{\lambda\beta\mu\nu} F^{\alpha\beta}\dfrac{\partial}{\partial p^\alpha} f_5^\lambda = 0, \\ -2ip_{[\mu} f_{\nu]} + 2imf_{\mu\nu} - 2\varepsilon_{\alpha\lambda\mu\nu} p^\alpha f_5^\lambda - eF^\alpha{}_\mu \dfrac{\partial}{\partial p^\alpha} f_{\nu]} \\ \qquad - ie\varepsilon_{\lambda\beta\mu\nu} F^{\alpha\beta}\dfrac{\partial}{\partial p^\alpha} f_5^\lambda = 0, \end{cases} \tag{12.62}$$

$$\begin{cases} -\varepsilon_{\beta\sigma\lambda\mu} p^\beta f^{\sigma\lambda} - 2ip_\mu f_5 - 2imf_{5\mu} \\ \qquad - \dfrac{i}{2} e\varepsilon_{\beta\sigma\lambda\mu} F^{\alpha\beta}\dfrac{\partial}{\partial p^\alpha} f^{\sigma\lambda} + eF^\alpha{}_\mu \dfrac{\partial}{\partial p^\alpha} f_5 = 0, \\ +\varepsilon_{\beta\sigma\lambda\mu} p^\beta f^{\sigma\lambda} - 2ip_\mu f_5 + 2imf_{5\mu} \\ \qquad - \dfrac{i}{2} e\varepsilon_{\beta\sigma\lambda\mu} F^{\alpha\beta}\dfrac{\partial}{\partial p^\alpha} f^{\sigma\lambda} - eF^\alpha{}_\mu \dfrac{\partial}{\partial p^\alpha} f_5 = 0. \end{cases} \tag{12.63}$$

This system has a more useful form by adding and subtracting the equations of each couple, and it reads

$$\begin{cases} p_\mu f^\mu - mf = 0, \\ eF^\alpha{}_\mu \dfrac{\partial}{\partial p^\alpha} f^\mu = 0, \end{cases} \tag{12.64}$$

$$\begin{cases} p_\mu f_5^\mu = 0, \\ eF^\alpha{}_\mu \dfrac{\partial}{\partial p^\alpha} f_5^\mu - 2imf_5 = 0, \end{cases} \tag{12.65}$$

$$\begin{cases} 2ip_\mu f - 2imf_\mu - eF^{\alpha\beta} \dfrac{\partial}{\partial p^\alpha} f_{\beta\mu} = 0, \\ -2ip^\beta f_{\beta\mu} + eF^\alpha{}_\mu \dfrac{\partial}{\partial p^\alpha} f = 0, \end{cases} \tag{12.66}$$

$$\begin{cases} 2p_{[\mu} f_{\nu]} - e\varepsilon_{\beta\lambda\mu\nu} F^{\alpha\beta} \dfrac{\partial}{\partial p^\alpha} f_5^\lambda = 0, \\ 2\varepsilon_{\beta\lambda\mu\nu} p^\beta f_5^\lambda + 2imf_{\mu\nu} + eF^\alpha{}_{[\mu} \dfrac{\partial}{\partial p^\alpha} f_{\nu]} = 0, \end{cases} \tag{12.67}$$

$$\begin{cases} 4p_\mu f_5 + e\varepsilon_{\sigma\lambda\beta\mu} F^{\alpha\beta} \dfrac{\partial}{\partial p^\alpha} f^{\sigma\lambda} = 0, \\ \varepsilon_{\beta\sigma\lambda\mu} p^\beta f^{\sigma\lambda} + 2imf_{5\mu} - eF^\alpha{}_\mu \dfrac{\partial}{\partial p^\alpha} f_5 = 0. \end{cases} \tag{12.68}$$

12.3.1. *Solution of the inhomogeneous equation*

The solution to the inhomogeneous Liouville equation

$$\begin{cases} \left\{ i\gamma \cdot \partial + 2(\gamma \cdot p - m) + ieF^\alpha{}_\mu \gamma^\mu \dfrac{\partial}{\partial p^\alpha} \right\} F(x,p) = S_1, \\ F(x,p) \left\{ i\gamma \cdot \overleftarrow{\partial} - 2(\gamma \cdot p - m) + ieF^\alpha{}_\mu \gamma^\mu \dfrac{\overleftarrow{\partial}}{\partial p^\alpha} \right\} = S_2 \end{cases} \tag{12.69}$$

is of importance when one is dealing with some kinetic equation or the linearized Vlasov equation, for instance. The source terms S_1 and S_2 cannot *a priori* be chosen arbitrarily and they must be consistent with each other, as explained in Chaps. 8 and 10. A simple example is provided elsewhere [R. Dominguez-Tenreiro and R. Hakim (1977)]. Given S_1 and S_2, it is not difficult to find whether they are consistent or not (see Chap. 10).

It can be shown that the general solution $F(x,p)$ depends on the two functions $f(x,p)$ and $f_5(x,p)$ only. This can be done by expanding the above system on the basis of the Dirac algebra and after simple calculations [for details see R. Hakim and H. Sivak (1982)]; it is therefore sufficient to solve the inhomogeneous equations

$$\begin{cases} p \cdot \partial f(x,p) + eF^{\mu\nu} p_\nu \dfrac{\partial}{\partial p^\mu} f(x,p) = Y, \\ p \cdot \partial f_5(x,p) + eF^{\mu\nu} p_\nu \dfrac{\partial}{\partial p^\mu} f_5(x,p) = Y_5, \end{cases} \tag{12.70}$$

which can both be written in the generic form

$$\frac{d}{d\tau} f = Y, \tag{12.71}$$

since these equations are the usual Liouville equations; τ is the proper time. The solution is thus of the form

$$f(x,p) = f_0(x,p) + \int^\tau ds\, Y[x(s), p(s)] \tag{12.72}$$

where the integration lies on the classical trajectories.[10] Such an integration has already been done in the relativistic nonquantum case by B. Kursunoglu (1966) and by O. Buneman (1968) in an investigation of the collective modes of a relativistic classical magnetized plasma. Using covariant cylindrical coordinates

$$\begin{cases} p \cdot r = -p_\perp \cos\theta, \\ p \cdot s = -p_{||} \sin\theta, \end{cases} \quad \begin{cases} p \cdot u = p_u, \\ p \cdot n = -p_{||}, \end{cases} \tag{12.73}$$

$$\begin{cases} k \cdot r = -k_\perp \cos\varphi, \\ k \cdot s = -k_{||} \sin\varphi, \end{cases} \quad \begin{cases} k \cdot u = \omega, \\ k \cdot n = -k_{||}, \end{cases} \tag{12.74}$$

where the four-vectors r^μ and s^μ have been discussed at the beginning of this chapter, and Fourier–transforming the Liouville equation, one is led to the equation

$$\left\{ \frac{\partial}{\partial\theta} + i\lambda^2(\omega p_u - p_{||}k_{||}) - i(\lambda^2 p_\perp k_\perp \cos\varphi)\cos\theta \right. \\ \left. - \lambda^2 p_\perp k_\perp \sin\varphi \sin\theta \right\} f = \lambda^2 Y \tag{12.75}$$

($\lambda^2 \equiv -1/eh$), whose solution is of the form [B. Kursunoglu (1966)]

$$f = \lambda^2 \exp[-\Lambda(\theta)] \int_{-\varepsilon.\infty}^{\theta} d\theta' \exp(\Lambda(\theta')) Y(\theta'), \tag{12.76}$$

[10] The word "classical" does not imply in any way a possible classical limit; the above equations are fully quantal and "classical" merely refers to a mathematical problem where it happens that the equation under study *looks* like a classical equation.

where[11]

$$\Lambda(\theta) \equiv i\lambda^2(\omega p_u - p_{||}k_{||})\theta - i\lambda^2 p_\perp k_\perp \sin(\theta - \varphi). \tag{12.77}$$

Details of the calculation can be found in B. Kursunoglu (1966) and O. Buneman (1968).

It remains for one to give the explicit expression of $F(x, p)$ in terms of $f(x, p)$ and $f_5(x, p)$; this has been done elsewhere [R. Hakim and H. Sivak (1982)], and such an expression is quite long and so will not be given here.

This achieves the solution to the inhomogeneous Liouville quantum system.

12.3.2. *The initial value problem*

Given the covariant Wigner function $F_\Sigma(x, p)$ on a specific spacelike surface Σ, what is the solution in the future of Σ? In other words, we look for a kernel $K^\mu(x, p; x', p')$ such that

$$F(x,p) = \int_\Sigma d\Sigma'_\mu \, d^4p' \, K^\mu(x,p;x'p')F_\Sigma(x',p') \tag{12.78}$$

reduces to $F_\Sigma(x, p)$ whenever x lies on Σ.

When it is observed that the Cauchy problem for the Dirac spinors is solved, this can easily be extended to the Wigner function. One has in fact

$$\psi(x) = \int_\Sigma d\Sigma'_\mu S(x - x')\gamma^\mu \psi_\Sigma(x'), \tag{12.79}$$

where $S(x - x')$ is given by[12]

$$S(x - x') = \sum_r \bar{\psi}_r(x) \otimes \psi_r(x'). \tag{12.80}$$

Next, using the definition of F_Σ,

$$F_\Sigma(x,p) = \frac{1}{(2\pi)^4} \int d^4R \exp(-i\pi \cdot R) \left\langle \bar{\psi}_\Sigma\left(x + \frac{1}{2}R\right) \otimes \psi_\Sigma\left(x - \frac{1}{2}R\right)\right\rangle, \tag{12.81}$$

[11] ε is the sign of $p \cdot u$; this means that particles turn from the infinite past to θ while antiparticles turn to the infinite future to θ in the opposite direction.

[12] It is not difficult to check that this $S(x - x')$ actually solves the Cauchy problem for the Dirac equations.

one is led directly to the kernel

$$K^{\mu}_{\alpha\beta\alpha'\beta'}(x,p;x',p') = \int d^4z\, d^4R \int_{\Sigma} d\Sigma''_{\nu} \exp[-i\pi\cdot R + ip'\cdot(x'-x'')]$$
$$\times\, \delta^{(4)}\left(z - \frac{x'+x''}{2}\right) \gamma^{\mu}_{(\beta'\rho)} \bar{S}_{(\rho\beta)}\left(x'-x-\frac{1}{2}R\right)$$
$$\times \left(x'-x-\frac{1}{2}R\right) S_{(\alpha\lambda)}\left(x-x''-\frac{1}{2}R\right)\gamma^{\nu}_{(\lambda\alpha')}, \tag{12.82}$$

which solves the Cauchy problem of the relativistic quantum Liouville equation. Note that the indices between parentheses refer to spinor indices.

12.4. The Equilibrium Wigner Function for Noninteracting Electrons

Let us begin with the density operator of such a system. It possesses three additive first integrals, the total charge Q, the energy $u\cdot P$, the momentum parallel to the magnetic field $P_{||}$ and the projection of the total angular momentum on the magnetic field axis $F^{\mu\nu}J_{\mu\nu}$, with

$$J^{\mu\nu} = L^{\mu\nu} + S^{\mu\nu}. \tag{12.83}$$

Finally, the grand-canonical density operator is written as

$$\rho = \frac{1}{Z}\exp(-\beta u\cdot P + \beta\mu Q - \beta A F_{\mu\nu}J^{\mu\nu}), \tag{12.84}$$

where μ is the chemical potential and A is given by

$$A = \frac{\omega}{B}, \tag{12.85}$$

where B is the magnetic field induction and ω the angular velocity of the system around the magnetic field. An alternative way to write the density operator is

$$\rho = \frac{1}{Z}\exp(-\beta_{\mu}\Omega^{\mu\nu}P_{\nu} + \beta\mu Q - \beta A F_{\mu\nu}J^{\mu\nu}), \tag{12.86}$$

where β^{μ} is an arbitrary four-vector of length β. When β^{μ} is chosen to be a timelike eigenvector of $\Omega^{\mu\nu}$, the invariant $\beta_{\mu}\Omega^{\mu\nu}P_{\nu}$ reads βP^0 in its rest frame and there is no collective motion parallel to the magnetic field. It should be noted that $\Omega^{\mu\nu}P_{\nu}$ actually represents *two*, and not four, constants of the motion since $\Omega^{\mu\nu}$ is a rank 2 matrix (remember that it is a projection on the two-plane spanned by the four-vectors n^{μ} and u^{μ}). In what follows, the collective motions of the magnetized electron gas will not be considered — neither those parallel to the magnetic field nor the possible rotations around it; and use will be made of

$$\rho = \frac{1}{Z}\exp(-\beta u \cdot P + \beta\mu Q). \tag{12.87}$$

For future use, let us calculate the partition function Z of such a system or any system of free fermions. As mentioned in Chap. 7, it always possesses the general form

$$Z = \sum_{r,\pm} \ln\{1 + \exp(-\beta[E_r \mp \mu])\}, \tag{12.88}$$

where r is the set of those quantum numbers that characterize the state of the system and $\pm$ refers to the electrons and the positrons respectively, whose energy is E_r. The sum over the quantum numbers nonexplicitly contained in the energy eigenvalue provides the degeneracy level, i.e.

$$d_n = \frac{1}{2\pi}\frac{|e|B}{m^2}. \tag{12.89}$$

In the magnetized electron gas case, one has

$$\begin{cases} r \equiv \{a, s, n, p_{||}\}, a \in \Re; s = \pm 1; n = 0, 1, 2, \ldots; p_{||} \in \Re, \\ \sum\limits_r \to \sum\limits_{n=0}^{\infty} \sum\limits_{s=\pm 1} \frac{1}{2\pi} \int dp_{||} \frac{1}{2\pi\lambda^2} \int da, \\ E_r = \sqrt{m^2 + p_{||}^2 + \frac{B}{B_{\text{crit}}}(2n + s + 1)}, \end{cases} \tag{12.90}$$

and one then gets

$$\ln Z = \sum_{n=0,\pm}^{\infty} \frac{1}{(2\pi)^2}\frac{m^2}{B_{\text{crit}}} \int_{-\infty}^{+\infty} dp_{||} \ln(1 + \exp\{-\beta[E_n(p_{||}) \mp \mu]\}). \tag{12.91}$$

In this last equation a vacuum term has been dropped. Note that a careless application of the usual formulae of statistical thermodynamics would give incorrect results, as is the case of the pressure which is not proportional to the derivative of Z with respect to the volume (see below).

12.4.1. *Thermodynamic quantities*

The first quantity of interest is the average charge of the electron system

$$\langle Q \rangle = \frac{1}{\beta}\frac{\partial}{\partial \mu} \ln Z, \tag{12.92}$$

or

$$\langle Q \rangle = \int_\Sigma d\Sigma_\mu \langle J^\mu \rangle, \tag{12.93}$$

with

$$\langle J^\mu \rangle = \int d^4p\, f^\mu(p), \tag{12.94}$$

and from the only timelike available four-vector u^μ, one necessarily has

$$\langle J^\mu \rangle = \int d^4p\, f^\mu(p) = n_{\text{eq}} u^\mu. \tag{12.95}$$

The next important quantity, which involves the energy density and the pressures within the system, is the average energy–momentum tensor. Its most general form is

$$T^{\mu\nu} = \rho u^\mu u^\nu - P_\perp \Pi^{\mu\nu} + P_{||} n^\mu n^\nu, \tag{12.96}$$

so that the energy density and the perpendicular and the parallel pressures are given by

$$\begin{cases} \rho = u_\mu u_\nu T^{\mu\nu}, \\ P_\perp = -\dfrac{1}{2}\Pi_{\mu\nu} T^{\mu\nu}, \\ P_{||} = (u_\mu u_\nu - \Omega_{\mu\nu}) T^{\mu\nu} = n_\mu n_\nu T^{\mu\nu}. \end{cases} \tag{12.97}$$

The anisotropy of the electrons' energy–momentum tensor is the main feature of such a system. However, in the thermodynamics of the system, it is the whole energy–momentum tensor which must be taken into account, including the magnetic field one:

$$\begin{cases} T^{\mu\nu}_{\text{tot}} = \rho u^\mu u^\nu - P_\perp \Pi^{\mu\nu} + P_{||} n^\mu n^\nu + T^{\mu\nu}_{\text{magn}}, \\ T^{\mu\nu}_{\text{magn}} = \dfrac{1}{4\pi}\left(F^{\mu\alpha} F^\nu{}_\alpha - \dfrac{1}{4}\eta^{\mu\nu} F_{\alpha\beta} F^{\alpha\beta} \right), \end{cases} \tag{12.98}$$

and not that of electrons only.

12.5. The Wigner Function of the Ideal Magnetized Electron Gas

From the available tensors in the theory and the form of the quantum Liouville equation, it can be shown that the equilibrium Wigner function can be expressed in terms of $f_{eq}(p)$ only.[13] A direct calculation confirms this property. The direct calculation can be achieved in several ways. For instance, the average value of $\langle\bar{\psi}(x+\frac{1}{2})\cdot\psi(x-\frac{1}{2})\rangle$ can be computed as

$$\left\langle\bar{\psi}\left(x+\frac{1}{2}\right)\cdot\psi\left(x-\frac{1}{2}\right)\right\rangle=\sum_{r,\pm}\varpi_{r,\pm}\bar{\psi}_{r,\pm}\left(x+\frac{1}{2}\right)\psi_{r,\pm}\left(x-\frac{1}{2}\right) \tag{12.98}$$

in order to get $f_{eq}(p)$, with

$$\varpi_{r,\pm}\equiv\frac{1}{\exp(\beta E_r\mp\beta\mu)+1}. \tag{12.99}$$

$\varpi_{r,\pm}$ is the statistical weight of the state ($\pm$, r) and the sum over these states is intended to mean the average value. As to

$$\sum_r\to\sum_{n=0}^{\infty}\sum_{s=\pm1}\frac{1}{2\pi}\int dp_{||}d_n\int da, \tag{12.100}$$

it represents a sum over *all* quantum numbers representing a Landau level [see Eq. (12.7)]. After a straightforward calculation, one obtains for $f_{eq}(p)$

$$f_{eq}(p)=\frac{2m}{(2\pi)^3}\exp(w)\times\left\{\frac{1}{E_0}\frac{\delta(p\cdot u-E_0)}{\exp(\beta[E_0-\mu])+1}\right.$$
$$\left.+\sum_{n=1}^{\infty}\frac{(-1)^n}{E_n}\frac{\delta(p\cdot u-E_n)}{\exp(\beta[E_n-\mu])+1}[L_n(-2w)-L_{n-1}(-2w)]\right\}, \tag{12.101}$$

where w is the variable

$$w\equiv d_n^{-1}\Pi^{\mu\nu}p_\mu p_\nu, \tag{12.102}$$

which is roughly the (squared) length of the four-momentum perpendicular to the magnetic field. L_n are the Laguerre polynomials.

The other components of the equilibrium Wigner function can be calculated likewise and one finds that

$$f_{eq}^{\mu}(p)=\left\{\frac{p^\mu}{m}-\frac{\lambda^2}{mw}(p^2-m^2)\Pi^{\mu\nu}p_\nu\right\}f_{eq}(p), \tag{12.103}$$

[13] This has been shown in R. Hakim and H. Sivak (1982); however, it was argued by A.E. Shabad (private communication) that a pseudoscalar was "missed" in our reasoning. When one takes account of the *PC* invariance of the system, this pseudoscalar vanishes identically, and the original reasoning becomes correct.

$$f_{\rm eq}^{\mu\nu}(p) = F^{\mu\nu}\frac{i}{h}\frac{2m}{(2\pi)^3}\exp(w)\times\left\{\frac{1}{E_0}\frac{\delta(p\cdot u - E_0)}{\exp(\beta[E_0-\mu])+1}\right.$$

$$\left.+\sum_{n=1}^{\infty}\frac{(-1)^n}{E_n}\frac{\delta(p\cdot u - E_n)}{\exp(\beta[E_n-\mu])+1}[L_n(-2w)+L_{n-1}(-2w)]\right\}, \tag{12.104}$$

$$f_{5\rm eq}^{\mu}(p) = -{}^*F^{\mu}_{\nu}p^{\nu}\frac{1}{h(2\pi)^3}\exp(w)\times\left\{\frac{1}{E_0}\frac{\delta(p\cdot u - E_0)}{\exp(\beta[E_0-\mu])+1}\right.$$

$$\left.+\sum_{n=1}^{\infty}\frac{(-1)^n}{E_n}\frac{\delta(p\cdot u - E_n)}{\exp(\beta[E_n-\mu])+1}[L_n(-2w)+L_{n-1}(-2w)]\right\}, \tag{12.105}$$

$$f_{5\rm eq}(p) \equiv 0. \tag{12.106}$$

12.5.1. *The nonmagnetic field limit*

Finally, it can be shown that the above "magnetized" Wigner function reduces to the "nonmagnetic" one of Chap. 8, in the limit $F_{\mu\nu}\to 0$. Let us do that. The essential trick consists in making the replacement

$$\sum_n \to \int dn, \tag{12.107}$$

so that it follows that $f_{\rm eq}(p)$ takes the form

$$f_{\rm eq}(p) = \frac{2m}{(2\pi)^3}\exp(-w)\left\{\int_0^{\infty} dn\frac{\delta(p_0-E_n)L_n(2w)}{\exp(\beta[E_n-\mu]+1)}\right.$$

$$\left.-\int_0^{\infty} dn\frac{\delta(p_0-E_n)L_{n-1}(2w)}{\exp(\beta[E_n-\mu])+1}\right\}. \tag{12.108}$$

Using the property

$$\delta(p_0-E_n) = \frac{\lambda^2}{E_\ell}\delta(n-\ell)$$

$$\ell \equiv \frac{b^2\lambda^2}{2}, b = [p_0^2 - p_{||}^2 - m^2]^{1/2}, \lambda^2 = \frac{1}{|eB|}, \tag{12.109}$$

$f_{\rm eq}$ can be rewritten as

$$f_{\rm eq}(p) = \frac{2m\lambda^2}{(2\pi)^3}\frac{\exp(-w)L_\ell^{-1}(2w)}{\exp(\beta[E_\ell-\mu])+1}, \tag{12.110}$$

where L_k^q is an associated Laguerre polynomial [I.S. Gradshtein and I.W. Ryzhik (1965)]. On the other hand, one also has

$$\exp(-w)L_\ell^{-1}(2w) = -2^{\ell+\frac{1}{2}}\left(\frac{2}{\pi}\right)^{1/2}\left(\frac{1}{\ell!}\right)2w^{1/2}\mathcal{F},$$

$$\mathcal{F} = \int_0^\infty dx\exp(-x^2)\sin(2w^{1/2}x)H_\ell(x)H_{\ell-1}(x). \tag{12.111}$$

From the asymptotic value of the Hermite polynomial H_ℓ and from the fact that when $B \to 0$, then $\lambda \to \infty$ and hence $\ell \to \infty$, one gets

$$\mathcal{F} \simeq -\pi(\ell-1)!!2^{\ell-7/2}(\ell-1)^{1/2}\left(\frac{1}{\lambda}\right)\delta(p_\perp - b). \tag{12.112}$$

Going back to $f_{\text{eq}}(p)$, one obtains

$$\exp(-w)L_\ell^{-1}(2w) \simeq \left(\frac{\pi}{2}\right)^{1/2}\frac{(\ell-1)!!(\ell-3)!!}{\ell!}(\ell-1)^{1/2}p_\perp^2\delta(p^2-m^2), \tag{12.113}$$

which, for large ℓ, reduces to

$$\exp(-w)L_\ell^{-1}(2w) \simeq \left(\frac{2}{\lambda^2}\right)\delta(p^2-m^2), \tag{12.114}$$

where use has been made of

$$\ell!! \simeq \Gamma\left(\frac{\ell}{2}+1\right)\frac{2^{(\ell+1)/2}}{\sqrt{\pi}}. \tag{12.115}$$

Finally, one gets the expression of f_{eq}:

$$\lim f_{\text{eq}}(p) = \frac{4m}{(2\pi)^3}\theta(p_0)\frac{\delta(p^2-m^2)}{\exp[\beta(E_p-\mu)]+1}. \tag{12.116}$$

The limits of the other f_A's are obtained in the same way.

12.5.2. *Equations of state*

Our first task is now to normalize the Wigner function, i.e. find the connection between the chemical potential μ and the charge density n_{eq}.

From the four-current

$$J^\mu = \int d^4p\, f_{\text{eq}}^\mu(p) \tag{12.117}$$

one obtains successively

$$\begin{aligned} n_{\text{eq}} &= \frac{1}{m}\int d^4p\, p\cdot u\, f_{\text{eq}}(p) \\ &= \frac{1}{(2\pi)^3}\int dp^0 dp_{||} d(-wd_n)d\theta\, p^0 f_{\text{eq}}(p), \end{aligned} \tag{12.118}$$

which finally leads to

$$n_{\rm eq} = \frac{1}{(2\pi\lambda)^2}\int dp_{||}\left\{\frac{1}{\exp(\beta[E_0-\mu])+1}+2\sum_{n=1}^{\infty}\frac{1}{\exp(\beta[E_n-\mu])+1}\right\}, \tag{12.119}$$

which is the result obtained by V. Canuto and H.Y. Chiu (1968). Note that the factor 2 which occurs in front of the sum is due to the spin degeneracy, which does not occur for the $n=0$ level: the energy of the level $(n+1)-1$ (spin down) is the same as for $(n-1)+1$ (spin up). In the derivation of this "normalization" condition, use was made of[14]

$$\int_0^{\infty} dt \exp\left(-\frac{1}{2}t\right) tL_n(t) = (2n+1)(-1)^n. \tag{12.120}$$

12.5.3. *Is the pressure isotropic?*

If one starts from the thermodynamic potential, the pressure is given by

$$P = n_{\rm eq}^2 \frac{\partial}{\partial n_{\rm eq}} \ln Z \tag{12.121}$$

and thus it appears to be isotropic, unlike the results obtained by V. Canuto and H.Y. Chiu (1968) from calculations based on the density operator or those obtained from the Wigner function. Several authors think that this last expression for P is the correct one, the validity of thermodynamics being absolute. They argued — we quote Dong Lai (2001), who explained the point of view of R.D. Blandford and L. Hernquist (1982) — that "When we compress the electron gas perpendicular to B we must also do work against the Lorentz force density $(\nabla \times \mathbf{M}) \times \mathbf{B}$ involving the magnetization current. Thus there is a magnetic contribution to the perpendicular pressure of magnitude $\mathbf{M}\cdot\mathbf{B}$. The composite pressure tensor is therefore isotropic, in agreement with the thermodynamic result $P = \Omega/V$".

On the other hand, the calculations leading to an anisotropic energy–momentum tensor are correct and they start from a universally admitted microscopic viewpoint and lead to an anisotropic stress tensor, exactly as do radiative corrections, so that the pressure appears to be anisotropic. Note also that these anisotropies constitute a typical quantum effect.

[14] I.S. Gradshteyn, I.W. Ryzhyk, *Tables of Integrals, Series and Products* (Academic, New York, 1965).

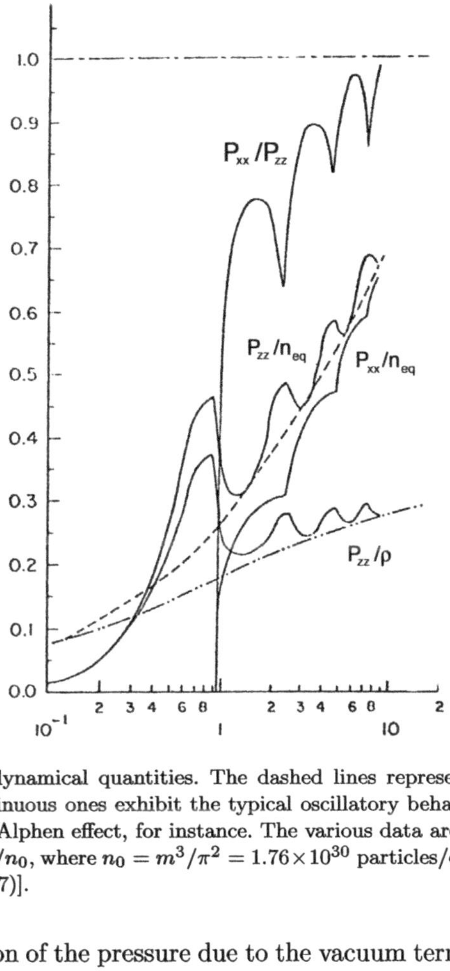

Fig. 12.1 Thermodynamical quantities. The dashed lines represent the nonmagnetic case, while the continuous ones exhibit the typical oscillatory behavior which one finds in the de Haas–van Alphen effect, for instance. The various data are plotted against the electron density n_{eq}/n_0, where $n_0 = m^3/\pi^2 = 1.76\times10^{30}$ particles/cm^3 [after V. Canuto and J. Ventura (1977)].

The calculation of the pressure due to the vacuum terms shows that the system is anisotropic: there exist a parallel and a perpendicular pressure. Finally, the various thermodynamic properties are shown in Fig. 12.1. We note the oscillations of the various variables corresponding to a jump of one Landau level to a consecutive one.

12.5.4. *The completely degenerate case*[15]

In the complete degenerate case, at $T = 0$ K, the Landau levels are uniformly occupied untill the Fermi level, i.e. the last occupied one. This

[15]H.Y. Chiu and V. Canuto (1968); see also V. Canuto and J. Ventura (1977).

occurs for a Landau level indexed by a quantum number n^* which depends on the electronic density, which is

$$n_{\text{eq}} = \frac{Bm^3}{\pi^2 B_{\text{crit}}} \left\{ \frac{1}{2}\sqrt{\varepsilon_f^2 - 1} + \sum_{n=1}^{n=n^*} \left(\varepsilon_f^2 - 1 - 2n\frac{B}{B_{\text{crit}}} \right)^{1/2} \right\}, \tag{12.122}$$

with, of course,

$$n < (\varepsilon_f^2 - 1)\frac{B_{\text{crit}}}{2B}. \tag{12.123}$$

The energy density ρ is given by

$$\begin{aligned} \rho = \frac{m^4}{\pi^2}\frac{B}{B_{\text{crit}}} \Bigg\{ & \frac{1}{4}(\varepsilon_f^2 - 1)^2 - \frac{1}{4}\ln\left[\varepsilon_f + \sqrt{\varepsilon_f^2 - 1}\right] \\ & + \sum_{n=1}^{n=n^*} \frac{1}{2}\varepsilon_f \left(\varepsilon_f^2 - 1 - 2n\frac{B}{B_{\text{crit}}} \right)^{1/2} + \sum_{n=1}^{n=n^*} \frac{1}{2}\left(1 + 2n\frac{B}{B_{\text{crit}}}\right) \\ & \times \ln\left(\frac{\varepsilon_f + \left(\varepsilon_f^2 - 1 - 2n\frac{B}{B_{\text{crit}}}\right)^{1/2}}{1 + 2n\frac{B}{B_{\text{crit}}}} \right) \Bigg\}, \end{aligned} \tag{12.124}$$

while the pressures are

$$\left\{ \begin{aligned} P_\perp &= \left(\frac{B}{B_{\text{crit}}}\right)^2 \sum_{n=1}^{n=n^*} n \ln\left(\frac{\varepsilon_f + \left(\varepsilon_f^2 - 1 - 2n\frac{B}{B_{\text{crit}}}\right)^{1/2}}{1 + 2n\frac{B}{B_{\text{crit}}}} \right), \\ P_{||} &= \frac{m^4}{\pi^2}\frac{B}{B_{\text{crit}}} \Bigg\{ \frac{1}{4}(\varepsilon_f^2 - 1)^2 - \frac{1}{4}\ln\left[\varepsilon_f + \sqrt{\varepsilon_f^2 - 1}\right] \\ &\quad + \sum_{n=1}^{n=n^*} \frac{1}{2}\varepsilon_f \left(\varepsilon_f^2 - 1 - 2n\frac{B}{B_{\text{crit}}} \right)^{1/2} - \sum_{n=1}^{n=n^*} \frac{1}{2}\left(1 + 2n\frac{B}{B_{\text{crit}}}\right) \\ &\quad \times \ln\left(\frac{\varepsilon_f + \left(\varepsilon_f^2 - 1 - 2n\frac{B}{B_{\text{crit}}}\right)^{1/2}}{1 + 2n\frac{B}{B_{\text{crit}}}} \right) \Bigg\}. \end{aligned} \right. \tag{12.125}$$

Note that the above condition on n implies that n^* is given by

$$n^* + \frac{s+1}{2} = \left(\frac{E}{m}\right)^2 \frac{B_{\text{crit}}}{2B}. \tag{12.126}$$

This equation comes from the expression of E and from minimizing the energy with respect to $p_{||}$. This explains the oscillations in the various thermodynamical quantities. Suppose that the electron has the highest level and we add some more energy. Then the electron energy begins not at the level n but rather at $p_{||} \neq 0$; and this occurs untill the level $n+1$ is reached, at which $p_{||}$ comes again, with the value $p_{||} = 0$. Thus, there exists an oscillation between $p_{||}$ and the increase of the quantum n.

This means that the stronger the magnetic field is, the lower n^* will be; also, the higher the density is, the higher n^* will be. Therefore, the typical oscillatory behavior of the thermodynamical data is expected for low values of n^* and, accordingly, at low density and/or high magnetic fields. As remarked by V. Canuto and J. Ventura (1977), in a typical white dwarf (where $n_{\rm eq} \approx 10^{30}$ particles/cm^3 and $B \approx 10^8\,G$) one finds that $n^* \approx 10^6$, while for a typical neutron star one still finds that $n^* \approx 10^6$. However, the quantizing effects of the strong magnetic field can show up in the much less dense plasma surrounding a pulsar.

Finally, let us note that in an interesting article C.O. Dib and O. Espinosa (2001), expressing various thermodynamical quantities in terms of Hurwitz functions (see their article), gave their temperature dependence starting from the fully degenerate case.[16] For instance, they gave for the grand potential at finite temperature

$$\Omega(T,\mu) = \int_{-(\mu-1)/T}^{\infty} dx\, \Omega_0(\mu + Tx)\frac{\exp(x)}{[\exp(x)+1]^2}, \tag{12.127}$$

where Ω_0 corresponds to the zero temperature case.

12.5.5. *Magnetization*

Here we follow the approach of V. Canuto and J. Ventura (1977) and start with the usual thermodynamic definition of the magnetization[17]:

$$M = \frac{1}{\beta}\frac{\partial}{\partial B}\ln Z. \tag{12.128}$$

The partition function Z of the system has already been calculated as

$$\ln Z = \sum_r \ln\{1 + \exp(-\beta[E_r - \mu])\}, \tag{12.129}$$

where r still represents the ensemble of all quantum numbers that define the state of an electron, and where the sum also involves an integral. Remember

[16]This result, although apparently correct, seems a bit strange, the more so since it is known that the zero temperature case often contains pathologies.

[17]See e.g. H.B. Callen, *Thermodynamics and an Introduction to Thermostatistics* (Wiley, New York, 1985).

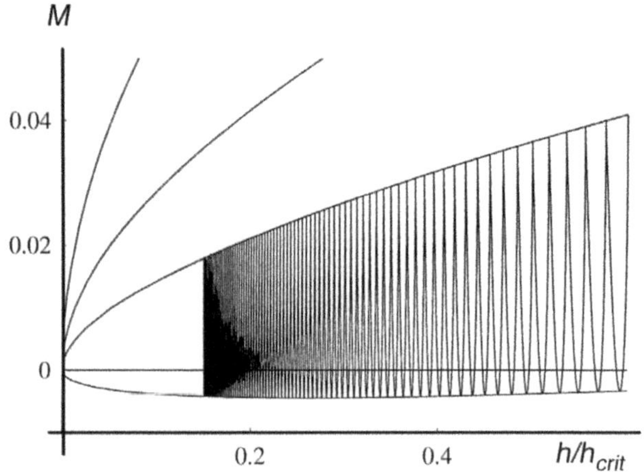

Fig. 12.2 The magnetization as a function of the magnetic field (in units of $h_{\rm crit}$) for the Fermi energy $\varepsilon_f = 4$. The other curves correspond to $\varepsilon_f = 8$, 16 [after C.O. Dib and O. Espinosa (2001)].

that the degeneracy of the Landau levels does depend on the magnetic field itself and hence is taken into account[18] in the derivation of M. Deriving $\ln Z$ with respect to β, one finds that

$$M = \frac{1}{\pi^2}\frac{e}{D^2_{\rm Compton}}\left(-\frac{mB}{B_{\rm crit}}\sum_{n=0}^{\infty} n\int_0^{\infty}\frac{dx}{\varepsilon_n(x)}\frac{1}{\exp\{m\beta[\varepsilon_n(x)-\mu]\}+1}\right.$$
$$+\frac{1}{2\beta}\int_0^{\infty} dx \ln\{1+\exp(-m\beta[\varepsilon_0-\mu])\}$$
$$\left.+\frac{1}{\beta}\sum_{n=1}^{\infty}\int_0^{\infty} dx \ln\{1+\exp(-m\beta[\varepsilon_n-\mu])\}\right), \tag{12.130}$$

with the notations

$$x \equiv \frac{p}{m}, \varepsilon_n(x) \equiv \frac{E_n(x)}{m}. \tag{12.131}$$

An integration by parts of the last two terms and a comparison with previous results led these authors to the following final expression for the magnetization of the system:

$$M = \frac{P_{||} - P_{\perp}}{B}. \tag{12.132}$$

[18] V. Canuto and J. Ventura (1977) gave several examples of incorrect calculations performed when one does not take this B dependence into account.

Thus the magnetization of the system appears to be proportional to the degree of anisotropy of the electron pressure. It is interesting to note that this relation is also valid for the magnetized vacuum [H. Sivak (1986)].

12.5.6. *Landau orbital ferromagnetism: LOFER states*

An interesting way to generate magnetic fields has been suggested by V. Canuto, H.Y. Chiu and C. Chiuderi (1969), which they called *Landau orbital ferromagnetism*, since the Landau levels play an essential role in their mechanism. They noticed that, in the above considerations, one has to make the usual distinction between the external magnetic field H and the magnetic induction B, to which the particle are sensitive through their equations of motion, B and H being connected by the relation

$$B = H + 4\pi M(B), \tag{12.133}$$

where $M(B)$ is the magnetization of the medium. The question they raised is then: Is it possible that when the external field H is switched off, the above equation has a nonvanishing solution for B? The answer can be given numerically only, owing to the involved character of $M(B)$, and the result is shown in Fig. 12.3, at $T = 0\,\mathrm{K}$. There exist several possible solutions. The $T \neq 0\,\mathrm{K}$ case can be expected to smoothen the oscillatory curves, which presents more and more oscillations with less and less intersections with the straight line in the figure.

Unfortunately, this interesting mechanism is eliminated by Coulomb interactions, as shown by J. Schmidt-Burgk (1973). However, it could still be valid in other circumstances (magnetars) and should be studied in the case of colored fields.

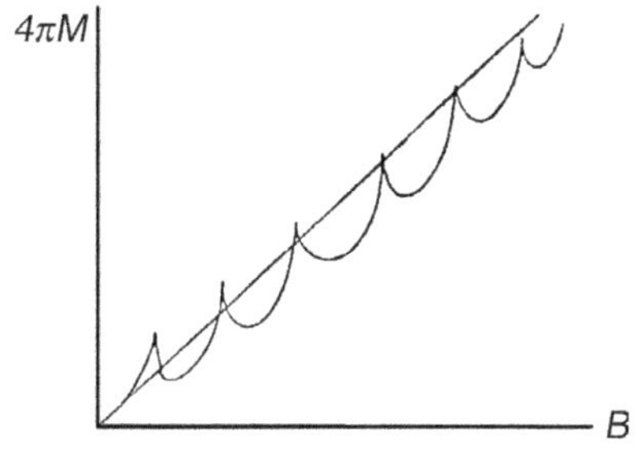

Fig. 12.3 Graphical representation of the possible solutions to the equation $M(B) = B$. The oscillatory line is $M(B)$ [after V. Canuto and J. Ventura (1977)].

12.6. The Magnetized Vacuum

In the presence of a strong magnetic field, the vacuum becomes magnetized and hence acquires anisotropy.[19] As a consequence, the vacuum Wigner function of this magnetized vacuum also reflects this anisotropy, and the magnetized Dirac ocean is different from that of ordinary electrons. The vacuum Wigner function plays an important role in the process of renormalization.

12.6.1. *The general structure of the vacuum Wigner function*

One could calculate the vacuum Wigner function directly; it is, however, more instructive to determine *a priori* its general structure from the quantum Liouville equation it obeys and the available tensor in the magnetized vacuum, namely p^μ, $\eta^{\mu\nu}$ and $F^{\mu\nu}$. From the fact that a pseudoscalar cannot be built up from these tensors, we deduce that $f^5_{\text{vac}} \equiv 0$. Also, since ${}^*F^{\mu\nu}p_\nu$ is the only pseudovector at our disposal, we have

$$f^\mu_{5\text{vac}}(p) = a(p)^*F^{\mu\nu}p_\nu. \tag{12.134}$$

Similarly, f^μ_{vac} assumes the general form

$$f^\mu_{\text{vac}} = b(p)p^\mu + c(p)F^{\mu\nu}p_\nu. \tag{12.135}$$

Note that $b(p)$ and $c(p)$ are actually functions of p since $p^2 \neq m^2$. We now have to specify $f^{\mu\nu}_{\text{vac}}$. To this end, the equation

$$-2ip^\beta f_{\beta\mu} + eF^\alpha{}_\mu \frac{\partial}{\partial p^\alpha} f = 0 \tag{12.136}$$

can be rewritten as

$$p^\beta \left[if_{\beta\mu}(p) - eF_{\beta\mu}\frac{\partial}{\partial p^2} f_{\text{vac}}(p) \right] = 0. \tag{12.137}$$

It follows that this last equation has the form

$$p^\mu A_{\mu\nu}(p) = 0$$

---

[19] A. Minguzzi, *Nuovo Cimento* **4**, 476 (1956); *ibid.* **6**, 501 (1957); J.J. Klein and B.P. Nigam, *Phys. Rev.* **B135**, 1279 (1964); H. Constantinescu, *Nucl. Phys.* **B36**, 121 (1972); D.B. Melrose and R.J. Stoneham, *Nuovo Cimento* **A32**, 435 (1976; *ibid., J. Phys.* **A10**, 1211 (1977); A.E. Shabad, *Lett. Nuovo Cimento* **3**, 457 (1972); *ibid., Ann. Phys. (N.Y.)* **90**, 166 (1975); I.A. Batalin and A.E. Shabad, *Sov. Phys. JETP* **33**, 483 (1971); Y.T. Wu, *Phys. Rev.* **D10**, 2699 (1974).

and hence

$$A_{\mu\nu}(p) = d(p)\Delta_{\mu\nu}(p) \tag{12.138}$$

or, equivalently,

$$f_{\beta\mu}(p) = eF_{\beta\mu}\frac{\partial}{\partial p^2}f_{\text{vac}}(p) = d(p)\Delta_{\mu\nu}(p) \tag{12.139}$$

where the right-hand side is symmetric ans the left-hand one is antisymmetric. The can occur only when $d(p) = 0$. Finally we have

$$f_{\beta\mu}(p) = eF_{\beta\mu}\frac{\partial}{\partial p^2}f_{\text{vac}}(p) \tag{12.140}$$

Finally, with use of the second equation (12.68)

$$f_5^{\mu}(p) = -\frac{e}{2m}{}^*F^{\mu\nu}p_v\frac{\partial}{\partial p^2}f_{\text{vac}}(p). \tag{12.141}$$

From Eqs. (66a) and (12.141) the equation for $f^{\mu}(p)$ is obtained:

$$f^{\mu}(p) = \frac{p^{\mu}}{m}f(p) + \frac{e}{2im}F^{\alpha}{}_{\beta}F^{\beta\mu}\frac{\partial}{\partial p^{\alpha}}\left[-ie\frac{\partial}{\partial p^2}f(p)\right] \tag{12.142}$$

(e is the electron charge). The right hand side of this last equation can be rewritten as

$$+\left(\frac{e^2}{m}\right)H^2\Pi^{\mu\alpha}p_{\alpha}\frac{\partial^2}{\partial(p^2)^2}f(p) \tag{12.143}$$

Also, the second derivative in this last equation can be evaluated with the help of Eq. (64a) as

$$\frac{\partial^2}{\partial(p^2)^2}f(p) = \left(\frac{1}{e^2H^2\Pi^{\alpha\beta}p_{\alpha}p_{\beta}}\right)(p^2 - m^2)f(p), \tag{12.144}$$

so that, finally, the general form of the vacuum Wigner function is found to be

$$F_{\text{vac}}(p) = \frac{1}{4}\left\{I + \gamma_{\mu}\left[\eta^{\mu}{}_{\nu} - \left(\frac{d_n}{|w|}\right)(p^2 - m^2)\Pi^{\mu}{}_{\nu}\right]\frac{p^{\nu}}{m}\right.$$
$$\left. + \frac{1}{2}\sigma_{\mu\nu}F^{\mu\nu}\frac{i}{H}\frac{\partial}{\partial w} - \gamma^5\gamma^{\mu}\frac{1}{2Hm}{}^*F_{\mu\nu}p^{\nu}\frac{\partial}{\partial w}\right\}f_{\text{vac}}(p).$$

12.6.2. *The wigner function of the magnetized vacuum*

Since a vacuum is invariant under space–time translations, the general form of its Wigner function $F_{\text{vac}}(p)$ is the one already obtained above and its determination reduces to that of $f_{\text{vac}}(p)$. Expanding the fields $\psi(x)$ and $\bar{\psi}(x)$ as

$$\begin{cases} \psi(x) = \sum\limits_{n,\sigma} \psi_{\sigma,n+} a_{\sigma,n} + \psi_{\sigma,n-} d^+_{\sigma,n}, \\ \bar{\psi}(x) = \sum\limits_{n,\sigma} \bar{\psi}_{\sigma,n+} a^+_{\sigma,n} + \bar{\psi}_{\sigma,n-} d_{\sigma,n}, \end{cases} \tag{12.145}$$

using the vacuum density operator

$$\rho_{\text{vac}} = |\text{vac}\rangle\langle\text{vac}|$$

and the definition of $f(p)$, it appears that the only surviving term is the one that involves products of the form $d_{n,\sigma} d^+_{n,\sigma}$, so that

$$\left\langle \text{vac} \left| \bar{\psi}\left(x + \frac{1}{2}R\right) \psi\left(x - \frac{1}{2}R\right) \right| \text{vac} \right\rangle = \sum_r \bar{\psi}_r\left(x + \frac{1}{2}R\right) \psi_r\left(x - \frac{1}{2}R\right), \tag{12.146}$$

where $\psi_r \equiv \psi_{n,\sigma-}$ and $r = \{n, p_{||}, (-a), \sigma\}$ is the set of the quantum numbers necessary for specifying the state of the electron. One then obtains [H. Sivak (1986)][20]

$$f_{\text{vac}}(p) = -\left[\frac{4m}{(2\pi)^3} \exp(w)\theta(-p\cdot u) \times \left\{\sum_{n=0}^{\infty} \frac{(-1)^n}{E_n} \delta(\Omega^{\alpha\beta} p_\alpha p_\beta - m^2 - 2n|e|B) L_n^{-1}(-2w)\right\}\right], \tag{12.147}$$

which indicates that the negative energy states are uniformly occupied — an expected result.

Finally, the Wigner function of the magnetized vacuum reads

$$F_{\text{vac}}(p) = \frac{1}{4m}\left\{m + \gamma_\mu \left[p^\mu - \Pi^{\mu\nu} p_\nu \frac{p^2 - m^2}{\Pi^{\alpha\beta} p_\alpha p_\beta}\right] + \frac{im}{2H} \sigma_{\mu\nu} F^{\mu\nu} \frac{\partial}{\partial w} - \frac{1}{2H} \gamma^5 \gamma^{\mu *} F_{\mu\nu} p^\nu \frac{\partial}{\partial w}\right\} f_{\text{vac}}(p), \tag{12.148}$$

which, of course, leads to divergent expressions for the average four-current or the energy–momentum tensor.

[20] Another equivalent expression has been given in R. Hakim and H. Sivak (1982).

12.6.3. *Renormalization of the vacuum Wigner function*

A renormalized expression for $F_{\text{vac}}(p)$ — say, $F^R_{\text{vac}}(p)$ — that leads to finite physical results can be obtained by subtracting from $F_{\text{vac}}(p)$ the first two terms of the McLaurin series as

$$F^R_{\text{vac}}(p,H) = F_{\text{vac}}(p,H) - F_{\text{vac}}(p,0) - \left.\frac{\partial F_{\text{vac}}(p,h)}{\partial h^2}\right|_{h=0} H^2. \qquad (12.149)$$

This is based on the fact that our basic reference is the unmagnetized vacuum and the need for a finite electron charge[21]; this leads to redefining both the magnetic field and the electron charge as

$$\begin{cases} H \equiv H_0\left\{1 + 8\pi \left.\dfrac{\partial}{\partial H_0^2}\rho\right|_{H_0=0}\right\}, \\ e \equiv e_0\left\{1 + 8\pi \left.\dfrac{\partial}{\partial H_0^2}\rho\right|_{H_0=0}\right\}^{-1}, \end{cases} \qquad (12.150)$$

where ρ is the vacuum energy density.

After some calculations [see H. Sivak (1986)], one finds that

$$f^R_{\text{vac}}(p) = -\frac{8m}{(2\pi)^4}\theta(-p\cdot u)Re\int_0^\infty ds \exp(i[\Omega^{\mu\nu}p_\mu p_\nu - m^2 + i\varepsilon]s)$$
$$\times\left\{\exp(iw\tan s|e|H) - \exp(iws|e|H)\left[1 + i\frac{w}{3}(s|e|H)^3\right]\right\}. \qquad (12.151)$$

With this renormalized Wigner function one can calculate various radiative corrections and, in particular, as to the energy–momentum tensor of the magnetized vacuum

$$T^{\mu\nu}_{\text{vac}} = \text{Tr}\int d^4p\, p^\mu\gamma^\nu F^R_{\text{vac}}(p), \qquad (12.152)$$

which leads to

$$T^{\mu\nu}_{\text{vac}} = \rho\Omega^{\mu\nu} - P_\perp\Pi^{\mu\nu}, \qquad (12.153)$$

with

$$\begin{cases} \rho = \dfrac{e^2H^2}{8\pi^2}\displaystyle\int\frac{ds}{s^2}\exp\left(-\frac{sm^2}{|e|H}\right)\cdot\left(\coth s - \frac{1}{s} - \frac{s}{3}\right), \\ P_\perp = -\dfrac{e^2H^2}{8\pi^2}\displaystyle\int\frac{ds}{s}\exp\left(-\frac{sm^2}{|e|H}\right)\cdot\left(\coth^2 s - \frac{1}{s^2} - \frac{2}{3}\right), \end{cases} \qquad (12.154)$$

[21] E. Lifschitz and L. Pitayevski, *Relativistic Quantum Theory* (Mir, Moscow, 1973).

where $P_\perp$ is the orthogonal pressure of the magnetized vacuum. Finally, the total energy–momentum tensor is obtained by adding to $T^{\mu\nu}_{\rm vac}$ the energy–momentum tensor of the magnetic field itself,

$$T^{\mu\nu}_H = \frac{H^2}{8\pi}(\Omega^{\mu\nu} - \Pi^{\mu\nu}), \tag{12.155}$$

where $T^{\mu\nu}_H$ is the well-known Euler–Heisenberg expression for the energy density of the magnetized vacuum, while $P_\perp$ represents the radiative corrections to the pressure orthogonal to the magnetic field.

12.7. Fluctuations

The importance of the equilibrium fluctuations has already been emphasized; in particular, it provides — via the use of the inverse of the fluctuation–dissipation theorem — a way to derive the plasma modes. Furthermore, the knowledge of the spectrum of the four-current fluctuations provides a direct calculation to the noninteracting Fermi gas.

From Ω, the usual thermodynamic potential

$$\Omega = -\beta^{-1}\ln[\mathrm{Tr}\exp(-\beta[H_0 + H_{\rm int}] + \beta\mu Q)], \tag{12.156}$$

we have

$$\frac{\partial\Omega}{\partial e} = \frac{1}{e}\frac{\mathrm{Tr}\{\exp[-\beta(H_0 + H_{\rm int} - \mu Q)]H_{\rm int}\}}{\mathrm{Tr}\{\exp[-\beta(H_0 + H_{\rm int} - \mu Q)]\}} \tag{12.157}$$

or

$$\frac{\partial}{\partial e}\Omega = \frac{1}{e}\langle H_{\rm int}\rangle. \tag{12.158}$$

H_0 and $H_{\rm int}$ are the free and the interaction[22] Hamiltonians, respectively. Using now Maxwell's equations in the form

$$-k^2 A^\mu(k) = 4\pi J^\mu(k), \tag{12.159}$$

it follows that

$$\Omega = \Omega_0 - \frac{2}{(2\pi)^3}\int_0^e \frac{de'}{e'}\int d^4k \left\langle \frac{J^\lambda_{\rm op} J_{\lambda\rm op}}{k^2}\right\rangle_k, \tag{12.160}$$

where Ω_0 is the noninteracting case. Similarly, the knowledge of the polarization tensor (see next section) gives rise to corrections[23] to the thermal

[22]That is to say, $H_{\rm int} = eJ_{\rm op}\cdot A$; $H_{\rm int}$ is a Hamiltonian and not a magnetic field.
[23]See e.g. E.S. Fradkin, *Nucl. Phys.* **12**, 455 (1955); H. Perez-Rojas and A.E. Shabad, *Ann. Phys.* (*N.Y.*) **121**, 432 (1979).

properties of the noninteracting electron gas through

$$\Omega = \Omega_0 - \frac{V}{(2\pi)^3\beta}\sum_{\omega,i}\int d^4k \int_0^e \frac{de'}{e'}\frac{\chi_i(\omega,\mathbf{k})}{k^2 - \chi_i(\omega,\mathbf{k})}, \tag{12.161}$$

where $\chi_i(\omega, \mathbf{k})$ ($i = 1, 2, 3$) are the three nonzero eigenvalues of the polarization tensor[24] (see next section).

Besides the four-current fluctuations, others are worth studying (energy–momentum tensor, etc.) and they can all be obtained from the following "fluctuation" of the covariant Wigner function:

$$\mathcal{F}(x,p;x',p') = \langle F_{\rm op}(x,p)F_{\rm op}(x',p')\rangle - F(x,p)F(x',p'). \tag{12.162}$$

For instance, the four-current fluctuations

$$\delta J^{\mu\nu}(x,x') \equiv \langle J^\mu(x)J^\nu(x')\rangle - \langle J^\mu(x)\rangle\langle J^\nu(x')\rangle \tag{12.163}$$

can be obtained from $\mathcal{F}$ as

$$\delta J^{\mu\nu}(x,x') = \mathrm{Sp}\int d^4p\, d^4p'\gamma^\mu \otimes \gamma^\nu \mathcal{F}(x,p;x',p'). \tag{12.164}$$

$\mathcal{F}$ has been calculated elsewhere [R. Hakim and H. Sivak (1982)] and the result is a bit too long to be reproduced here.

12.7.1. *Fluctuations of the four-current*

From the definition of the four-current operator and the development of the electron–positron field

$$\begin{cases} \psi(x) = \sum_r a_r\psi_r^{(+)}(x) + d_r^+\bar{\psi}_r^{(-)}(x), \\ \psi(x) = \sum_r a_r^+\bar{\psi}_r^{(+)}(x) + d_r\psi_r^{(-)}(x), \end{cases} \tag{12.165}$$

in which the ψ's are the covariant Johnson–Lippman spinors, one makes the explicit calculation of four-current operator fluctuations using Wick's theorem,

$$\langle a_1a_2a_3a_4\rangle = \langle a_1a_2\rangle\langle a_3a_4\rangle - \langle a_1a_3\rangle\langle a_2a_4\rangle + \langle a_1a_4\rangle\langle a_2a_3\rangle, \tag{12.166}$$

where the a's are either a, d or a^+, d^+. Setting

$$\delta J^{\mu\nu}(x) \equiv \delta J^{\mu\nu}(0, x_2 - x_1), \tag{12.167}$$

[24] Because of the transverse character of the polarization, zero is an eigenvalue and k^μ the corresponding eigenvector.

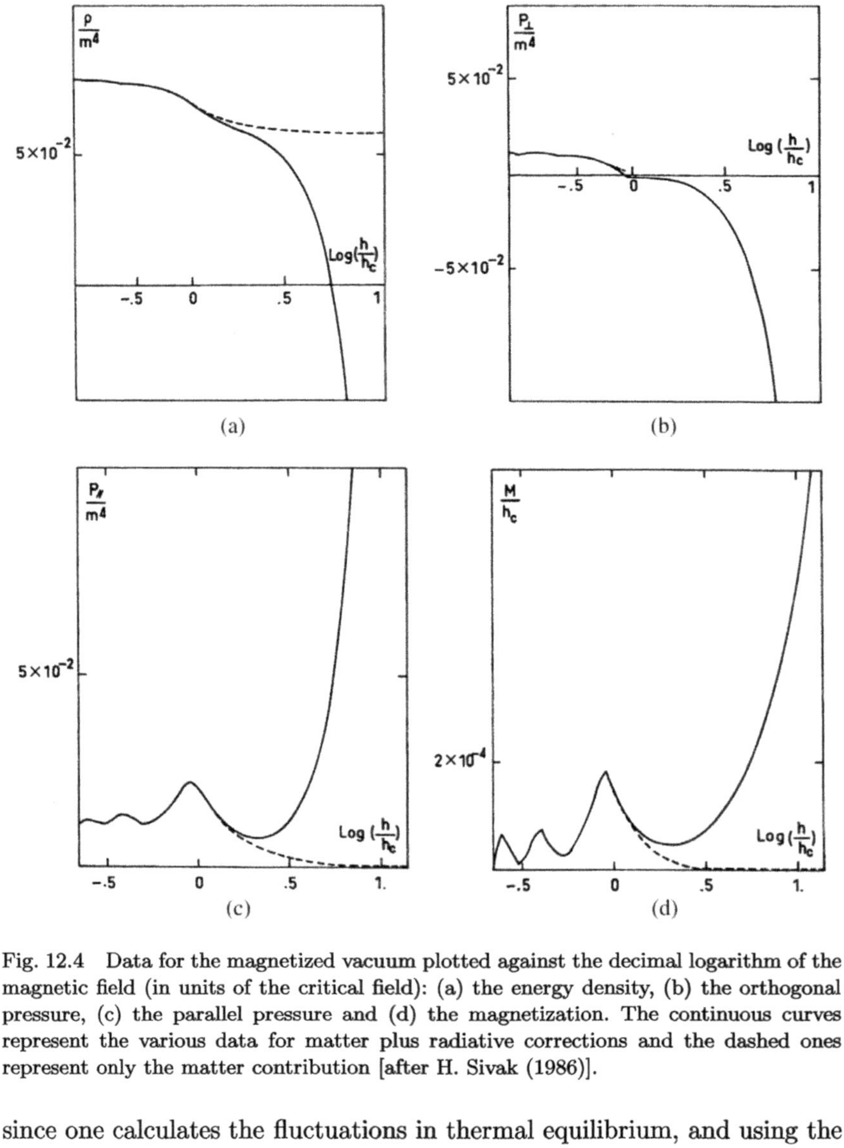

Fig. 12.4 Data for the magnetized vacuum plotted against the decimal logarithm of the magnetic field (in units of the critical field): (a) the energy density, (b) the orthogonal pressure, (c) the parallel pressure and (d) the magnetization. The continuous curves represent the various data for matter plus radiative corrections and the dashed ones represent only the matter contribution [after H. Sivak (1986)].

since one calculates the fluctuations in thermal equilibrium, and using the average occupation numbers of positive and negative energy electrons in the quantum state r,

$$\begin{cases} n_r^+ \equiv \langle a_r^+ a_r \rangle = \dfrac{1}{\exp(\beta[E_r - \mu]) + 1}, & E_r > 0, \\ n_r^- \equiv \langle d_r^+ d_r \rangle = \dfrac{1}{\exp(-\beta[E_r - \mu]) + 1}, & E_r < 0, \end{cases} \tag{12.168}$$

one obtains

$$\delta J^{\mu\nu}(x) = \left\langle \sum_{r_1,r_2} \{ n^+_{r_1}(1-n^+_{r_2})\bar{\psi}^{(+)}_{r_1}(x_1)\gamma^\mu\psi^{(+)}_{r_2}(x_1) \times \bar{\psi}^{(+)}_{r_2}(x_2)\gamma^\nu\psi^{(+)}_{r_1}(x_2) \right.$$
$$+ (1-n^-_{r_1})n^-_{r_2}\bar{\psi}^{(-)}_{r_1}(x_1)\gamma^\mu\psi^{(-)}_{r_2}(x_1) \times \bar{\psi}^{(-)}_{r_2}(x_2)\gamma^\nu\psi^{(-)}_{r_1}(x_2)$$
$$+ n^+_{r_1}n^-_{r_2}\bar{\psi}^{(+)}_{r_1}(x_1)\gamma^\mu\psi^{(-)}_{r_2}(x_1) \times \bar{\psi}^{(-)}_{r_2}(x_2)\gamma^\nu\bar{\psi}^{(+)}_{r_1}(x_2)$$
$$\left. + (1-n^-_{r_1})(1-n^+_{r_2})\bar{\psi}^{(-)}_{r_1}(x_1)\gamma^\mu\psi^{(+)}_{r_2}(x_1)\bar{\psi}^{(+)}_{r_2}(x_2)\gamma^\nu\psi^{(-)}_{r_1}(x_2)\} \right\rangle. \tag{12.169}$$

The first term on the right hand side of this last equation represents the contributions of the electrons only, while the second term is that of the positrons. The last two terms are connected with the possibility of pair creations or annihilations by electromagnetic waves propagating through the plasma.[25] After some long calculations [see some details in R. Hakim and H. Sivak (1982)], one finds that

$$\delta J^{\mu\nu}(k) = \frac{\exp\left(-\frac{1}{2}d_n^2 k_\perp^2\right)}{4\pi d_n^2[\exp(\beta\omega)-1]} \sum_{n,n'} \sum_{\ell,j=\pm 1} j\ell(-1)^{n+n'}$$
$$\times \int \frac{dp_{||}}{E_n E'_{n'}}(j_S^{\mu\nu} + j_A^{\mu\nu}) \times \delta(E_n + \ell E'_{n'} + j\omega), \tag{12.170}$$

where $j^{\mu\nu}_{S,A}$ designates a symmetrical (S) or antisymmetrical (A) tensor given by

$$j_S^{\mu\nu} = (\delta_{\ell 1} - [n_n^+ + n_n^-])$$
$$\times[-\ell E_n E'_{n'} + p_{||}(p_{||}-k_{||})]\cdot[L_n^{n'-n}L_{n'}^{n-n'} + L_{n-1}^{n'-n}L_{n'-1}^{n-n'}]\Omega_+^{\mu\nu}$$
$$+ \left[m^2(L_n^{n'-n}L_{n'}^{n-n'} + L_{n-1}^{n'-n}L_{n'-1}^{n-n'}) + \frac{4n}{d_n^2}L_n^{n'-n}L_{n'-1}^{n-n'}\right]\Omega^{\mu\nu}$$
$$-j[E_n(p_{||}-k_{||}) - \ell E'_{n'}p_{||}]\cdot[L_n^{n'-n}L_{n'}^{n-n'} + L_{n-1}^{n'-n}L_{n'-1}^{n-n'}]u^{(\mu}n^{\nu)}$$
$$-[\ell E_n E'_{n'} + m^2 + p_{||}(p_{||}-k_{||})]\cdot[L_n^{n'-n-1}L_{n'-1}^{n-n'+1} + L_{n-1}^{n'-n+1}L_{n'}^{n-n'-1}]\Pi^{\mu\nu}$$

[25]These two cases correspond either to a damping of these waves or to Cerenkov emission [see V.N. Tsytovich (1961)].

$$
\begin{aligned}
&- 2k_\alpha k_\sigma L_{n-1}^{n'-n+1} L_{n'}^{n-n'-1} (2\Pi^{\mu\sigma}\Pi^{\nu\alpha} - \Pi^{\alpha\sigma}\Pi^{\mu\nu}) \\
&- j[E_n(L_{n-1}^{n'-n+1} L_{n'-1}^{n-n'} + L_n^{n'-n} L_{n'-1}^{n-n'+1}) \\
&- \ell E'_{n'}(L_{n-1}^{n'-n} L_{n'-1}^{n-n'+1} - L_{n-1}^{n'-n+1} L_{n'}^{n-n'})] u^{(\mu}\Pi^{\nu)\sigma} k_\sigma \\
&+ [p_{||}(-L_{n-1}^{n'-n+1} L_{n'-1}^{n-n'} + L_n^{n'-n} L_{n'-1}^{n-n'+1}) \\
&+ (p_{||} - k_{||})(L_{n-1}^{n'-n} L_{n'-1}^{n-n'+1} - L_{n-1}^{n'-n+1} L_{n'}^{n-n'})] n^{(\mu}\Pi^{\nu)\sigma} k_\sigma,
\end{aligned}
\tag{12.171}
$$

$$
\begin{aligned}
j_A^{\mu\nu} = \frac{i}{h}(n_n^- - n_n^+) \\
&\times \{-j(L_n^{n'-n-1} L_{n'-1}^{n-n'+1} - L_{n-1}^{n'-n+1} L_{n'}^{n-n'-1}) \\
&\times [\ell E_n E'_{n'} + m^2 + p_{||}(p_{||} - k_{||})] F^{\mu\nu} \\
&+ [E_n(L_{n-1}^{n'-n+1} L_{n'-1}^{n-n'} + L_n^{n'-n} L_{n'-1}^{n-n'+1}) \\
&- \ell E'_{n'}(L_{n-1}^{n'-n} L_{n'-1}^{n-n'+1} + L_{n-1}^{n'-n+1} L_{n'}^{n-n'})] u^{[\mu} F^{\nu]\sigma} k_\sigma \\
&- j[p_{||}(L_{n-1}^{n'-n+1} L_{n'-1}^{n-n'} + L_n^{n'-n} L_{n'-1}^{n-n'+1}) \\
&+ [(p_{||} - k_{||})(L_{n-1}^{n'-n} L_{n'-1}^{n-n'+1} + L_{n-1}^{n'-n+1} L_{n'}^{n-n'})] n^{[\mu} F^{\nu]\sigma} k_\sigma,
\end{aligned}
\tag{12.172}
$$

where use has been made of the notations[26]

$$
\begin{cases}
E_n \equiv \sqrt{m^2 + p_{||}^2 + 2n/d_n^2}, \\
E'_{n'} \equiv \sqrt{m^2 + (p_{||} - k_{||})^2 + 2n'/d_n^2}, \\
L_a^b \equiv L_a^b\left(\frac{1}{2}d_n^2 k_\perp^2\right) \text{ (associated Laguerre polynomials)}, \\
\Omega_+^{\mu\nu} \equiv u^\mu u^\nu + n^\mu n^\nu.
\end{cases}
\tag{12.173}
$$

The above results for $\delta J^{\mu\nu}(k)$ require a few comments. First, the various values of ℓ and $j(\pm)$ (a separation between j and $(\pm)$) and the δ-term indicate that the fluctuations of the electron (or positron) four-current are nonvanishing only when

$$
\omega = \pm(E_n - E'_{n'}),
\tag{12.174}
$$

[26]We adopt the convention $L_n^a = 0$ whenever $n < 0$.

and these fluctuations correspond to transitions between Landau levels for either electrons or positrons. On the other hand, the fluctuations between positrons and electrons are possible only when

$$\omega = \pm(E_n + E'_{n'}) \tag{12.175}$$

and correspond to either pair creation or anihilation by (or into) an electromagnetic wave of frequency ω.

Another remark deals with the charge conservation, a property implying that

$$k_\mu \delta J^{\mu\nu}(k) = 0. \tag{12.176}$$

Although not obvious on the explicit form of $\delta J^{\mu\nu}(k)$, this can be proven by using various properties of the associated Laguerre polynomials (see App. A).

Finally, in order to complete the calculation of $\delta J^{\mu\nu}(k)$, the integration on $p_{||}$ can be performed and this can be achieved with the use of the well-known formula

$$\delta(g(x)) = \sum_\ell \frac{\delta(x - x_\ell)}{|g'(x_\ell)|} \tag{12.177}$$

where x_ℓ is a simple zero of $g(x)$, which is the case here.

The two figures below represent the spectra of longitudinal (Fig. 12.5) and transverse (Fig. 12.6) fluctuations of the density, at $T = 0\,\mathrm{K}$, and for the critical value of the magnetic field. We have chosen $\varepsilon_f = 1.6\,\mathrm{m}$. A common feature is the existence of transparency regions where $\langle \delta n^2 \rangle_k = 0$.

Let us begin with the longitudinal fluctuation spectrum, i.e. those fluctuations with $k_\perp = 0$. Two curves are presented (Fig. 12.5), each for one specific value of $k_{||}$, namely $k_{||} = 1$ or 4. The first branch on the left of the figure is related to the fluctuations' density of matter only (i.e. without any vacuum contribution), while the other branches contain the vacuum fluctuations occurring because of possible transitions from the vacuum to positive energy Landau levels (such as pair creation electromagnetic waves, a phenomenon that gives rise to a damping of electromagnetic waves) not already occupied (the presence of matter inhibits such fluctuations, owing to the exclusion principle), and vice versa. The purely material part, i.e. the first branch, appears at a given frequency and disappears at another: this effect comes from energy conservation, as can be seen from a closer inspection of $\delta J^{00}(k)$, where δ factors occur. As a matter of fact, the main effect of a longitudinal wave on the electrons of the Fermi sea is to accelerate them along the magnetic field without ever exciting transitions $n \to n'$. Unlike the first

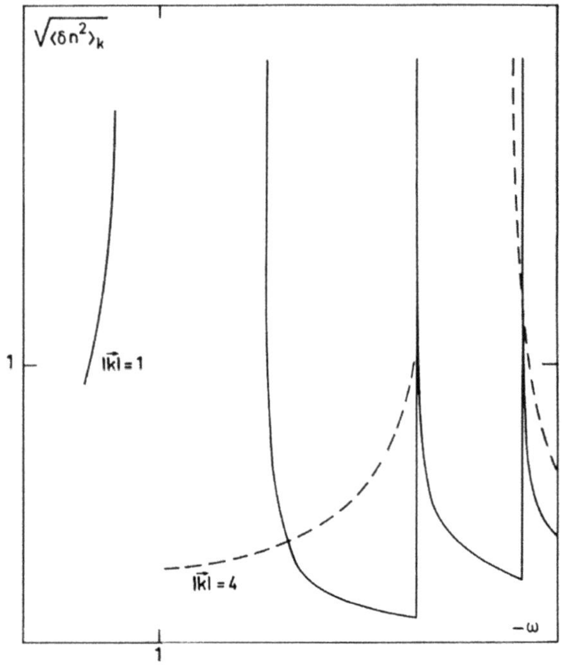

Fig. 12.5 Spectrum of the longitudinal density fluctuations versus $-\omega$, for two values of $k_{||}$ (continuous line — $k_{||} = 1$; dashed line — $k_{||} = 4$). This curve represents the case of the critical field and is plotted for a value of the Fermi energy $\varepsilon_f = 1.6\,m(T = 0\,\text{K})$.

branch, the others — related to the vacuum fluctuations — exhibit *resonances*, corresponding to pair creations or annihilations on those Landau levels such that $|\omega| = 2E$ and $k_{||} = 2p_{||}$.

Between the "matter part" and the "vacuum part" of the figure lies a *transparency region*, in the sense that the imaginary part of the polarization tensor[27] (see next section) vanishes. Physically, this is due to the fact that the frequency of the electromagnetic wave is not sufficient to excite vacuum fluctuations. Let us also add that this brief discussion is valid for all values of $k_{||}$ as well.

As to the case of transverse fluctuations of density ($k_{||} = 0$), it also presents some branches which are related to matter and others to the vacuum (see Fig. 12.6). A plot of $\langle \delta n^2 \rangle_k^{1/2}$ has been made on the figure

[27] See e.g. F. Bakshi, R.A. Cover and G. Kalman (1976, 1980); H. Perez-Rojas and A.E. Shabad (1979); H. Sivak (1985); D.B. Melrose and R.J. Stoneham (1977).

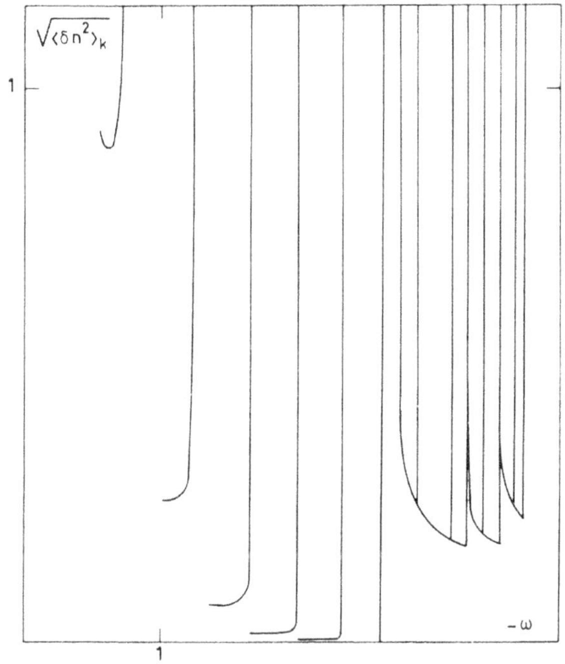

Fig. 12.6 Spectrum of the transverse density fluctuations versus $-\omega$, for $k_\perp = 0$. This curve represents the case of the critical field and is plotted for a value of the Fermi energy $\varepsilon_f = 1.6\,\mathrm{m}\,(T = 0\,\mathrm{K})$.

(still versus $-\omega$) for $k_\perp = 0.35$, and it presents similar features.[28] First, the branches on the left of the figure begin at specific frequencies for the same reasons as before (energy conservation) and all present resonances corresponding to

$$\frac{1}{2}|\omega| = \sqrt{m^2 + \frac{2nB}{B_{\text{crit}}}} \pm \sqrt{m^2 + \frac{2n'B}{B_{\text{crit}}}}. \tag{12.178}$$

Between two successive branches, there exists a gap still corresponding to a transparency region; it occurs when ω is not yet sufficient to excite a Landau level n to another one n'. These left hand side branches still refer to matter only. The last such branch is represented by a vertical line on the figure, and on its right are the vacuum contributions *except for* the vertical lines, which are also due to matter.

[28]For other values of $k_\perp$, the curves present the same characteristics.

12.8. Polarization Tensors of the Magnetized Electron Gas and of the Magnetized Vacuum

The polarization tensor of a neutralized magnetized electron gas has the form

$$\Pi^{\mu\nu}(k) = \Pi^{\mu\nu}_{\text{mat}}(k) + \Pi^{\mu\nu}_{\text{vac}}(k), \tag{12.179}$$

and its matter part $\Pi^{\mu\nu}_{\text{mat}}(k)$ is studied independently of its vacuum part $\Pi^{\mu\nu}_{\text{vac}}(k)$. The latter is infinite and must be renormalized. From the four-current fluctuations tensor, one has (see Chap. 15)

$$\Pi^{\mu\nu}(k) = -\frac{1}{2\pi}\int d\omega' \frac{\exp(\beta\omega') - 1}{\omega' - \omega - i\varepsilon}\langle J^\mu J^\nu\rangle_{\text{eq}}(\omega', \mathbf{k}), \tag{12.180}$$

in which the above expression for $\langle J^\mu J^\nu\rangle_{\text{eq}}$ has to be substituted. The result is an expression of the above general form where $\Pi^{\mu\nu}_{\text{mat}}$ refers to the (finite) matter part of the polarization tensor while $\Pi^{\mu\nu}_{\text{vac}}$ is its infinite vacuum contribution, to be renormalized. The results obtained by H. Sivak (1985), which are quite reliable, are presented below without any discussion and the reader should refer to the original articles for more details.

In Chap. 15, it will be shown how the inverse of the fluctuation–dissipation theorem can be used to obtain the dispersion properties of the plasma, with *no* magnetic field.

The matter part of the polarization tensor is then given by

$$\Pi^{\mu\nu}_{\text{mat}} = e^2\frac{\exp(-x^2)}{2(2\pi)^2 d_n} \times \sum_{n,n'}(-1)^{n+n'}\sum_{\ell,a}\int \frac{dp_{||}}{E_n E_{n',a}}\frac{J^{\mu\nu}(k)}{E_n + \ell E_{n',a} + a(\omega + i\varepsilon)}, \tag{12.181}$$

with the notations[29]

$$\begin{cases} x^2 \equiv -\frac{1}{2}d_n \Pi^{\mu\nu}k_\mu k_\nu, \\ E_n \equiv \sqrt{m^2 + p_{||}^2 + \frac{2n}{d_n}}, \\ E_{n',a} \equiv \sqrt{m^2 + (p_{||} + ak_{||})^2 + \frac{2n'}{d_n}}. \end{cases} \tag{12.182}$$

[29] Here $\Pi^{\mu\nu}$ is the projection operator on the two-plane orthogonal to the four-vectors u^μ and n^ν.

It can be decomposed into a Hermitian part, responsible for the propagation of the waves, and an anti-Hermitian one, responsible for the damping of waves. In any case, it is too long to be written explicitly here and the interested reader should go back to the original article of H.D. Sivak (1985).

12.8.1. *The vacuum polarization tensor*

The vacuum polarization tensor can be decomposed into its Hermitian and anti-Hermitian parts:

$$\begin{cases} \Pi^{\mu\nu}_{A,\text{vac}} = -\dfrac{i}{2} \lim\limits_{\substack{\mu\to 0\\ \beta\to\infty}} [\exp(\beta\omega) - 1] J^{\nu\mu}(k), \\ \Pi^{\mu\nu}_{H,\text{vac}} = -\dfrac{1}{2\pi} \displaystyle\int \frac{d\omega'}{\omega' - \omega} \lim\limits_{\substack{\mu\to 0\\ \beta\to\infty}} [\exp(\beta\omega') - 1] J^{\nu\mu}(\omega', \mathbf{k}). \end{cases} \tag{12.183}$$

The anti-Hermitian part is finite, while the Hermitian one is divergent and must be renormalized. The renormalized polarization tensor $\bar{\Pi}^{\mu\nu}_{H,\text{vac}}$ is obtained by

$$\bar{\Pi}^{\mu\nu}_{H,\text{vac}}(F^{\alpha\beta}, k) = \Pi^{\mu\nu}_{H,\text{vac}}(F^{\alpha\beta}, k) - \Pi^{\mu\nu}_{H,\text{vac}}(0, k) + \bar{\Pi}^{\mu\nu}_{H,\text{vac}}(0, k). \tag{12.184}$$

Let us remember that $\bar{\Pi}^{\mu\nu}_{H,\text{vac}}(0, k)$ can be obtained by subtracting from $\Pi^{\mu\nu}_{H,\text{vac}}(0, k)$ the first two terms of the McLaurin series

$$\bar{\Pi}^{\mu\nu}_{H,\text{vac}}(0, k) = \Pi^{\mu\nu}_{H,\text{vac}}(0, k) - \Pi^{\mu\nu}_{H,\text{vac}}(0, 0) - \frac{1}{2} \frac{\partial^2}{\partial k^\sigma \partial k^\lambda} \Pi^{\mu\nu}_{H,\text{vac}}(0, k) \bigg|_{k=0} k^\sigma k^\lambda \tag{12.185}$$

and after some calculations [see H. Sivak (1985)] one finds that

$$\bar{\Pi}^{\mu\nu}_{H,\text{vac}}(F^{\alpha\beta}, k) = \sum_{\ell=1}^{\ell=3} \bar{\Pi}_\ell(k) \frac{b^\mu_\ell b^\nu_\ell}{b^2_\ell}, \tag{12.186}$$

where $\bar{\Pi}_\ell(k)$ are the three nonvanishing eigenvalues of $\bar{\Pi}^{\mu\nu}_{H,\text{vac}}(F^{\alpha\beta}, k)$:

$$\bar{\Pi}_\ell(k) = \frac{e^2}{(2\pi)^2} \int_0^\infty dy \int_0^1 d\xi \left\{ \frac{C_\ell \exp(\chi)}{\lambda^2 \sinh(y/d_n^2)} - \frac{2k^2\xi(1-\xi)}{y} \exp(-ym^2) \right\}, \tag{12.187}$$

with

$$\begin{cases} C_1 = k^2(1-\xi)\dfrac{\sinh(2y\xi/d_n^2)}{\sinh(y/d_n^2)}, \\ C_2 = \dfrac{\Omega^{\mu\nu}k_\mu k_\nu}{k^2}C_1 + 2\Pi^{\mu\nu}k_\mu k_\nu \dfrac{\sinh(y\xi/d_n^2)\sinh(y[1-\xi]/d_n^2)}{\sinh^2(y/d_n^2)}, \\ C_3 = \dfrac{\Omega^{\mu\nu}k_\mu k_\nu}{k^2}C_1 + 2\Omega^{\mu\nu}k_\mu k_\nu \xi(1-\xi)\cosh(y/d_n^2), \end{cases} \tag{12.188}$$

$$\begin{cases} \chi = -\Pi^{\mu\nu}k_\mu k_\nu d_n^2 s \dfrac{\sinh(y\xi/d_n^2)\sinh(y[1-\xi]/d_n^2)}{\sinh(y/d_n^2)}, \\ s = \mathrm{sgn}(\Omega^{\mu\nu}k_\mu k_\nu \xi[1-\xi] - m^2). \end{cases} \tag{12.189}$$

The three symmetric tensors $b_\ell^\mu b_\ell^\nu$ are constructed from

$$\begin{cases} b_1^\mu = \left\{\Pi^{\mu\nu}k_\nu - \dfrac{\Pi^{\alpha\beta}k_\alpha k_\beta}{k^2}k^\mu\right\} = -\left\{\Omega^{\mu\nu}k_\nu - \dfrac{\Omega^{\alpha\beta}k_\alpha k_\beta}{k^2}k^\mu\right\}, \\ b_2^\mu = F^{\mu\nu}k_\nu, \\ b_3^\mu = {}^*F^{\mu\nu}k_\nu. \end{cases} \tag{12.190}$$

These results coincide with those obtained by I.A. Batalin and A.E. Shabad (1971) and A.E. Shabad (1972, 1975).

12.9. Remarks on the Transport Coefficients of the Magnetized Electron Gas

The transport coefficients are evaluated from a kinetic equation, as has been done several times in this book. Here we only want to show how modifications occur because of the magnetic field.

Let us begin with the kinetic equation. In an arbitrary gauge, the BGK equations read

$$\begin{aligned} &\{i\gamma\cdot\partial + 2[\gamma\cdot p - m]\}F(x,p) \\ &\quad + 2e\int \frac{d^4R}{(2\pi)^4}d^4\xi \exp[-i(p-\xi)\cdot R]F(x,\xi)\gamma\cdot A_{\text{ext}}\left(x-\frac{1}{2}R\right) \\ &= -i\gamma\cdot u\frac{F(x,p)-F_{\text{eq}}(p)}{\tau} \end{aligned} \tag{12.191}$$

$$F(x,p)\{i\gamma\cdot\partial-2[\gamma\cdot p-m]\}$$
$$-2e\int\frac{d^4R}{(2\pi)^4}d^4\xi\exp[-i(p-\xi)\cdot R)]F(x,\xi)\gamma\cdot A_{\text{ext}}\left(x-\frac{1}{2}R\right)$$
$$=-\frac{F(x,p)-F_{\text{eq}}(p)}{\tau}i\gamma\cdot u, \tag{12.192}$$

which reduces to

$$\{i\gamma\cdot\partial+2[\gamma\cdot p-m]\}F(x,p)+eF^{\mu\nu}p_\nu\frac{\partial}{\partial p^\mu}F(x,p)$$
$$=-i\gamma\cdot u\frac{F(x,p)-F_{\text{eq}}(p)}{\tau} \tag{12.193}$$

in the Lorentz gauge and the other equation.

In the absence of any magnetic field, it was recognized that there exists another expansion parameter — besides the ratio of the relaxation time and a macroscopic hydrodynamical time — connected in the quantum domain to the existence of the Compton wavelength. When a magnetic field is present, a new scale of length does exist, the electron *Larmor radius*,

$$r_L=\frac{eBv_{\text{th}}}{m}, \tag{12.194}$$

where the thermal velocity v_{th} should be replaced by 1, the velocity of light, in a truly relativistic regime.

In order to get an idea of the various possible regimes, the above equation is first rewritten in terms of the dimensionless quantities[30]

$$x=\ell_1\bar{x},R=\ell_1\bar{R},A=\left(\frac{eB}{\ell_2}\right)\bar{A},p=\frac{\bar{p}}{\ell_3}, \tag{12.195}$$

where $\{\dots\}$ is one of the possible available lengths; and one obtains

$$\left\{i\ell_1^{-1}\gamma\cdot\tilde{\partial}+2\left[\ell_3^{-1}\gamma\cdot\bar{p}-\frac{1}{\lambda_{\text{Compton}}}\right]\right\}F(x,p)$$
$$+2\ell_2^{-1}\int\frac{d^4\bar{R}}{(2\pi)^4}d^4\bar{\xi}\exp[-i(\bar{p}-\bar{\xi})\cdot\bar{R}]F(\bar{s},\bar{\xi})\gamma\cdot\bar{A}_{\text{ext}}\left(\ell_1\bar{x}-\frac{1}{2}\bar{R}\ell_1\right)$$
$$=-i\gamma\cdot u\frac{F(\bar{x}\ell_1,\bar{p}\ell_3)-F_{\text{eq}}(\bar{p}\ell_3)}{\tau} \tag{12.196}$$

[30]Note that since the BGK equation is linear in F, it is not necessary to replace this quantity by a dimensionless one.

(and the other dimensionless companion equation), which allows a comparison of the orders of magnitude of the different terms and hence does indicate the possible regimes, after a choice for the various lengths has been made. Since there are *a priori* three possible length scales, this gives rise to nine choices and a few dozen possible regimes. Only one case has been treated [R. Dominguez Tenreiro and R. Hakim (1977b)], i.e. the one in which

$$\tau \ll \ell_1^{-1} \approx eB, \tag{12.197}$$

so that the left hand sides of the equations are of order $O(\tau)$. This is a relatively simple situation.

The results for the various transport coefficients are essentially an oscillatory behavior around the values obtained (see Chaps. 2 and 10) in the absence of a magnetic field [R. Dominguez Tenreiro (1978)]. In this regime, however, not all transport coefficients are nonvanishing.

The next problem deals with the definition of the transport coefficients in the presence of a magnetic field. This has been done by J.L. Anderson (1976), in an unpublished work, on the basis of group-theoretical arguments. It is, however, possible to extend nonrelativistic results by S.I. Braginskii[31] to the relativistic case [R. Dominguez Tenreiro and R. Hakim (1977b)], although there is no unique way to reach this aim.[32] These different possibilities are largely a matter of adaptation to experimental situations, or to the symmetries of the problem under consideration. Nevertheless, it should be noted that the physically meaningful quantity is the viscous stress tensor and not necessarily its decomposition. We have already given the covariant generalization of Braginskii's results, so we shall not repeat them (see Chap. 2).

When one calculates the off-equilibrium part of the energy–momentum tensor, one can encounter the nonsymmetric tensor

$$\Pi^{(\alpha\mu}n^{\beta)}n^{\rho}[\partial_\rho u_\mu - \partial_\mu u_\rho], \tag{12.198}$$

which is a *vorticity* tensor expressing the rotation of the charged particles around the magnetic field axis. This term also exists in the nonrelativistic

[31] S.I. Braginskii, Review of Plasma Physics, Vol. 1 (1965).

[32] See e.g. A.N. Kaufman, *Phys. Fluids* **3**, 610 (1960); J.A.R. Coope and R.F. Snider, *J. Chem. Phys.* **56**, 2056 (1970); S.R. de Groot and P. Mazur, *Non-equilibrium Thermodynamics* (North-Holland, Amsterdam, 1962); P.C. Clemmow and J.P. Dougherty, *Electrodynamics of Particles and Plasmas* (Addison-Wesley, Reading, 1969).

and nonquantum case.[33] As to the various viscosity coefficients, they correspond to a transverse coefficient, a parallel one, a crossed parallel/transverse one, etc.

Let us show briefly the off-equilibrium part of the four-current and the heat flow; the latter can be decomposed as

$$Q^\mu = Q^\mu_\perp + Q^\mu_{||}, \tag{12.199}$$

with

$$\begin{cases} Q^\mu_\perp = \Pi^{\mu\nu} Q_\nu = +\lambda_\perp \Pi^{\mu\nu}(\beta^{-1}\dot{u}_\nu - \partial_\nu \beta^{-1}), \\ Q^\mu_{||} = n^\mu n^\nu Q_\nu = -\lambda_{||} n^\mu n^\nu (\beta^{-1}\dot{u}_\nu - \partial_\nu \beta^{-1}), \end{cases} \tag{12.200}$$

in agreement with Eckart's definition. However, the calculation of J^μ_{off} also contains terms proportional to $\partial_\mu n_{\text{eq}}$. The latter can be eliminated with the help of the zeroth order conservation equations but terms involving gradients of the magnetic field do remain. Conversely, if these terms are not eliminated but only the gradients of the magnetic fields, then we should resort to a Robinson–Bernstein[34] version of the entropy considerations, namely that the entropy obeys a minimum–maximum principle, i.e. it is not a minimum but a saddle point. Therefore, this question should be considered as open.

12.10. Astrophysical Aspects

In several astrophysical quantities, the issues of the occurrence and the properties of strong magnetic fields have been questioned. We mention the influence of strong magnetic fields on the β decay of some models of disintegration, the neutronization reactions, the mass–radius relation of a star (magnetic white dwarfs or neutron stars), the stability of the star, the possibility of an *a priori* strong magnetic field in the primeval universe, etc.

Let us briefly review some of these modifications brought about by the strong magnetic fields, and let us first examine the neutron disintegration of the neutron

$$N \to P + e^- + \bar{\nu}_e,$$

which is the simplest one. The essentials of this disintegration lie in the modification of its phase space; in neutron stars, the intensity of the magnetic field is of the order of the critical one, so that the proton is not affected.

[33] See S.R. de Groot and P. Mazur, *loc. cit.*
[34] B.B. Robinson and I.B. Bernstein, *Ann. Phys.* (*N.Y.*) **18**, 110 (1962).

In terms of the disintegration lifetime, it reads

$$\frac{1}{\tau} = \frac{2\pi^2 G_V^2}{E_N} \int \frac{d^2p}{E_P} \frac{d^3q}{E_\nu} \{f^\mu(q_N - p - q) - f_5^\mu(q_N - p - q)\}$$
$$\times\{(1+\lambda^2) q_N \cdot q p_\mu + (1-\lambda^2)\, p \cdot q_{N\mu} - (1-\lambda^2) m_P m_N q_\mu\}, \tag{12.201}$$

where q_N is the neutron four-momentum, p the proton four-momentum, q the neutrino, G_V the weak coupling constant and λ the axial vector coupling constant. As expected, the magnetic field occurs only in f^μ and f_5^μ, i.e. in $F(q)$.

As to the neutronization, occurring for example in neutron stars,

$$P + e^- \rightleftarrows N + \nu_e,$$

it is governed by the equality of chemical potentials

$$\mu_P + \mu_e = \mu_N + \mu_\nu, \tag{12.202}$$

where μ_ν vanishes almost completely since the neutrinos escape the star. From these equalities between the μ's, from their expressions as functions of B, n_{eq}, etc., one obtains relations for the nucleons, etc. (see e.g. G.A. Schulman (1974ff)).

As to white dwarfs, for instance, J.P. Ostriker and F.D.A. Hartwick[35] made detailed calculations on magnetic white dwarfs, showing that inside the magnetic field it could be as high as 10^{11}–10^{13} G. We look briefly at the impact of magnetic fields on white dwarfs. We limit our considerations to the ones presented in the book by S.L. Shapiro and S.A. Teukolski (1983).[36] Starting with the *virial* theorem

$$W + 3\Pi + \mathrm{M} = 0, \tag{12.203}$$

where W is the gravitational energy, Π the work done by the pressure and M the magnetic energy; we write explicitly the various data as

$$W = \frac{1}{2}\int d^3x\, \rho(x)\phi(x), \mathrm{M} = \frac{1}{8\pi}\int d^3x\, B^2, \Pi = \int d^3x P, \tag{12.204}$$

or

$$W = -aG\frac{M^2}{R}, \Pi = bM\left\langle\frac{P}{\rho}\right\rangle, \mathrm{M} = c\left\langle\frac{B^2}{8\pi}\right\rangle\frac{4}{3}\pi R^3 \tag{12.205}$$

---

[35] J.P. Ostriker and F.D.A. Hartwick, *Astrophys. J.* **153**, 105 (1968).
[36] S.I. Shapiro and S.A. Teukolski, *Black Holes, White Dwarfs and Neutron Stars* (J. Wiley and Sons, New York, 1983).

(the constants a, b and c are of order unity), so that the virial theorem can be written as

$$-a\frac{GM^2}{R} + bM\left\langle\frac{P}{\rho}\right\rangle + c\frac{\psi^2}{R} = 0, \tag{12.206}$$

where the last term is the magnetic flux

$$\psi = \frac{B}{R^2} \tag{12.207}$$

and use has been made of an ultrarelativistic and nonmagnetic limit for the equation of state of electrons. The nonmagnetic field limit is taken since the various calculations indicate that only the "weight" of the magnetic field makes sense. Therefore, the effect of the magnetic field is — crudely speaking — to decrease the effect of gravity:

$$G' \to G - \frac{c}{a}\frac{\phi^2}{M^2} = G\left(1 - \frac{\mathrm{M}}{|W|}\right). \tag{12.208}$$

This means that, for a given mass of the star, its radius can be substantially larger. This has also been found by D. Adam,[37] in a model less rigorous than that of Ostriker and Hartwick, but more rigorous than the crude model. In fact, he also took account of the modifications of the equations of state of the electrons brought about by the magnetic field. In addition, he calculated the various pycnonuclear reactions.

[37] D. Adam, *Astron. Astrophys.* **160**, 95 (1986).

Chapter 13

Statistical Mechanics of Relativistic Quasiparticles*

The concept of the *quasiparticle* appears in a large number of nonrelativistic physical situations, where it brings substantial technical simplifications and also far-reaching clarifications of the problems under consideration. Quasiparticles are widely used in the framework of solid state physics[1] and, consequently, have been studied in detail. Another case where quasiparticles occur, namely plasmons, has been systematically considered in plasma physics.[2]

However, in a relativistic context, in spite of numerous studies on QED plasmas[3,4] and of an exponentially growing interest in the quark–gluon plasma expected to be observed in heavy ion collisions,[5] no systematic considerations of *relativistic* quasiparticles have been undertaken. It should also be mentioned that such theoretical objects are of much interest in the context of nuclear matter whether relativistic [B.D. Serot and J.D. Walecka (1986)] or not and in the study of Bose condensates[6] occurring in dense matter. Very simple relativistic quasiparticles (i.e. "free" quasiparticles endowed with an effective mass) are also considered in a

---

*This chapter was partially done with our late friend Horacio D. Sivak (1946–2000) and has been partially published in *Class. Quantum Grav.*: **10** (Suppl.), 223 (1993).

[1] See e.g. C. Kittel, *Quantum Theory of Solids* (J. Wiley, New York, 1963).

[2] T. Kihara, O. Aono and T. Dodo, *Nucl. Fusion* **2**, 66 (1962); A. I. Alekseev and Yu. P. Nikitin, *Sov. Phys. JETP* **23**, 608 (1966); E. G. Harris, *Adv. Plasma Phys.* **3**, 157 (1969).

[3] See e.g. N.P. Landsman and Ch. G. van Weert, *Phys. Rep.* **145**, 141 (1987).

[4] One should also include the "phenomenological QED" of J.M. Watson, *Phys. Rev.* **74**, 950 (1948); *ibid.* **74**, 1485 (1948); *ibid.* **75**, 1249 (1949).

[5] See e.g. L.P. Csernai, *Introduction to Relativistic Heavy Ion Collisions* (Wiley, Chichester, 1994) or *Hot Hadronic Matter: Theory and Experiment*, eds. J. Letessier and J. Rafelski (Plenum, New York, 1994).

[6] A.B. Migdal, *Rev. Mod. Phys.* **50**, 107 (1978).

toy model constituted by the $\lambda\varphi^4$ theory in Gaussian (or other nonperturbative) approximations[7,8] or in the case of the so-called relativistic "scalar plasma."[9] Apart from the early attempts by J.M. Jauch and K. Watson[10] (1948, 1949), the most advanced study of relativistic quasiparticles seems to be the one by A.B. Migdal (1978) in connection with considerations of boson condensates and, in what follows, his results are found anew in a more general context and also are extended and specified more precisely.

The reasons why there do not exist systematic studies are multiple: (i) many people seem to consider that a relativistic extension of the concept of the quasiparticle is trivial (it will be realized, below, that this is certainly not the case); (ii) the need for relativistic quasiparticles was not really urgent, because of the relatively recent emergence of statistical considerations for relativistic quantum dense matter. Apart from the interest of their own, relativistic quasiparticles can be a source of new progress when we are dealing with dense matter and, in particular, with the QCD plasma; nevertheless, it should be clear that only new physical inputs (ideas, approximations, etc.) can bring new developments and, in this respect, quasiparticles can only represent a technical instrument that may express such an input.

It is, of course, not possible to deal with all cases and hence we limit ourselves to a sufficiently general one so as to accommodate many situations: it is assumed that the polarization tensor (for bosons) or the mass operator (for fermions) is given as a function not only of the four-momentum but also of *macroscopic* quantities such as the average four-velocity of the system, its temperature, its density, etc. Accordingly, and at least as a first step, only free quasiparticles are dealt with. Whether the system under consideration is correctly described by an assembly of (free or interacting) quasiparticles or not is another question depending on the specific problem at hand. So is the case regarding the approximation method that leads to the polarization tensor (or the mass operator) at hand.

---

[7]Among numerous articles on the subject, we quote only the following: G. Baym and G. Grinstein, *Phys. Rev.* **D15**, 2897 (1977); T. Barnes and G.I. Ghandour, *Phys. Rev.* **D22**, 924 (1980); W.A. Bardeen and M. Moshe, *Phys. Rev.* **D28**, 1372 (1983); P.M. Stevenson, *Phys. Rev.* **D33**, 2305 (1985); M. Ciancitto, *Nucl. Phys.* **B254**, 653 (1985); etc. See Ref. 9 for more references.

[8]F. Grassi, R. Hakim and H. Sivak, *Int. J. Mod. Phys.* **A6**, 4579 (1991).

[9]G. Kalman, *Phys. Rev.* **D9**, 1656 (1974).

[10]J.M. Jauch and K. Watson, *Phys. Rev.* **74**, 950 (1948); *ibid.* **74**, 1485 (1949).

Therefore, our first assumption is concerned with the general form of the equations of motion obeyed by the quasiparticle fields. For quasibosons, it reads

$$\Box\varphi(x) + \int d^4x\ \Pi(x, x';\ \text{macroscopic quantities})\, \varphi(x') = 0, \qquad (13.1)$$

while for quasifermions one has

$$i\gamma \cdot \partial\Psi(x) - \int d^4x'\ \Sigma(x, x';\ \text{macroscopic quantities})\, \Psi(x') = 0. \qquad (13.2)$$

In the following, we deal rather with the space–time translational invariant case, where

$$\Pi(x, x') = \Pi(x - x', 0) \equiv \Pi(x - x') \qquad (13.3)$$

for bosons, and

$$\Sigma(x, x') = \Sigma(x - x', 0) \equiv \Sigma(x - x') \qquad (13.4)$$

for fermions. Therefore, we treat mainly the case of Eqs. (1.1) and (1.2), which are rewritten in Fourier space as

$$\begin{cases} [k^2 - \Pi(k)]\varphi(k) = 0 \quad \text{(quasibosons)}, \\ [\gamma \cdot p - \Sigma(p)]\Psi(p) = 0 \quad \text{(quasifermions)}, \end{cases} \qquad (13.5)$$

and these equations also signify that the quasibosons obey the dispersion equation

$$D(k) \equiv [k^2 - \Pi(k)] = 0 \qquad (13.6)$$

while the quasifermions obey

$$\text{Det}[\gamma \cdot p - \Sigma(p)] = 0. \qquad (13.7)$$

From this "dynamical" starting point, the usual quantum field theory path is followed. First, the equations for the "classical" field Φ or Ψ are cast into a Lagrangian form; this allows the derivation of the usual conservation laws, *whenever valid* (this is discussed later on), and hence explicit expressions for the four-current, energy–momentum tensor, angular momentum, etc. These expressions are needed not only for the macroscopic description of the system at hand but also for the quantization of the "classical" quasiparticle field. Indeed, explicit normalization of those quasiparticle plane waves, in which the field is expanded, is a necessity since a covariant expression for the scalar product of two free wave packets is needed to that end. Furthermore, from the energy–momentum tensor the quasiparticle Hamiltonian is obtained. Once quantization is achieved, statistical mechanics (for both equilibrium and nonequilibrium situations) is almost a matter of routine. Next, applications can follow. The above dispersion equations lead to nonlocal equations for the quasiparticle fields.

Finally, a strong emphasis should be put on the comparison with the nonrelativistic case since, essentially, two new classes of problems show up. First, one knows that negative energy states lead to the concept of antiparticles: for quasiparticles, the question is much more delicate and subtle and "antiquasiparticles" depend strongly on the medium of which they are supposed to be excitations. It appears that this question is entangled with the problem of the thermodynamic stability of the system under consideration. This is briefly discussed in Sec. 13.3. A second class of problems is connected with causality. In general, nonlocal field theories are plagued with troubles arising from a "lack of causality." This can be understood in an intuitive way: the function $\Pi(x)$ [or $\sum(x)$], besides its order of magnitude (Π), introduces implicitly a new dimensioned scale, connected with its spatial/temporal extension, i.e. the "width" of these functions. This means that the approximations contained in the calculation of Π generally involve a change of scales for time and space that does not preserve the light cone: this either becomes sharper or gets flattened so that, for example, a timelike four-vector becomes spacelike, implying a certain lack of causality. A well-known example of such a lack of causality is provided by the propagation of heat, whether in a nonrelativistic or a relativistic context.

13.1. Classical Fields

In this section, we limit ourselves to the case of complex scalar fields, the extension to other possibilities being either straightforward or explicitly dealt with. These fields obey the equation

$$\Box\varphi(x) + \int d^4y\ \Pi(y)\varphi(x-y) = 0. \tag{13.8}$$

A possible Lagrangian for this equation is

$$L = \partial_\mu\varphi(x)\partial^\mu\varphi(x) - \int d^4y\ \Pi(y)\varphi^*\left(x+\frac{1}{2}y\right)\varphi\left(x-\frac{1}{2}y\right), \tag{13.9}$$

as can be checked by minimizing the action integral, while the variation with respect to Φ

$$\frac{\delta}{\delta\Phi(x)}\int d^4x'\ L\left[\{\Phi^*\},\{\Phi\}\right] = 0$$

yields

$$\Box\Phi^*(x) + \int d^4x\ \Pi(y)\Phi^*(x+y) = 0 \tag{13.10}$$

with $\Pi^*(y) = \Pi(-y)$ in order that both equations may be consistent; the Lagrangian is thus Hermitian. On the other hand, this condition joined to Onsager relations implies that, in Fourier space,

$$\Pi(k) = \Pi^*(k) = \Pi(-k). \tag{13.11}$$

This condition comes from our results in the case of long-lived quasiparticles. Such a case can be found in many physical situations or problems: plasmons in QED plasmas, quasinucleons of the Walecka model, quasibosons of the $\lambda\varphi^4$ theory in Gaussian approximation, quasiparticles involved in some regimes of hard thermal loops, etc. Note also that dissipative contributions to these long-lived quasiparticles arise from mode–mode interactions[11] or from a more or less phenomenological collision term like the Boltzmann or the relativistic BGK one.

The equations of motion are obtained by using the well-known relation

$$\frac{\delta}{\delta\Phi(x)}\Phi(x') = \delta^{(4)}(x-x'). \tag{13.12}$$

However, such a derivation of the equations of motion (13.1) and (13.2) from the Lagrangian (13.3) possesses the inconvenience of hiding some conditions to be satisfied by Φ and Φ^* and, furthermore, it needs a separate deduction of the various currents associated with symmetries, whether space–time or possible internal ones.

13.1.1. *Internal symmetries and conserved currents*

From the Lagrangian, one can obtain (such as from a minimal coupling assumption) the Lagrangian of the system in the presence of an "electromagnetic" field $A^\mu(x)$ — say, $L\left[\{\Phi^*\},\{\Phi\},\{A^\lambda\}\right]$. In such a case, the charge current $J^\mu(x)$ is nothing but the functional derivative of the action integral with respect to the electromagnetic field and considered for $A^\mu \equiv 0$.

---

[11]See e.g. R.Z. Sagdeev and A.A. Galeev, *Nonlinear Plasma Theory* (Benjamin, New York, 1969).

Such a gauge-invariant Lagrangian is given by

$$L\left[\{\Phi^*\},\{\Phi\},\{A^\lambda\}\right] = D_\mu^*\Phi^*(x)\cdot D^\mu\Phi(x) + \int d^4y\,\Pi(y)$$
$$\times\,\Phi^*\left(x+\frac{1}{2}y\right)\exp(-iey^\mu)\int_0^1 ds\;A^\mu\left[x+y\left(s-\frac{1}{2}\right)\right]\Phi\left[x-\frac{1}{2}y\right], \tag{13.13}$$

with

$$D^\mu \equiv \partial^\mu + ieA^\mu(x), \tag{13.14}$$

and where the integral over the parameter s is nothing but an integral along the straight line joining the space–time points $x-y/2$ and $x+y/2$. Notice that, for quantum fields, the exponential occurring in this last equation must involve an ordering with respect to the parameter s. The above Lagrangian represents the dynamics of fields Φ and Φ^* minimally coupled to the electromagnetic field A^μ and is inspired by J. Schwinger's.[12] Similar problems (of gauge invariance) connected with covariant Wigner functions can be found elsewhere [E.A. Remler (1977); J. Winter (1984); H. Th. Elze, M. Gyulassi and D. Vasak (1986, 1987); U. Heinz (1985); H. Th. Elze and U. Heinz (1989); etc.]. The basic idea is that quantities like $\Phi(x-y)$ can be rewritten as exp $(y\cdot\partial)$. $\Phi(x)$ and, accordingly, the gauge-invariant form are obtained by replacing ∂ with $\partial + ieA$. The fact that the path integral occurring in this last equation is taken along a straight line was first proven by D.G. Boulware[13] on the basis of both Lorentz and space–time translation invariance. Another argument has been given in another context by R. Marnelius (1973).[14]

Let us form the functional derivative $\delta S/\delta A^\mu$, where S is the action integral $S=\int Ld^4x$. We get

$$\left.\frac{\delta S}{\delta A_\mu(x)}\right|_{A^\lambda\equiv 0}$$
$$= J^\mu(x) = i\Phi^*\overleftrightarrow{\partial}^\mu\Phi - \int d^4x'\,d^4y'\,\Pi(y)\Phi^*\left(x+\frac{1}{2}y\right)$$
$$\times\,\frac{\delta}{\delta A_\mu(x)}\exp\left\{-iy^\lambda\int_0^1 ds\;A_\lambda\left[x'+y\left(s-\frac{1}{2}\right)\right]\right\}_{A^\lambda\equiv 0}\Phi\left(x'-\frac{1}{2}y\right)$$

[12]J. Schwinger, *Phys. Rev.* **125**, 397 (1962); *ibid.* **128**, 2425 (1962).
[13]D.G. Boulware, *Phys. Rev.* **151**, 1024 (1966).

$$= i\Phi^* \overleftrightarrow{\partial}^\mu \Phi + i \int d^4x'\, d^4y'\, \Pi(y) y^\nu \Phi^* \left(x + \frac{1}{2}y\right)$$

$$\times \int_0^1 ds \frac{\delta A_\nu \left[x' + y\left(s' - \frac{1}{2}\right)\right]}{\delta A_\mu(x)} \exp\left\{ - iy^\lambda \right.$$

$$\left. \times \int_0^1 ds A_\lambda \left[x' + y\left(s - \frac{1}{2}\right)\right]\right\}_{A^\lambda \equiv 0} \Phi\left(x' - \frac{1}{2}y\right). \tag{13.15}$$

Using now

$$\frac{\delta A_\lambda(x')}{\delta A^\mu(x)} = \delta_{\lambda\mu}\delta^{(4)}(x - x'), \tag{13.16}$$

the above equation for the four-current reduces to

$$J^\mu(x) = i\Phi^* \overleftrightarrow{\partial}^\mu \Phi + i \int d^4y \int_{-1/2}^{+1/2} ds \Pi(y) y^\mu \Phi^* \left[x + y\left(s + \frac{1}{2}\right)\right]$$

$$\times \Phi\left[x + y\left(s - \frac{1}{2}\right)\right] \tag{13.17}$$

and is similar to the expression given by R. Marnelius[14] after a Fourier transformation and an integration over the parameter s.

Such a derivation is valid for any gauge field whatsoever and allows one, accordingly, to obtain any macroscopic four-current, such as an isospin four-current. For instance, in the gauge-invariant Lagrangian used in the derivation of J^μ the exponential has to be changed as follows:

$$\exp\left\{-iey^\lambda \int_0^1 ds\, A_\lambda \left[x + y\left(s - \frac{1}{2}\right)\right]\right\} \to P \exp\left\{-igy^\lambda \int_0^1 ds \right.$$

$$\left. \times\, \mathbf{T} \cdot \mathbf{A} \left[x + y\left(s - \frac{1}{2}\right)\right]\right\}, \tag{13.18}$$

where the matrices $\mathbf{T}$ are the infinitesimal generators of the group under consideration g is the coupling constant of the gauge field and P is a chronological ordering of the integration path. Then the functional derivative with respect to the gauge field $\mathbf{A}_\lambda$, specialized to $\mathbf{A}_\lambda \equiv 0$, provides the corresponding four-currents. One can easily check that J^μ is conservative: $\partial_\mu J^\mu = 0$.

[14]R. Marnelius, *Phys. Rev.* **D8**, 2472 (1973).

13.1.2. *Space–time symmetries*

The conservation laws for space–time symmetries obviously depend on the transformation properties of the polarization tensor $\Pi(x)$ under the Poincaré group. Clearly, for equations of motion as given above, one expects the conservation of the energy–momentum tensor: Π depends on the difference $x - x'$ and is thus invariant under space–time translations. However, this is not so for the general equations of motion. Also, it is clear that Π is not necessarily invariant under spatial translations; therefore, even though the energy–momentum tensor of the quasiparticles is conserved, this is not necessarily the case for the general angular momentum tensor.

Let us look for the conserved — or nonconserved — energy–momentum and angular tensors of quasiparticles by means of Noether's theorem. To this end, let us consider an infinitesimal x-dependent coordinate change,[15]

$$x \to x + \varepsilon(x), \tag{13.19}$$

in the expression of the action integral $S = \int d^4xL$. Owing to the fact that the Jacobian of this transformation is $1 + \partial\varepsilon(x)$, up to first order, and that — still to this order — the derivatives become

$$\partial_\mu \to \partial_\mu - \partial_\mu \varepsilon^\lambda(x) \cdot \partial_\lambda, \tag{13.20}$$

we find that

$$\begin{aligned} \delta S = 0 = \int d^4x \, \{ & L\partial \cdot \varepsilon(x) + \partial^\mu \delta\Phi^* \partial_\mu \Phi + \partial^\mu \Phi^* \partial_\mu \delta\Phi \\ & - \partial^\mu \varepsilon^\lambda(x) \partial_{(\lambda} \Phi^* \partial_{\mu)} \Phi + \delta L_\Pi \} , \end{aligned} \tag{13.21}$$

with

$$L_\Pi = -\int d^4y \, \Pi(y) \Phi^* \left(x + \frac{1}{2} y \right) \Phi \left(x - \frac{1}{2} y \right) \tag{13.22}$$

and $\delta\Phi = \varepsilon(x)\partial\Phi$. After an integration by parts, the equation for δS yields

$$\begin{aligned} 0 &= \int d^4x \left(-\varepsilon\partial L + \varepsilon^\lambda \partial^\mu [\partial_{(\lambda} \Phi^* \partial_{\mu)} \Phi] + \delta L_\Pi - \delta\Phi^* \partial^2 \Phi - \partial^2 \Phi^* \delta\Phi \right) \\ &= \int d^4x \bigg\{ - \varepsilon\partial L + \varepsilon^\lambda \partial^\mu [\partial_{(\lambda} \Phi^* \partial_{\mu)} \Phi] + \delta L_\Pi \\ &\quad + \int d^4y \, \Pi(y) \left[\delta\Phi^* \Phi(x - y) + \Phi^*(x + y) \delta\Phi \right] \bigg\} , \end{aligned} \tag{13.23}$$

---

[15] See the excellent article by E.L. Hill, *Rev. Mod. Phys.* **23**, 253 (1951).

where use has been made of the equations of motion. This last equation can be rewritten as

$$\begin{aligned}
0 = \int d^4x \Bigg(& -\varepsilon\partial L + \varepsilon^\lambda \partial^\mu [\partial_{(\lambda}\Phi^* \partial_{\mu)}\Phi] + \int d^4y \int_0^{1/2} ds\, \Pi(y) \\
& \times \frac{\partial}{\partial s}\left\{\Phi^*\left[x + y\left(s + \frac{1}{2}\right)\right] \delta\Phi\left[x + y\left(s - \frac{1}{2}\right)\right]\right. \\
& \left.+\ \delta\Phi^*\left[x - y\left(s - \frac{1}{2}\right)\right] \Phi\left[x - y\left(s + \frac{1}{2}\right)\right]\right\}\Bigg) \\
= \int d^4x \Bigg(& -\varepsilon\partial L + \varepsilon^\lambda \partial^\mu [\partial_{(\lambda}\Phi^* \partial_{\mu)}\Phi] + \varepsilon_\lambda \int d^4y \int_0^{1/2} ds\, \Pi(y) \\
& \times y \cdot \partial \left\{\Phi^*\left[x + y\left(s + \frac{1}{2}\right)\right] \partial^\lambda\Phi\left[x + y\left(s - \frac{1}{2}\right)\right]\right. \\
& \left.-\ \partial^\lambda\Phi^*\left[x - y\left(s - \frac{1}{2}\right)\right] \Phi\left[x - y\left(s + \frac{1}{2}\right)\right]\right\}\Bigg)
\end{aligned} \tag{13.24}$$

or

$$0 = \int d^4x\, \varepsilon^\lambda(x) \partial^\mu T_{\mu\lambda}(x), \tag{13.25}$$

where $T^{\mu\nu}$ is the energy–momentum tensor we were looking for; explicitly, it is given by

$$\begin{aligned}
T^{\mu\nu} = {} & \partial^{(\mu}\Phi^* \partial^{\nu)}\Phi - \eta^{\mu\nu} L + \int_0^{1/2} ds \int d^4y\ \Pi(y) y^\mu \\
& \times \left\{\Phi^*\left[x + y\left(s + \frac{1}{2}\right)\right] \partial^\nu\Phi\left[x + y\left(s - \frac{1}{2}\right)\right]\right. \\
& \left.-\ \partial^\nu\Phi^*\left[x - y\left(s - \frac{1}{2}\right)\right] \Phi\left[x - y\left(s + \frac{1}{2}\right)\right]\right\}.
\end{aligned} \tag{13.26}$$

Note also that, as already remarked for the four-current J^μ, this $T^{\mu\nu}$ is similar to the form given by R. Marnelius. From the vanishing of δS for arbitrary $\varepsilon^\lambda(x)$, it is deduced that the energy–momentum flow is conservative: $\partial_\mu T^{\mu\nu} = 0$. This can also be checked directly by using the equations of motion, and the specific form of $T^{\mu\nu}$, as expected for a system invariant under space–time translations. However, it should be borne in mind that this conservation property holds as far as the *first* index is concerned: $\partial_\mu T^{\mu\nu} = 0$, and generally $\partial_\nu T^{\mu\nu} \neq 0$. On the other hand, the energy–momentum tensor is even not symmetric, as is often the case for the *canonical* one.[16]

These last points have to be discussed a little bit further on. Usually, a symmetric and conserved energy–momentum tensor can be obtained by the Belinfante–Rosenfeld method[17]: in the Lagrangian all Lorentz indices are replaced by generally covariant ones, while the Lorentz metric $\eta^{\mu\nu}$ is replaced by a general one, $g^{\mu\nu}$; then $T^{\mu\nu}_{\mathrm{BR}}$ is obtained via the following functional derivation of the action integral then specialized to $g^{\mu\nu} = \eta^{\mu\nu}$:

$$T^{\mu\nu}_{\mathrm{BR}} = \frac{\delta}{\delta g_{\mu\nu}(x)} S\left(\{\Phi^*\},\{\Phi\},\{g_{\alpha\beta}\}\right)|_{g_{\mu\nu}=\eta_{\mu\nu}} . \tag{13.27}$$

This expression is the exact analog of the derivative functional with respect to the electromagnetic field given above for the four-current: gravitation [i.e. $g_{\mu\nu}(x)$] is minimally coupled to matter via the source $T^{\mu\nu}_{\mathrm{BR}}$, itself a conservative and symmetric tensor by construction. Unfortunately, while for fields obeying local equations of motion this last construction method can be worked out more or less easily, in our nonlocal case the situation is technically rather involved and delicate problems of bitensors, transport, etc. show up.[18] Furthermore, at the moment, there does not seem to exist a gravitational analog of the so-called "link operator,"[19]

$$U(x,y) = \exp\left[ie\int_x^y A^\mu(z)dz_\mu\right] \tag{13.28}$$

(used in the calculation of the four-current), in order to get a covariant [under the gauge group U(1), in that case] space–time translation operator acting on Φ.

In fact, the nonsymmetric character of $T^{\mu\nu}$ is by no means surprising since the polarization tensor $\Pi(x)$ *a priori* contains macroscopic tensors that might break the Lorentz (and also the rotational) invariance of the basic system; as a consequence, the angular momentum tensor is in general no longer conserved and hence the usual symmetrization procedure of the energy–momentum tensor no longer applies. Therefore, even though the

---

[16] D.G. Boulware, *Phys. Rev.* **151**, 1024 (1966); C. Itzykson and J.B. Zuber, *Quantum Field Theory* (Mc Graw-Hill, New York, 1980); Ch. W. Misner, K.S. Thorne and J.A. Wheeler, *Gravitation* (Freeman, San Francisco, 1973).

[17] See preceding footnote.

[18] See e.g. S. Dowker, *J. Phys.* **A7**, 1256 (1974).

[19] See e.g. [17]. H.-Th. Elze, M. Gyulassi and D. Vasak, *Nucl. Phys.* **B276**, 706 (1986); D. Vasak, M. Gyulassi and H.-Th. Elze, *Ann. Phys. (N.Y.)* **173**, 462 (1987); U. Heinz, *Ann. Phys. (N.Y.)* **161**, 48 (1985); H.-Th. Elze and U. Heinz, *Phys. Rep.* **183**, 81 (1989).

technical problems mentioned above were solved, the intrinsic lack of symmetry of the energy–momentum tensor would be inconsistent with the symmetric form obtained via the Belinfante–Rosenfeld method. These technical problems also appear, in our case, as reflecting the possible breaking of rotational symmetry.

In order to get further insights, let us now consider the invariance of the system under the infinitesimal rotations

$$\begin{cases} \varepsilon^{\lambda}(x) = \omega^{\lambda\sigma}(x)x_{\sigma}, \\ \omega^{\lambda\sigma}(x) = -\omega^{\sigma\lambda}(x). \end{cases} \tag{13.29}$$

In this particular case of arbitrary infinitesimal variations, the basic relation $\delta S = 0$ yields

$$\begin{aligned} 0 &= \int d^4x\, \omega^{\lambda\sigma}(x)\left\{\partial^{\mu}\left[x_{\sigma}T_{\lambda\mu}(x)\right] - \eta^{\mu}_{\sigma}T_{\lambda\mu}(x)\right\} \\ &= \int d^4x\, \omega^{\lambda\sigma}(x)\{\partial^{\mu}\left[x_{[\sigma}T_{\lambda]\mu}(x)\right] - \eta^{\mu}_{[\sigma}T_{\lambda]\mu}(x)\}, \end{aligned} \tag{13.30}$$

showing that the usual definition of the bosonic angular momentum tensor

$$J^{\mu\nu\lambda}(x) = T^{\mu\nu}x^{\lambda} - T^{\mu\lambda}x^{\nu} \tag{13.31}$$

is still valid; nevertheless, $J^{\mu\nu\lambda}(x)$ is generally *not conservative* and its divergence is

$$\partial_{\mu}J^{\mu\nu\lambda}(x) = T^{[\nu\lambda]}(x), \tag{13.32}$$

where $T^{[\nu\lambda]}(x)$ is the antisymmetric part of the energy–momentum tensor.

A more transparent form of the energy–momentum tensor for discussing symmetries can be obtained and it reads

$$\begin{aligned} T^{\mu\nu}(x) &= \partial^{(\mu}\phi^*(x)\partial^{\nu)}\phi(x) - \eta^{\mu\nu}\{(\partial\phi(x))^2 + \phi^*(x)\Box\phi(x) \\ &\quad + \Box\phi^*(x)\phi(x)\} + \int_{-1/2}^{+1/2} ds \int d^4y\, y^{\mu}\frac{\partial}{\partial y_{\nu}}\Pi(y) \\ &\quad \times \phi^*\left[x + y\left(s + \frac{1}{2}\right)\right]\phi\left[x + y\left(s - \frac{1}{2}\right)\right], \end{aligned} \tag{13.33}$$

from which it follows that whenever $\Pi(y)$ is a function of $y^2 \equiv y^{\lambda}y_{\lambda}$ only, $T^{\mu\nu}(x)$ is symmetric and $J^{\mu\nu\lambda}(x)$ is conservative.

These apparently strange results occur, as already mentioned above, because of the noninvariance of $\Pi(y)$ under the Lorentz group, although the basic dynamics (i.e. before one uses approximations involving quasiparticles) used to describe the system are invariant. This does not mean that the consideration of quasiparticles does not make sense in relativity or even that they should be restricted to fully Poincaré-invariant cases, but only that, being part of an approximation scheme, they bring a number of apparent pathologies when they are considered independently of their physical context: not only are they mutually interacting but they are also interacting with the ground state (i.e. the "vacuum"); in turn, the latter acts upon them more or less as an external force field and the subsequent exchange of energy, momentum and angular momentum has also to be taken into account.

Another important case is found whenever $\Pi(y)$ depends on two variables, namely y^2 and $y \cdot u$, where u^μ is a timelike unit four-vector, as the average four-velocity of the system. Then, there exists a reference frame — precisely the one in which u^μ reduces to $(1, \mathbf{0})$ — where the system is isotropic and hence where the *spatial part*,

$$J^{\mu\alpha\beta}_{\text{spat}} \equiv J^{\mu\alpha\beta}\Delta^\alpha_\rho(u)\Delta^\beta_\sigma(u), \tag{13.34}$$

of $J^{\mu\alpha\beta}(x)$ is conservative: $\partial_\mu J^{\mu\alpha\beta}_{\text{spat}} = 0$.

13.1.3. *A general remark*

The above results were obtained with the help of functional methods which greatly simplified their derivation. However, the same results can also be obtained with more "pedestrian" means. Starting again from the Lagrangian

$$L = \partial_\mu\varphi^*(x)\partial^\mu\varphi(x) - \int d^4y\,\Pi(y)\varphi^*\left(x + \frac{1}{2}y\right)\varphi\left(x - \frac{1}{2}y\right) \tag{13.35}$$

and expanding the fields φ and φ^* into Taylor's series, one gets

$$L = \partial_\mu\varphi(x)\partial^\mu\varphi(x) - \sum_{\ell,n}(-)^n\left(\frac{1}{2}\right)^{\ell+n}$$
$$\times \int d^4y\,\Pi(y)y^{\mu(\ell)+\nu(n)}\partial_{\mu(\ell)}\varphi^*(x)\partial_{\nu(n)}\varphi(x), \tag{13.36}$$

where use has been made of A.O. Barut's notations:[20]

$$\begin{cases} y^{\mu(n)} \equiv y_1^{\mu_1}y_2^{\mu_2}...y_n^{\mu_n}, \\ y^{\mu(n)}\partial_{\mu(n)} \equiv y_1^{\mu_1}\partial_{\mu_1}\,y_2^{\mu_2}\partial_{\mu_2}...y_n^{\mu_n}\partial_{\mu_n}. \end{cases} \tag{13.37}$$

[20] A.O. Barut and G. Mullen, *Ann. Phys. (N.Y.)* **20**, 184 (1962); *ibid.* **20**, 205 (1962); A.O. Barut, *ibid.* **5**, 95 (1958).

It follows that we have to deal with a Lagrangian with derivatives of arbitrary orders. It can be treated as usual and, at the end of the calculation, the complete series can be summed and it provides the above equations of motion. The same is true of the symmetries of the system. The four-current and the energy–momentum tensors are obtained as double series which must first be rearranged before being summed.

As an example, we evaluate the energy–momentum tensor of quasi-photons whose action reads (see Chap. 15)

$$S = \int d^4x \left\{ -\frac{1}{4} F^{\alpha\beta} F_{\beta\alpha} - \frac{\lambda}{2} (\partial A)^2 \right\} + \frac{1}{2} \int d^4y \, A_\mu \left(x + \frac{1}{2}y \right) \Pi^{\mu\nu}(y) A_\nu \left(x - \frac{1}{2}y \right), \tag{13.38}$$

where $\Pi^{\mu\nu}$ is the polarization tensor of the medium and λ is a gauge-fixing parameter. The first term is the usual action for the electromagnetic field which gives rise to the ordinary energy–momentum tensor and we concentrate on the second part only, which we call S_Π.

The infinitesimal space–time traslations

$$\begin{cases} \delta A^\mu(x) = \varepsilon^\alpha(x) \partial_\alpha A^\mu(x), \\ \delta \partial^\nu A^\mu(x) = \partial^\nu \left(\varepsilon^\alpha(x) \partial_\alpha A^\mu(x) \right) \end{cases} \tag{13.39}$$

provide

$$\delta S_\Pi = \frac{1}{2} \int d^4y \left\{ \varepsilon^\alpha \left(x + \frac{1}{2}y \right) \partial_\alpha A_\mu \left(x + \frac{1}{2}y \right) \Pi^{\mu\nu}(y) A_\nu \left(x - \frac{1}{2}y \right) + A_\mu \left(x + \frac{1}{2}y \right) \Pi^{\mu\nu}(y) \, \varepsilon^\alpha \left(x - \frac{1}{2}y \right) \partial_\alpha A_\nu \left(x - \frac{1}{2}y \right) \right\}, \tag{13.40}$$

which, after addition and subtraction of a derivative of the term involving the polarization tensor, yields

$$\begin{aligned} \delta S_\Pi = \int d^4x \Bigg\{ \varepsilon^\alpha(x) \partial_\alpha \mathcal{L}_\Pi + \Bigg\{ \frac{1}{2} \varepsilon^\alpha(x) \int d^4y \, \partial_\alpha A_\mu(x) \Pi^{\mu\nu}(y) A_\nu(x-y) \\ + A_\mu(x+y) \Pi^{\mu\nu}(y) \partial_\alpha A_\nu(x) - \partial_\alpha A_\mu \left(x + \frac{1}{2}y \right) \Pi^{\mu\nu}(y) A_\nu \left(x - \frac{1}{2}y \right) \\ - A_\mu \left(x + \frac{1}{2}y \right) \Pi^{\mu\nu}(y) \partial_\alpha A_\nu \left(x - \frac{1}{2}y \right) \Bigg\} \Bigg\}. \end{aligned} \tag{13.41}$$

The last two terms of this equality represent the term $-\partial_\alpha \mathcal{L}_\Pi$, and they are now expanded in a Taylor series,

$$A_\mu \left(x \pm \frac{1}{2}y \right) = \sum_{n=0}^{+\infty} \frac{(\pm y)^{\lambda(n)}}{2^n n!} \partial_{\lambda(n)} A_\mu(x), \tag{13.42}$$

where use has been made of A.O. Barut's notations. Accordingly, one rewrittes

$$\partial_\alpha \mathcal{L}_\Pi = \frac{1}{2}\int d^4y \sum_{n=1}^{+\infty}\sum_{k=0}^{k=n-1}\left(\frac{y}{2}\right)^{\lambda(n)}\frac{(-1)^k}{n!}\partial_{\lambda(\ell)}$$
$$\times\left[\partial_{\lambda(n-k-1)}\partial_\alpha A_\mu(x)\Pi^{\mu\nu}(y)\partial_{\lambda(k)}A_\nu\left(x-\frac{1}{2}y\right)\right]$$
$$+\partial_\alpha A_\mu(x)\Pi^{\mu\nu}(y)A_\nu\left(x-\frac{1}{2}y\right)$$
$$+\sum_{\ell=1}^{+\infty}\left(\frac{y}{2}\right)^{\lambda(\ell)}\frac{(-1)^\ell}{\ell!}\partial_\alpha A_\mu(x)\Pi^{\mu\nu}(y)\partial_{\lambda(\ell)}A_\nu\left(x-\frac{1}{2}y\right)$$
$$+\sum_{\ell=1}^{+\infty}\sum_{k=0}^{n-1}\left(-\frac{y}{2}\right)^{\lambda(n)}\frac{(-1)^k}{n!}\partial_{\lambda(\ell)}$$
$$\times\left[\partial_{\lambda(k)}\partial_\alpha A_\mu\left(x+\frac{1}{2}y\right)\Pi^{\mu\nu}(y)\partial_{\lambda(n-k-1)}A_\nu(x)\right]$$
$$+A_\mu\left(x+\frac{1}{2}y\right)\Pi^{\mu\nu}(y)\partial_\alpha A_\nu\left(x-\frac{1}{2}y\right)$$
$$+\sum_{\ell=1}^{+\infty}\left(\frac{-y}{2}\right)^{\lambda(\ell)}\frac{(-1)^\ell}{\ell!}\partial_{\lambda(\ell)}A_\mu\left(x+\frac{1}{2}y\right)\Pi^{\mu\nu}(y)\partial_\alpha A_\nu(x)\,. \tag{13.43}$$

Using the following expression for the function $B(m+1,\ell+1)$,

$$B(m+1,\ell+1) = \frac{m!\ell!}{(m+\ell+1)!} = \int_0^1 ds\; s^m(1-s)^\ell \tag{13.44}$$

and changing the indices as

$$\begin{cases} m = n-k-1, \\ l = k \end{cases} \tag{13.45}$$

in the expressions of $T^{\mu\nu}$, where $m = 0,1,2,\ldots$ and $l = 0,1,2,\ldots$, one finally sees that

$$T_\gamma^{\mu\nu} = -\frac{1}{2}\partial^\nu A_\alpha F^{\mu\alpha} - \frac{1}{2}F^{\mu\alpha}\partial^\nu A_\alpha - \frac{\lambda}{2}\partial^\nu A^\mu(\partial\cdot A) - \frac{\lambda}{2}(\partial\cdot A)\partial^\nu A^\mu$$
$$+\frac{1}{2}\int d^4y\; y^\mu\int_0^{1/2} ds\left\{\partial^\nu A_\alpha\left(x-y\left[s-\frac{1}{2}\right]\right)\Pi^{\alpha\beta}(y)\right.$$

$$\times A_\beta\left(x - y\left[s+\frac{1}{2}\right]\right) - A_\alpha\left(x+y\left[s+\frac{1}{2}\right]\right)\Pi^{\alpha\beta}(y)$$
$$\times\, \partial^\nu A_\beta\left(x+y\left[s-\frac{1}{2}\right]\right)\Bigg\} - \eta^{\mu\nu}\mathcal{L}_\gamma, \tag{13.46}$$

where $\mathcal{L}_\gamma$ is the Lagrangian of the quasiparticles (quasiphotons),

$$\mathcal{L}_\gamma = -\frac{1}{4}F^{\mu\nu}F_{\mu\nu} - \frac{\lambda}{2}(\partial\cdot A)^2$$
$$+\frac{1}{2}\int d^4y\, A_\mu\left(x+\frac{1}{2}y\right)\Pi^{\mu\nu}(y)A_\nu\left(x-\frac{1}{2}y\right), \tag{13.47}$$

and the Hermitian character of the polarization tensor implies that

$$\Pi^{\mu\nu +}(y) = \Pi^{\mu\nu}(-y), \tag{13.48}$$

after analysis of the various terms of these expressions.

13.2. Quantum Quasiparticles

In the preceding section, the "classical" quasiparticle fields have been studied and their conservation laws derived; the latter are to be used in what follows. Before one quantizes these fields, it must be recalled that, in actual practice, one starts with a given physical system obeying an involved quantum field theory at finite density and/or temperature; thence approximations of various types (perturbation, one-loop, random phase, etc.) must be performed and they finally lead to equations of motion for the excitations propagating within the system. It is the quantization of these excitations, of these *approximate* "free" fields, which we want to consider in this section. The important point, here, is that our starting point is an approximation and hence it should not be surprising if pathologies appear. For instance, since our equations are nonlocal in essence, problems of causality might be expected. However, it should also be emphasized that, within the framework of a specific approximation, only particular scales (of time, length, energy, etc.) are to be dealt with and, most generally, pathologies appear at smaller scales: when one is dealing with scales of $\sim 10^{-10}$ cm, a possible lack of causality occurring on the scale of 10^{-14} cm should not really be considered as a problem.

In this section, the quantization of the "classical" quasiparticle fields is first performed in a *formal* manner and next discussed in connection with specific problems such as the existence and interpretation of antiquasiparticles. Such a quantization is not very complicated, the more so since only free quasiparticle fields are dealt with here.

A last remark is, however, in order. The quantization performed below parallels the standard quantum field quantization. It has, however, been argued that quasiparticles behave *as if* they would obey *parastatistics*.[21] In such a case, there exist peculiarities of the quantization procedure which should be taken into account.

13.2.1. *Formal quantization*

First, we assume that the Φ vacuum is normal, in the sense that $\langle \text{vac} | \Phi | \text{vac} \rangle = 0$, so that the field Φ can be expanded into *normalized* plane waves. Whenever $\langle \text{vac} | \Phi | \text{vac} \rangle \neq 0$, the vacuum average value of Φ has just to be subtracted out.

The vector space of the solutions of the (linear) quasiparticle field can be attributed to the Hermitian product

$$\begin{aligned}\langle \Phi_1 | \Phi_2 \rangle = \int_\Sigma d\Sigma_\mu \Bigg\{ & \Phi_1^*(x) \overleftrightarrow{\partial}^\mu \Phi_2(x) \\ & + \int_{-1/2}^{+1/2} \int ds\, d^4y\, \Pi(y) y^\mu \Phi_1^* \left[x + y \left(s + \frac{1}{2} \right) \right] \\ & \times \Phi_2 \left[x + y \left(s - \frac{1}{2} \right) \right] \Bigg\}, \end{aligned} \tag{13.49}$$

where Σ is an arbitrary spacelike three-surface and $\Phi_i(x)$ $(i = 1, 2)$ are two solutions to the equations of motion. It is not difficult to see, that the integrant of this last expression is divergent-free and hence that this Hermitian product does not actually depend on Σ (use the equations of motion). It reduces to the usual one (the first term on the right hand side) when $\Pi(y) = -\mu^2 \delta^{(4)}(y)$ (Klein–Gordon equation) and is suggested by the form of the four-current and, as in the latter case, is not definite positive since "charge" can have either sign. It allows the normalization of plane wave solutions to the equations of motion

$$\Phi_k(x) = N_k \exp(-ik \cdot x) \tag{13.50}$$

through the usual condition

$$\langle \Phi_{\mathbf{k}} | \Phi_{\mathbf{k}'} \rangle = \delta^{(3)}(\mathbf{k} - \mathbf{k}'), \tag{13.51}$$

---

[21] See e.g. O.W. Greenberg and A. Messiah, *Phys. Rev.* **138**, 1155 (1965).

and one finally obtains

$$\phi_k(x) = \frac{1}{(2\pi)^{1/2}\left(2\omega_{\mathbf{k}} - \frac{\partial\Pi(k)}{\partial\omega_{\mathbf{k}}}\right)^{1/2}} \exp(-ikx), \tag{13.52}$$

where $\omega_{\mathbf{k}} \equiv k^0$ is a solution to the dispersion equation given at the beginning of the first section. Such a normalization obviously implies that the quantity $2\omega_{\mathbf{k}} - (\partial/\partial\omega_{\mathbf{k}})\Pi(k)$ is positive: this is, however, the case for particles [see the discussion below and A.D. Migdal (1978)] and, presumably, when the medium is thermodynamically stable. This property is discussed later in a subsection. Note that the above normalization was already obtained in another way by Migdal.

In order to be specific, let us consider the case of a real scalar field, the extension to other fields (complex, with internal degrees of freedom, fermion, etc.) being straightforward. It is expanded into normalized plane waves as

$$\phi(x) = \sum_\ell \frac{1}{(2\pi)^{3/2}} \int \frac{d^3k}{\left|\frac{\partial D}{\partial\omega}\right|^{1/2}_{\omega=\omega_\ell(\mathbf{k})}}$$
$$\times \left\{a_\ell(\mathbf{k})\exp[-ik\cdot x] + a_\ell^+(\mathbf{k})\exp[+ik\cdot x]\right\} \tag{13.53}$$

where $D(k) \equiv k^2 - \Pi(k)$ and where the sum over l refers to a sum over all possible modes, solutions to $D(k) = 0$. Of course, when $\Pi(k) = -\mu^2$, one recovers the usual free field expansion; $a_\ell(\mathbf{k})$ and $a_\ell^+(\mathbf{k})$ are the annihilation and creation operators, respectively, of the quasiparticles under consideration and obey the conventional commutation relations

$$\left[a_\ell(\mathbf{k}), a_\ell^+(\mathbf{k}')\right] = \delta_{\ell\ell'}\delta^{(3)}(\mathbf{k}-\mathbf{k}'). \tag{13.54}$$

Covariant annihilation/creation operators $A_{\mathbf{k}}$ and $A_{\mathbf{k}}^+$ are connected to the above $\{\mathrm{a}(\mathbf{k}), \mathrm{a}^+(\mathbf{k})\}$ through

$$a_\ell(\mathbf{k}) = \left[\frac{\partial D}{\partial\omega}\right]^{1/2} A_{\mathbf{k}} \tag{13.55}$$

and obey the commutation relation

$$\left[A_{\mathbf{k}}, A_{\mathbf{k}'}^+\right] = \left[\frac{\partial D}{\partial\omega}\right] \delta^{(3)}(\mathbf{k}-\mathbf{k}'). \tag{13.56}$$

Furthermore, $\phi(x)$ can be rewritten in a manifestly covariant form as

$$\phi(x) = \frac{1}{(2\pi)^{3/2}} \int d^4k\ \delta(D)\left\{A_{\mathbf{k}}\exp[-ik\cdot x] + A_{\mathbf{k}}^+\exp[+ik\cdot x]\right\}. \tag{13.57}$$

After a little algebra and using the basic commutation relations for the creation/annihilation operators, one gets the commutation relations obeyed by $\Phi(x)$, i.e.

$$[\phi(x), \phi(y)] = -\frac{2i}{(2\pi)^3} \int d^4k \; \delta\left[D(k)\right] \sin\left[k \cdot (x-y)\right] \theta(k^0), \tag{13.58}$$

or in the form

$$[\phi(x), \phi(y)] = -\frac{1}{(2\pi)^3} \int d^4k \; \varepsilon(k^0)\delta\left[D(k)\right] \exp\left[ik \cdot (x-y)\right]. \tag{13.59}$$

In this last equation, and in the transition from the noncovariant to the covariant forms, use has been made of the well-known formula

$$\delta\left[D(k)\right] = \sum_\ell \delta\left[\omega - \omega_\ell(\mathbf{k})\right] \left(\left|\frac{\partial D}{\partial \omega}\right|^{1/2}_{\omega=\omega_\ell(\mathbf{k})} \right)^{-1}, \tag{13.60}$$

where $\omega_\ell(\mathbf{k})$ is a simple root of the dispersion equation $D(k) = 0$.

In a vacuum, the field $\Phi(x)$ has fluctuations given by

$$\langle \mathrm{vac}| \, \phi(x)\phi(y) \, |\mathrm{vac}\rangle = -\frac{2i}{(2\pi)^3} \int d^4k \; \delta\left[D(k)\right] \exp\left[ik \cdot (x-y)\right] \tag{13.61}$$

so that, at point $x = y$, they are

$$\langle \mathrm{vac}| \, \phi^2(x) \, |\mathrm{vac}\rangle = -\frac{2i}{(2\pi)^3} \sum_\ell \int \frac{d^3k}{\left|\frac{\partial D}{\partial \omega}\right|^{1/2}_{\omega=\omega_\ell(\mathbf{k})}}. \tag{13.62}$$

This last quantity is certainly infinite: the propagating modes are likely to behave as $\omega \approx \mathbf{k}$ for large $\mathbf{k}$'s (this is the case in known relativistic systems such as QED or QCD plasmas), so that this last integral would diverge. Neveretheless, there exists a counterexample with the longitudinal modes of a QED plasma. Although this question is interesting in itself, it should be borne in mind that, for $(\omega, \mathbf{k}) \to \infty$, the notion of the quasiparticle can be questioned. On the other hand, tachyonic modes could lead to finite fluctuations of the quasiparticle vacuum and also to finite renormalizations, although they are expected to possess very short lifetimes.

From the expansion of the field $\Phi(x)$ in plane waves and the expression of the energy–momentum tensor, one immediately obtains the Hamiltonian of the quasiparticles as

$$\begin{aligned} H &= \int_{t=\mathrm{const}} d^3x \; T^{00} \\ &= \sum_\ell \int_{t=\mathrm{const}} \frac{d^3k}{(2\pi)^3} \omega_\ell(\mathbf{k}) \left[a_\ell^+(\mathbf{k}) a_\ell(\mathbf{k}) + \frac{1}{2}\right], \end{aligned} \tag{13.63}$$

as it should be in the usual cases (see below, however). Of course, from this Hamiltonian and Schwinger's action principle, one could recover the assumed commutation relations of the creation/destruction operators given above.

Unfortunately, actual physical situations are not so simple and they lead to difficult problems, which we now discuss.

13.3. Problems with the Quantization of Quasiparticles

13.3.1. *A first example*

In order to be specific and yet remain at a level where calculations do not hide the problems, the following equation of motion is first chosen:

$$\left(\Box + m_1^2\right)\left(\Box + m_2^2\right)\Phi(x) = 0, \tag{13.64}$$

where $\Phi(x)$ is still a real scalar field. Equations of that form, or more general ones, have been discussed a few decades ago in connection with the hope of resolving the question of the divergences of quantum electrodynamics.[22]

This particular equation of motion corresponds to a "polarization tensor" Π(k) given by

$$\Pi(k) = \frac{k^4 + m_1^2 m_2^2}{m_1^2 + m_2^2} \tag{13.65}$$

and to the dispersion equation

$$D(k) = (k^2 - m_1^2)(k^2 - m_2^2) = 0, \tag{13.66}$$

so that the quantity

$$2\omega_{\mathbf{k}} - \left.\frac{\partial \Pi}{\partial \omega_k}\right|_{\omega_\ell = \omega_{\mathbf{k}}} = 2\omega\left[\frac{m_1^2 + m_2^2 - 2k^2}{m_1^2 + m_2^2}\right] \tag{13.67}$$

is positive for one root of the dispersion equation and negative for the other. It follows immediately that the plane wave normalization is no longer possible and must be replaced by

$$\langle \Phi_{\mathbf{k}} | \Phi_{\mathbf{k}'} \rangle = \pm \delta^{(3)}(\mathbf{k} - \mathbf{k}'); \tag{13.68}$$

[22]See e.g. B.T. Darling, *Phys. Rev.* **92**, 1547 (1953); P.T. Mathews, *Proc. Cambridge Philos. Soc.* **45**, 441 (1949); W. Pauli and F. Villars, *Rev. Mod. Phys.* **21**, 434 (1949); W. Pauli, *Rev. Mod. Phys.* **15**, 175 (1943); R.J.N. Phillips, *Nuovo Cimento* **1**, 823 (1955); L.K. Pandit, *Suppl. Nuovo Cimento* **11**, 157 (1959); K.L. Nagy, *Suppl. Nuovo Cimento* **17**, 92 (1960); T.W.B. Kibble and J.C. Polkinghorne, *Nuovo Cimento* **8**, 74 (1958); H.M. Fried and J. Plebanski, *Nuovo Cimento* **18**, 884 (1960); A.O. Barut and G.H. Mullen, *Ann. Phys. (N.Y.)* **20**, 184 (1962); *ibid.* **20**, 203 (1962); A. Pais and G.E. Uhlenbeck, *Phys. Rev.* **79**, 145 (1950); E.C.G. Sudarshan, *Phys. Rev.* **123**, 2183 (1961).

in other words, the Hilbert space of the solutions to the equations of motion is now endowed with an *indefinite metric*. In this last normalization relation, ±s refers to the sign of

$$2\omega_{\mathbf{k}} - \left.\frac{\partial \Pi}{\partial \omega_k}\right|_{\omega_\ell=\omega_{\mathbf{k}}} \equiv \left.\frac{\partial}{\partial \omega}\left[\omega^2 - \Pi(k)\right]\right|_{\omega=\omega_\ell(\mathbf{k})} = \left.\frac{\partial D(k)}{\partial \omega}\right|_{\omega=\omega_\ell(\mathbf{k})}. \quad (13.69)$$

With such an indefinite metric, the Hamiltonian operator now reads

$$H = \sum_\ell \int \frac{d^3k}{(2\pi)^3}\,\mathrm{sgn}\left(\left.\frac{\partial D(k)}{\partial \omega}\right|_{\omega=\omega_\ell(\mathbf{k})}\right)\omega_\ell(\mathbf{k}) a_\ell^+(\mathbf{k}) a_\ell(\mathbf{k}), \quad (13.70)$$

where the creation/destruction operators now obey the commutation relations

$$\left[a_\ell(\mathbf{k}), a_\ell^+(\mathbf{k})\right] = \mathrm{sgn}\left(\left.\frac{\partial D(k)}{\partial \omega}\right|_{\omega=\omega_\ell(\mathbf{k})}\right)\delta_{\ell\ell'}\delta^{(3)}(\mathbf{k}-\mathbf{k}') \quad (13.71)$$

and where sgn is the sign function. In fact, rather than using a plane wave expansion of $\Phi(x)$ in the above form, it is preferable to write it as[23]

$$\Phi(x) = \int \frac{d^3k}{(2\pi)^{3/2}} \times \left\{\frac{a_\ell(\mathbf{k})\exp(-ik\cdot x)}{\left|\frac{\partial D}{\partial \omega}\right|^{1/2}_{\omega=\omega_1(\mathbf{k})}} + \frac{b_\ell^+(\mathbf{k})\exp(+ik\cdot x)}{\left|\frac{\partial D}{\partial \omega}\right|^{1/2}_{\omega=\omega_2(\mathbf{k})}}\right\}, \quad (13.72)$$

so that the commutation relations now reads

$$\begin{aligned} [a(\mathbf{k}), a^+(\mathbf{k}')] &= \delta^{(3)}(\mathbf{k}-\mathbf{k}'), \\ [b(\mathbf{k}), b^+(\mathbf{k}')] &= -\delta^{(3)}(\mathbf{k}-\mathbf{k}'), \end{aligned} \quad (13.73)$$

where the a's correspond to the plus sign of ε and the b's to the minus sign. The Hamiltonian then reads

$$H = \int \frac{d^3k}{(2\pi)^3}\left\{\sqrt{\mathbf{k}^2 + m_1^2}\,a^+(\mathbf{k})a(\mathbf{k}) - \sqrt{\mathbf{k}^2 + m_2^2}\,b^+(\mathbf{k})b(\mathbf{k})\right\}, \quad (13.74)$$

assuming that $k^2 = m_1^2$ is the positive energy mode or, equivalently, that $m_1^2 < m_2^2$. This expression shows clearly that the Hamiltonian operator is not positive definite, with the disastrous consequence that there no longer exists a vacuum state defined as being a minimum energy state.

Many solutions to these drawbacks have been proposed and discussed in the literature: transformation of the real field Φ into a purely imaginary one; introduction of supplementary conditions; projection onto positive energy

[23] In order to simplify the notation, we assumed only one positive energy and only one negative energy mode, which is the case of our example. Of course, the various other commutation relations are vanishing as $[a, b] = 0$, etc.

states regarded as sole physical states; etc. Unfortunately, none of them can be considered as being really satisfactory.

Moreover, one has to distinguish between theories supposed to be fundamental and phenomenological ones, describing partly a given physical system in a definite class of states. While in the first case, known properties, such as causality or Lorentz invariance, have to be imperatively obeyed, in the second one, these basic properties might be relaxed at given scales. In order to gain some further insights, several other examples will now be discussed.

13.3.2. *Another example: the QED plasma*

A more physical situation can be found in the case of a QED plasma, and even in the case of a nonquantum and/or a nonrelativistic one. For such systems (ions plus electrons plus modes or radiation) the existence of what is known as "negative energy modes" has been recognized for a long time[24]: these modes are precisely those which correspond to

$$2\omega_{\mathbf{k}} - \frac{\partial \Pi(k)}{\partial \omega} < 0. \tag{13.75}$$

Usually they are interpreted via the sign of the imaginary part of the frequency of an eigenmode of the plasma

$$\gamma = - \left. \frac{\mathrm{Im}\Pi(\omega, k)}{2\omega_{\mathbf{k}} - \frac{\partial \mathrm{Re}\Pi(k)}{\partial \omega}} \right|_{\omega = \omega_\ell(\mathbf{k})}, \tag{13.76}$$

where $|\gamma| \ll \omega$ and $\omega_\ell(\mathbf{k})$ is a solution to the dispersion equation. In this expression for γ, Π is to be understood as being either the transverse or the longitudinal part of the polarization tensor. It can also be shown to represent essentially the ratio of the energy dissipated by the propagating wave and of its *total* energy[25]; and it can be rewritten as

$$\gamma = - \left. \frac{\mathrm{Re}\mathbf{E}^* \cdot \mathbf{J}}{\frac{\mathbf{E}^2}{4\pi} \frac{\partial}{\partial \omega} (\omega \varepsilon_L)} \right|_{\omega = \omega_L(\mathbf{k})}, \tag{13.77}$$

where $\mathbf{E}$ is the electric field, $\mathbf{J}$ the three-current and ε_L the longitudinal dielectric function. For transverse waves there also exists a quite similar expression. The negative energy modes then appear when the signs of these

---

[24] T. Kihara, O. Aono and T. Dodo, *Nucl. Fus.* **2**, 66 (1962); A.I. Alexseev and Yu. P. Nikitin, *Sov. Phys. JETP* **23**, 608 (1966); E.G. Harris, *Adv. Plas. Phys.* **3**, 157 (1969).
[25] G. Bekefi, *Radiation Processes in Plasmas* (Wiley, New York, 1966).

two energy differ: when the total energy is positive and the dissipated energy is negative, the particles of the plasma provide energy to the wave so that the latter is amplified; the opposite case leads to a similar conclusion.[2] Detailed features of these modes, in connection with mode–mode interactions, have been discussed by V.N. Tsytovich.[26]

A first conclusion of this short analysis is that when one considers modes, whether quantized as quasiparticles or not, it is essential to take the energy (and possibly quantum numbers) exchanges with the basic system or, in other words, with a prescribed vacuum supposed to provide a satisfactory description of it. This shows that when one quantizes the eigenmodes of a system, not only negative energies may appear but also they are *as physical as positive ones*. Unlike the case of a system obeying a fourth order equation (or a higher order one) like in the example briefly studied in the preceding section, where there is no obvious way to treat the negative energy states (which lead to a basic instability of the system since, in such a case, there is no vacuum in the theory), the case of a plasma (whether classical or not, whether relativistic or not) is quite different: in a real physical system, there always exists a ground state; this ground state is, of course, the lowest possible energy (or free energy) state and the negative energy of the modes is physically bound from below.

To these remarks, it should be added that for a nonquantum and nonrelativistic plasma, a necessary and sufficient condition for the existence of negative energy modes, has been given by several authors.[27] It is based on the analysis of the second variation of the free energy of "equilibrium" (i.e. steady states) distributions, whose sign determines its linear stability. Unfortunately, such an analysis has not yet been performed in the relativistic quantum case.

Finally, one should conclude that these strange modes have to be dealt with and must not be eliminated *a priori*. Furthermore, a full discussion on the stability of the original system (considered in the approximation at hand) is certainly required.

13.3.3. *Migdal's approach*

In a general review of pion condensation, A.D. Migdal (1978) gave a few basic tools for the quantization of quasiparticle fields. With our notations,

---

[26]V.N. Tsytovich, *Non-linear Effects in Plasma* (Plenum, New York, 1970).

[27]H. Weitzner and D. Pfirsch, *Phys. Rev.* **A43**, 4532 (1991); P. J. Morrison and D. Pfirsch, *Phys. Rev.* **A40**, 3898 (1989); *Phys. Fluids* **B2**, 1105 (1990).

for a complex scalar field $\Phi(x)$, he wrote (we limit ourselves to the case of only one mode, for brevity)

$$\Phi(x) = \int \frac{d^3k}{(2\pi)^{3/2}} \times \left\{ \frac{a(\mathbf{k}) \exp(-ik^+ \cdot x)}{\left|\frac{\partial D}{\partial \omega}\right|^{1/2}_{\omega=\omega^+(\mathbf{k})}} + \frac{b^+(\mathbf{k}) \exp(+ik^- \cdot x)}{\left|\frac{\partial D}{\partial \omega}\right|^{1/2}_{\omega=\omega^-(\mathbf{k})}} \right\}, \tag{13.78}$$

where the + and − signs refer to the positive and the negative energy mode, respectively. Accordingly, the corresponding Hamiltonian reads

$$H = \int \frac{d^3k}{(2\pi)^3} k^0 \operatorname{sgn} \left(\frac{\partial D(k)}{\partial \omega} \bigg|_{\omega=\omega(\mathbf{k})} \right) \left[a^+(\mathbf{k}) a(\mathbf{k}) - b^+(\mathbf{k}) b(\mathbf{k}) \right], \tag{13.79}$$

where, for this discussion, an irrelevant vacuum term has been dropped and $\varepsilon(x)$ is still the sign function. Furthermore, the factor $k^0\text{sgn}(\ldots)$ should be understood as being taken for $k^0 = w^+(\mathbf{k})$ when acting on the term $a^+(\mathbf{k})a(\mathbf{k})$, and as being taken for $k^0 = -\omega^-(k)$ when acting on $b^+(\mathbf{k})b(\mathbf{k})$.

For a normal stable vacuum, whose quantum numbers are essentially zero, one has $D(k) = D(-k)$ and hence $\partial D/\partial k^0$, as a function of k^0, is odd. Consequently, the quantity $k^0\text{sgn}(\partial D/\partial k^0)$ is always positive and the Hamiltonian H is definite positive. This is the case when, for instance, one deals with symmetric nuclear matter: the vacuum does not carry any nonvanishing quantum number (for instance, the equilibrium state of the system has a vanishing isotopic spin) and the quasiparticle spectrum is symmetric with respect to the change $\mathbf{k} \to -\mathbf{k}$ [see A.D. Migdal (1978) for some examples]. Unlike symmetric nuclear matter, neutron matter gives rise to a "vacuum" state carrying isospin and the relation $D(k) = D(-k)$ is no longer valid. It follows that the change of sign of k^0 and of $\partial D/\partial k^0$ generally do not occur at the same time. In this last case, the Hamiltonian is nondefinite positive: there is no vacuum and the system *as quantized*, as above, *is unstable*. Several examples of such spectra are provided by A. D. Migdal (1978).

This brief discussion shows that Migdal's approach is not always adequate. Moreover, it is easy to see that when $\partial D/\partial k^0$ changes its sign for $k^0 \neq 0$, the plane wave expansion of the "quasifield" does not provide a complete basis in the Hilbert space of the solutions to the equations of motion since Migdal suppresses two terms in the plane wave expansion of the field.

Although this important problem of negative energy modes cannot be considered as being solved, an alternative, but not unique, interpretation is briefly outlined and discussed in the next section. Note also that a quantization of quasiparticle fields according to parastatistics would perhaps provide an interesting way out.

13.4. The Covariant Wigner Function

In this section and in the following, a stable "vacuum" or equilibrium state is assumed and hence no negative energy modes can be excited within the medium. When these conditions are realized, it is not very difficult to obtain the basic statistical properties of those quasiparticles propagating in the system, whether in thermal equilibrium or not. In this section we still limit ourselves to the case of a complex scalar field without any internal symmetry, the extension to this case (or to other Bose fields) being straightforward.

When quasiparticles do interact, or when a number of problems such as the calculation of the fluctuations of various physical quantities are dealt with, the use of the covariant Wigner function presents a certain interest. As before, it is defined as the average value of the Wigner operator

$$f_{\rm op}(x,p) = \frac{1}{(2\pi)^4}\int d^4R \exp(-ip\cdot R)\,\tilde{\phi}^*\left(x+\frac{1}{2}R\right)\tilde{\phi}\left(x-\frac{1}{2}R\right), \tag{13.80}$$

where, as mentioned in Chaps. 8 and 9, the average value of the field has been subtracted:

$$\tilde{\phi}(x) \equiv \phi(x) - \langle\phi(x)\rangle. \tag{13.81}$$

Such a definition should generally be preferred, in contrast with those given by other authors, since quasiparticles are associated with the field *fluctuations* above a "vacuum."

Using now the equations of motion obeyed by the fields $\phi^*(x)$ and $\phi(x)$, a straightforward calculation similar to those already used[28] in preceding chapters leads to the transport equation

$$\begin{aligned} p\cdot\partial f(x,p) &- \frac{1}{2}\int d^4p\, d^4y\, \Pi(y)\exp(ip\cdot y) \\ &\times\left[f\left(x+\frac{1}{2}y,p\right) - f\left(x-\frac{1}{2}y,p\right)\right] = 0, \end{aligned} \tag{13.82}$$

[28]See also P. Carruthers and F. Zachariasen (1976); F. Cooper and D. Sharp (1975); F. Cooper and M. Feigenbaum (1976); P. Carruthers and F. Zachariasen (1983).

valid for *free* quasiparticles only. This equation is the analog of the relativistic Liouville equation for a classical relativistic system of free particles. In fact, there exists another equation obeyed by $f(x,p)$ which plays the role of a mass shell: this property is more obvious in Fourier space and hence it will be shown below. This "Liouville equation" is a basic equation when one is dealing with transport properties of the system; in such a case it must be supplemented by a collision term $C(f)$ which might be of the Uhlenbeck–Uehling form or another phenomenological nature. As a matter of fact, as has been emphasized several times, the most reasonable one is probably a relaxation time approximation where

$$C(f) = -u^{\mu}\left(p_{\mu} - \frac{\partial}{\partial p^{\mu}}\Pi(p)\right)\frac{f - f_{\text{eq}}}{\tau}. \tag{13.83}$$

In this collision term, τ is the relaxation time and the factor

$$p_{\mu} - \frac{\partial}{\partial p^{\mu}}\Pi(p) \tag{13.84}$$

has been chosen so that the Landau matching conditions

$$u_{\mu}J^{\mu}_{\text{off}} = 0, \quad u_{\mu}T^{\mu\nu}_{\text{off}} = 0 \tag{13.85}$$

are satisfied. This can be checked by integrating the two sides of the consequent kinetic equation for the "small" *off*-equilibrium part of the Wigner function.

In terms of the above Wigner function, the four-current operator $J^{\mu}_{\text{op}}(x)$ can be written as

$$\begin{aligned} J^{\mu}_{\text{op}}(x) = \int d^4p\, d^4y\, p^{\mu} f_{\text{op}}(x,p) + \frac{1}{2}\int d^4y \int_{-1/2}^{+1/2} ds \\ \times \int d^4\xi \Pi(y)\exp(i\xi\cdot y) f_{\text{op}}(x+ys,\xi), \end{aligned} \tag{13.86}$$

while the energy–momentum tensor operator $T^{\mu\nu}_{\text{op}}(x)$ reads

$$\begin{aligned} T^{\mu\nu}_{\text{op}}(x) = \int d^4p \Bigg(\left\{2p^{\mu}p^{\nu} - \eta^{\mu\nu}\left[p^2 - \frac{1}{4}\partial^2 - \Pi(p)\right]\right\} f_{\text{op}}(x,p) \\ - i\int_0^{1/2} ds \int d^4y \exp(ip\cdot y)\Pi(y)y^{\mu} \\ \times \left[\left(p^{\nu} + \frac{i}{2}\partial^{\nu}\right) f(x+ys,p) + \left(p^{\nu} - \frac{i}{2}\partial^{\nu}\right) f(x-ys,p)\right]\Bigg), \end{aligned} \tag{13.87}$$

an unpleasant expression[29] quite different from what was assumed in different relativistic versions[30] of the Landau theory of normal Fermi liquids.

Let us now go over to the Fourier space expressions, since much can be learned from them. We still denote by the same symbol a function and its Fourier transform, the variables k and x being sufficient to make the distinction. First, the Fourier transform of $f(x,p)$ reads

$$f(k,p) = \frac{1}{(2\pi)^4}\left\langle \tilde{\phi}^*\left(p-\frac{1}{2}k\right)\tilde{\phi}\left(p+\frac{1}{2}k\right)\right\rangle, \tag{13.88}$$

so that the equation of motion for $\phi(k)$,

$$\left[k^2 - \Pi(k)\right]\phi(k) = 0, \tag{13.89}$$

yields

$$\left[\left(p \pm \frac{1}{2}k\right)^2 - \Pi\left(p \pm \frac{1}{2}k\right)\right] f(k,p) = 0, \tag{13.90}$$

which finally gives rise to the equations for $f(k,p)$

$$\begin{cases} \left\{p\cdot k - \frac{1}{2}\left[\Pi\left(p+\frac{1}{2}k\right) - \Pi\left(p-\frac{1}{2}k\right)\right]\right\} f(k,p) = 0, \\ \left\{\left(p^2 + \frac{1}{4}k^2\right) - \frac{1}{2}\left[\Pi\left(p+\frac{1}{2}k\right) + \Pi\left(p-\frac{1}{2}k\right)\right]\right\} f(k,p) = 0, \end{cases} \tag{13.91}$$

obtained by subtracting or adding the preceding equations. The first of these last equations is nothing but the Fourier transform of the above "Liouville equation." In the long wavelength and low frequency limit $k \approx 0$, it reads

$$k \cdot v f(k,p) = 0 \tag{13.92}$$

where v^μ can be considered as the four-velocity of the (free) quasiparticles. The second equation is directly connected with the "mass shell" of the quasiparticles and this can be seen in the case of a stationary and homogeneous state — like an equilibrium state — since, in such a case, one has

$$f(k,p) = \delta^{(4)}(k) f(0,p) \equiv \delta^{(4)}(k) f(p), \tag{13.93}$$

[29] An irrelevant divergence-free term has been omitted in the expression of the energy–momentum tensor.

[30] G. Baym and S.A. Chin (1976); T. Matsui (1981); Ch. G. van Weert and M.C.J. Leermakers (1984); M.C.J. Leermakers and Ch. G. van Weert (1984); Ch. G. van Weert and M.C.J. Leermakers (1985a,b,c); M.C.J. Leermakers Ch. G. van Weert (1986); M.C.J. Leermakers, Ch. G. van Weert and A.M.J. Schakel (1986).

so that the second equation reduces to

$$[p^2 - \Pi(p)]f(p) = 0, \tag{13.94}$$

whose solution has the general form

$$f(p) = \delta\left[p^2 - \Pi(p)\right]\chi(p), \tag{13.95}$$

where $\chi(p)$ is an arbitrary function.

In Fourier space, the Lagrangian reads

$$L(k) = \int d^4p \left[p^2 - \frac{1}{4}k^2 - \Pi(k)\right] f(k,p), \tag{13.96}$$

so that the total action I is

$$I = L\,(k = 0)\,. \tag{13.97}$$

The average four-current $J^\mu(k)$ (see App. E for useful formulae) is easily found to be

$$\langle J^\mu(k)\rangle = 2\int d^4p\; p^\mu f(k,p) - \int_{-1/2}^{+1/2} ds \int d^4p\; p^\mu f(k,p)\frac{\partial}{\partial p_\mu}\Pi(k+ps), \tag{13.98}$$

while the energy–momentum tensor is given by

$$\langle T^{\mu\nu}(k)\rangle = 2\int d^4p \left[p^\mu p^\nu - \frac{1}{4}k^\mu k^\nu\right] f(k,p) - a^{\mu\nu}(k) - \eta^{\mu\nu}L(k), \tag{13.99}$$

with

$$\begin{aligned} a^{\mu\nu}(k) = \int_0^{1/2} ds \int d^4p\; p^\mu f(k,p) \Big\{ & p^\nu \frac{\partial}{\partial p_\mu}\left[\Pi(p+ks) + \Pi(p-ks)\right] \\ & - \frac{1}{2}k^\nu \frac{\partial}{\partial p_\mu}\left[\Pi(p+ks) + \Pi(p-ks)\right]\Big\}. \end{aligned} \tag{13.100}$$

It is obviously not symmetric as expected; however, when one deals with an isotropic equilibrium system, it is symmetric, as can easily be shown. Note the general form below for $\langle T^{\mu\nu}(k)\rangle$ in thermal equilibrium.

13.5. Equilibrium Properties

From the knowledge of the quasiparticle Hamiltonian and of the "charge" (i.e. the conserved property whose four-current is given above)

$$Q = \int d^3x\; J^0(x), \tag{13.101}$$

the equilibrium density operator is given, as usual, by

$$\rho = \frac{1}{Z}\exp\left(-\beta\left[H - \mu Q\right]\right), \tag{13.102}$$

with

$$Z = \mathrm{Tr}\left\{\exp\left(-\beta\left[H - \mu Q\right]\right)\right\}. \tag{13.103}$$

In manifestly covariant form ρ can be rewritten as

$$\rho = \frac{1}{Z}\exp\left(-\beta_\nu \int_\Sigma d\Sigma_\lambda \left[T^{\nu\lambda} - \mu J^\lambda\right]\right), \tag{13.104}$$

where μ is the chemical potential associated with the conserved charge Q and β_ν is the average velocity four-vector of the system times the inverse temperature β (in the rest frame of the system); this last equation reduces to the preceding one.

Since either expression for the density operator has the same structure as usual,[31] this leads to quite similar expressions for the physical quantities such as the average charge density, energy–momentum tensor (or energy density ρ and pressure P), entropy four-current, or average occupation number.

In manifestly covariant form these latter quantities respectively read

$$n(k) = \frac{1}{\exp\left(\beta_\mu k^\mu - \beta\mu\right) - 1} \tag{13.105}$$

for the average occupation number,

$$J^\mu = \frac{1}{(2\pi)^3}\int d^4k\,\mathrm{sgn}(k^0)n(k)\delta\left[D(k)\right]\frac{\partial D(k)}{\partial k_\mu} \tag{13.106}$$

for the "charge" four-current, and

$$\begin{aligned} S^\mu = -k_B\frac{1}{(2\pi)^3}&\int d^4k\,\mathrm{sgn}(k^0)n(k)\delta\left[D(k)\right]\frac{\partial D(k)}{\partial k_\mu} \\ &\times\left\{n(k)\log[n(k)] - [n(k)+1]\log[n(k)+1\right\} \end{aligned} \tag{13.107}$$

for the entropy four-current. When the system does not involve any other four-vector than u^μ, $\langle T^{\mu\nu}\rangle$ has the general form

$$\langle T^{\mu\nu}\rangle = (\rho + P)\,u^\mu u^\nu - P\eta^{\mu\nu}, \tag{13.108}$$

[31] See e.g. K. Huang, *Statistical Mechanics* (Wiley, New York, 1963).

with

$$\rho = \frac{1}{(2\pi)^3} \int d^4k \; \varepsilon(k^0)\delta\left(D(k)\right) k_\nu \frac{\partial D(k)}{\partial k^\mu} u^\mu u^\nu n(k) \quad \text{(energy density)}, \tag{13.109}$$

$$P = \frac{1}{3\,(2\pi)^3} \int d^4k \; \varepsilon(k^0)\delta\left(D(k)\right) k_\nu \frac{\partial D(k)}{\partial k^\mu} \Delta^{\mu\nu}(u) n(k) \quad \text{(pressure)}. \tag{13.110}$$

In more usual notations, i.e. in a Lorentz frame where $u^\mu = (1, \mathbf{0})$, one has

$$n_{\text{eq}} \equiv \frac{\langle Q \rangle}{V} = J^0 = \sum_{\ell,\pm} \frac{1}{(2\pi)^3} \int d^3k \frac{\pm 1}{\exp\left[\beta\left(\omega_\ell(k) \mp \mu\right)\right] - 1}$$

$$\rho = \sum_{\ell,\pm} \frac{1}{(2\pi)^3} \int d^3k \frac{\omega_\ell(k)}{\exp\left[\beta\left(\omega_\ell(k) \mp \mu\right)\right] - 1}$$

$$P = \frac{1}{3} \sum_{\ell,\pm} \frac{1}{(2\pi)^3} \int \frac{d^3k}{\left.\frac{\partial D(k)}{\partial \mathbf{k}}\right|_{\omega=\omega_\ell(\mathbf{k})}} \left. \mathbf{k} \frac{\partial D(k)}{\partial \mathbf{k}} \right|_{\omega=\omega_\ell(\mathbf{k})}$$

$$\times \frac{\omega_\ell(k)}{\exp\left[\beta\left(\omega_\ell(k) \mp \mu\right)\right] - 1}. \tag{13.111}$$

The only differences from the usual case[32] are (i) the summation over the various modes, denoted by $|$, and over the "anti"-quasiparticles (summation over $\pm$), and (ii) the expression for the pressure, which, in the case where $k^2 = m^2$, reduces to the known expression.

These equations indicate that the role of the four-velocity of the quasiparticles is played by the quantity

$$v^\mu(k) = \frac{1}{2} \frac{\partial D(k)}{\partial k_\mu}, \tag{13.112}$$

while their four-momentum is k^μ. It follows from this equation that the three-velocity $\mathbf{w}$ of a quasiparticle excited in the ℓth mode is given by

$$\mathbf{w} = \left. \frac{\partial D(k)}{\partial \mathbf{k}} \times \left(\frac{\partial D(k)}{\partial \omega} \right)^{-1} \right|_{\omega=\omega_\ell(\mathbf{k})} \tag{13.113}$$

so that the pressure P retains its customary form:

$$P = \frac{1}{3} \langle \mathbf{w} \cdot \mathbf{k} \rangle. \tag{13.114}$$

[32]See e.g. J. Ehlers (1971).

This is in contrast with an earlier incorrect result[33] but in accordance with Ch. G. Van Weert and M.C.J. Leemakers (1984) when one realizes that their spectral function $A(k)$ is related to our $D(k)$ through $A(k) = \theta[D(k)]$, where θ is the Heaviside step function. Note also that the above definition of $\mathbf{w}$ is identical to the definition $\mathbf{w} = \mathbf{v}/v^0$.

Other derived thermodynamic quantities can be obtained via the customary relations. For instance, the heat capacity of the quasiparticles is given by

$$C_V = \frac{\partial \rho}{\partial T}; \tag{13.115}$$

it should be remembered that an expression like this contains the temperature not only through the average occupation number parameter β but it also occurs implicitly *in* the mode $\omega_\ell(\mathbf{k})$. One has to bear in mind too that this is only the contribution of the quasiparticles and that the medium must be taken into account. This is briefly developed below.

13.6. A Simple Example: The $\lambda\phi^4$ Model

The previous concepts are now illustrated in the case of a real scalar field obeying equations of motion derived from the Lagrangian

$$L = \frac{1}{2}\left(\partial\phi\right)^2 - \frac{1}{2!}\mu_0^2\phi^2 - \frac{1}{4!}\lambda_0\phi^4. \tag{13.116}$$

This simple model is, of course, not intended to represent an actual physical situation whatsoever, even though scalar fields play an important role in particle physics via the Higgs mechanism and, consequently, in the primordial universe.[34]

From the equation of motion

$$\left(\Box + \mu_0^2\right)\phi + \frac{1}{3!}\lambda_0\phi^3 = 0 \tag{13.117}$$

and the definition of the Wigner operator, after an average one obtains the two equations

$$2ip\cdot\partial f(x,p) - \frac{\lambda_0}{3!}\int \frac{d^4R}{(2\pi)^4}\exp(-ip\cdot R)\left\langle \phi\left(x+\frac{1}{2}R\right)\phi^3\left(x-\frac{1}{2}R\right)\right.$$
$$\left. - \phi\left(x+\frac{1}{2}R\right)\phi^3\left(x-\frac{1}{2}R\right)\right\rangle = 0, \tag{13.118}$$

[33]R. Hakim (1978).

[34]See e.g. G.W. Gibbons, S.W. Hawking and S.T.C. Siklos, *The Very Early Universe* (Cambridge University Press, 1983).

$$2\left[p^2 - \Box - \mu_0^2\right] f(x,p) - \frac{\lambda_0}{3!}\int \frac{d^4R}{(2\pi)^4}\exp(-ip\cdot R)\left\langle \phi\left(x+\frac{1}{2}R\right)\right.$$
$$\left.\times\,\phi^3\left(x-\frac{1}{2}R\right) + \phi\left(x+\frac{1}{2}R\right)\phi^3\left(x-\frac{1}{2}R\right)\right\rangle = 0, \qquad (13.119)$$

where, for simplicity, we have assumed — but this is not an essential restriction — that $\langle\phi\rangle = 0$. The integral terms of these equations could be expressed in terms of the two-particle covariant Wigner function; however, this is not necessary, since they will be expressed here in terms of $f(x,p)$ with the help of the Gauss (Hartree–Vlasov pairing) approximation. This latter is expressed by

$$\langle\phi(1)\,\phi(2)\,\phi(3)\,\phi(4)\rangle \approx \sum_{\text{all pairs}} \langle\phi(i)\,\phi(j)\rangle\langle\phi(k)\,\phi(l)\rangle, \qquad (13.120)$$

where the labels $1, 2, \ldots, i, \ldots$ indicate space–time points. Inserted into the above equation, the Gaussian approximation leads to

$$\begin{cases} ip\cdot\partial f(x,p) - \dfrac{\lambda_0}{4}\displaystyle\int \frac{d^4R}{(2\pi)^4} d^4p'\,d^4p''\exp(-ip\cdot R)\,f(x,p') \\ \quad\times\left[f\left(x+\dfrac{1}{2}R,p''\right) - f\left(x-\dfrac{1}{2}R,p''\right)\right] = 0, \\ \left[p^2 - \dfrac{1}{4}\Box - \mu_0^2\right] f(x,p) - \dfrac{\lambda_0}{4}\displaystyle\int \frac{d^4R}{(2\pi)^4} d^4p'\,d^4p''\exp(-ip\cdot R) f(x,p') \\ \quad\times\left[f\left(x+\dfrac{1}{2}R,p''\right) + f\left(x-\dfrac{1}{2}R,p''\right)\right] = 0, \end{cases} \qquad (13.121)$$

the solution of which must be mutually consistent. The first equation expresses the statistical evolution of the system of quasiparticles, while the second one is connected to their "mass shell."

Of course, other kinds of approximations would lead to other kinetic equations. In particular, this is the case when, instead of terms like $\approx\ \phi^3\phi$, one introduces a two-body Wigner function $f_2(x,\,p;\,x',\,p')$; then new functional expressions of $f_2[\{f\}]$ yield new kinetic equations and it should be emphasized that the choice of this functional greatly depends on the physical problem under study and also on the available scales of energy, length, time, etc. Here, use was made of the simplest possible choice, where only *collective* effects are taken into account so that the kinetic equations obtained are strongly reminiscent of the ordinary Vlasov equation of the usual plasma physics.

Let us now briefly provide some elements on thermodynamic equilibrium for this system and in this approximation. Since $f(x,p) = f_{\rm eq}(p)$, the first of the above equations is trivially satisfied while the second one reduces to

$$\left\{\left[p^2 - \mu_0^2\right] - \frac{\lambda_0}{2}\int d^4p'' f_{\rm eq}(p'')\right\} f_{\rm eq}(p) = 0, \tag{13.122}$$

which can be rewritten as

$$\left[p^2 - M^2\right] f_{\rm eq}(p) = 0, \tag{13.123}$$

where the constant *effective* mass M satisfies the transcendental equation

$$M^2 = \mu_0^2 + \frac{\lambda_0}{2}\int d^4p\, f_{\rm eq}(p). \tag{13.124}$$

These equations indicate that, in Gaussian approximation, the system can be considered as being composed of *free* quasiparticles endowed with the effective mass M. This means that $f_{\rm eq}(p)$ is the usual Bose–Einstein function — plus the vacuum term — and hence this last equation is a self-consistent equation for M that controls the whole thermodynamics of the system. This equation reads

$$M^2 = \mu_0^2 + \frac{\lambda_0}{2}\int \frac{d^3p}{\sqrt{\mathbf{p}^2 + M^2}}\left[\frac{1}{\exp\left(\beta\sqrt{\mathbf{p}^2 + M^2}\right) - 1} + \frac{1}{2}\right], \tag{13.125}$$

which has yet to be renormalized. This result is by no means new and can be found many times in the literature, where it was derived with several techniques.[35]

There exist many possible cases of thermodynamic states for this system, which are analyzed in detail elsewhere. The above figure, for the effective mass plotted against the temperature, indicates a phase transition and the existence of a critical temperature above which $M = 0$. As mentioned earlier, the thermodynamic properties of the system are controlled by the effective mass and they can be obtained through the free energy of the system $\mathcal{F} = -P$, which reads in the case of Fig. 13.1,

$$\mathcal{F} = \frac{M^4}{128\pi^2}\left(2\ln\left[\frac{M^2}{\Lambda}\right] - 1\right) + \frac{M^2}{2}\left[\langle\phi\rangle^2 - \frac{m_1}{8\pi^2}\right], \tag{13.126}$$

where m_1 and Λ are arbitrary constants which can be related to the renormalized quantities λ_R and μ_R^2. The figure below shows the *effective potential*

[35]For a more complete list of references, see F. Grassi, R. Hakim and H. Sivak (1991).

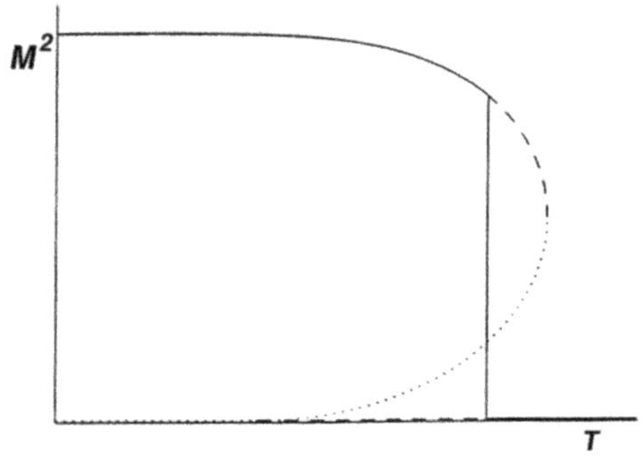

Fig. 13.1 A typical behavior of the renormalized effective mass as a function of temperature for $\langle\phi\rangle = 0$, in the case of the so-called "precarious" solution [after F. Grassi, R. Hakim and H. Sivak (1991)]. Continuous line — stable solution; dotted line — unstable; dashed line — metastable.

(free energy) corresponding to the effective mass of the preceding figures as a function of $\langle\phi\rangle$ (see Fig. 13.2).

Let us briefly mention the off-equilibrium properties of this system and, in particular, its transport properties. They can be calculated from the relativistic version of the Bathnagar–Gross–Krook equation, which, in Fourier space, can be written in the particularly simple form

$$ip \cdot kf(k,p) = -p \cdot u \frac{f(k,p) - f_{\rm eq}(p)}{\tau} \tag{13.127}$$

and hence is identical to J.L. Anderson and H.R. Witting's form (1974). Accordingly, the *general form* of the various transport coefficients of this system is identical to those given by these authors, although their dependence on the thermodynamic properties (such as from the temperature) is not the same owing to the presence of the effective mass. Note also that some care is needed because of the absence of a conserved current in this particular system.

13.7. Remarks on the Thermodynamics of Quasiparticles

The expressions obeyed earlier for P, ρ and Q do not generally obey the laws of thermodynamics unless strong conditions are imposed on the polarization Π. This can conveniently be seen in the rest frame where the *ordinary* forms of thermodynamics apply, at least for ordinary systems. One might indeed think that an assembly of free quasiparticles obeys the law of thermodynamics. However, things are not that simple. For instance, while in the $\lambda\phi^4$

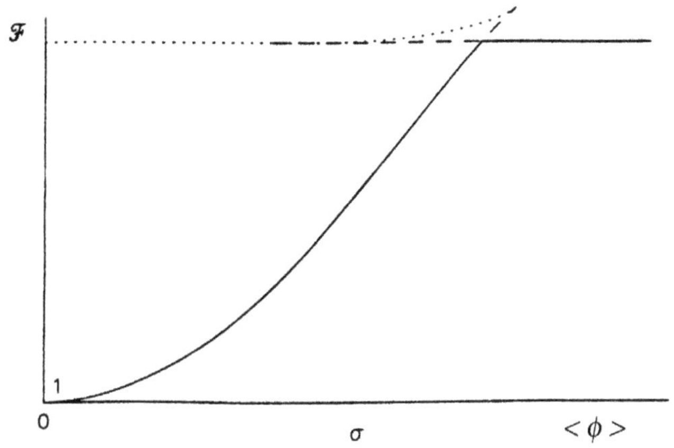

Fig. 13.2 The free energy as a function of the average scalar field at a given temperature. The various branches correspond to those indicated on the effective mass curve [after F. Grassi, R. Hakim and H. Sivak (1991)].

theory at finite temperature [F. Grassi, R. Hakim and H. Sivak (1991)] the laws of thermodynamics are not satisfied by the quasiparticles alone, they are valid for the quasiparticles of the Walecka model of relativistic nuclear matter [B.D. Serot and J.D. Walecka (1986)]. Also, in the $\lambda\phi^4$ theory at finite temperature, when the interactions between the quasiparticles are taken into account, one finds that the thermodynamics is recovered.

The reasons why thermodynamics are obeyed in these two examples lie in the following remarks. In both cases the macroscopic parameters involved in the excitation spectrum of the quasiparticles are determined in a variational manner: in the Walecka model by minimizing the grand potential Ω, and in the $\lambda\phi^4$ theory by minimizing the free energy $\mathcal{F}$. In either model there exists an effective mass M obeying a self-consistent equation whose general form is

$$\begin{cases} \dfrac{\partial \Omega}{\partial M} = 0 \text{ (Walecka model)}, \\ \dfrac{\partial \mathcal{F}}{\partial M} = 0 \text{ } (\lambda\phi^4 \text{ theory}). \end{cases} \tag{13.128}$$

As a result, since the various thermodynamic quantities are essentially derivatives of either Ω or $\mathcal{F}$, their dependence on M can be ignored.[36]

[36]B.D. Serot and J.D. Walecka (1986); F. Grassi, R. Hakim and H. Sivak (1991); M.I. Gorenstein and S.N. Yang, *J. Phys.* **G21**, 1053 (1995); *Phys. Rev.* **D52**, 5206 (1995).

These variational properties were noticed by M.I. Gorenstein and Shin Nan Yang (1995), and they suggested another solution to this problem: only the whole system, composed of the quasiparticles *and* the background (i.e. the vacuum), must obey the laws of thermodynamics and not the system reduced to the free quasiparticles. As a matter of fact, they assumed that

$$T^{\mu\nu}_{\text{system}} = T^{\mu\nu}_{\text{quasiparticles}} + T^{\mu\nu}_{\text{vacuum}}. \tag{13.129}$$

Accordingly, they assumed that the pressure and the energy density should be modified as

$$\begin{cases} \rho_{\text{system}} = \rho_{\text{quasiparticles}} + B, \\ P_{\text{system}} = P_{\text{quasiparticles}} - B, \\ B: \ \text{contribution of the vacuum.} \end{cases} \tag{13.130}$$

However, their approach rests on the implicit assumption that the vacuum contribution is as usual, i.e. such that $\rho + P = 0$. In general, a material medium, like the quasiparticle vacuum, possesses an energy–momentum tensor of the form

$$T^{\mu\nu}_{\text{vacuum}} = (A + B)\, u^\mu u^\nu - B\eta^{\mu\nu} \tag{13.131}$$

(or even a more involved form), so that, on the line of M.I. Gorenstein and Shin Nan Yang (1995), one should perform the changes

$$\begin{cases} \rho_{\text{system}} = \rho_{\text{quasiparticles}} + A, \\ P_{\text{system}} = P_{\text{quasiparticles}} - B, \end{cases} \tag{13.132}$$

in order to recover the thermodynamics, and this is precisely what was noticed above.

In order to be more specific, their approach is rephrased with our notations and in accordance with the preceding remarks. Let us call L_{system} the Lagrangian of the system under consideration, averaged with the density operator of the quasiparticles; this means that, in L_{system}, the fields are considered as being assimilated into the quasiparticle fields and then averaged. Assume also that this is a correct approximation in some range of the macroscopic parameters. In addition, let us denote by $L_{\text{quasiparticle}}$ the Lagrangian of the quasiparticles, not to be confused with L_{system}. Also, let $T^{\mu\nu}_{\text{system}}$ and $T^{\mu\nu}_{\text{quasiparticle}}$ be the corresponding energy–momentum tensors. They are respectively given by

$$\begin{cases} T^{\mu\nu}_{\text{system}} = t^{\mu\nu} - \eta^{\mu\nu} L_{\text{system}}, \\ T^{\mu\nu}_{\text{quasiparticle}} = t^{\mu\nu} + a^{\mu\nu} - \eta^{\mu\nu} L_{\text{quasiparticle}}, \end{cases} \tag{13.133}$$

where $t^{\mu\nu}$ is the kinetic part of the energy–momentum tensor, essentially that part involving first order derivatives; where $a^{\mu\nu}$ is defined above. Note that $t^{\mu\nu}$ for the system and for the quasiparticles are identical since the true Lagrangian and energy–momentum tensor are approximated by quasiparticle fields. Thence one can write

$$T^{\mu\nu}_{\text{system}} = T^{\mu\nu}_{\text{quasiparticle}} - [a^{\mu\nu} + \eta^{\mu\nu}(L_{\text{system}} - L_{\text{quasiparticle}})] \tag{13.134}$$

where $t^{\mu\nu}$ has been eliminated. The term $a^{\mu\nu}$ contains in general a term proportional to $u^\mu u^\nu$ and also to $\eta^{\mu\nu}$.

Following M.I. Gorenstein and S.N. Yang (1995), the term in the brackets $[\cdots]$, on the right hand side of the above equation, was interpreted as the medium (or the vacuum) contribution to be added to the quasiparticles' macroscopic data; and this seems quite natural. However, a glance at this term indicates that such an interpretation is hardly tenable. All that can be said is that, indeed, this term contains a contribution of the medium and of the interactions with the quasiparticles, but only in a general and loose sense.

This can best be realized by looking at specific examples. In the J.D. Walecka model of relativistic nuclear matter, one finds that

$$T^{\mu\nu}_{\text{system}} = T^{\mu\nu}_{\text{quasiparticle}} \tag{13.135}$$

and the brackets vanish identically. In such a case the system is approximated by an assembly of free quasiparticles and obeys the law of thermodynamics. A second example is provided by the $\lambda\phi^4$ theory in Gaussian approximation; in this model, one finds that

$$[a^{\mu\nu} + \eta^{\mu\nu}(L_{\text{system}} - L_{\text{quasiparticle}})] = \frac{\lambda}{8}\langle\phi^2\rangle^2\eta^{\mu\nu}, \tag{13.136}$$

which shows that (i) the bracket term is certainly connected with the interactions and that (ii) it cannot be identified with the contribution of the medium. Furthermore, one can verify that the laws of thermodynamics are still obeyed by the quantities derived from $T^{\mu\nu}_{\text{system}}$ and not from $T^{\mu\nu}_{\text{quasiparticle}}$.

Finally, one must conclude that each case requires a delicate discussion.

13.8. Equilibrium Fluctuations

In many physical situations the equilibrium fluctuations of some observables are required. For instance, this is the case for a QED plasma where the fluctuations of the four-current $\langle\tilde{J}^\mu(x)\tilde{J}^\nu(y)\rangle$ induce the electromagnetic modes propagating in the system via the inverse of the fluctuation–dissipation theorem (see Chap. 15). Another example of interest is the one provided by the pion correlations that occur in heavy ion collisions[37]: although these correlations are well represented by those given by a free pion gas, there exist some discrepancies which might be interpreted as being due to the fact that one should rather consider a gas madeup of quasipions.

Here the *equilibrium* expression for the correlations of the Wigner function operator, namely

$$\langle \tilde{f}(y,p)\tilde{f}(x,p')\rangle = \langle \tilde{f}(0,p)\tilde{f}(x-y,p')\rangle \equiv \mathcal{F}(x-y;p,p'), \tag{13.137}$$

or, more precisely, the Fourier transform of $\mathcal{F}$, is evaluated below — say, $\mathcal{F}(k;\, p,\, p')$. From this quantity one can easily calculate the fluctuations of all one-quasiparticle observables. For instance, the equilibrium four-current fluctuation tensor is given by

$$\delta J^{\mu\nu}(k) \equiv \int d^4(x-y)\exp\left[ik\cdot(x-y)\right]\left\langle \tilde{J}^{\mu}(x)\tilde{J}^{\nu}(y)\right\rangle. \tag{13.138}$$

First, we define the "adjoint" Wigner function operator

$$^*f_{\rm op} \equiv \frac{1}{(2\pi)^4}\int d^4R\exp\left(-ip\cdot R\right)\tilde{\phi}\left(x-\frac{1}{2}R\right)\tilde{\phi}^*\left(x+\frac{1}{2}R\right). \tag{13.139}$$

With this definition and Wick's theorem, which is valid since only free quasiparticles are considered, we obtain successively

$$\begin{aligned}
\mathcal{F}(k;p,p') &\equiv \int d^4x\exp(ik\cdot x)\left\langle \tilde{f}(0,p)\tilde{f}(x,p')\right\rangle \\
&= \frac{1}{(2\pi)^8}\int d^4x\, d^4R\, d^4R'\exp\left[ik\cdot x - p\cdot R - p'R'\right] \\
&\quad\times\left\{\left\langle \tilde{\phi}^*\left(\frac{1}{2}R\right)\tilde{\phi}\left(-\frac{1}{2}R\right)\tilde{\phi}^*\left(x+\frac{1}{2}R'\right)\tilde{\phi}\left(x-\frac{1}{2}R'\right)\right\rangle\right. \\
&\quad\left.-\left\langle \tilde{\phi}^*\left(\frac{1}{2}R\right)\tilde{\phi}\left(-\frac{1}{2}R\right)\right\rangle\left\langle \tilde{\phi}^*\left(x+\frac{1}{2}R'\right)\tilde{\phi}\left(x-\frac{1}{2}R'\right)\right\rangle\right\} \\
&= \int d^4x\, d^4R\, d^4R'\exp\left[i(k\cdot x - p\cdot R - p'R'\right] \\
&\quad\times\left\langle \tilde{\phi}^*\left(\frac{1}{2}R\right)\tilde{\phi}\left(x-\frac{1}{2}R'\right)\right\rangle\left\langle \tilde{\phi}\left(-\frac{1}{2}R'\right)\tilde{\phi}^*\left(x+\frac{1}{2}R'\right)\right\rangle,
\end{aligned} \tag{13.140}$$

[37] J. Rafelski and J. Letessier (eds.), *Hot Hadronic Matter* (Plenum, New York, 1995).

and finally we get

$$\mathcal{F}(k;p,p') = (2\pi)^4\,\delta^{(4)}(p-p')f\left(p+\frac{1}{2}k\right)^* f\left(p-\frac{1}{2}k\right). \tag{13.141}$$

On the other hand, the Fourier transform of the commutator of the fields ϕ and ϕ^* and the fact that $[\phi(x),\phi(y)] = [\phi^*(x),\phi^*(y)] = 0$ yield

$$^*f(k,p) - f(k,p) = \Delta(p), \tag{13.142}$$

so that the Wigner function for the fluctuations reads

$$\mathcal{F}(k;p,p') = (2\pi)^4\,\delta^{(4)}(p-p')\left[f\left(p+\frac{1}{2}k\right)f\left(p-\frac{1}{2}k\right)\right.$$
$$\left. + f\left(p+\frac{1}{2}k\right)\Delta\left(p-\frac{1}{2}k\right)\right]. \tag{13.143}$$

While the first term of this expression represents the fluctuations of matter (under the form of quasiparticles), the last one contains a "vacuum" contribution or, more precisely, it is a term arising from the equilibrium state of the system.

This expression can be rewritten in another form, by using the identity

$$\frac{1}{x-1}\frac{y}{y-1} = \left\{\frac{1}{y-1} - \frac{1}{x-1}\right\}\frac{1}{x/y-1} \tag{13.144}$$

and the equality

$$f(p) = n(p)\Delta(p), \tag{13.145}$$

where $n(p)$ is the Bose–Einstein factor. One finds that

$$n\left(p+\frac{1}{2}k\right)\left[n\left(p-\frac{1}{2}k\right)+1\right] = \frac{n\left(p-\frac{1}{2}k\right) - n\left(p+\frac{1}{2}k\right)}{\exp(\beta\omega)-1} \tag{13.147}$$

and thus

$$\mathcal{F}(k;p,p') = (2\pi)^4\,\delta^{(4)}(p-p')\frac{\Delta\left(p+\frac{1}{2}k\right)\Delta\left(p-\frac{1}{2}k\right)}{\exp(\beta\omega)-1}$$
$$\times\left\{n\left(p-\frac{1}{2}k\right) - n\left(p+\frac{1}{2}k\right)\right\}; \tag{13.148}$$

in these last expressions one recognizes the Bose–Einstein factor of the excitations of frequency ω. Note that the commutor has already been calculated in a preceding section as

$$[\phi(x),\phi(y)] = -\frac{1}{(2\pi)^3}\int d^4k\ \mathrm{sgn}(k^0)\delta[D(k)]\exp[ik\cdot(x-y)], \tag{13.149}$$

and hence

$$\Delta(p) = -\mathrm{sgn}(k^0)\delta\left[D(p)\right]. \tag{13.150}$$

13.9. Remarks on the Negative Energy Modes

Let us now briefly examine how the negative energy modes behave in thermal equilibrium. When they are taken into account in the equilibrium density operator, they give a contribution of the form

$$\rho \approx \exp\left[-\beta \sum_{\mathbf{p}} b_{\mathbf{p}}^{+} b_{\mathbf{p}}(-\omega_{\mathbf{p}} + \mu)\right], \tag{13.151}$$

which is quite similar to the usual case and hence gives rise to the average occupation number for these modes

$$\langle b_{\mathbf{p}}^{+} b_{\mathbf{p}}\rangle = \frac{1}{\exp\left[-\beta(\omega_{\mathbf{p}} - \mu)\right] - 1} = -\frac{\exp\left[\beta(\omega_{\mathbf{p}} - \mu)\right]}{\exp\left[\beta(\omega_{\mathbf{p}} - \mu)\right] - 1}, \tag{13.152}$$

obtained by applying without any precaution the usual procedures of quantum statistical mechanics. In other words, these modes give a wrong sign for the Bose–Einstein factor; however, the average occupation number is still positive, owing to the vacuum term $+1$.

Let us look at the consequences of this "wrong sign," and let us begin with the average energy of these modes. Surprisingly enough, the wrong sign of the Bose–Einstein factor compensates for the negative sign of the energy. On the other hand, the sign of the four-current is the same as that of the normal modes. These two properties, joined to the fact that $-\mu$ and not $+\mu$ appears in the Bose–Einstein factor, do not justify A.D. Migdal's claim (1978) that the negative energy modes are the antiparticles of the normal modes. Moreover, the vacuum term — which gives a negative energy — cannot be forgotten.

However, when we go back to the usual technicalities involved in the implicit derivation of $\langle b_{\mathbf{p}}^{+} b_{\mathbf{p}}\rangle$, it is easy to see that we have actually summed a divergent geometrical series, and hence that the above expression is not valid as such. Nevertheless, this suggests that a kind of renormalization has to be performed although its physical basis is yet unclear in the absence of a specific problem. In any case, the vacuum should play an important role.

We now give a merely qualitative example that shows how the negative energy modes could be generated and what could be the role of the vacuum. Let us assume that the system under consideration possesses a free energy

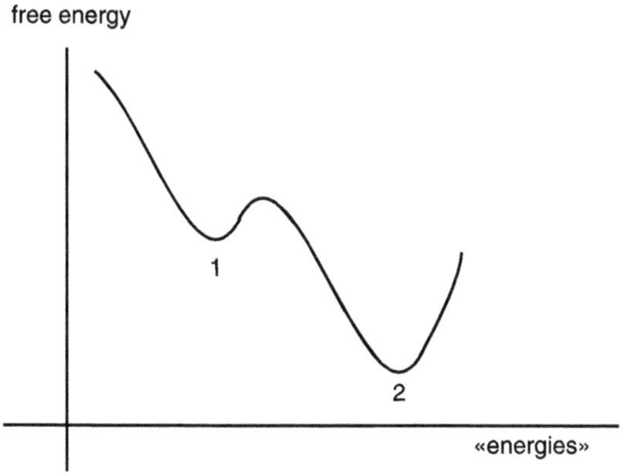

Fig. 13.3 A possible model for the generation of negative energy modes (see text).

of the shape shown in Fig. 13.3, i.e. it consists of a metastable state, 1, and a stable state, 2. This also means that the system is capable of excitations above the relative minima 1 and 2. Suppose now that the system is condensed on the state 1, usually termed the "false vacuum." Then the system decays gradually to the "true vacuum," 2. When undergoing this phase transition, the system releases energy: modes are excited above the state 2. These modes are amplified at the expense of the "false vacuum," 1, and possibly — but this is not essential here — also at the expense of the "true vacuum," 2. They are thus unstable and have the "wrong" sign for the energy.

This situation is, of course, suggested by the usual interpretation of the negative energy modes of an ordinary plasma. Such a situation is hidden when one is simply given the dispersion equation for the quasiparticles: the system at hand must imperatively be specified in a precise way. Let us also emphasize that this example does not eliminate other possible explanations for the "wrong" modes; it is based only on what can be learnt from the usual plasma physics.

13.10. Interacting Quasibosons

There are many examples in the current literature of quasiparticles whose associated polarization operator is not a mere function $\Pi(p)$ but an integral operator. For instance, in Chap. 9, when investigating the Hartree–Fock approximation, a more general "mass operator" was met which did not enter into the scheme studied above. A closer inspection of the transport equations obeyed by the quasiparticles considered so far shows that they are not subjected to any force whatsoever; this is in fact quite natural, since our kinetic equation contains only the kinematics of noninteracting quasiparticles. Also, when one tries to obtain the relativistic analog of the Landau theory of the normal Fermi liquid[38,39] from the above results,[40] it appears that this absence of a force term is a serious drawback. Therefore, the study of quasiparticles obeying more general equations of motion is absolutely necessary. Another reason can be found in the following considerations.

In this section the various modifications of the preceding results are briefly given, without any comment except whenever necessary, since the basic arguments and calculations are essentially similar.

The equations of motion are now written as

$$\Box\Phi(x) + \int d^4y\, \Pi(x,y)\Phi(y) = 0, \tag{13.153}$$

which can be derived from the Lagrangian

$$L[\{\Phi\},\{\Phi^*\}] = \partial\Phi^*(x)\cdot\partial\Phi(x) - \int d^4y\, \Pi\left(x+\frac{1}{2},x-\frac{1}{2}y\right) \times \Phi^*\left(x+\frac{1}{2}y\right)\Phi\left(x-\frac{1}{2}y\right) \tag{13.154}$$

or from the action

$$S[\{\Phi\},\{\Phi^*\}] = \int d^4y\; \partial\Phi^*(y)\cdot\partial\Phi(y) - \int d^4x\; d^4y\; \Pi(x,y)\Phi^*(y)\Phi(x), \tag{13.155}$$

where the fact that the action must be real implies that

$$\Pi(x,y) = \Pi^*(y,x), \tag{13.156}$$

which is nothing but a statement of the Hermiticity of the operator Π. After some elementary calculations the covariant Wigner function $f(x,p)$ can be shown to obey the equations

$$\left[p+\frac{1}{2}k\right]^2 f(k,p) = \frac{1}{(2\pi)^4}\int d^4k'\, \Pi\left(p+\frac{1}{2}k; -p+\frac{1}{2}k-k'\right) \times f\left(k'; p-\frac{1}{2}[k-k']\right), \tag{13.157}$$

---

[38]L.P. Kadanoff and G. Baym, *Quantum Statistical Mechanics* (Addison-Wesley, Reading, 1989).

[39]G. Baym and S.A. Chin, *Nuclear Physics* (1976).

[40]...Either from equations for quasifermions (see Chap. 14) or from the ones above, which amounts to neglecting the spin, a nonessential point as to the present discussion.

$$\left[p-\frac{1}{2}k\right]^2 f(k,p) = \frac{1}{(2\pi)^4}\int d^4k'\,\Pi\left(p-\frac{1}{2}k;-p-\frac{1}{2}k+k'\right)$$
$$\times f\left(k';p-\frac{1}{2}k+\frac{1}{2}k'\right), \tag{13.158}$$

where use has been made of the property

$$\Pi\left(p+\frac{1}{2}k;-p+\frac{1}{2}k\right) = \Pi^+\left(p-\frac{1}{2}k;-p-\frac{1}{2}k\right). \tag{13.159}$$

From the action integral, one obtains the (conserved) four-current

$$J^\mu(x) = \Phi^*(x)\overleftrightarrow{\partial}^\mu\Phi(x) + i\int_{-1/2}^{+1/2} ds \int d^4y\, y^\mu$$
$$\times \Pi\left[x+y\left(s+\frac{1}{2}\right);x+y\left(s-\frac{1}{2}\right)\right]$$
$$\times \Phi^*\left[x+y\left(s+\frac{1}{2}\right)\right]\Phi\left[x+y\left(s-\frac{1}{2}\right)\right] \tag{13.160}$$

and the (nonconserved) energy–momentum tensor

$$T^{\mu\nu} = \partial^{(\mu}\Phi^*(x)\cdot\partial^{\nu)}\Phi(x) + i\int_{-1/2}^{+1/2} ds\int d^4y\, y^\mu$$
$$\times\left\{\Pi\left[x+y\left(s+\frac{1}{2}\right);x+y\left(s-\frac{1}{2}\right)\right]\right.$$
$$\times\Phi^*\left[x+y\left(s+\frac{1}{2}\right)\right]\partial^\nu\Phi\left[x+y\left(s-\frac{1}{2}\right)\right]$$
$$-\Pi\left[x-y\left(s-\frac{1}{2}\right);x-y\left(s+\frac{1}{2}\right)\right]$$
$$\left.\times\partial^\nu\Phi^*\left[x-y\left(s-\frac{1}{2}\right)\right]\Phi\left[x-y\left(s+\frac{1}{2}\right)\right]\right\}$$
$$-\eta^{\mu\nu}L. \tag{13.161}$$

The divergence of $T^{\mu\nu}$ is now nonvanishing and is given by

$$\partial_\mu T^{\mu\nu}(x) = -\int d^4y\,\partial^\nu\Pi\left[x+\frac{1}{2}y;x-\frac{1}{2}y\right]\Phi^*\left(x+\frac{1}{2}y\right)\Phi\left(x-\frac{1}{2}y\right). \tag{13.162}$$

The fact that the energy–momentum tensor is not conserved can be understood if one remembers that the polarization operator is not invariant under space–time translations. As a matter of fact, this lack of conservation amounts to considering quasiparticles in a force field representing an interaction or/and an external field (see below).

13.10.1. *The long wavelength and low frequency limit*

This can be seen as follows. Introducing now the notation

$$\Pi(k,p) = \tilde{\Pi}\left(k-p, \frac{k+p}{2}\right), \tag{13.163}$$

the function $f(k, p)$ can easily be shown to obey the equations

$$\left(p \pm \frac{1}{2}k\right)^2 f(k,p) = \int \frac{d^4k'}{(2\pi)^4}\tilde{\Pi}\left[k-k';p\pm\frac{1}{2}k'\right] f\left(k',p\mp\frac{1}{2}(k-k')\right), \tag{13.164}$$

and by looking at the weak gradient approximation of the latter equations or, equivalently, in the long wavelength and low frequency approximation. In this approximation the difference between these equations reduces to

$$v(x,p)\cdot\partial f(x,p) + F(x,p)\cdot\frac{\partial}{\partial p}f(x,p) = 0, \tag{13.165}$$

where $v^{\mu}(x, p)$ and $F^{\mu}(x, p)$ are given by the Hamiltonian-like equations

$$\begin{cases} v^{\mu}(x,p) &= \nabla^{\mu}H(x,p), \\ F^{\mu}(x,p) &= -\partial^{\mu}H(x,p), \end{cases} \tag{13.166}$$

with

$$H(x,p) = \frac{1}{2}\left[p^2 - \tilde{\Pi}(x,p)\right],$$

where $\tilde{\Pi}(x,p)$ is the covariant Wigner transform of $\tilde{\Pi}(x,x')$:

$$\tilde{\Pi}\,(x,p) = \frac{1}{(2\pi)^4}\int d^4R \exp(-ip\cdot R)\,\tilde{\Pi}\left(x+\frac{1}{2}R;x-\frac{1}{2}R\right). \tag{13.167}$$

These properties are proven below. The transport equation then appears as the ordinary relativistic Liouville equation and F^{μ} as an external force field which is not present when Π is invariant under space–time translations, while v^{μ} is the quasiparticle four-velocity.

Let us now briefly indicate how the above "Liouville equation" can be obtained. The expansions

$$\begin{aligned} \tilde{\Pi}\left(x\pm\frac{1}{2}R;y\right) &\approx \tilde{\Pi}\,(x;y) \pm \tfrac{1}{2}R\cdot\partial_x\tilde{\Pi}\,(x;y)\,, \\ f\left(x\pm\frac{1}{2}y;p'\right) &\approx f\,(x;p') \pm \tfrac{1}{2}y\cdot\partial_x\tilde{\Pi}\,(x;y)\,, \end{aligned} \tag{13.168}$$

introduced in the Fourier transform of the equations of motion for $f(x, p)$, i.e. in

$$\left(p \pm \frac{1}{2}\partial\right)^2 f(x,p) = \int \frac{d^4R}{(2\pi)^4} d^4y\, d^4p'\, e^{ip.y} e^{-i(p-p').R} \times \tilde{\Pi}\left[x \mp \frac{1}{2}R; y\right] f\left(x \mp \frac{1}{2}y, p'\right), \tag{13.169}$$

lead to

$$\begin{cases} p \cdot \partial f(x,p) - \frac{1}{2}\nabla_\mu \tilde{\Pi}(x,p)\partial^\mu f(x,p) + \frac{1}{2}\partial_\mu \tilde{\Pi}(x,p) \cdot \frac{\partial}{\partial p} f(x,p) = 0, \\ p^2 f(x,p) - \tilde{\Pi}(x,p) f(x,p) = 0. \end{cases} \tag{13.170}$$

The last equation represents the "mass shell" where the quasiparticles live, while the first one is nothing but the above "Liouville equation."

It should be noted that the calculation of transport coefficients via the use of for example the Chapman–Enskog method contains an assumption of weak gradients; it follows that such a calculation can be based on a transport equation of the form

$$v(x,p) \cdot \partial f(x,p) + F(x,p) \cdot \frac{\partial}{\partial p} f(x,p) = -u_\lambda v^\lambda(x,p) \frac{f(x,p) - f_{\text{eq}}(p)}{\tau} \tag{13.171}$$

or with any other collision term on the right hand side. This has been done in order to evaluate the transport coefficient of nuclear matter within the Walecka model at finite temperature (see Chap. 11).

Chapter 14

The Relativistic Fermi Liquid

Landau's theory of the Fermi liquid[1] possesses a large domain of applications, ranging from quantum plasmas to solid state physics, without forgetting nuclear matter. A relativistic generalization has been proposed by G. Baym and S.A. Chin (1976) with a close parallel to the usual Newtonian theory. Several other authors, like T. Matsui (1981), have used the conventional Landau's theory by replacing essentially the nonrelativistic expression of the energy by the relativistic one, although covariant forms are sometimes conserved. The only tentative work which is manifestly covariant is the one by Ch. G. van Weert and M.C.J. Leermakers (1984); they applied their theory to the QED plasma (1985). In this chapter, relativistic concepts, equivalent to the phenomenological ones commonly used, are considered *ab initio* and developed step by step as far as necessary for a phenomenological theory.

14.1. Independent Quasifermions

In this section, one starts again from the equations of motion of quasifermions, written in the form

$$\begin{cases} i\gamma \cdot \partial\Psi(x) - \displaystyle\int d^4x'\, \Sigma(x-x')\Psi(x') = 0, \\ i\bar{\Psi}(x)\gamma \cdot \partial - \displaystyle\int d^4x'\, \bar{\Psi}(x')\bar{\Sigma}(x-x') = 0, \end{cases} \tag{14.1}$$

where the mass operator $\sum$ is assumed to be invariant under space–time translations — an important property when one is dealing with equilibrium,

[1]See e.g. G. Baym and C. Pethick, *Landau Fermi Liquid Theory: Concepts and Applications* (Wiley, New York, 1991).

for instance. $\bar{\Sigma}$ is defined as

$$\bar{\Sigma}(x-x') \equiv \gamma^0 \Sigma^+(x'-x)\gamma^0, \tag{14.2}$$

where the cross indicates the Hermitian conjugation. These equations of motion provide the following system for the covariant Wigner function:

$$\begin{cases} \left[\gamma \cdot \left(p - \frac{1}{2}k\right) - \Sigma\left(p - \frac{1}{2}k\right)\right] F(k,p) = 0, \\ F(k,p)\left[\gamma \cdot \left(p + \frac{1}{2}k\right) - \Sigma\left(p + \frac{1}{2}k\right)\right] = 0. \end{cases} \tag{14.3}$$

In the preceding chapters, several examples have been considered for the "mass operator" Σ. For the scalar plasma in the Hartree approximation, one had a very simple case with $\Sigma = M$. For the Walecka model of nuclear matter (see Chap. 13), in the same approximation, the following relation was obtained:

$$\Sigma(k) = \left[\gamma \cdot \left(k - \frac{g_V^2 n_{\text{eq}}}{m_V^2}\right) - M\right], \tag{14.4}$$

where M is the effective mass. However, a much more involved expression was obtained for the scalar plasma in the Hartree–Fock approximation (see Chap. 9). *A priori* the mass operator $\Sigma(k)$ can be expanded on the algebra of the 16 Dirac matrices

$$\Sigma(p) = \frac{1}{4} \sum_{A=1}^{A=16} \Sigma_A(p)\gamma^A; \tag{14.5}$$

however, the most common case usually encountered is the one where only its components on I and γ^μ are present, as is the case for the examples given above:

$$\Sigma(p) = \Sigma_s(p)I + \Sigma_\mu(p)\gamma^\mu. \tag{14.6}$$

This corresponds to an *unpolarized* medium, owing to the absence of the other terms (see Chap. 8). In this chapter, we shall limit ourselves to this case, the extension to other possibilities being generally straightforward albeit it contains sometimes-involved calculations. In an unpolarized medium where the only macroscopic four-vector u_μ exists, $\Sigma(p)$ has the general form

$$\Sigma(p) = \Sigma_s(p)I + [\Sigma_p(p)p_\mu + \Sigma_u(p)u_\mu]\gamma^\mu, \tag{14.7}$$

to which we shall restrict ourselves. Note also that the Hermitian character of $\Sigma(p)$ implies the fact that $\Sigma_s(p)$, $\Sigma_u(p)$ and $\Sigma_p(p)$ are all real.

From the Fourier transform of the equations of motion and this latter form for $\Sigma(k)$, one obtains the mass shell on which the quasiparticles live as

$$\mathrm{Det}[\gamma \cdot p - \Sigma(p)] = 0, \tag{14.8}$$

or

$$[1 - u \cdot p\Sigma_p(p) - \Sigma_u(p)]^2 = [1 - \Sigma_p(p)]^2 p^2 + [\Sigma_s(p)]^2. \tag{14.9}$$

Note that the quantity

$$\alpha(p) \equiv [1 - u \cdot p\Sigma_p(p) - \Sigma_u(p)]^2 - [1 - \Sigma_p(p)]^2 p^2 - (\Sigma_S(p))^2 \tag{14.10}$$

is directly connected with the eigenvalues of the matrix $D(p)$

$$D(p) \equiv \gamma \cdot p - \Sigma(p) \tag{14.11}$$

or

$$\begin{cases} D(p)u_\sigma(p) = \alpha(p)u_\sigma(p), \\ D(-p)v_\sigma(p) = -\alpha(-p)v_\sigma(p). \end{cases} \tag{14.12}$$

14.1.1. *Quantization and observables*

As was done for the quasibosons in the preceding chapter, the field ψ can be quantized *mutatis mutandis* as

$$\psi(x) = \sum_{\ell} \sum_{\sigma=1,2} \int \frac{d^3p}{(2\pi)^{3/2}} \left\{ \left(\frac{\partial \bar{D}(p)}{\partial p_0} \right)^{-1/2} \Bigg|_{k_0=E_p} u_{\ell\sigma}(p)\, a_{\ell\sigma}(p) \exp(-ip\cdot x) \right.$$
$$\left. + \left(\frac{\partial \bar{D}(-p)}{\partial p_0} \right)^{-1/2} \Bigg|_{k_0=E_p} \bar{v}_{\ell\sigma}(p)\, d^+_{\ell\sigma}(p) \exp(+ip\cdot x) \right\}, \tag{14.13}$$

where ℓ indicates the excited mode, σ is the spin index and $\bar{D}$ are the eigenvalue of D. The a's and the d's are the creation/annihilation operators of the particles and antiparticles, respectively; and they obey the canonical anticommutation relations

$$\begin{cases} \{a_{\ell\sigma}(\mathbf{p}), a^+_{\ell'\sigma'}(\mathbf{p}')\} = \delta_{\ell\ell'}\delta_{\sigma\sigma'}\delta^{(3)}(\mathbf{p} - \mathbf{p}'), \\ \{d_{\ell\sigma}(\mathbf{p}), d^+_{\ell'\sigma'}(\mathbf{p}')\} = \delta_{\ell\ell'}\delta_{\sigma\sigma'}\delta^{(3)}(\mathbf{p} - \mathbf{p}'), \\ \{a_{\ell\sigma}(\mathbf{p}), d_{\ell'\sigma'}(\mathbf{p}')\} = 0, \text{etc.}, \end{cases} \tag{14.14}$$

which specify the normalization of the spinors u and v as

$$\begin{cases} \bar{u}_{\ell\sigma}(p)\dfrac{\partial}{\partial p_0}D(p)u_{\ell\sigma}(p) = \delta_{\sigma\sigma'}\text{sgn}\left(R_+[p]\dfrac{\partial D(p)}{\partial p_0}\right)\left|\dfrac{\partial D(p)}{\partial p_0}\right|, \\ \bar{v}_{\ell\sigma}(p)\dfrac{\partial}{\partial p_0}D(-p)v_{\ell\sigma'}(p) = -\delta_{\sigma\sigma'}\text{sgn}\left(R_-[p]\dfrac{\partial D(-p)}{\partial p_0}\right)\left|\dfrac{\partial D(p)}{\partial p_0}\right|, \end{cases} \tag{14.15}$$

for one mode. Remember that $D(p)$ is the 4×4 matrix

$$D(p) = \gamma \cdot p - \Sigma(p), \tag{14.16}$$

while $R_+[k]$ is given by

$$\begin{cases} R_+(p) = \omega[1 - \Sigma_p(p)] - \Sigma_u(p) + \Sigma_s(p), \\ R_-(p) = -R_+(-p). \end{cases} \tag{14.17}$$

The spinors u and v can be calculated without any difficulty and one finds that

$$\begin{aligned} u_\sigma(p) &= \sqrt{|R_+(\mathbf{p})|}\begin{bmatrix} \chi_\sigma \\ \dfrac{1-\Sigma(p)}{R_+(p)}\mathbf{p}\cdot\sigma\,\chi_\sigma \end{bmatrix}, \\ v_\sigma(p) &= \sqrt{|R_-(\mathbf{p})|}\begin{bmatrix} \chi_\sigma \\ \dfrac{1-\Sigma(-p)}{R_-(p)}\mathbf{p}\cdot\sigma\,\chi_\sigma \end{bmatrix}, \end{aligned} \tag{14.18}$$

where

$$\chi_1 = \begin{bmatrix} 1 \\ 0 \end{bmatrix}, \quad \chi_2 = \begin{bmatrix} 0 \\ 1 \end{bmatrix} \tag{14.19}$$

represent spinors for spin up and spin down, respectively.

In order to obtain the anticommutation relations at equal time, one imposes the following normalization (on the mass shell):

$$\begin{cases} \bar{u}_{\ell\sigma\alpha}(p)\dfrac{\partial D^{\alpha\beta}(p)}{\partial p_0}u_{\ell\sigma'\beta}(p) = \delta_{\sigma\sigma'}\varepsilon(p)\left|\dfrac{\partial \bar{D}(p)}{\partial p_0}\right|, \\ \bar{v}_{\ell\sigma\alpha}(p)\dfrac{\partial D^{\alpha\beta}(-p)}{\partial p_0}u_{\ell\sigma'\beta}(p) = \delta_{\sigma\sigma'}\varepsilon(-p)\left|\dfrac{\partial \bar{D}(-p)}{\partial p_0}\right|. \end{cases} \tag{14.20}$$

Note that $\bar{D}(p)$ is the eigenvalue of $D^{\alpha\beta}(p)$. Note also that these normalizations of u and v can be checked on Eq. (14.15).

The charge (or baryonic, etc.) four-current and the energy–momentum tensor are easily found to be

$$J^{\mu}_{\rm op}(x) = \bar{\psi}(x)\gamma^{\mu}\psi(x)$$
$$- i\int d^4y\, y^{\mu}\int_{-1/2}^{+1/2} ds\bar{\psi}\left(x + y\left[s + \frac{1}{2}\right]\right)\Sigma(y)\psi\left(x + y\left[s - \frac{1}{2}\right]\right), \tag{14.21}$$

$$T^{\mu\nu}_{\rm op}(k) = \int d^4p\, p^{\mu}\gamma^{\nu}F_{\rm op}(k,p) - \int_0^{1/2} ds\left\{\left(p^{\mu} - \frac{1}{2}k^{\mu}\right)\nabla^{\nu}\Sigma(p+ks)\right.$$
$$\left. + \left(p^{\mu} + \frac{1}{2}k^{\mu}\right)\nabla^{\nu}\Sigma(p-ks)\right\}F_{\rm op}(k,p)$$
$$- \eta^{\mu\nu}\{\gamma\cdot p - \Sigma(p)\}F_{\rm op}(k,p), \tag{14.22}$$

where the Fourier transform of the energy–momentum tensor has been given for future use and where

$$\nabla^{\lambda} \equiv \frac{\partial}{\partial p_{\lambda}}. \tag{14.23}$$

The Hamiltonian of the quasiparticles is the zeroth component of the energy–momentum tensor taken for $k = 0$ (this amounts to integrating $T^{\mu\nu}(x)$ over a three-plane $t = \text{const}$). After using the plane wave expansion of the quasiparticles' field, one obtains

$$H \equiv T^{00}(k=0)$$
$$= \sum_{\ell,\sigma}\int d^4p\, p^0\{\varepsilon(p)a^{+}_{\ell,\sigma}(\mathbf{p})a_{\ell,\sigma}(\mathbf{p}) - \varepsilon(-p)d_{\ell,\sigma}(\mathbf{p})d^{+}_{\ell,\sigma}(\mathbf{p})\}. \tag{14.24}$$

As for quasibosons, this Hamiltonian can be negative, a pathology indicating a possible instability of the system, a phase transition, etc.

Similarly, the expression for the Wigner function operator reads, for $k = 0$,

$$F_{\beta\alpha}(0,p) = 2\pi\sum_{\ell,\sigma,\sigma'}\left\{\delta[p^0 - \omega_{\ell+}(p)]\left|\frac{\partial\bar{D}(-p)}{\partial p_0}\right|^{-1}\right.$$
$$\times\, \bar{u}_{\alpha\ell\sigma}(p)u_{\beta\ell\sigma'}(p)b^{+}_{\ell\sigma}(p)b_{\ell\sigma'}(p) + \delta[p^0 + \omega_{\ell-}(p)]\left|\frac{\partial\bar{D}(-p)}{\partial p_0}\right|^{-1}$$
$$\left.\times\, \bar{v}_{\alpha l\sigma}(-p)v_{\beta\ell\sigma'}(-p)d_{\ell\sigma}(-p)d^{+}_{\ell\sigma'}(-p)\right\}, \tag{14.25}$$

which is to be used below.

14.1.2. *Statistical expressions*

Let us now turn to the main statistical relations used in what follows. After taking the trace of sums and the difference of the resulting equations of motion (with the γ's), they read

$$\begin{cases} \left\{k^\lambda - \left[\Sigma^\lambda\left(p+\frac{1}{2}k\right) - \Sigma^\lambda\left(p-\frac{1}{2}k\right)\right]\right\} f(k,p) \\ \quad = \left[\Sigma_S\left(p+\frac{1}{2}k\right) - \Sigma_S\left(p-\frac{1}{2}k\right)\right] f^\lambda(k,p), \\ \left\{2p^\lambda - \left[\Sigma^\lambda\left(p+\frac{1}{2}k\right) + \Sigma^\lambda\left(p-\frac{1}{2}k\right)\right]\right\} f(k,p) \\ \quad = \left[\Sigma_S\left(p+\frac{1}{2}k\right) + \Sigma_S\left(p-\frac{1}{2}k\right)\right] f^\lambda(k,p). \end{cases} \tag{14.28}$$

Note that we have set

$$\Sigma^\lambda(p) \equiv p^\lambda \Sigma_p(p) + u^\lambda \Sigma_u(p) \tag{14.29}$$

in order to make the correspondence with further notations. The consistency with results of Chaps. 9 and 11 can easily be verified.

The four-current and the energy–momentum tensor are expressed as

$$\begin{cases} J^\mu(x) = \langle J^\mu_{\rm op}(x)\rangle, \\ T^{\mu\nu}(x) = \langle T^{\mu\nu}_{\rm op}(x)\rangle, \end{cases} \tag{14.30}$$

and their general form is the same as the one above for the corresponding operators. $J^\mu(x)$ and $T^{\mu\nu}(x)$ assume an interesting form when the system is invariant under space–time translations. In such a case, they are given by

$$J^\nu = {\rm Sp}\int d^4p\, \nabla^\nu D(p) F(p), \tag{14.31}$$

$$T^{\mu\nu} = {\rm Sp}\int d^4p\, p^\mu \nabla^\nu D(p) F(p), \tag{14.32}$$

which shows — as was the case for quasibosons — that in *this* case the role of the four-velocity of a quasifermion is provided by

$$v^\mu(p) \equiv \nabla^\nu D(p) = \gamma^\nu - \nabla^\nu \Sigma(p) \tag{14.33}$$

This remark will be exploited in a later section. Note that when

$$\Sigma(p) = m = {\rm const}, \tag{14.34}$$

one recovers the usual Dirac result. Also, it should be realized that this quasiparticle four-velocity is now a 4×4 matrix. Note that d^4p is a shorthand for

$$d^4p = 2\theta(p^0)\delta[p^2 - \bar{D}^2(p)]. \tag{14.35}$$

14.1.3. *Thermal equilibrium*

We notice first that the algebric structure of the quasiparticle Hamiltonian is similar to that of the free particle case. Next, it should be remarked that the structure chosen for $\Sigma(p)$ possesses the same general form as that of the Walecka model in thermal equilibrium. Accordingly, one has

$$F_{\text{eq}}(p) = \frac{\gamma \cdot [p - \Sigma_V(p)] + \Sigma_S(p)}{4\Sigma_S(p)} f_{\text{eq}}(p), \tag{14.36}$$

with

$$\begin{aligned} f_{\text{eq}}(p) = \frac{d}{(2\pi)^3}\delta[p^{*2} - \Sigma_s(p)^2]\bigg\{&\frac{\theta(p^* \cdot u)}{\exp(\beta p^* \cdot u - \mu^*) + 1} \\ &+ \frac{\theta(-p^* \cdot u)}{\exp(-\beta p^* \cdot u - \mu^*) + 1} - \theta(-p^* \cdot u)\bigg\}, \end{aligned} \tag{14.37}$$

where use has been made of the notations

$$\begin{cases} p^{*\mu} = p^\mu[1 - \Sigma_p(p)] - \Sigma_u(p)u^\mu, \\ \mu^* = \mu - \Sigma_S(p), \end{cases} \tag{14.38}$$

and where the last term, $\theta(-p^* \cdot u)$ is a "vacuum" (ground state) term. The main observables are finally obtained as

$$\begin{cases} n_{\text{eq}} = d\sum_{\ell,\pm} \displaystyle\int \frac{d^3p}{(2\pi)^3} \frac{\pm 1}{\exp\{\beta[\omega_\ell(p) \mp \mu]\} + 1}, \\ \rho = d\sum_{\ell,\pm} \displaystyle\int \frac{d^3p}{(2\pi)^3} \frac{\omega_\ell(p)}{\exp\{\beta[\omega_\ell(p) \mp \mu]\} + 1}, \\ P = \dfrac{d}{3}\sum_{\ell,\pm} \displaystyle\int \frac{d^3p}{(2\pi)^3} \frac{\mathbf{p} \cdot \nabla\omega_\ell(p)}{\exp\{\beta[\omega_\ell(p) \mp \mu]\} + 1}, \end{cases} \tag{14.39}$$

where the vacuum term has been omitted. It will be briefly discussed later on.

14.2. Interacting Quasifermions

We begin by extending briefly the results of the last section to interacting quasifermions, which are supposed to obey quasifield equations of the forms

$$\begin{cases} i\gamma\cdot\partial\psi(x) - \int d^4x'\,\Sigma(x,x')\psi(x') = 0, \\ \bar\psi(x)i\gamma\cdot\overleftarrow{\partial} + \int d^4x'\,\bar\psi(x')\Sigma(x',x) = 0, \end{cases} \tag{14.40}$$

which derive from the Lagrangian

$$L = \frac{i}{2}\bar\psi(x)\gamma\cdot\partial\psi(x) - \int d^4y\,\bar\psi\left(x+\frac{1}{2}y\right)\Sigma\left(x+\frac{1}{2}y, x-\frac{1}{2}y\right)\psi\left(x-\frac{1}{2}y\right). \tag{14.41}$$

These equations, together with the definition of the Wigner function, provide the following system for $F(x,p)$:

$$\begin{cases} [i\gamma\cdot\partial + 2\gamma\cdot p]F(x,p) \\ \quad -2\int \frac{d^4R}{(2\pi)^4} d^4y\,d^4p' \exp\left(-ip\cdot R + ip'\cdot\left[x+\frac{1}{2}R-y\right]\right) \\ \quad \times\Sigma\left(x-\frac{1}{2}R, y\right) F\left(\frac{x+y+\frac{1}{2}R}{2}, p'\right) = 0, \\ F(x,p)[i\gamma\cdot\partial - 2\gamma\cdot p] \\ \quad +2\int \frac{d^4R}{(2\pi)^4} d^4y\,d^4p' \exp\left(-ip\cdot R + ip'\cdot\left[y+\frac{1}{2}R-x\right]\right) \\ \quad \times F\left(\frac{x+y-\frac{1}{2}R}{2}, p'\right)\Sigma\left(y, x+\frac{1}{2}R\right) = 0. \end{cases} \tag{14.42}$$

These equations can be rewritten in Fourier space a quite useful form:

$$\begin{cases} \gamma\cdot\left(p+\frac{1}{2}k\right)F(k,p) \\ \quad -\int \frac{d^4k'}{(2\pi)^4}\Sigma\left(p+\frac{1}{2}k, -p+\frac{1}{2}k-k'\right)F\left(k', p-\frac{1}{2}(k-k')\right) = 0, \\ F(k,p)\gamma\cdot\left(p-\frac{1}{2}k\right) \\ \quad -\int \frac{d^4k'}{(2\pi)^4}F\left(k', p+\frac{1}{2}(k-k')\right)\bar\Sigma\left(p-\frac{1}{2}k, -p-\frac{1}{2}k+k'\right) = 0. \end{cases} \tag{14.43}$$

From the above Lagrangian, the four-current is obtained as

$$J^\mu(x) = \bar{\psi}(x)\gamma^\mu\psi(x) + i\int d^4y \int_{-1/2}^{+1/2} ds\, y^\mu \bar{\psi}\left(x + y\left[s + \frac{1}{2}\right]\right)$$
$$\times \Sigma\left(x + y\left[s + \frac{1}{2}\right]; x + y\left[s - \frac{1}{2}\right]\right)\psi\left(x + y\left[s - \frac{1}{2}\right]\right), \tag{14.44}$$

which, in terms of the covariant Wigner function, can be written as

$$J^\mu(x) = \mathrm{Sp}\int d^4p\Big\{\gamma^\mu F(x,p) + i\int d^4y\; y^\mu$$
$$\times \int_{-1/2}^{+1/2} ds \exp(ip\cdot y)\tilde{\Sigma}(x + ys; y)F(x + ys, p)\Big\}, \tag{14.45}$$

where use of the convenient definition

$$\Sigma(x;y) \equiv \tilde{\Sigma}\left(\frac{x+y}{2}; x - y\right) \tag{14.46}$$

has been made. Note the useful relation

$$\Sigma(k;k') = \tilde{\Sigma}\left(k + k'; \frac{1}{2}(k - k')\right), \tag{14.47}$$

and also

$$\Sigma(k;k') = \delta^{(4)}(k + k')\Sigma(k') \tag{14.49}$$

for a system invariant under space–time translations.

Similarly, the energy–momentum tensor reads

$$T^{\mu\nu} = \frac{i}{2}\bar{\psi}(x)\gamma^\mu\partial^\nu\psi(x) - \int d^4y\, y^\mu \int_0^{1/2} ds \left[\partial^\nu\bar{\psi}\left(x - y\left[s - \frac{1}{2}\right]\right)\right.$$
$$\times \Sigma\left(x - y\left[s - \frac{1}{2}\right]; x + y\left[s + \frac{1}{2}\right]\right) \times \psi\left(x - y\left[s + \frac{1}{2}\right]\right)$$
$$- \bar{\psi}\left(x + y\left[s + \frac{1}{2}\right]\right)\Sigma\left(x + y\left[s + \frac{1}{2}\right]; x + y\left[s - \frac{1}{2}\right]\right)$$
$$\left.\times \partial^\nu\psi\left(x - y\left[s + \frac{1}{2}\right]\right)\right] - \eta^{\mu\nu}\left\{\frac{i}{2}\bar{\psi}(x)\gamma\cdot\partial\psi(x) - \int d^4y\, \bar{\psi}\left(x + \frac{1}{2}y\right)\right.$$
$$\left.\times \Sigma\left(x + \frac{1}{2}y, x - \frac{1}{2}y\right)\psi\left(x - \frac{1}{2}y\right)\right\}. \tag{14.50}$$

This energy–momentum tensor is not conserved as expected since the system is not invariant under space–time translations, and one has

$$\partial_\mu T^{\mu\nu}(x) = \int d^4x'\, \bar{\psi}\left(x + \frac{1}{2}x'\right) \partial^\nu \Sigma\left(x + \frac{1}{2}x'; x - \frac{1}{2}x'\right) \psi\left(x - \frac{1}{2}x'\right). \tag{14.51}$$

This equation shows that, after averaging, and in equilibrium where

$$\Sigma\left(x + \frac{1}{2}x'; x - \frac{1}{2}x'\right) = \Sigma\left(\left[x + \frac{1}{2}x'\right] - \left[x - \frac{1}{2}x'\right]; 0\right) = \Sigma(x'; 0), \tag{14.52}$$

the right hand side vanishes and $\partial_\mu T^{\mu\nu}(x) = 0$, as it should be. In terms of the Wigner function, it reads

$$\begin{aligned} T^{\nu\mu}(x) = \mathrm{Sp}\Bigg\{ & \int d^4p\, (p^\mu \gamma^\nu - \eta^{\mu\nu}[\gamma \cdot p - \tilde{\Sigma}(x,p)]) F(x,p) \\ & + \int d^4y\, d^4p \int_0^{1/2} ds\; y^\mu \exp(ip \cdot y) \\ & \times \tilde{\Sigma}(x + ys; y) \left(p^\nu + \frac{i}{2}\partial^\nu\right) F(x + ys, p) \\ & + \tilde{\Sigma}(x - ys; y) \left(p^\nu - \frac{i}{2}\partial^\nu\right) F(x - ys, p) \Bigg\}. \end{aligned} \tag{14.53}$$

14.2.1. *The long wavelength and low frequency limit*

The approximation of the long wavelength and low-frequency limit is obtained by retaining the lowest order terms in the expression

$$\begin{cases} \gamma \cdot \left(p + \frac{1}{2}k\right) F(k,p) - \int \frac{d^4k'}{(2\pi)^3} \left[1 - \frac{(k - k') \cdot \nabla}{2} \cdots\right] \\ \quad \times \Sigma\left(p + k - \frac{1}{2}k', -p - \frac{1}{2}k'\right) F(k',p) = 0, \\ F(k',p)\gamma \cdot \left(p - \frac{1}{2}k\right) - \int \frac{d^4k'}{(2\pi)^3} F(k',p) \\ \quad \times \Sigma\left(p - \frac{1}{2}k', -p + k - \frac{1}{2}k'\right) \left[1 + \left(\frac{k - k'}{2}\right) \cdot \nabla \cdots\right] = 0. \end{cases} \tag{14.54}$$

Taking the trace and making the difference of these two equations, one obtains

$$k_\mu f^\mu - \int \frac{d^4k'}{(2\pi)^3}[(k-k')\cdot\nabla]\left\{\left[\Sigma_s\left(p+k-\frac{1}{2}k', -p-\frac{1}{2}k'\right)f(k',p)\right]\right.$$
$$+\left[\Sigma_u\left(p+k-\frac{1}{2}k', -p-\frac{1}{2}k'\right)u_\mu f^\mu(k',p)\right]$$
$$\left.+\left[\Sigma_p\left(p+k-\frac{1}{2}k', -p-\frac{1}{2}k'\right)p_\mu f^\mu(k',p)\right]\right\} = 0. \qquad (14.55)$$

When spin does not play an important role, $f^\mu \approx p^\mu f/\Sigma_S(p)$ and the above equation reduces to

$$k\cdot p\ f - \int \frac{d^4k'}{(2\pi)^3}[(k-k')\cdot\nabla]\left\{\left[\Sigma_s\left(p+k-\frac{1}{2}k', -p-\frac{1}{2}k'\right)f(k',p)\right]\right.$$
$$+\left[\Sigma_u\left(p+k-\frac{1}{2}k', -p-\frac{1}{2}k'\right)u\cdot p\ f(k',p)\right]$$
$$\left.+\left[\Sigma_p\left(p+k-\frac{1}{2}k', -p-\frac{1}{2}k'\right)u\cdot p\ f(k',p)\right]\right\} = 0, \qquad (14.56)$$

which is a closed equation for f, similar to a relativistic Liouville equation, once we go back to configuration space; it can be rewritten as

$$k\cdot p\ f - \int \frac{d^4k'}{(2\pi)^3}[(k-k')\cdot\nabla][K(k',p)f(k',p)] = 0, \qquad (14.57)$$

where $K(k',p)$ is given by

$$K(k',p) \equiv \Sigma_S\left(p+k-\frac{1}{2}k', -p-\frac{1}{2}k'\right) + \Sigma\left(p+k-\frac{1}{2}k', -p-\frac{1}{2}k'\right). \qquad (14.58)$$

14.3. Kinetic Equation for Quasiparticles

Let us now treat the kinetic equation for the quasiparticles; however, in order not to mix up all the indices — spin, internal numbers, etc. — we shall treat the problem of an unimportant spin. Therefore, only the "boson" part of Σ has to be considered and $\tilde{\Sigma}(k,p)$, $\tilde{\Pi}(k,p)$. Let us also recall that the tilde over $\tilde{\Sigma}(k,p)$ or $\tilde{\Pi}(k,p)$ denotes a Wigner transformation. In addition, we should recall that we have

$$H(x,p) = \frac{1}{2}[p^2 - \tilde{\Pi}(x,p)] \qquad (14.59)$$

$$\begin{cases} v^\mu(x,p) = \nabla^\mu H(x,p), \\ F^\mu(x,p) = -\partial^\mu H(x,p) \end{cases} \qquad (14.60)$$

[see Eqs. (13.167) and the following]. Now, an important point is to be emphasized. While in the usual Landau theory of Fermi liquids what plays the role of a Hamiltonian is simply the energy of a quasiparticle, here in the relativistic context, no Hamiltonian at all with the meaning of an energy (or an energy density) exists. If we take the above Hamiltonian in the case of free particles, we have only

$$H(x,p) = \frac{1}{2}(p^2 - m^2), \tag{14.61}$$

which is formally a Hamiltonian but has no particular meaning. Of course, in the case of a nonmanifestly covariant theory, one can find a Hamiltonian.

Finally, the relativistic kinetic equation for the quasiparticle reads

$$v^\nu(x,p)\partial_\nu f(x,p) + F^\mu(x,p)\nabla_\mu f(x,p) = C[p, f(p)], \tag{14.62}$$

where we have added a collision term which might be a BGK one,

$$C[p, f(p)] = -u_\lambda v^\lambda(x,p)\frac{f(x,p) - f_{\text{eq}}(x,p)}{\tau}, \tag{14.63}$$

or any other. At this stage, the collision term is left unspecified and satisfies only the standard conservation relations and H theorem conditions:

$$\begin{cases} \displaystyle\int d^4p \; C[p, f(x,p)] = 0, \\ \displaystyle\int d^4p \; p^\beta C[p, f(x,p)] = 0, \\ \displaystyle -k_B \int d^4p \; v^\alpha(x,p) \log\left[\frac{f(x,p)}{1 - f(x,p)}\right] C[p, f(x,p)] \geq 0. \end{cases} \tag{14.64}$$

While an integration of the above Liouville equation (or, equivalently, over the relativistic kinetic equation) over the four-momenta immediately provides the four-current conservation, the energy–momentum tensor conservation law is not obtained, as expected from the results of the last chapter. This means that in the whole system — which is conservative — energy and momentum are not conserved. Such a pathology occurs mainly because of the implicit use of two different definitions of the energy–momentum tensor: that of the quasiparticles and that used for the system.

Let us now study a bit further the energy–momentum tensor that arises from the Liouville equation; we first multiply it by p^μ and then integrate over d^4p. We get

$$\partial_\nu T^{\mu\nu}_{\text{quasi}}(x) - \int d^4p \; \partial^\nu H(x,p) f(x,p) = 0. \tag{14.65}$$

This means that the "dressing" of the quasiparticles (implied by the second term) does contain some energy and momentum. Actually, a solution to this problem has been obtained by C.G. van Weert and M.C.J. Leermakers (1984) in the context of the Landau Fermi liquid theory.

14.4. Remarks on the Relativistic Landau Theory

The Landau theory[2] of the Fermi liquid was originally a phenomenological theory designed to describe a weakly excited fermion system in a normal state, where the word "normal" means that no phase transition occurs whatsoever and that the basic ground state is not symmetry-breaking, for instance. Loosely speaking, Landau's idea is essentially that of a continuity and a one-to-one correspondence between the nonexcited states of the system and the low-lying excited ones. This means that starting from the noninteracting Fermi system, the distribution function preserves its general shape when the interaction is adiabatically switched on.

Although we do not give a complete Landau theory, we present a number of remarks on the relativistic case:

(1) The first idea is to choose the noninteracting-looking Fermi distribution; this is simply the usual thing to do, i.e. as the Fermi–Dirac function

$$n_{\text{eq}}(p) = \frac{1}{\exp[\beta(p_\mu u^\mu - \mu)] + 1} \tag{14.66}$$

where $p \cdot u$ is the quasiparticle energy $\varepsilon(\mathbf{p})$, which is a complicated function of $\mathbf{p}$, and this function is by no means trivial. The density tensor of the quasiparticles is

$$T^{\mu\nu} = \int d^4p\, p^\mu v^\nu(p) n_{\text{eq}}(p), \tag{14.67}$$

where $v^\mu(p)$ is the velocity of the quasiparticle. The next step is to expand (functionally) the energy density of the system until the second order of a possible variation δn of n_{eq}:

$$\begin{aligned} \delta E &\equiv \delta(T^{\mu\nu} u_\mu u_\nu) \\ &= \int d^4p\, p^\mu u_\mu v^\nu(p) u_\nu \delta n(p) + \int d^4p\, dp'^4 f(p,p') \delta n(p) \delta n(p') \\ &\quad + \cdots \int dp\, p^\mu u_\mu a^\nu(p) u_\nu \delta n(p) = \int d^3p\, p^\mu u_\mu \delta n(p). \end{aligned} \tag{14.68}$$

[2]It is useful to consult the article of G. Baym and C. Pethick, *loc. cit.*

We shall thus try a more natural way in which the manifest covariance will be part of the theory.

Therefore, let us use the distribution function developed above. This means that it still assumes its equilibrium form,

$$F_{\text{eq}}(p) = \frac{\gamma \cdot [p - \Sigma_0(p)] + \Sigma_S(p)}{4\Sigma_S(p)} f_{\text{eq}}(p), \tag{14.69}$$

with

$$f_{\text{eq}}(p) = \frac{d}{(2\pi)^3}\delta[p^{*2} - \Sigma_s(p)^2] \times \left\{\frac{1}{\exp(\beta p^* \cdot u - \mu) + 1} + \frac{1}{\exp(\beta p^* \cdot u + \mu) + 1}\right\} \tag{14.70}$$

and

$$\Sigma_0(p) \equiv p \cdot \Sigma_p + u \cdot \Sigma_u, \tag{14.71}$$

where the vacuum term should be omitted, since we deal with a phenomenological theory, and where positrons also have to be omitted. Although in a complete theory positrons must be dealt with, here we assume that the temperature is relatively low and hence there are only a few antiparticles present. Note, however, that the second Fermi factor refers to the quasiholes and not to the particles themselves: indeed, they cannot be separated from the excitations of the system. As in the nonrelativistic case, this expression for $F_{\text{eq}}(p)$ is by no means trivial since it depends on itself through the $p^0(p)$ which occurs in the explicit expression.

(2) In the usual Landau theory, the distribution function $F_{\text{eq}}(p)$ is derived from the basic principles; note that, although simple, it is not trivial from the point of view of physics. This derivation consists in minimizing the entropy (density) while taking account of the total energy (density) and the total number (density) within the system. However, the entropy of a Wigner function is by no means clear. We could define it as

$$S^\mu = -\int d^4p\, v^\mu(p)\{f_{\text{eq}}(p)\log f_{\text{eq}}(p) + [1 - f_{\text{eq}}(p)]\log[1 - f_{\text{eq}}(p)]\}, \tag{14.72}$$

which is the common entropy of $f_{\text{eq}}(p)$ for Fermi particles, and minimize the result. This would lead to the usual part of $F_{\text{eq}}(p)$, i.e. the expression for $f_{\text{eq}}(p)$, and also we could add a nonessential term in order to take the coefficient

$$\frac{\gamma \cdot [p - \Sigma_0(p)] + \Sigma_S(p)}{4\Sigma_S(p)} \tag{14.73}$$

into account. Therefore, we shall follow the usual way and thus admit that the basic function is that given by Eq. (14.69). However, the fact that the entropy is the one drawn above comes from the fact that we deal with *free* quasiparticles: the derivation of the entropy is the same as for free particles; it comes from $S = -\mathrm{Tr}(\rho \log \rho)$. The other part of $F_{\mathrm{eq}}(p)$ comes from the equation of motion (see Chap. 7).

In the above expressions, there is still something that should be explained. If we look at the symbols of the equilibrium quantities [namely Eqs. (14.39)], they are all of the form

$$\langle g(p) \rangle = \int d^3p \frac{g(p)}{\exp\{\beta[\omega(p) - \mu]\} + 1}, \tag{14.75}$$

and if we want to vary the quantity $\langle g(p) \rangle$, we must vary also $\omega(p) = p^0$. This is quite awkward from a covariant point of view and it would certainly be more valuable to keep the function $f_{\mathrm{eq}}(p)$ fixed and thus use the equivalent form

$$\langle g(p) \rangle = \int d^4p\, \delta[p^{*2} - \Sigma_S(p)] \frac{g(p)}{\exp(\beta[p \cdot u - \mu]) + 1}, \tag{14.76}$$

so that the variation depends not on that of $f_{\mathrm{eq}}(p)$ but on that of $\omega(p)$ only.

The starting point is now the equilibrium expression for the energy–momentum tensor,

$$\begin{aligned} T^{\mu\nu} &= \mathrm{Sp} \int d^4p\; p^\mu \nabla^\nu D(p)\; F_{\mathrm{eq}}(p) \\ &= \mathrm{Sp} \int d^4p\; p^\mu \nabla^\nu S_\Sigma^{-1}(p)\; F_{\mathrm{eq}}(p), \end{aligned} \tag{14.77}$$

where $S_\Sigma^{-1}(p)$ is the propagator of the free quasiparticle:

$$S_\Sigma^{-1}(p) = \gamma \cdot p - \Sigma(p). \tag{14.78}$$

(3) Usually, the first step of the Landau theory consists of adding a particle to the medium and investigating the subsequent variation of the energy (or energy density) of the system, after the added particle has been "dressed" by the interaction with the quasifermions in the medium. Such an additional particle thus modifies the energy — and the energy density — of the system.

Following first G. Baym and S.A. Chin (1976) or T. Matsui (1981), we consider the variation of the energy density,[3] with no modification of the *general form* of the equilibrium distribution.

[3] This is the same thing as considering the variation of the energy: we look at the density times a volume of value 1.

Therefore, we have

$$\rho = T^{\mu\nu} u_\mu u_\nu \tag{14.79}$$

and hence

$$\delta E \equiv \delta\rho = \delta(T^{\mu\nu} u_\mu u_\nu). \tag{14.80}$$

Thus, we have two possible variations: (i) the usual one, as dealt with by G. Baym and S.A. Chin (1976) — $\delta T^{\mu\nu} \cdot u_\mu u_\nu$; and (ii) the variation of the timelike u^μ, which is typically relativistic — in the Newtonian theory all timelike four-vectors are parallel and there is not any possibility of varying u^μ.

Let us begin with the second point. Since $T^{\mu\nu}$ is dependent on u^μ and on $\eta^{\mu\nu}$, it has the general form

$$T^{\mu\nu} = A u^\mu u^\nu + B\eta^{\mu\nu}, \tag{14.81}$$

and since $u^\mu \delta u_\mu = 0$, the variation is zero. Note that this is only for an equilibrium energy–momentum tensor.

Let us now briefly look at the first variation; it reads[4]

$$\begin{aligned}\delta T^{\mu\nu} u_\mu u_\nu = {} & \mathrm{Sp} \int d^4p\, p^\mu u_\mu\, v^\nu(p)|_{n=n_{\mathrm{eq}}} u_\nu\, \delta F(p) \\ & + \frac{1}{2}\mathrm{Sp} \int d^4p\, d^4p'\, f(p,p')|_{n=n_{\mathrm{eq}}} \delta F(p)\delta F(p') + O\{(\delta F)^3\},\end{aligned} \tag{14.82}$$

where[5] $v^\nu(p)$ is as given by Eq. (14.33); note that

$$v^\nu(p) \equiv \gamma^\nu - \nabla^\nu \Sigma(p), \tag{14.83}$$

and that $\bar{v}^\mu(p) \neq v^\mu(p)$.

In addition, note that if the system involves three- or four-body interactions, as is the case for quark matter, the above form should be supplemented by the terms

$$\frac{1}{3!}\mathrm{Sp} \int d^4p\, d^4p'\, d^4p''\, f(p,p',p'')|_{n=n_{\mathrm{eq}}} \delta f_{\mathrm{eq}}(p)\delta f_{\mathrm{eq}}(p')\delta f_{\mathrm{eq}}(p'') \tag{14.84}$$

---

[4] $f(p,p')$ has the meaning, in this section, of the usual "interacting" factor of the Landau theory: it should not be mistaken for other $f(p,p')$'s which occur in other chapters.

[5] It should be noted that assuming a tensor $f_{\mu\nu}(p,p')$ gives rise to a scalar $f(p,p')$ owing to the product by $u^\mu u^\nu$.

for three-body interactions, and

$$\frac{1}{4!}\mathrm{Sp}\int d^4p\, d^4p'\, d^4p''\, d^4p'''\; f(p,p',p'',p''')|_{n=n_{\mathrm{eq}}}$$
$$\times\, \delta f_{\mathrm{eq}}(p)\delta f_{\mathrm{eq}}(p')\delta f_{\mathrm{eq}}(p'')\delta f_{\mathrm{eq}}(p''') \tag{14.85}$$

in the case of four-body forces. These terms should be added in the theory unless there exist arguments as to their possible small size.

It should be noted that two relations hold:

$$u_\mu u_\nu \frac{\delta T^{\mu\nu}}{\delta F(p)} = p^\mu u_\mu v^\nu(p)\cdot u_\nu = p^\mu u_\mu[\gamma^\nu - \nabla^\nu\Sigma(p)]u_\nu, \tag{14.86}$$

$$u_\mu u_\nu \frac{\delta T^{\mu\nu}}{\delta F(p)\delta F(p')} = f(p,p'). \tag{14.87}$$

Let us rewrite the basic variation of the Landau assumptions, but with the spin indices

$$\delta T^{\mu\nu}\cdot u_\mu u_\nu = \mathrm{Sp}\int d^4p\, p^\mu u_\mu \nu^\nu(p) u_\nu\, \delta F_{\alpha\beta}(p)$$
$$+\frac{1}{2}\mathrm{Sp}\int d^4p\, d^4p'\, f^{\alpha\beta\chi\gamma}(p,p')\delta F_{\alpha\beta}(p)\delta F_{\chi\gamma}(p'), \tag{14.88}$$

where the spin indices obey the usual tensorial rules. We shall now analyze a little bit this expression. The first term indicates that it is only $\delta f_{\mathrm{eq}}(p)$ that occurs in the equation. The second one necessitates a more detailed analysis. $f_{\alpha\beta\chi\gamma}(p,p')$ depends on p and p' only. It is thus twice a four-vector: $\bar{f}_{\mu\nu}(p,p')$. The second term has the form

$$\delta F_{\alpha\beta}(p)\delta F_{\chi\gamma}(p') = \delta([\,]|_{\alpha\beta} f_{\mathrm{eq}}(p))\delta([\,]|_{\chi\gamma} f_{\mathrm{eq}}(p')), \tag{14.89}$$

where the expressions between brackets are

$$\frac{\gamma^\mu_{\alpha\beta}\cdot[p_\mu - \Sigma_{0\mu}(p)] + I_{\alpha\beta}\Sigma_S(p)}{\Sigma_S(p)}. \tag{14.90}$$

This term has the value

$$\bar{f}_{\mu\nu}(p,p')\{[\,]_{\alpha\beta}[\,]_{\beta\gamma}\}\delta f_{\mathrm{eq}}(p)\delta f_{\mathrm{eq}}(p') \tag{14.91}$$

or

$$f^{\alpha\beta\chi\gamma}(p,p')\,\delta F_{\alpha\beta}(p)\,\delta F_{\chi\gamma}(p')$$
$$= [p^\mu - \Sigma_0^\mu(p)][p'^\nu - \Sigma_0^\nu(p')]\bar{f}_{\mu\nu}(p,p')\delta f_{\mathrm{eq}}(p)\delta f_{\mathrm{eq}}(p'). \tag{14.92}$$

It should be borne in mind that in the Landau assumptions [Eqs. (14.69), (14.84), (14.86) and (14.87)], use was made of the fact that the energy density is a functional series, each term of it being taken at $f_{\rm eq}(p) = \text{const}$; this is the reason why there was no differentiation of the term $[\,]_{\alpha\beta}$.

Finally, the basic Landau assumption reads

$$\delta T^{\mu\nu}u_\mu u_\nu = \mathrm{Sp}\int d^4p\, p^\mu u_\mu v^\nu(p)u_\nu\,\delta f_{\rm eq}(p)$$
$$+\frac{1}{2}\mathrm{Sp}\int d^4p\, d^4p'\,\bar{f}_{\mu\nu}(p,p')\{[p^\mu - \Sigma_0^\mu(p)][p^\nu - \Sigma_0^\nu(p)]\}$$
$$\times\,\delta f_{\rm eq}(p)\delta f_{\rm eq}(p') \tag{14.93}$$

and hence we can safely set

$$f(p,p') \equiv \bar{f}_{\mu\nu}(p,p')\;[p^\mu - \Sigma_0^\mu(p)][p^\nu - \Sigma_0^\nu(p)], \tag{14.94}$$

where we have given the same name for the initial $f(p,p')$ and the final one.

(4) Finally, we have the two equations at the basis of the theory

$$\begin{cases} \dfrac{\delta(T^{\mu\nu}u_\mu u_\nu)}{\delta f_{\rm eq}(p)} = p_\mu u^\mu v^\nu(p)u_\nu, \\[2ex] \dfrac{\delta^2(T^{\mu\nu}u_\mu u_\nu)}{\delta f_{\rm eq}(p)\delta f_{\rm eq}(p')} = f(p,p'). \end{cases} \tag{14.95}$$

The first equation is shown to be an exact generalization of the nonrelativistic one: in the rest frame it reduces to p^0v^0, and thus to p^0 in the Galilean limit. The second equation, once integrated (functionally) over $\delta f_{\rm eq}(p)$, yields

$$p_\mu u^\mu v^\nu(p)u_\nu = p_\mu u^\mu v^\nu(p)u_\nu|_0 + \int d^4p'\,f(p,p')\delta f_{\rm eq}(p'), \tag{14.96}$$

where the reference to the implicit dependence on $f_{\rm eq}$ has been omitted and the index 0 in the first term on the right hand side refers to a *noninteracting* quasiparticle at $T = 0$ K. Explicitly, one has

$$[p^0v^0(p)]_0 = \sqrt{[\mathbf{p} - \Sigma_0(\varepsilon_F)]^2 + m_{\rm eff}^2}\,[v^0(p)]_0, \tag{14.97}$$

where $m_{\rm eff}^2$ is the effective mass of the quasiparticle and $\Sigma_0(\varepsilon_F)$ comes from the mass shell of the quasiparticle. It is defined, as usual, through

$$\begin{cases} m_{\rm eff} \equiv \dfrac{|\mathbf{p}_F|}{|\mathbf{v}_F|} = \dfrac{|\mathbf{p}_F|}{v_F}, \\[2ex] \mathbf{v} \equiv \nabla p^0(\mathbf{p}) = \dfrac{dp^0(\mathbf{p})}{d\mathbf{p}}, \end{cases} \tag{14.98}$$

where the index F refers to the value at the Fermi surface. In the Landau theory of the Fermi liquid, the system is indeed weakly excited above its Fermi surface, where the quasiparticles have a sufficiently long lifetime. Near the Fermi surface, one can always expand $p^0(\mathbf{p})$ into a Taylor series and write

$$p^0(\mathbf{p}) \approx \varepsilon_F + (|\mathbf{p}| - p_F)v_F, \tag{14.99}$$

as in the nonrelativistic case.

Furthermore, we shall also neglect spin effects in the following way. A glance at the equation for the effective four-velocity

$$\bar{v}^\mu(p) = \gamma^\nu - \nabla^\nu \Sigma(p) \tag{14.100}$$

indicates that, in the case considered, the only 4×4 matrix dependence occurs only through the usual term introduced by the four-current. This latter implicitly contains two terms: a spin term — the Gordon term — which we neglect, and a convective one which has the form $\frac{i}{2}\overleftrightarrow{\partial}$. All this leads to the substitution

$$\gamma^\nu \to \frac{p^\nu - \frac{1}{2}\nabla^\nu \Pi(p)}{m_{\text{eff}}} \tag{14.101}$$

and hence to a quasifermion four-velocity that reads

$$v^\mu(p) = \frac{p^\nu - \frac{1}{2}\nabla^\nu \Pi(p)}{m_{\text{eff}}}. \tag{14.102}$$

(5) In this approach to Landau–Fermi relativistic liquids, the knowledge of the effective mass is immediate since the dispersion equations of the quasiparticles are already known as

$$p^2 = \Pi(p), \tag{14.103}$$

and since they "live" in the vicinity of the Fermi surface, one has

$$p^2 = \Pi(p_F) \equiv m_{\text{eff}}^2 \tag{14.104}$$

or, if one has

$$[p - V(p)]^2 = \tilde{\Pi}(p), \tag{14.105}$$

then one also gets

$$\Pi(p_F) \equiv m_{\text{eff}}^2. \tag{14.106}$$

Note that the term with the index 0 does not depend on $f_{\text{eq}}(p)$ taking account of the fact that it is related to free quasiparticles. This equation being valid for arbitrary δf_{eq}, performing the δ operation and taking into account the symmetry of $f(p,p')$, one has

$$p^\mu u_\mu = p_\lambda u^\lambda \frac{v^\mu(p)u_\mu|_0}{v^\nu(p)u_\nu} + \frac{1}{v\cdot u}\int d^4p' f(p,p')\delta f_{\text{eq}}(p'). \tag{14.107}$$

Note the difference from the nonrelativistic equation

$$\varepsilon = \varepsilon_0 + \int d^4p' f(p,p')\delta f_{\text{eq}}(p'). \tag{14.108}$$

(6) A final remark is that spin cannot be dealt with here separately: it is not a constant of the motion; only the total kinetic momentum is a first integral, or rather its projection over an axis. Hence, we shall first look at the case where spin is not explicitly taken into account. Later we shall deal with the case of a magnetic field interacting with spin; in such a case, the additional Hamiltonian,

$$H_1 = \mu_N F^{\mu\nu} J_{\mu\nu} = \mu_N F^{\mu\nu}\cdot\left[L_{\mu\nu} + \frac{1}{2}\sigma_{\mu\nu}\right], \tag{14.109}$$

also commutes with the ordinary Hamiltonian (μ_N is the magnetic moment of the quasiparticle) and we have to consider spin. In this case, since $\mu_N F^{\mu\nu} J_{\mu\nu}$ is a first integral of the motion, we have to make the substitution

$$f_{\text{eq}}(p) \to f_{\text{eq}B}(p) = \frac{1}{\exp(\beta H\cdot u + \beta\mu_N F^{\mu\nu} J_{\mu\nu}) + 1}, \tag{14.110}$$

where $F^{\mu\nu} J_{\mu\nu}$ commutes with $H\cdot u$, the latter quantity being the Hamiltonian of the system *with* a magnetic field (see Chap. 12). However, the Wigner function with a magnetic field is certainly not of the above form and is extremely complicated. Nevertheless, when the magnetic field is weak enough not to perturb the equilibrium distribution — i.e. when the above $f_{\text{eq}B}$ is still valid — spin effects can be studied.

(7) Let us now present the way the distribution function is varied under Landau's assumption. This variation consists of two types: that of f_{eq} and the other. Let us first consider the first variation. This distribution function can vary (i) from δT, (ii) from $\delta\mu$, and finally from δu^μ. Let us look at these

variations. They are given by

$$\begin{aligned}\delta f_{\rm eq}(p) &= \delta\frac{1}{\exp[\beta p\cdot u-\beta\mu]+1}\\ &= -\frac{p\cdot u-\mu+\frac{\partial\mu}{\partial T}+\frac{\partial(p\cdot u)}{\partial T}}{(\exp[\beta p\cdot u-\beta\mu]+1)^2}\delta T-\frac{\frac{\partial(p\cdot u)}{\partial\mu}-T}{(\exp[\beta p\cdot u-\beta\mu]+1)^2}\delta\mu\\ &\quad -\frac{\beta p_\mu}{(\exp[\beta p\cdot u-\beta\mu]+1)^2}\delta u^\mu. \end{aligned}\tag{14.110}$$

The other parts of $F(p)$, i.e.

$$\frac{\gamma\cdot[p-\Sigma(p)]+\Sigma_S(p)}{\Sigma_S(p)},\tag{14.111}$$

are subject to

$$\begin{aligned}\delta\left\{\frac{\gamma\cdot[p-\Sigma(p)]+\Sigma_S(p)}{\Sigma_S(p)}\right\} &= -\delta\left\{\frac{\gamma\cdot u\Sigma_u(p)+\gamma\cdot p\Sigma_S(p)}{\Sigma_S}\right\}\\ &= -\gamma\cdot u\left\{\frac{\partial}{\partial T}\left(\frac{\Sigma_u}{\Sigma_S}\right)\delta T+\frac{\partial}{\partial\mu}\left(\frac{\Sigma_u}{\Sigma_S}\right)\right\}\\ &\quad -\gamma\cdot p\left\{\frac{\partial}{\partial T}\left(\frac{\Sigma_p}{\Sigma_S}\right)\delta T+\frac{\partial}{\partial\mu}\left(\frac{\Sigma_p}{\Sigma_S}\right)\right\}, \end{aligned}\tag{14.112}$$

where the variation of u^μ is zero because of $u^\mu\delta u_\mu=0$.

(8) Let us now go beyond the equilibrium and start with the idea of C.G. van Weert and M.C.J. Leermakers (1984) to get a conservative energy–momentum tensor. They used the following remark: if there exists a primitive P of the functional equation

$$\delta P(x)=\int d^4p\; H(x,p)\delta f(x,p),\tag{14.113}$$

then the tensor

$$T^{\mu\nu}=T^{\mu\nu}_{\rm quasi}-\eta^{\mu\nu}\left[P-\int d^4p\; H(x,p)f(x,p)\right]$$

is conservative and thus represents the energy–momentum tensor of the system. Before discussing this expression, let us show explicitly how it occurs. A variation of the position $x\to x+\delta x$ yields

$$\delta P(x)=\partial_\mu P(x)\delta x^\mu\tag{14.114}$$

and hence it gives

$$\partial_\mu P(x)=\int dp\; H(x,p)\partial_\mu f(x,p).\tag{14.115}$$

From this relation, it follows that the conservation equation now takes on the announced form. It remains for one to identify the specific value of the functional P. Those authors found that P can be identified with the usual pressure. To see this, let us evaluate the above energy–momentum tensor in thermal equilibrium; we then obtain

$$P_{\rm eq} = \int dp\ H(p) f_{\rm eq}(p), \tag{14.116}$$

so that from Landau's hypothesis — i.e. the off-equilibrium one has the same form as the equilibrium one — it follows that

$$P(x) = \int dp\ H(x,p) f(x,p). \tag{14.117}$$

Chapter 15

The QED Plasma

The quantum-electrodynamical plasma is a theoretical object of particular importance in astrophysics and, as already mentioned, in white dwarfs, where it is degenerate, and in the magnetosphere of pulsars, where it is strongly magnetized. Below, some physical properties where it occurs for white dwarfs are briefly mentioned. The QED plasma has been studied by numerous authors, in the pioneering works of E.S. Fradkin (1959a,b, 1960, 1965) — extended by I.A. Akhiezer and S.V. Peletminskii (1960), A.I. Akhiezer, I.A. Akhiezer and A.G. Sitenko (1962) — and of V.N. Tsytovich (1961), B. Bezzerides and D.F. DuBois (1972). Finally, in an article by H.A. Weldon (1982), an analysis of this kind of plasmas has been made from the viewpoint of a non-Abelian system.

15.1. Basic Equations

The electron field obeys the Dirac equations

$$\begin{cases} [i\gamma \cdot (\partial + ieA) - m]\psi(x) = 0, \\ \bar{\psi}(x)[i\gamma \cdot (\overleftarrow{\partial} - ieA) + m] = 0, \end{cases} \tag{15.1}$$

and the electromagnetic field Maxwell equations, which are written for the four-potentials as

$$\{\eta_{\mu\nu}\Box - \partial_\mu\partial_\nu\}A^\nu(x) = 4\pi e\bar{\psi}(x)\gamma_\mu\psi(x), \tag{15.2}$$

to which the Lorentz condition

$$\partial_\nu A^\nu(x) = 0 \tag{15.3}$$

should be added. Note that the gauge could be fixed by adding to the Lagrangian a term proportional to $(\partial_\nu A^\nu(x))^2$ or any other nonlinear term: the modes and other physical results must not depend on the proportionality coefficient, exhibiting thereby their gauge invariance.[1]

These equations are expressed in terms of the Wigner function operator:

$$\begin{cases} \{i\gamma\cdot\partial + 2[\gamma\cdot p - m]\}F_{\text{op}}(x,p) \\ \quad = 2e\displaystyle\int \frac{d^4x'}{(2\pi)^4} d^4p' \exp[-ip'\cdot(x-x')]\gamma_\mu F_{\text{op}}\left(x, p - \frac{1}{2}p'\right) A^\mu(x') \\ F_{\text{op}}(x,p)\{i\gamma\cdot\overleftarrow{\partial} - 2[\gamma\cdot p - m]\} \\ \quad = -2e\displaystyle\int \frac{d^4x'}{(2\pi)^4} d^4p' \exp[-ip'\cdot(x-x')]A^\mu(x')F_{\text{op}}\left(x, p + \frac{1}{2}p'\right)\gamma_\mu \\ \Box A^\mu(x) - (1-\lambda)\partial^\mu\partial_\nu A^\nu(x) = 4\pi e \mathrm{Sp}\displaystyle\int d^4p\ \gamma^\mu F_{\text{op}}(x,p), \end{cases} \tag{15.4}$$

where λ is the gauge-fixing parameter. From this generating equation and the Hartree–Vlasov ansatz, one gets the relativistic quantum Vlasov equation

$$\begin{aligned} &\{i\gamma\cdot\partial + 2[\gamma\cdot p - m]\}F(x,p) \\ &\quad = 2e\int \frac{d^4x'}{(2\pi)^4} d^4p' \exp[-ip'\cdot(x-x')]\gamma_\mu F\left(x, p - \frac{1}{2}p'\right) \langle A^\mu(x')\rangle, \\ &F(x,p)\{i\gamma\cdot\overleftarrow{\partial} - 2[\gamma\cdot p - m]\} \\ &\quad = -2e\int \frac{d^4x'}{(2\pi)^4} d^4p' \exp[-ip'\cdot(x-x')]\langle A^\mu(x')\rangle F\left(x, p + \frac{1}{2}p'\right)\gamma_\mu, \\ &\Box\langle A^\mu(x)\rangle = 4\pi e \mathrm{Sp}\int d^4p\ \gamma^\mu F(x,p), \end{aligned} \tag{15.5}$$

where we have chosen $\lambda = 0$.

15.2. Plasma Collective Modes

We could follow exactly the same path as for the "scalar plasma" or for the classical electrodynamic plasma: the Hartree–Vlasov equation is first linearized around the equilibrium state and next solved in a Fourier transform so as to find a homogeneous equation for the electromagnetic field and

[1]This is done, as an example, at the end of this chapter.

finally the dispersion relations obeyed by the plasma modes [R. Hakim and J. Heyvaerts (1978)].

The Vlasov–Hartree ansatz for solving the equation is

$$\langle F(x,p) \otimes A(x') \rangle \sim F(x,p) \otimes \langle A(x') \rangle \tag{15.6}$$

and, in order to obtain the dispersion relation for electromagnetic waves propagating through the plasma, it is linearized about an equilibrium state,

$$\begin{cases} \langle A^\mu(x) \rangle_{\text{eq}} \equiv 0, \\ F_{\text{eq}}(p) = \dfrac{\gamma \cdot p + m}{4m} f_{\text{eq}}(p), \end{cases} \tag{15.7}$$

where $f_{\text{eq}}(p)$ is given, as usual, by

$$f_{\text{eq}}(p) = \frac{2m}{(2\pi)^3} \sum_{\pm} \frac{\theta(\pm p^0)\delta(p^2 - m^2)}{\exp[\pm\beta(u \cdot p - \mu)] + 1}. \tag{15.8}$$

Notice that the linearization procedure

$$\begin{cases} F(x,p) \sim F_{\text{eq}}(p) + F^{(1)}(p), \\ \langle A^\mu(x) \rangle \sim A^{\mu(1)}(x) \end{cases} \tag{15.9}$$

is essentially equivalent to the random phase approximation, as has been noted long ago [(see e.g. B. Jancovici (1962)]. Once they have been linearized and Fourier-transformed, the resulting equations read

$$\begin{cases} \left[\gamma \cdot \left(p - \dfrac{1}{2}k\right) - m\right] F^{(1)}(k,p) = eA^{(1)}(k) \cdot \gamma F_{\text{eq}}\left(p + \dfrac{1}{2}k\right), \\ F^{(1)}(k,p) \left[\gamma \cdot \left(p + \dfrac{1}{2}k\right) - m\right] = eA^{(1)}(k) F_{\text{eq}}\left(p - \dfrac{1}{2}k\right) \cdot \gamma. \end{cases} \tag{15.10}$$

Particular solutions to these equations are easily found to be

$$\begin{cases} F_a^{(1)}(k,p) = \frac{eA_\mu^{(1)}(k)}{4m} \left(\frac{\left[\gamma\cdot\left(p-\frac{1}{2}k\right)+m\right]\gamma^\mu\left[\gamma\cdot\left(p+\frac{1}{2}k\right)+m\right]}{\left(p-\frac{1}{2}k\right)^2-m^2} \right) f_{\text{eq}}\left(p + \dfrac{1}{2}k\right), \\ F_a^{(1)}(k,p) = \frac{eA_\mu^{(1)}(k)}{4m} \left(\frac{\left[\gamma\cdot\left(p-\frac{1}{2}k\right)+m\right]\gamma^\mu\left[\gamma\cdot\left(p+\frac{1}{2}k\right)+m\right]}{\left(p+\frac{1}{2}k\right)^2-m^2} \right) f_{\text{eq}}\left(p - \dfrac{1}{2}k\right), \end{cases} \tag{15.11}$$

where the necessary $i\varepsilon$ terms have been provisionally omitted. The most general solutions to these equations are respectively of the general forms

$$\begin{cases} F^{(1)} = F_a^{(1)} + \left[\gamma \cdot \left(p - \frac{1}{2}k\right) + m\right] G_1 \left(p - \frac{1}{2}k\right), \\ F^{(2)} = F_b^{(1)} + G_2 \left(p + \frac{1}{2}k\right) \left[\gamma \cdot \left(p + \frac{1}{2}k\right) + m\right], \end{cases} \tag{15.12}$$

where the last terms on the right hand sides represent the *arbitrary* solutions to the inhomogeneous equations (15.11). In the last equations G_1 and G_2 are arbitrary 4×4 matrices of p and are respectively on the mass shell

$$\begin{cases} \left(p - \frac{1}{2}k\right)^2 = m^2 \quad \text{for } i = 1, \\ \left(p + \frac{1}{2}k\right)^2 = m^2 \quad \text{for } i = 2. \end{cases} \tag{15.13}$$

The reason why $G_1(p - k/2)$, for instance, contains a $\delta[(p - k/2)^2 - m^2]$ factor can be seen by applying the operator $[\gamma \cdot (p - k/2) - m]$ to Eq. (15.11) from the left. The first term vanishes since $F_a^{(1)}$ is a solution, while the second vanishes only if $G_1(p - k/2)$ also contains a $\delta[(p - k/2)^2 - m^2]$ factor.

From the necessary identity of these equations, one concludes that

$$\begin{aligned} F^{(1)} &= F_a^{(1)} + F_b^{(2)} \\ &= -\frac{e}{8m} \left[\frac{\left[\gamma \cdot (p - \frac{1}{2}k) + m\right] \gamma \cdot A^{(1)}(k) \left[\gamma \cdot (p + \frac{1}{2}k) + m\right]}{k \cdot p} \right] \\ &\quad \times \left[f_{\text{eq}}\left(p + \frac{1}{2}k\right) - f_{\text{eq}}\left(p - \frac{1}{2}k\right) \right], \end{aligned} \tag{15.14}$$

where use was made of the fact that

$$\left(p \pm \frac{1}{2}k\right)^2 - m^2 = \pm 2k \cdot p \quad \text{when} \quad \left(p \pm \frac{1}{2}k\right)^2 = m^2, \tag{15.15}$$

valid only when both of the Equations (15.15) hold.

Using

$$4\pi J_{(1)}^{\lambda}(k) = \Pi^{\lambda\mu}(k) A_{\mu}^{(1)}(k) \tag{15.16}$$

finally, one then finds that

$$\Pi^{\mu\nu}(k) = -\omega_P^2 K^{\mu\nu}(k) - \Omega_P^2 \eta^{\mu\nu} - k^2 \Delta^{\mu\nu}(k) \frac{\omega_P^2}{4n} I \tag{15.17}$$

for the polarization tensor, where one has set

$$K^{\mu\nu}(k) = \frac{1}{n}\int d^4p\, p^\mu p^\nu \left(\frac{f\left(p+\frac{1}{2}k\right) - f\left(p-\frac{1}{2}k\right)}{k\cdot p + i\varepsilon}\right), \tag{15.18}$$

$$I = \int d^4p \left(\frac{f\left(p+\frac{1}{2}k\right) - f\left(p-\frac{1}{2}k\right)}{k\cdot p + i\varepsilon}\right), \tag{15.19}$$

$$\Omega_P^2 = \frac{4\pi e^2}{m}\int d^4p\, f_{\text{eq}}(p). \tag{15.20}$$

In these equations the $i\varepsilon$ of the resonant denominator $(1/k\cdot p)^{-1}$ have been re-established — they correspond to the usual Landau prescription; as a result these equations acquire an imaginary part that is vanishing as long as the waves are superluminous.

From the general equation for collective modes

$$\text{Det}\left[k^2\eta^{\mu\nu} - \Pi^{\mu\nu}(k)\right] = 0, \tag{15.21}$$

and the Lorentz gauge condition

$$k\cdot A_{(1)}(k) = 0, \tag{15.22}$$

the transverse modes are found to be

$$-\frac{\Omega_P^2}{\omega^2 - k^2} + \frac{\omega_P^2}{\omega^2 - k^2}K^{11}(k) - \frac{\omega_P^2}{4n}I = 0, \tag{15.23}$$

while the longitudinal modes are

$$1 - \frac{\Omega_P^2}{\omega^2 - k^2} - \frac{\omega_P^2}{\omega^2 - k^2}K^{00}(k) + \frac{\omega}{|\mathbf{k}|}\frac{\omega_P^2}{\omega^2 - k^2}K^{30}(k) - \frac{\omega_P^2}{4n}I = 0. \tag{15.24}$$

Several remarks are now in order. First, these last equations for the longitudinal and transverse modes reduce to the classical relativistic equations [R. Hakim and A. Mangeney (1968, 1971)] when $\hbar \to 0$. To see this, (i) neglect spin, i.e. I, (ii) suppress the contributions of the positrons and (iii) suppress the $+1$ of the Fermi factor in Eq. (15.14); finally, take the long wavelength limit.

A second remark deals with the absence of vacuum contributions: indeed, in the absence of matter $f_{\text{eq}}(p)$ goes to zero. This is due to the fact that we have implicitly used a normal ordering of our field operators, thereby killing all vacuum contributions. Actually, if we do not omit the vacuum contribution to $f_{\text{eq}}(p)$, we have to replace $f_{\text{eq}}(p)$ by its expression

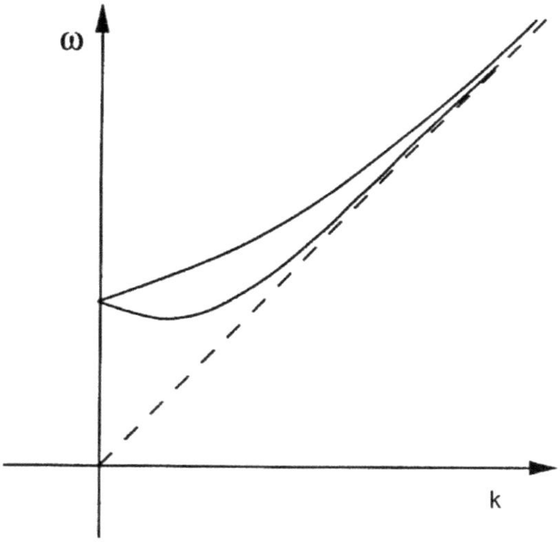

Fig. 15.1 Two typical curves of the collective oscillation of a plasma; It means "longitudinal" and T "transverse."

plus the vacuum Wigner funtion

$$f_{\text{vac}}(p) = -\frac{2m}{(2\pi)^3}\theta(-p^0)\delta(p^2 - m^2). \tag{15.25}$$

Inserting this expression into Eq. (15.17), for instance, gives rise to the usual polarization tensor at order e^2. Notice that $F_{\text{vac}}(p)$ is given by an expression quite similar to that of $F_{\text{eq}}(p)$; the calculation of $f_{\text{vac}}(p)$ is performed with $\rho_{\text{vac}} \equiv |\text{vac}\rangle\langle\text{vac}|$. The last equation expresses the fact that the Dirac ocean of negative energy electrons is uniformly filled.

These relations have been derived with a great variety of methods by several authors [V. Tsytovich (1961), D. Biskamp (1961), B. Bezzerides and D.F. Dubois (1972), R. Hakim and J. Heyvaerts (1978), H. Sivak (1984), B. Jancovici (1962) (at $T = 0\,\text{K}$)] and we refer to V.N. Tsytovich (1960) for a discussion.

It is, however, quite instructive to look for these modes via the use of plasma fluctuations[2] [H. Sivak (1984)].

---

[2]See the general study of plasma fluctuations in A.G. Sitenko, *Electromagnetic Fluctuations in Plasmas* (Academic, New York, 1967).

15.3. The Fluctuation–Dissipation Theorem and Its Inverse

As an example of the role played by fluctuations, let us consider the thermodynamic potential Ω,

$$\Omega = -\frac{1}{\beta}\ln\{\mathrm{Tr}\exp(-\beta[H_0 + H_{\mathrm{int}} - \mu B])\}, \tag{15.26}$$

where B is the baryonic number (or charge) operator, μ the chemical potential, H_0 the free Hamiltonian and H_{int} the interaction Hamiltonian,

$$H_{\mathrm{int}} = \int d^3x\, J_{\mathrm{op}}(x) \cdot A(x), \tag{15.27}$$

in the case of a relativistic quantum plasma (J_{op} is the four-current operator which is proportional to e; A is the electromagnetic four-potential). One can show that[3]

$$\Omega = \Omega_{\mathrm{free}} - \frac{2}{(2\pi)^3}\int_0^e de \int \frac{d^4k}{k^2}\langle J_{\mathrm{op}} \cdot J_{\mathrm{op}}\rangle_{(k)}, \tag{15.28}$$

an expression that exhibits the role of the four-current fluctuations in the calculation of the thermodynamics of an electromagnetic plasma (see also Chap. 12).

A formal connection can exist between the linear response to a given excitation of a system and a related quantity, and this connection is known as the *fluctuation–dissipation theorem*.[4] For instance, if a plasma is excited by an external electromagnetic field $A_{\mathrm{ext}}(x)$, the perturbation being linear as

$$\int d^3x\, O_{\mathrm{op}} \cdot A_{\mathrm{ext}}(x), \tag{15.29}$$

where O_{op} is the operator representing some physical quantity, then the spectrum of O_{op} in thermal equilibrium is provided by

$$\langle O^{\mu}_{\mathrm{op}} O^{\nu}_{\mathrm{op}}\rangle_{(k)} = -\frac{i}{\exp(\beta\omega) - 1}[\alpha^{*\mu\nu}(k) - \alpha^{\mu\nu}(k)], \tag{15.30}$$

where $\alpha^{\mu\nu}(k)$ characterizes the linear response of O_{op} to the external disturbance $A_{\mathrm{ext}}(x)$:

$$\langle O_{\mathrm{op}}\rangle_{(k)} = \alpha^{\mu\nu}(k) A_{\nu\mathrm{ext}}(k). \tag{15.31}$$

---

[3] A. Fetter and J. Walecka, *Quantum Theory of Many Particle System* (McGraw-Hill, New York, 1971).

[4] See e.g. L.E. Reichl, *A Modern Course in Statistical Physics* (Arnold, London, 1980).

Therefore, the spectrum of the fluctuations of $O_{\rm op}$ is determined when the anti-Hermitian part of the response function is known. Conversely, causality implies the Kramers–Kronig relations, which relate the Hermitian and anti-Hermitian parts of $\alpha^{\mu\nu}(k)$. Accordingly, the full response function $\alpha^{\mu\nu}(k)$ can be obtained from the knowledge of the spectrum of the fluctuations of $O_{\rm op}$.

Let us now be more specific. First, we split the response function $\alpha^{\mu\nu}(k)$ into its symmetric and antisymmetric parts,

$$\alpha^{\mu\nu}(k) = \alpha_S^{\mu\nu}(k) + \alpha_A^{\mu\nu}(k), \tag{15.32}$$

so that the fluctuation–dissipation theorem can be rewritten as

$$\langle O_{\rm op}^{\mu} O_{\rm op}^{\nu} \rangle_{(k)} = -\frac{2}{\exp(\beta\omega) - 1}\left[i{\rm Re}\, \alpha_A^{\mu\nu}(k) + {\rm Im}\, \alpha_S^{\mu\nu}(k) \right]. \tag{15.33}$$

Next, with the use of the Kramers–Kronig relations[5]

$$\begin{cases} {\rm Re}\, \alpha^{\mu\nu}(\omega, \mathbf{k}) = \alpha^{\mu\nu}(\omega \to \infty, \mathbf{k}) - \dfrac{1}{\pi} \displaystyle\int d\omega' \frac{{\rm Im}\, \alpha^{\mu\nu}(\omega', \mathbf{k})}{\omega' - \omega}, \\ {\rm Im}\, \alpha^{\mu\nu}(\omega, \mathbf{k}) = \dfrac{1}{\pi} \displaystyle\int d\omega' \frac{\alpha^{\mu\nu}(\omega \to \infty, \mathbf{k}) - {\rm Re}\, \alpha^{\mu\nu}(\omega', \mathbf{k})}{\omega' - \omega}, \end{cases} \tag{15.34}$$

the above expression can be inverted as

$$\alpha^{\mu\nu}(k) = \alpha^{\mu\nu}(\omega \to \infty, \mathbf{k}) - \frac{1}{2\pi} \int d\omega' \frac{\exp(\beta\omega') - 1}{\omega' - \omega - i\varepsilon} \langle O_{\rm op}^{\mu} O_{\rm op}^{\nu} \rangle_{\rm eq}(\omega, \mathbf{k}), \tag{15.35}$$

allowing thereby the full determination of the response function from the equilibrium fluctuation spectrum. The quantity $\alpha^{\mu\nu}(\omega \to \infty, \mathbf{k})$, which does not appear in the nonrelativistic case, must be determined by other physical considerations and has been discussed and found by H. Sivak (1984), whom we have followed here.

Finally, note that this derivation of the inversion of the fluctuation–dissipation formula is quite standard and that relativity does not enter into it.

15.4. Four-Current Fluctuations and the Polarization Tensor

Let us now use Kubo's formula and its inverse to study some fluctuations in the QED plasma.

[5]See e.g. L. Landau, E. Lifschitz, *Statistical Physics* (Pergamon, Oxford, 1980).

(1) We first study the fluctuations of the four-current J^μ and their connections with energy–momentum fluctuations $T^{\mu\nu}$. Using the fluctuation of the Wigner function $F(k;p,p')$ and after a straightforward calculation, we obtain [H. Sivak (1984)]

$$\begin{aligned}\langle \tilde{J}^\mu \tilde{J}^\nu \rangle_{\text{eq}}(k) &= \frac{e^2}{(2\pi)^2}\frac{1}{\exp(\beta\omega)-1}\\ &\times \sum_{\ell,\ell'} \ell\ell' \int \frac{d^3p}{EE_{\ell'}}(\delta_{\ell 1} - N_E)\delta(E+\ell E_{\ell'} + \ell'\omega)\\ &\times \left[\frac{1}{2}k^2\eta^{\mu\nu} + 2p^\mu p^\nu + \ell' p^{(\mu}k^{\nu)}\right],\end{aligned} \tag{15.37}$$

$$\begin{aligned}\langle \tilde{T}^{\mu\nu} \tilde{T}^{\alpha\beta} \rangle_{\text{eq}}(k) &= \frac{1}{(2\pi)^2}\frac{1}{\exp(\beta\omega)-1}\\ &\times \sum_{\ell,\ell'} \ell\ell' \int \frac{d^3p}{EE_{\ell'}}(\delta_{\ell 1} - N_E)\delta(E+\ell E_{\ell'} + \ell'\omega)\\ &\times \left(p^\nu + \frac{\ell'}{2}k^\nu\right)\left(p^\beta + \frac{\ell'}{2}k^\beta\right)\\ &\times \left[\frac{1}{2}k^2\eta^{\mu\alpha} + 2p^\mu p^\alpha + \ell' p^{(\mu}k^{\alpha)}\right],\end{aligned} \tag{15.38}$$

$$\begin{aligned}\langle \tilde{T}^{\mu\nu} \tilde{J}^{\alpha} \rangle_{\text{eq}}(k) &= -\frac{e}{(2\pi)^2}\frac{1}{\exp(\beta\omega)-1}\\ &\times \sum_{\ell,\ell'} \ell\ell' \int \frac{d^3p}{EE_{\ell'}}Q_E\delta(E+\ell E_{\ell'} + \ell'\omega)\\ &\times \left(p^\nu + \frac{\ell'}{2}k^\nu\right)\left[\frac{1}{2}k^2\eta^{\mu\alpha} + 2p^\mu p^\alpha + \ell' p^{(\mu}k^{\alpha)}\right],\end{aligned} \tag{15.39}$$

where the following notations have been used:

$$\begin{cases} E = p^0 = \sqrt{\mathbf{p}^2 + m^2}, \\ E_{\ell'} = \sqrt{(\mathbf{p}+\ell'\mathbf{k})^2 + m^2}, \end{cases} \tag{15.40}$$

$$\begin{bmatrix} N_E \\ Q_E \end{bmatrix} = \frac{1}{\exp(\beta[E-\mu])+1} \pm \frac{1}{\exp(\beta[E+\mu])+1}. \tag{15.41}$$

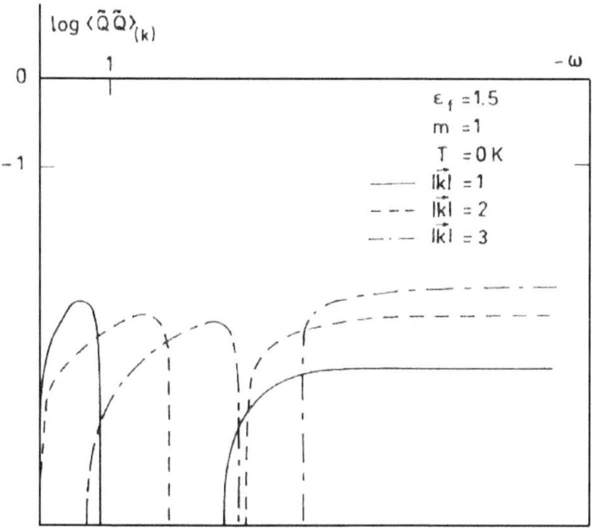

Fig. 15.2 *Spectrum of density fluctuations* (parameters are indicated), after H. Sivak, *Ann. Phys. (N.Y.)* **159**, 351 (1984).

Let us note that these three spectra are all orthogonal to k^μ owing to the four-current and energy–momentum conservation, as can directly be checked on Figs. 15.2–15.4. An explicit expression for them at zero temperature can be found in the article by H. Sivak (1984). The figures below refer to this case where the fluctuations are plotted versus ω for only one value of $\mathbf{k}$ [other values give rise to similar curves and can be found in H. Sivak (1984)].

These three figures present common characteristics; there exist two branches for each curve, except for the energy density/charge density spectrum, which has no right branch. The left branch is connected with the process of electron–Fermi-hole pair production and the right one with electron–positron pair production. This can be seen from the δ terms which occur in the formulae for the various spectra, which express the energy conservation when the pairs of whatever kind are excited by a plasma wave of frequency ω. The gap between the two branches corresponds to a frequency region where a plasma wave cannot excite pairs of whatever type — a transparency region.

(2) Conversely, let us study the electromagnetic fluctuations from the polarization tensor $\Pi^{\mu\nu}(k)$. We first relate $\Pi^{\mu\nu}(k)$ and $\alpha^{\mu\nu}(k)$. From the definition of $\Pi^{\mu\nu}(k)$ and $\alpha^{\mu\nu}(k)$, and from Maxwell's equations together with

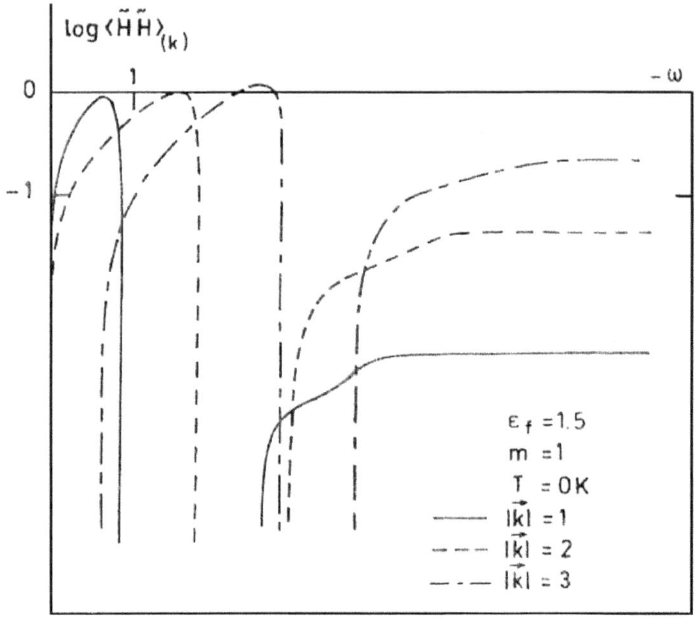

Fig. 15.3 *Spectrum of energy density fluctuations* (parameters are indicated), after H. Sivak, *Ann. Phys. (N.Y.)* **159**, 351 (1984).

the Lorentz condition

$$\begin{cases} -k^2 \tilde{A}^{\mu}_{\text{int}}(k) = 4\pi J^{\mu}(k), \\ k \cdot \tilde{A}^{\mu}_{\text{int}}(k) = 0, \end{cases} \tag{15.41}$$

we get

$$J_{\lambda}(k) = k^2 \{k^2 \eta^{\alpha\beta} + 4\pi \Pi^{\alpha\beta}(k)\}^{-1}_{\lambda\mu} \Pi^{\mu\nu}(k) A_{\nu\text{ext}}(k) \tag{15.42}$$

and hence

$$\alpha^{\nu}_{\lambda}(k) = k^2 \{k^2 \eta^{\alpha\beta} + 4\pi \Pi^{\alpha\beta}(k)\}^{-1}_{\lambda\mu} \Pi^{\mu\nu}(k). \tag{15.43}$$

Finally, the four-current fluctuations are given by

$$\begin{aligned} \langle \tilde{J}_{\mu} \tilde{J}_{\nu} \rangle_{(k)} = & -ik^2 \frac{1}{\exp(\beta\omega) - 1} [\{k^2 \eta^{\alpha\beta} + 4\pi \Pi^{\alpha\beta}(k)\}^{*-1}_{\mu\lambda} \Pi^{*\lambda}_{\nu}(k) \\ & - \{k^2 \eta^{\alpha\beta} + 4\pi \Pi^{\alpha\beta}(k)\}^{-1}_{\nu\lambda} \Pi^{\lambda}_{\mu}(k)]. \end{aligned} \tag{15.44}$$

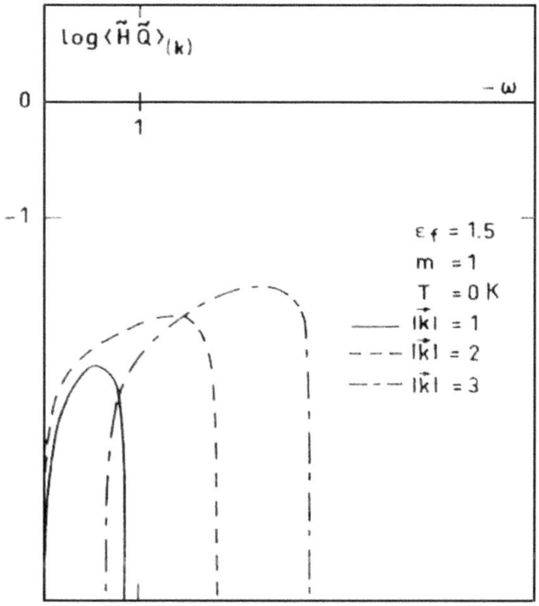

Fig. 15.4 *Spectrum of energy density/charge density fluctuations* (parameters are indicated), after H. Sivak, *Ann. Phys. (N.Y.)* **159**, 351 (1984).

From Maxwell's equations, the fluctuations of the electromagnetic field are easily obtained as

$$\langle \tilde{A}_\mu \tilde{A}_\nu \rangle_{(k)} = -i\frac{(4\pi)^2}{k^2}\frac{1}{\exp(\beta\omega)-1}[\{k^2\eta^{\alpha\beta}+4\pi\Pi^{\alpha\beta}(k)\}^{*-1}_{\mu\lambda}\Pi^{*\lambda}_{\nu}(k)$$
$$-\{k^2\eta^{\alpha\beta}+4\pi\Pi^{\alpha\beta}(k)\}^{-1}_{\nu\lambda}\Pi^{\lambda}_{\mu}(k)]. \tag{15.45}$$

In matrix form, it looks a bit simpler:

$$\langle \tilde{A}\otimes\tilde{A}\rangle_{(k)} = -i\frac{(4\pi)^2}{k^2}\frac{1}{\exp(\beta\omega)-1}\frac{\Pi^*(k)-\Pi(k)}{k^2 I+4\pi\Pi(k)}. \tag{15.46}$$

15.5. The Polarization Tensor at Order e^2

We now compute the polarization tensor (i.e. the linear response to an external electromagnetic field) from the ideal gas four-current fluctuations with the use of the fluctuation–dissipation theorem, and we get

$$\Pi^{\mu\nu}(k) = -\frac{1}{2\pi}\int d\omega'\frac{\exp(\beta\omega')-1}{\omega'-\omega-i\varepsilon}\langle J^\mu J^\nu\rangle_{\rm eq}(\omega',\mathbf{k}). \tag{15.47}$$

In order to make an evaluation at order e^2, it is sufficient to replace the equilibrium fluctuation spectrum of the four-current by its ideal electron gas expression, which has been calculated above. We then find that

$$\Pi^{\mu\nu}(k) = \Pi^{\mu\nu}_{\text{vac}}(k) - \frac{e^2}{(2\pi)^3} \sum_{\ell,\ell'=\pm 1\ell} \int \frac{d^3p}{EE_{\ell'}} \times \left[\frac{1}{\exp(\beta[E-\mu])+1} - \frac{1}{\exp(\beta[E+\mu])+1}\right] \times \frac{\frac{1}{2}\eta^{\mu\nu}k^2_{\ell'} + 2p^\mu p^\nu + \ell' p^{(\mu}k^{\nu)}_{\ell'}}{E + \ell E_{\ell'} + \ell'(\omega + i\varepsilon)}, \tag{15.48}$$

where $\Pi^{\mu\nu}_{\text{vac}}(k)$ is the vacuum polarization tensor, calculated from the fluctuation–dissipation theorem, and in which the limit

$$\begin{cases} n \to 0, \\ \beta \to \infty \end{cases} \tag{15.49}$$

is taken. Also, the following notation has been used:

$$k^\mu_{\ell'} \equiv \{-\ell'(E + \ell E_{\ell'}),\ \mathbf{k}\}. \tag{15.50}$$

The vacuum polarization tensor is obviously infinite and must be renormalized; this has been done many times and the result is[6]

$$\bar{\Pi}^{\mu\nu}_{\text{vac}}(k) = \frac{e^2k^2}{3(2\pi)^2}\Delta^{\mu\nu}(k)\left\{\frac{5}{3} + \frac{4m^2}{k^2} - \left(1 + \frac{2m^2}{k^2}\right)\sqrt{1 - \frac{4m^2}{k^2}} \times \left[2\,\text{arth}\left(1 - \frac{4m^2}{k^2}\right)^{-1/2} - i\pi\varepsilon(\omega)\theta(k^2 - 4m^2)\right]\right.. \tag{15.51}$$

As to the matter part of the polarization tensor, it can be written in the form

$$\Pi^{\mu\nu}_{\text{mat}}(k) = P_1(k)\Delta^{\mu\nu}(k) + P_2(k)\Delta^{[\mu\lambda}(u)\Delta^{\nu]\sigma}(u)k_\lambda k_\sigma, \tag{15.52}$$

where $P_1(k)$ and $P_2(k)$ are given by

$$\begin{cases} P_1(k) = -\dfrac{k^2}{\mathbf{k}^2}\Pi^{\mu\nu}_{\text{mat}}(k)u_\mu u_\nu, \\ P_2(k) = \dfrac{1}{\mathbf{k}^2}\left\{\dfrac{1}{2\mathbf{k}^2}\Pi_{\mu\nu\text{mat}}(k)\Delta^{[\mu\lambda}(u)\Delta^{\nu]\sigma}(u)k_\lambda k_\sigma - P_1(k)\right\}, \end{cases} \tag{15.53}$$

[6] See e.g. C. Nash, *op. cit.*; or C. Itzykson and J.B. Zuber, *op. cit.* See also the interesting derivation by H. Sivak (1984).

which we now give at $T = 0\,\text{K}$. The real part of $\Pi^{\mu\nu}_{\text{mat}}(k)$ is obtained from the real parts of $P_1(k)$ and $P_2(k)$; $P_1(k)$ is found to be

$$\begin{aligned}\text{Re}P_1(k) = \frac{k^2}{12\pi}\Bigg\{ & 8\frac{\varepsilon_f p_f}{\mathbf{k}^2} - 2\text{arsh}\left(\frac{p_f}{m}\right) + \left(1 + \frac{2m^2}{k^2}\right)\sqrt{\left|1 - \frac{4m^2}{k^2}\right|}I(k) \\ & + \frac{2\varepsilon_f}{|\mathbf{k}|^3}\left(\frac{3}{4}k^2 + \varepsilon_f^2\right)\ln\left|\frac{(k^2 - 2p_f|\mathbf{k}|)^2 - 4\omega^2\varepsilon_f^2}{(k^2 + 2p_f|\mathbf{k}|)^2 - 4\omega^2\varepsilon_f^2}\right| \\ & + \frac{\omega}{4|\mathbf{k}|^3}(3\mathbf{k}^2 - 12\varepsilon_f^2 - \omega^2)\ln\left|\frac{k^4 - (p_f|\mathbf{k}| + \omega\varepsilon_f)^2}{k^4 - 4(p_f|\mathbf{k}| - \omega\varepsilon_f)^2}\right|\Bigg\},\end{aligned} \tag{15.54}$$

while $P_2(k)$ is given by

$$\begin{aligned}\text{Re}\,P_2(k) = \frac{1}{8\pi^2k^2}\Bigg\{ & -4p_f\varepsilon_f\left(1 + 2\frac{k^2}{\mathbf{k}^2}\right) \\ & + \frac{k^2}{4|\mathbf{k}|}\sum_{\ell'=\pm 1}\Bigg[\left(1 - \frac{4m^2}{k^2} - \frac{(2\varepsilon_f + \ell'\omega)^2}{\mathbf{k}^2}\right) \\ & \times(2\varepsilon_f + \ell'\omega)\ln\left|\frac{k^2 - 2|\mathbf{k}|p_f + 2\ell'\omega\varepsilon_f}{k^2 + 2|\mathbf{k}|p_f + 2\ell'\omega\varepsilon_f}\right|\Bigg]\Bigg\},\end{aligned} \tag{15.55}$$

where $I(k)$ is given by

$$I(k) = \frac{1}{2}\ln\left|\frac{\left(\varepsilon_f + p_f\sqrt{1 - \frac{4m^2}{k^2}}\right)^2 - \frac{4m^4\omega^2}{k4}}{\left(\varepsilon_f - p_f\sqrt{1 - \frac{4m^2}{k^2}}\right)^2 - \frac{4m^4\omega^2}{k4}}\right|, \tag{15.56}$$

for $1 - \frac{4m^2}{k^2} \geq 0$

$$I(k) = \text{artg}\left[\frac{2m^2\,|\mathbf{k}| + k^2p_f}{\varepsilon_f\sqrt{k^2(4m^2 - k^2)}}\right] - \text{artg}\left[\frac{2m^2|\mathbf{k}| - k^2p_f}{\varepsilon_f\sqrt{k^2(4m^2 - k^2)}}\right]. \tag{15.57}$$

for $1 - \frac{4m^2}{k^2} < 0$

This result is identical with B. Jancovici's (1961) except for a misprint: $I(k)$ has a negative sign for $k^2 > 0$.

At the frequencies of pair creations of whatever kind (electron–positron or electron–hole), the real part of $\Pi^{\mu\nu}_{\text{mat}}(k)$ presents singularities, i.e. whenever

$$\begin{cases} |\omega| = |\varepsilon_- \pm \varepsilon_f|, \\ |\omega| = \varepsilon_+ \pm \varepsilon_f. \end{cases} \tag{15.57}$$

15.6. Quasiparticles in the Relativistic Plasma

As mentioned at the beginning of this chapter, the thermal properties of the QED plasma play an important role in astrophysics. Here only two properties are considered. First, the modification of the blackbody spectrum due to the propagation of quasiphotons in the medium; next, the modifications brought by the quasielectrons in the QED plasma. While the first considerations apply mainly to white dwarfs and the second ones to the primeval cosmology, only very small variations were obtained [M. Lemoine (1995)]. In this section, we follow M. Lemoine's (1995) work. Note that this has been considered in the nonrelativistic case by R. Chappell.[7]

15.6.1. *Quasiphotons in thermal equilibrium*

Basic definitions and equations. The quasiphotons obey the field equation

$$[k^2 - (1-\lambda)k^\mu k^\nu - \Pi^{\mu\nu}(k)]A_\nu(k) = 0, \tag{15.57b}$$

where λ is still the gauge-fixing parameter and $\Pi^{\mu\nu}(k)$ is intended to be the *retarded* polarization operator. This field equation can be derived from the Lagrangian

$$\mathcal{L} = -\frac{1}{4}F^{\mu\nu}F_{\mu\nu} - \frac{\lambda}{2}(\partial A)^2 + \frac{1}{2}\int d^4y A_\mu\left(x+\frac{1}{2}y\right)\Pi^{\mu\nu}(y)A_\nu\left(x-\frac{1}{2}y\right). \tag{15.58}$$

The Hermitian character of $\mathcal{L}$ implies that

$$\Pi^{\mu\nu+}(y) = \Pi^{\mu\nu}(-y). \tag{15.59}$$

From these equations, the energy–momentum tensor is found as in the preceding chapter, and reads

$$\begin{aligned} T^{\mu\nu}_{\text{quasiphotons}} &= -\frac{1}{2}[\partial^\nu A_\alpha F^{\mu\alpha} + F^{\mu\alpha}\partial^\nu A_\alpha] - \frac{\lambda}{2}[\partial^\nu A^\mu(\partial A) + (\partial A)\partial^\mu A^\nu] \\ &\quad + \frac{1}{2}\int d^4y\, y^\mu \int_0^{1/2} ds\left\{\partial^\nu A_\alpha\left(x - y\left[s-\frac{1}{2}\right]\right)\Pi^{\alpha\beta}(y)\right. \\ &\quad \times A_\beta\left(x-y\left[s+\frac{1}{2}\right]\right) - A_\alpha\left(x+y\left[s+\frac{1}{2}\right]\right)\Pi^{\alpha\beta}(y) \\ &\quad \left.\times\partial^\nu A_\beta\left(x+y\left[s-\frac{1}{2}\right]\right)\right\} - \mathcal{L}\eta^{\mu\nu}. \end{aligned} \tag{15.60}$$

[7]R. Chappell, Thesis (1959).

This expression can be rewritten in terms of the covariant Wigner operator[8] of the quasiphoton field,

$$f_{\mu\nu}(x,k) \underset{\text{def}}{\equiv} \int \frac{dR}{(2\pi)^4} \exp(-ik\cdot R)\; A_\nu\left(x+\frac{1}{2}R\right) A_\mu\left(x-\frac{1}{2}R\right), \quad (15.61)$$

where the following useful properties are still valid:

$$f_{\mu\nu}(k,p) = \frac{1}{(2\pi)^4} A_\nu\left(\frac{1}{2}k-p\right) A_\mu\left(\frac{1}{2}k+p\right), \quad (15.62)$$

$$A_\mu(x)A_\nu(y) = \int \frac{d^4R}{(2\pi)^4} \exp[ip\cdot(x-y)] f_{\nu\mu}\left(\frac{x+y}{2},p\right), \quad (15.63)$$

$$\partial_\alpha A_\mu(x)\partial_\beta A_\nu(x) = -\int \frac{d^4k}{(2\pi)^4}\exp(-ik\cdot x)\int d^4p\left(\frac{1}{2}k_\alpha - p_\alpha\right) \times \left(\frac{1}{2}k_\beta + p_\beta\right) f_{\nu\mu}(k,p). \quad (15.64)$$

With these notations and definitions, the equation for the Wigner operator can be written as

$$\begin{cases} \left\{\left(p+\frac{1}{2}k\right)^2 \eta^{\alpha\nu} - (1-\lambda)\left(\frac{1}{2}k^\alpha + p^\alpha\right)\left(\frac{1}{2}k^\nu + p^\nu\right) \right. \\ \left. \quad - \Pi^{\alpha\nu}\left(\frac{1}{2}k+p\right)\right\} f_{\nu\beta\,\mathrm{op}}(k,p) = 0, \\ f_{\alpha\mu\,\mathrm{op}}(k,p)\left\{\left(p-\frac{1}{2}k\right)^2 \eta^{\mu\beta} - (1-\lambda)\left(\frac{1}{2}k^\mu - p^\mu\right)\left(\frac{1}{2}k^\beta - p^\beta\right) \right. \\ \left. \quad - \Pi^{\mu\beta}\left(\frac{1}{2}k-p\right)\right\} = 0, \end{cases} \quad (15.65)$$

while those for the energy–momentum tensor read

$$T^{\mu\nu}_{\text{quasiphoton}}(k) = \int d^4p \frac{1}{2}\left[\left(\frac{1}{2}k^\nu - p^\nu\right)\left(\frac{1}{2}k^\mu + p^\mu\right) f^\alpha_{\ \alpha}\right.$$

[8]This corresponds to the general definition of $f_{\mu\nu}$; however, in the case considered here, one has $\langle F_{\mu\nu}\rangle_{\text{eq}} = 0$, and $\langle A_\mu\rangle$ reduces to a pure gauge. Note that $f_{\mu\nu}$ is not $\mathrm{Sp}(F\sigma_{\mu\nu})$. For the sake of simplicity, we shall choose $\langle A_\mu\rangle = 0$, and we must show the gauge invariance of the final results.

$$-\left(\frac{1}{2}k^\nu - p^\nu\right)\left(\frac{1}{2}k^\alpha + p^\alpha\right) f^\mu{}_\alpha + \left(\frac{1}{2}k^\mu - p^\mu\right)\left(\frac{1}{2}k^\nu + p^\nu\right) f^\alpha{}_\alpha$$
$$-\left(\frac{1}{2}k^\alpha - p^\alpha\right)\left(\frac{1}{2}k^\nu + p^\nu\right) f^\mu{}_\alpha + \lambda\left(\frac{1}{2}k^\nu - p^\nu\right)\left(\frac{1}{2}k^\alpha + p^\alpha\right) f^\mu{}_\alpha$$
$$+\lambda\left(\frac{1}{2}k^\alpha - p^\alpha\right)\left(\frac{1}{2}k^\nu + p^\nu\right) f^\mu{}_\alpha\Big]$$
$$+\frac{1}{2}\int_0^{1/2} ds \left[\left(\frac{1}{2}k^\nu + p^\nu\right)\nabla_\mu \Pi^{\alpha\beta}(p - ks)\right.$$
$$\left. - \left(\frac{1}{2}k^\nu - p^\nu\right)\nabla_\mu \Pi^{\alpha\beta}(p + ks)\right] f_{\beta\alpha}$$
$$-\eta^{\mu\nu}\left[\frac{1}{2}\left(\frac{1}{4}k^2 - p^2\right) f^\alpha{}_\alpha - \frac{1}{2}\left(\frac{1}{2}k^\alpha - p^\alpha\right)\left(\frac{1}{2}k^\beta + p^\beta\right) f_{\beta\alpha}\right.$$
$$\left. + \frac{\lambda}{2}\left(\frac{1}{2}k^\alpha - p^\alpha\right)\left(\frac{1}{2}k^\beta + p^\beta\right) f_{\alpha\beta} + \frac{1}{2}\Pi^{\alpha\beta}(p) f_{\beta\alpha}\right]. \tag{15.66}$$

The second-quantized field $A^\mu(x)$ is expressed in terms of the plane wave solutions to the wave equation in the same way as in Chap. 13 and reads

$$A^\mu(x) = \sum_\ell \int \frac{d^4k}{(2\pi)^{3/2}} \frac{1}{\left|\frac{\partial D_\ell}{\partial k_0}\right|^{1/2}} \{\varepsilon^\mu_\ell(k) a_\ell(k) \exp(-ik\cdot x)$$
$$+ \varepsilon^{\mu*}_\ell(k) a^+_\ell(k) \exp(+ik\cdot x)\}\big|_{k^0=\omega_\ell(k)}. \tag{15.67}$$

As in Chap. 3, the dispersion relations read

$$\begin{cases} D_T(k) \equiv k^2 - \pi_T(k) = 0, \\ \text{(transverse modes)} \\ D_L(k) \equiv k^2 - \pi_L(k) = 0, \\ \text{(longitudinal modes)} \end{cases} \tag{15.68}$$

so that the sum over ℓ is a sum over T and L, T being twice-degenerated. We now quantize the field A^μ exactly as in Chap. 13 and, finally, the quasiphoton Hamiltonian can be written as

$$\mathcal{H}_{\text{quasiphoton}} = \frac{1}{2}\sum_{\ell=L,T} \int d^4p\; p_0 \delta[p_0 - \omega_\ell(\mathbf{p})] \operatorname{sgn}\left(\frac{\partial D_\ell(p)}{\partial p_0}\right)$$
$$\times \left\{a^+_\ell(p) a_\ell(p) + a_\ell(p) a^+_\ell(p)\right\}, \tag{15.69}$$

which is quite similar to the ordinary Hamiltonian of quantum field theory, but with two differences: (i) there exists an extra degree of freedom that concerns the longitudinal mode and (ii) according to the sign of the derivative of the dispersion relation, it is possible to get the so-called "negative energy modes" discussed in Chap. 13.

Statistical properties of the quasiphoton gas. In thermal equilibrium where the system is invariant under space–time translations, the energy–momentum tensor reads

$$T^{\mu\nu}_{\text{quasiphoton}} = -\frac{1}{2}\int d^4p\, p^\nu \frac{\partial D^{\alpha\beta}(p)}{\partial p_\mu} f_{\beta\alpha}(p), \tag{15.70}$$

where an irrelevant $\delta^{(4)}(k)$ factor has been eliminated. The calculation of $f_{\beta\alpha}(p)$ is straightforward, owing to the quasiusual form of the Hamiltonian operator and hence to the fact that

$$\langle a_\ell^+(p)a_{\ell'}(p')\rangle_{\text{eq}} = \delta^{(3)}(\mathbf{p}-\mathbf{p}')\frac{1}{\exp[\beta\omega_\ell(\mathbf{p})]-1}, \tag{15.71}$$

and one finds that

$$f^{\mu\nu}(p) = -2\pi\delta(p^2)\left\{\theta(p_0) + \frac{1}{\exp(\beta p\cdot u)-1}\right\}\left[\eta^{\mu\nu} - \left(1-\frac{1}{\lambda}\right)\frac{p^\mu p^\nu}{p^2}\right]. \tag{15.72}$$

Finally, the quasiphoton energy–momentum tensor has the perfect fluid form, where the energy density and the pressure are given by

$$\begin{cases} \rho_{\text{quasiphoton}} = \displaystyle\sum_{\ell=L,T}\int \frac{d^3p}{(2\pi)^3}\omega_\ell(\mathbf{p})\text{sgn}\left(\frac{\partial D_\ell(p)}{\partial p_0}\right) \\ \qquad\qquad \times \displaystyle\frac{1}{\exp(\beta p_0)-1}\bigg|_{p_0=\omega_\ell(|\mathbf{p}|)}, \\ P_{\text{quasiphoton}} = \displaystyle -\frac{1}{3}\sum_{\ell=L,T}\int \frac{d^3p}{(2\pi)^3}\frac{1}{\left|\frac{\partial D_\ell(p)}{\partial p_0}\right|}\mathbf{p}\cdot \\ \qquad\qquad \times \displaystyle\frac{\partial D_\ell(p)}{\partial \mathbf{p}}\frac{1}{\exp(\beta p_0)-1}\bigg|_{p_0=\omega_\ell(|\mathbf{p}|)}. \end{cases} \tag{15.73}$$

In these expressions a vacuum term — which has to be discussed — has been dropped. In the various figures below, some comparisons with the usual cosmological blackbody results are given.

The numerical results shown in Fig. 15.5, obtained by M. Lemoine (1994), indicate that the difference from the usual blackbody equation of state is very small. Lemoine has also calculated the effects of these modifications on the light element production in the early universe and, as expected, they are quite tiny.

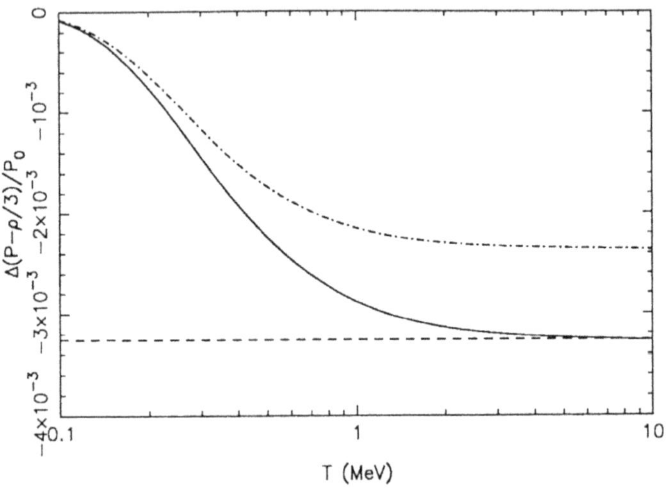

Fig. 15.5 The shift of the pressure, with respect to the ideal photon gas, as a function of the temperature and at zero chemical potential (after M. Lemoine).

15.6.2. *Gauge properties*

We now have to discuss the gauge properties of what we outlined from the statistical properties of quasiparticles. We are going to show that our results are gauge-invariant provided that they are considered in thermal equilibrium.

We first use the Wigner function

$$f_{\mu\nu}(x,k) \underset{\text{def}}{\equiv} \int \frac{d^4R}{(2\pi)^4} \exp(-ik\cdot R) A_\nu\left(x+\frac{1}{2}R\right) A_\mu\left(x-\frac{1}{2}R\right), \tag{15.74}$$

which is transformed as

$$f_{\mu\nu} \to f_{\mu\nu} + \frac{1}{2}\int \frac{d^4R}{(2\pi)^4} \exp(-ik\cdot R)\left[+A_\nu\left(x+\frac{1}{2}R\right)\partial_\mu\Lambda\left(x-\frac{1}{2}R\right)\right.$$
$$\left.+\,\partial_\nu\Lambda\left(x+\frac{1}{2}R\right)A_\mu\left(x-\frac{1}{2}R\right)+\partial_\nu\Lambda\left(x+\frac{1}{2}R\right)\partial_\mu\Lambda\left(x-\frac{1}{2}R\right)\right] \tag{15.75}$$

when performing a gauge transformation

$$A_\mu(x) \to A_\mu(x) + \partial_\mu\Lambda(x); \tag{15.76}$$

note that the derivatives act as

$$\partial_\mu \to \frac{\partial}{\partial(x\pm\frac{1}{2}R)^\mu}. \tag{15.77}$$

Let us use the Fourier transform of this last quantity,

$$f_{\mu\nu}(k,p) = \frac{1}{(2\pi)^4} A_\mu\left(p - \frac{1}{2}k\right) A_\nu\left(p + \frac{1}{2}k\right); \tag{15.78}$$

after a gauge transformation, we have

$$\begin{aligned} f_{\mu\nu} \to f_{\mu\nu} + \frac{1}{(2\pi)^4}\bigg\{&\left(\frac{1}{2}k_\mu - p_\mu\right)\Lambda\left(\frac{1}{2}k - p\right)A_\nu\left(\frac{1}{2}k + p\right) \\ &+ \left(\frac{1}{2}k_\nu + p_\nu\right)A_\mu\left(\frac{1}{2}k - p\right)\Lambda\left(\frac{1}{2}k + p\right) \\ &- p_\mu p_\nu \Lambda\left(\frac{1}{2}k - p\right)\Lambda\left(\frac{1}{2}k + p\right)\bigg\}. \end{aligned} \tag{15.79}$$

At this point, we have all that is necessary for proving the gauge invariance of our results.

In thermal equilibrium, $k = 0$, the Fourier transform of $f_{\mu\nu}$ is somewhat simpler. Let us look, however, at the energy–momentum tensor in thermal equilibrium; it reads

$$T^{\mu\nu}_{\text{quasiphoton}} = -\frac{1}{2}\int d^4p\, p^\nu \frac{\partial D^{\alpha\beta}(p)}{\partial p_\mu} f_{\beta\alpha}(p), \tag{15.80}$$

where

$$D^{\alpha\beta}(p) = p^2\eta^{\alpha\beta} - (1-\lambda)p^\alpha p^\beta - \Pi^{\alpha\beta}. \tag{15.81}$$

Discarding the λ-dependent terms for the moment, taking account of the transverse character of A^μ, i.e. $p_\alpha A^\alpha = 0$, using the fact that $k = 0$, and also the transverse character of $\Pi^{\mu\nu}$ (which makes the term including $\nabla_\mu \Pi^{\alpha\beta}$ vanish), we find that after a gauge transformation

$$-p_\alpha \Lambda(-p)A_\beta(+p) + p_\beta A_\alpha(-p)\Lambda(+p) - p_\alpha p_\beta \Lambda(-p)\Lambda(+p) \tag{15.82}$$

$$\times \frac{\partial}{\partial p^\mu}[p^2\eta^{\alpha\beta} - p^\alpha p^\beta - \Pi^{\alpha\beta}] = 0. \tag{15.83}$$

Let us now take the average value: it follows that the equilibrium energy–momentum tensor is itself gauge-invariant. Similarly, all the average values, i.e.

$$\langle B(p)\rangle_\mu = -\frac{1}{2}\int d^4p\, B(p)\frac{\partial D^{\alpha\beta}(p)}{\partial p^\mu}\langle f_{\alpha\beta}(p)\rangle, \tag{15.84}$$

are gauge-invariant. It remains for us observe that the remaining λ part does not play any role: this is due to the independence of the macroscopic quantities of this term; it contains no term in A^μ — only on $\Lambda(-p)\Lambda(p)$.

This can also be inferred from the fact that, in spite of the lack of gauge invariance of the equation

$$\Box A^\mu(x) - (1-\lambda)\partial^\mu\partial_\nu A^\nu(x) = 4\pi e \mathrm{Sp}\int d^4p\, \gamma^\mu F_{\mathrm{op}}(x,p), \tag{15.85}$$

its poles, i.e. $P_{||}(k)$ and $P_\perp(k)$, are independent of the gauge chosen; from the four equations leading to the poles, only one depends on the gauge [see Eq. (15.81)]. As a result, the calculation of the data of the quasiphotons — such as the pressure or the energy density — is actually gauge-independent. This also results from the use of only $P_{||}(k)$ and $P_\perp(k)$ in the thermal equilibrium Wigner function.

Finally, it should be noticed that the same results as those obtained with replacing in the equilibrium distribution the only gauge-invariant quantities, are obtained.

15.6.3. *Quasielectron modes in thermal equilibrium*

The quantum-electrodynamical plasma involves a phenomenon which does not occur in the relativistic classical case: the temperature- and density-dependent "dressing" of the electron mass. This can be understood easily, since the one-loop mass operator is proportional to $\hbar$. One expects that in astrophysical conditions — such as in the case of white dwarfs — this effect does not play an important role, owing to the energies involved; and this is what was actually found [M. Lemoine (1995)].

Let us rederive the electron mass operator $\Sigma(p)$ at order e^2, by using the Wigner function technique. We start again from the first equations of the quantum BBGKY hierarchy:

$$\begin{cases} \{i\gamma\cdot\partial + 2[\gamma\cdot p - m]\}F(x,p) \\ \quad = 2e\displaystyle\int \frac{d^4R}{(2\pi)^4} d^4p' \exp[-i(p'-p)\cdot R] \\ \quad\quad \times \left\langle F_{\mathrm{op}}(x,p')\gamma_\mu A^\mu\left(x-\frac{1}{2}R\right)\right\rangle, \\ F(x,p)\{i\gamma\cdot\overleftarrow{\partial} - 2[\gamma\cdot p - m]\} \\ \quad = -2e\displaystyle\int \frac{d^4x'}{(2\pi)^4} d^4p' \exp[-ip'\cdot(x-x')] \\ \quad\quad \times \left\langle \gamma_\mu A^\mu\left(x+\frac{1}{2}R\right)F(x,p')\right\rangle, \end{cases} \tag{15.86}$$

$$\begin{cases} \{i\gamma\cdot\partial+2[\gamma\cdot p-m]\}\langle F_{\rm op}(x,p)A_\alpha(x')\rangle \\ \qquad = 2e\displaystyle\int\frac{d^4R}{(2\pi)^4}d^4p'\exp[-i(p'-p)\cdot R] \\ \qquad\quad \times\left\langle F_{\rm op}(x,p')\gamma_\mu A^\mu\left(x-\frac{1}{2}R\right)A_\alpha(x')\right\rangle, \\ \langle A_\alpha(x')F_{\rm op}(x,p)\rangle\{i\gamma\cdot\overleftarrow{\partial}-2[\gamma\cdot p-m]\} \\ \qquad = -2e\displaystyle\int\frac{d^4x'}{(2\pi)^4}d^4p'\exp[-ip'\cdot(x-x')] \\ \qquad\quad \times\left\langle\gamma_\mu A_\alpha(x')A^\mu\left(x+\frac{1}{2}R\right)F(x,p')\right\rangle, \end{cases} \tag{15.87}$$

$$\begin{cases} \{\Box\eta^{\mu\nu}-(1-\lambda)\partial^\mu\partial^\nu\}\langle A_\nu(x)\rangle = 4\pi e{\rm Sp}\displaystyle\int d^4p'F(x,p'), \\ \{\Box\eta^{\mu\nu}-(1-\lambda)\partial^\mu\partial^\nu\}\langle A_\nu(x)A_\alpha(x')\rangle \\ \qquad = 4\pi e\,{\rm Sp}\displaystyle\int d^4p'\gamma^\mu\langle F_{\rm op}(x,p')A_\alpha(x')\rangle, \\ \{\Box\eta^{\mu\nu}-(1-\lambda)\partial^\mu\partial^\nu\}\langle F_{\rm op}(x',p)A_\nu(x)\rangle \\ \qquad = 4\pi e\,{\rm Sp}\displaystyle\int d^4p'\gamma^\mu\langle F_{\rm op}(x',p)F_{\rm op}(x,p')\rangle, \end{cases} \tag{15.88}$$

$$\begin{cases} \{i\gamma\cdot\partial+2[\gamma\cdot p-m]\}F(x,p) \\ \qquad = 2e\displaystyle\int\frac{d^4R}{(2\pi)^4}d^4p'\exp[-i(p'-p)\cdot R]\left\langle F_{\rm op}(x,p')\gamma_\mu A^\mu\left(x-\frac{1}{2}R\right)\right\rangle, \\ F(x,p)\{i\gamma\cdot\overleftarrow{\partial}-2[\gamma\cdot p-m]\} \\ \qquad = -2e\displaystyle\int\frac{d^4x'}{(2\pi)^4}d^4p'\exp[-ip'\cdot(x-x')]\left\langle\gamma_\mu A^\mu\left(x+\frac{1}{2}R\right)F(x,p')\right\rangle, \end{cases} \tag{15.89}$$

which are formally inverted as

$$\langle\gamma^\mu F_{\rm op}(x,p')A_\mu(x')\rangle_{\rm eq} = e\int d^4y\;d^4p''\{\Box\eta^{\mu\nu}-(1-\lambda)\partial^\mu\partial^\nu\}^{-1}(y) \\ \times\langle\gamma_\mu F_{\rm op}(x,p')\gamma_\nu F_{\rm op}(x'-y,p'')\rangle, \tag{15.90}$$

$$\langle\gamma^\mu F_{\rm op}(x,p')A_\mu(x')\rangle_{\rm eq} = e\int\frac{d^4R'}{(2\pi)^4}d^4y\;d^4p''\exp[-i(p''-p')\cdot R'] \\ \times\gamma^\mu\{i\gamma\cdot\partial+2[\gamma\cdot p-m]\}^{-1}(y)\gamma^\nu \\ \times\left\langle F_{\rm op}(x-y,p'')A_\nu\left(x-y-\frac{1}{2}R'\right)A_\mu(x')\right\rangle_{\rm eq}. \tag{15.91}$$

In these expressions — which look quite different from each other — two uncompletely defined operators appear: the propagator

$$\{\Box\eta^{\mu\nu} - (1-\lambda)\partial^\mu\partial^\nu\}^{-1} \tag{15.92}$$

and

$$\{i\gamma\cdot\partial + 2[\gamma\cdot p - m]\}^{-1}. \tag{15.93}$$

Actually, they are defined up to a homogeneous solution and it can be checked that one form obtained above for $\langle F_{\text{op}}A\rangle$ is a homogeneous solution for the other equation. As a result, a general expression for this last quantity is the sum of the two solutions up to a common homogeneous solution which can be obtained from the initial conditions on the system, such as an adiabatic switching-on of the interactions.

In order to obtain a simple form for the mass operator $\Sigma(p)$, a few approximations must be made. First, use is made of the Hartree–Fock approximation of $\langle F_{\text{op}}F_{\text{op}}\rangle$. Its form has been given in Chap. 8.

$$\mathcal{F}_{\text{eq}}(k;p,p') \approx (2\pi)^4\delta^{(4)}(p-p')F_{\text{eq}}\left(p+\frac{1}{2}k\right)F_{\text{eq}}^T\left(p-\frac{1}{2}k\right), \tag{15.94}$$

and, next, the "photon gas" is considered as being uncorrelated (only collective effects of the latter are considered) with the electron matter

$$\langle F_{\text{op}}AA\rangle \approx F\times\langle AA\rangle; \tag{15.95}$$

$\langle AA\rangle$ is essentially the photons' Wigner function.

A calculation, made by D. Lemoine (1995), with the help of the fluctuation–dissipation theorem, yields

$$\begin{aligned}\langle A_\mu A_\nu\rangle\big|_k &= 2\left\{\theta(k_0) + \frac{1}{\exp(\beta k\cdot u)-1}\right\}\\ &\quad\times\Bigg(-\pi\Big\{\delta[k^2 - \operatorname{Re}\Pi_T(k)]\delta[\operatorname{Im}\Pi_L(k)]\Big\}P_{\mu\nu}\\ &\quad+\delta[k^2-\operatorname{Re}\Pi_L(k)]\delta(\operatorname{Im}\Pi_L)Q_{\mu\nu}\\ &\quad+\operatorname{sgn}(k_0)P\left[\frac{\operatorname{Im}\Pi_T(k)}{[k^2-\Pi_T(k)]^2}P_{\mu\nu} + \frac{\operatorname{Im}\Pi_L(k)}{[k^2-\Pi_L(k)]^2}Q_{\mu\nu}\right]\Bigg),\end{aligned} \tag{15.96}$$

where P is the principal value. With these approximations and after the replacement of the expression obtained for $\langle F_{\text{op}}A\rangle$, the first equation of the quantum hierarchy can be rewritten in the form

$$\{\gamma\cdot p - m - \Sigma(p)\}F_{\text{eq}}(p) = 0, \tag{15.97}$$

with

$$\Sigma(p) \equiv G_{(1)}(p) + G_{(2)}(p) \tag{15.98}$$

and

$$\begin{cases} G_{(1)}(p) = e^2 \int d^4k \dfrac{\gamma^\mu F^T_{\text{eq}}(p-k)\gamma^\nu}{k^2} \left\{ \eta_{\mu\nu} - \left(1 - \dfrac{1}{\lambda}\right) \dfrac{k_\mu k_\nu}{k^2} \right\}, \\ G_{(2)}(p) = -e^2 \int \dfrac{d^4k}{(2\pi)^4} \dfrac{\gamma^\mu[\gamma \cdot (p+k) + m]\gamma^\nu}{(p+k)^2 - m^2} \langle A_\mu A_\nu \rangle_{\text{eq}}(k). \end{cases} \tag{15.99}$$

Note that, in Chap. 14, for systems that contain only the tensor u^μ, $\Sigma(p)$ has the form

$$\Sigma(p) = a(p)I + b(p)\gamma \cdot u + c(p)\gamma \cdot p, \tag{15.100}$$

where $a(p)$, $b(p)$ and $c(p)$ are known functions. This immediately leads to

$$D(p) = [1 - c(p)]u \cdot p - [b(p)]^2 - [1 - c(p)]^2 \Delta^{\mu\nu}(u) p_\mu p_\nu - [m + a(p)]^2, \tag{15.101}$$

which in turn leads to the dispersion relation of the quasielectron mass, $D(p) = 0$.

The mass operator is then obtained at order e^2 by replacing F^T_{eq} and $\langle A_\mu A_\nu \rangle_{\text{eq}}$ with their ideal gas expressions:

$$\begin{cases} \langle A^\mu A^\nu \rangle_{(0)}(k) \\ \quad = -2\pi \left\{ \theta(k_0) + \dfrac{1}{\exp(-\beta k_0) - 1} \right\} \delta(k^2) \left\{ \eta^{\mu\nu} - \left(1 - \dfrac{1}{\lambda}\right) \dfrac{k^\mu k^\nu}{k^2} \right\}, \\ F^T_{\text{eq}(0)}(p) = \dfrac{1}{(2\pi)^3} [1 - f_{\text{eq}}(p)][\gamma \cdot p + m] \text{sgn}(p_0) \delta(p^2 - m^2). \end{cases} \tag{15.102}$$

Otherwise, the expression for $\Sigma(p)$ is a rather involved nonlinear expression which exhibits the nonperturbative nature of the Hartree–Fock approximation. Note that an expression for $\langle A^\mu A^\nu \rangle_{(k)}$ has already been obtained by H. Sivak (1984) in terms of the polarization tensor given above, and another form by D. Lemoine (1995).

Appendix A

A Few Useful Properties of Some Special Functions

A.1. Kelvin's Functions

Kelvin's functions must be considered when dealing with the Jüttner–Synge equilibrium distributions and in various approximation schemes. They are defined by

$$K_n(x) = \int_0^{\infty} ds\ \cosh(sn)\exp(-x\ \cosh s),$$
$$K_n(x) = \left(\frac{x}{2}\right)^n \frac{\Gamma\left(\frac{1}{2}\right)}{\Gamma\left(n+\frac{1}{2}\right)} \int_1^{\infty} ds\ \exp(-x\cosh s)\sinh^{2n} s. \tag{A.1}$$

The following relations are repeatedly used in relativistic statistical mechanics:

$$K_{n+1}(\xi) = \frac{2n}{\xi}K_n(\xi) + K_{n-1}(\xi), \tag{A.2}$$

$$\begin{cases} \dfrac{d}{d\xi}\left[\xi^n K_n(\xi)\right] = -\xi^n K_{n-1}(\xi), \\ \dfrac{d}{d\xi}\left[\xi^{-n} K_n(\xi)\right] = -\xi^{-n} K_{n+1}(\xi). \end{cases} \tag{A.3}$$

Equivalently, one has

$$\begin{cases} \xi K_n'(\xi) - nK_n(\xi) = -\xi K_{n+1}(\xi), \\ \xi K_n'(\xi) + nK_n(\xi) = -\xi K_{n-1}(\xi), \end{cases} \tag{A.4}$$

$$2K_n'(\xi) = -\left\{K_{n+1}(\xi) + K_{n-1}(\xi)\right\}, \tag{A.5}$$

$$\frac{2n}{\xi}K_n'(\xi) = \left\{K_{n+1}(\xi) - K_{n-1}(\xi)\right\}. \tag{A.6}$$

The Kelvin's functions obey the second-order differential equation

$$\frac{d^2}{d\xi^2}K_n(\xi) + \frac{1}{\xi}\frac{d}{d\xi}K_n(\xi) - \left(1 + \frac{n^2}{\xi^2}\right)K_n(\xi) = 0. \tag{A.7}$$

They have several integral representations:

$$\begin{aligned} K_n(\xi) &= \frac{\xi^n}{1 \cdot 3 \cdot 5 \cdots (2n-1)} \int_0^\infty d\zeta \sinh^{2n}\zeta \exp(-\xi\cosh\zeta) \\ &= \int_0^\infty d\zeta \cosh n\zeta \exp(-\xi\cosh\zeta) . \end{aligned} \tag{A.8}$$

For large values of argument ξ, the functions $K_n(\xi)$ behave like

$$K_n(\xi) = \left(\frac{\pi}{2\xi}\right)^{1/2} \exp(-\xi)\left\{1 + \frac{4n^2-1}{1!8\xi} + \frac{(4n^2-1^2)(4n^2-3^2)}{2!(8\xi)^2} + \cdots\right\}, \tag{A.9}$$

while for small values of ξ, one has

$$\begin{cases} K_0(\xi) = -0.5772 - \log\dfrac{\xi}{2} + \cdots , \\ K_n(\xi) = \dfrac{1}{2}\Gamma(n)\left(\dfrac{2}{\xi}\right)^n + \cdots . \end{cases} \tag{A.10}$$

More properties can be found in M. Abramovicz and I.A. Stegun (1965) or in I.S. Gradshteyn and I.W. Rizhik (1965).

A.2. Associated Laguerre Polynomials

They repeatedly enter into the quantum calculations that involve magnetic fields. We give only a few formulae that are used here:

$$\begin{cases} x\dfrac{d}{dx}L_n^\alpha = nL_n^\alpha - (n+\alpha)L_{n-1}^\alpha \\ \qquad\quad = (n+1)L_{n+1}^\alpha - (n+\alpha+1-x)L_n^\alpha, \\ xL_n^{\alpha+1} = (n+\alpha+1)L_n^\alpha - (n+1)L_{n+1}^\alpha \\ \qquad\quad = (n+\alpha)L_{n-1}^\alpha - (n-x)L_n^\alpha, \\ L_n^{\alpha-1} = L_n^\alpha - L_{n-1}^\alpha \end{cases} \tag{A.11}$$

Appendix B

γ Matrices

In the derivation of various formulae dealing with the covariant Wigner function for spin 1/2 particles, use was made of the following properties of Dirac's matrices. Besides the usual anticommutation relations

$$\gamma^\mu \gamma^\nu + \gamma^\nu \gamma^\mu = \eta^{\mu\nu} \tag{B.1}$$

and the definitions

$$\begin{cases} \gamma^5 = \dfrac{i}{4!} \varepsilon^{\mu\nu\alpha\beta} \gamma_\mu \gamma_\nu \gamma_\alpha \gamma_\beta \\ \sigma_{\mu\nu} = \dfrac{i}{2} \left[\gamma_\mu \gamma_\nu - \gamma_\nu \gamma_\mu \right] \end{cases}, \tag{B.2}$$

from which one has

$$\gamma^\mu \gamma^5 + \gamma^5 \gamma^\mu = 0, \tag{B.3}$$

use was made of the representation

$$\gamma^0 = \begin{pmatrix} I & 0 \\ 0 & -I \end{pmatrix}, \quad \gamma^i = \begin{pmatrix} 0 & \sigma_i \\ -\sigma_i & 0 \end{pmatrix}, \quad \gamma^5 = \begin{pmatrix} 0 & -I \\ -I & 0 \end{pmatrix} \tag{B.4}$$

where the σ_i's are the common Pauli matrices. From the commutation relations and the various definitions given, one easily gets the following formulae

$$\gamma^\mu \gamma^\nu = \sigma^{\mu\nu} + \eta^{\mu\nu}, \tag{B.5}$$

$$\gamma^\mu \gamma^\nu \gamma^\sigma = \left\{ \eta^{\mu[\nu} \eta^{\sigma]\alpha} + \eta^{\mu\alpha} \eta^{\nu\sigma} \right\} \gamma_\alpha + i\varepsilon^{\mu\nu\sigma\alpha} \gamma^5 \gamma_\alpha, \tag{B.6}$$

$$\begin{cases} \gamma^\nu \sigma^{\beta\lambda} = \eta^{\nu[\beta} \gamma^{\lambda]} - i\varepsilon^{\nu\beta\lambda\sigma} \gamma_\sigma \gamma^5, \\ \sigma^{\beta\lambda} \gamma^\nu = \eta^{\nu[\lambda} \gamma^{\beta]} - i\varepsilon^{\nu\beta\lambda\sigma} \gamma_\sigma \gamma^5, \end{cases} \tag{B.7}$$

$$\gamma^5 \gamma^\lambda = \frac{i}{3!} \varepsilon^{\mu\nu\alpha\lambda} \gamma_\mu \gamma_\nu \gamma_\alpha, \tag{B.8}$$

$$\begin{cases} \gamma^{\mu}\gamma^{\nu}\sigma^{\alpha\beta} = \eta^{\nu[\alpha}\eta^{\beta]\mu} + \left\{\eta^{\tau[\alpha}\eta^{\beta]\mu}\eta^{\lambda\nu} + \eta^{\lambda[\alpha}\eta^{\beta]\nu}\eta^{\mu\tau} \right. \\ \qquad \left. + \eta^{\alpha\lambda}\eta^{\beta\tau}\eta^{\mu\nu}\right\}\sigma_{\lambda\tau} + i\varepsilon^{\alpha\beta\mu\nu}\gamma^5, \\ \sigma^{\alpha\beta}\gamma^{\mu}\gamma^{\nu} = \eta^{\nu[\alpha}\eta^{\beta]\mu} + \left\{\eta^{\mu[\alpha}\eta^{\beta]\tau}\eta^{\lambda\nu} + \eta^{\nu[\alpha}\eta^{\beta]\lambda}\eta^{\mu\tau} \right. \\ \qquad \left. + \eta^{\alpha\lambda}\eta^{\beta\tau}\eta^{\mu\nu}\right\}\sigma_{\lambda\tau} + i\varepsilon^{\alpha\beta\mu\nu}\gamma^5, \end{cases} \tag{B.9}$$

$$\gamma^{\mu}\gamma^{\nu}\gamma^{\alpha}\gamma^5 = -i\varepsilon^{\mu\nu\alpha\beta}\gamma_\beta + \left\{\eta^{\alpha[\beta}\eta^{\mu]\nu} + \eta^{\mu\beta}\eta^{\nu\alpha}\right\}\gamma_\beta\gamma_5, \tag{B.11}$$

$$\left[\gamma^5, \sigma_{\mu\nu}\right] = 0,$$

$$\gamma_5\sigma^{\mu\nu} = -\frac{i}{2}\varepsilon^{\mu\nu\alpha\beta}\sigma_{\alpha\beta}. \tag{B.12}$$

Furthermore, any 4×4 matrix M can be decomposed on the basis of the 16 matrices γ_A of the Dirac's algebra

$$\begin{cases} \gamma^A = \left\{I, \gamma^{\mu}, \sigma^{\mu\nu}, \gamma^5, \gamma^5\gamma^{\lambda}\right\}, \\ \gamma_A = \left\{I, \gamma_{\mu}, \sigma_{\mu\nu}, \gamma^5, \gamma_\lambda\gamma_5\right\} \end{cases} \tag{B.13}$$

with

$$\gamma^A\gamma_A = I, \ \operatorname{Sp}\left\{\gamma^A\gamma_B\right\} = 4\delta^A_B, \ \operatorname{Sp}\left\{\gamma_A\right\} = 0. \tag{B.14}$$

Therefore, one has

$$\begin{cases} M = \dfrac{1}{4}\sum_A m_A\gamma^A, \\ m_A = \operatorname{Sp}\{M\gamma_A\}. \end{cases} \tag{B.15}$$

Let p^μ and u^μ be two time-like four-vectors such that

$$p^2 = m^2, \quad u^2 = 1,$$

then the following relations too are extremely useful:

$$(\gamma \cdot p \pm m)\,\gamma^{\mu}\,(\gamma \cdot p \pm m) = 2p^{\mu}(\gamma \cdot p \pm m), \tag{B.16}$$

$$(\gamma \cdot p \pm m)\,\sigma_{\alpha\beta}u^{\alpha}p^{\beta}\,(\gamma \cdot p \pm m) = 0, \tag{B.17}$$

$$(\gamma \cdot p \pm m)\,\sigma^{\mu\nu}\,(\gamma \cdot p \pm m) = 2\left\{p^{[\mu}\sigma^{\nu]\lambda}p_\lambda + m^2\sigma^{\mu\nu} \pm i\varepsilon^{\mu\nu\alpha\beta}p_\alpha\gamma_5\gamma_\beta\right\}, \tag{B.18}$$

$$\frac{(\gamma \cdot p \pm m)}{2m}\gamma_5\frac{(\gamma \cdot p \mp m)}{2m} = \gamma_5\frac{(\gamma \cdot p \mp m)}{2m}, \tag{B.19}$$

$$\frac{(\gamma\cdot p+m)}{2m}\gamma_\mu\frac{(\gamma\cdot p-m)}{2m}=\frac{1}{2m^2}\{p_\mu\gamma\cdot p+m\sigma_{\mu\nu}p^\nu-m^2\gamma_\mu\},\tag{B.20}$$

$$\frac{(\gamma\cdot p+m)}{2m}\gamma_5\gamma_\mu\frac{(\gamma\cdot p-m)}{2m}=-\frac{p^\mu}{m}\gamma_5\frac{(\gamma\cdot p-m)}{2m},\tag{B.21}$$

$$\gamma\cdot p\sigma^{\mu\nu}\gamma\cdot p=m^2\sigma^{\mu\nu}+2p^{[\mu}\sigma^{\nu]\lambda}p_\lambda,\tag{B.22}$$

$$\frac{\gamma\cdot p\pm m}{2m}\gamma^\mu\gamma^5\frac{\gamma\cdot p\pm m}{2m}=\frac{1}{2}\gamma^\mu\gamma^5\pm\frac{i}{4m}\varepsilon^{\mu\nu\alpha\beta}p_\nu\sigma_{\alpha\beta}-\frac{p^\mu}{2m^2}\gamma\cdot p\gamma^5,\tag{B.23}$$

$$\frac{\gamma\cdot p-m}{2m}\gamma\cdot u\frac{\gamma\cdot p+m}{2m}=\left\{\frac{p\cdot u}{m}-\gamma\cdot u\right\}\frac{\gamma\cdot p+m}{2m},\tag{B.24}$$

$$\frac{(\gamma\cdot p+m)}{2m}\sigma^{\mu\nu}\frac{(\gamma\cdot p-m)}{2m}=\frac{1}{2m^2}\left\{mp^{[\mu}\gamma^{\nu]}+p^{[\mu}\sigma^{\nu]\alpha}p_\alpha\right\},\tag{B.25}$$

$$(\sigma^{\mu\nu}p_\mu u_\nu)^2=-\Delta^{\mu\nu}(u)p_\mu p_\nu,\tag{B.26}$$

$$[\gamma\cdot p,\gamma\cdot u]_+=2p\cdot u,\tag{B.27}$$

$$[\gamma\cdot p,\sigma^{\mu\nu}u_\mu p_\nu]_+=0,\tag{B.28}$$

$$[\gamma\cdot u,\sigma^{\mu\nu}u_\mu p_\nu]_+=0.\tag{B.29}$$

Various formulae including the Levi-Civita tensor $\varepsilon^{\mu\nu\alpha\beta}$ are also used. This tensor is completely antisymmetric and is such that

$$\varepsilon^{0123}=-\varepsilon_{0123}=+1\tag{B.30}$$

and obeys the relations

$$\varepsilon_{\tau\lambda\mu\nu}\varepsilon^{\mu\nu\alpha\beta}=-2\left(\delta^\alpha_\tau\delta^\beta_\lambda-\delta^\alpha_\lambda\delta^\beta_\tau\right),\tag{B.31}$$

$$\varepsilon^{\sigma\beta\mu\nu}\varepsilon_{\sigma\rho\lambda\alpha}=-\begin{vmatrix}\delta^\beta_\rho & \delta^\mu_\rho & \delta^\nu_\rho\\ \delta^\beta_\lambda & \delta^\mu_\lambda & \delta^\nu_\lambda\\ \delta^\beta_\alpha & \delta^\mu_\alpha & \delta^\nu_\alpha\end{vmatrix}.\tag{B.32}$$

Appendix C

Outline of Functional Methods

Functional methods are nowadays absolutely basic in various branches of theoretical physics. Functional integration, for instance, has become essential when quantizing gauge theories and/or for the decription of their thermal properties. In this appendix,[1] mathematical problems are completely skipped and our approach is merely formal or intuitive; it is important that the reader gets familiar with these methods and can make some elementary manipulations.

In what follows, we start with a given functional, that is a mathematical object that depends on a function in its entirety as $\mathcal{E}(\{\varphi\}; a, b, f(x), \text{etc})$, where $f(x)$ is a given function, $a, b, \ldots$ are constants, etc.; $\mathcal{E}$ depends on the function φ in its globality as, e.g. in

$$\mathcal{E}(\{\varphi\}; a, b, f(x), \text{etc.}) = ag(x) \int dx\, \varphi(x) f(x) + b. \tag{C.1}$$

In a sense, a functional is a "function of functions." Of course, $\varphi(x)$ itself is a also functional.

Several definitions are of common use in theoretical problems.

One consists in making the line interval discrete, the interval of definition of the function φ is then divided into intervals Δx_i where the function φ is approximated by its average value φ_i, so that a functional $\mathcal{E}(\{\varphi\})$ reduces to

$$\mathcal{E}(\{\varphi\}) = \mathcal{E}(\varphi_1, \varphi_2, \ldots). \tag{C.2}$$

In this case, the functional appears to be approximated by a function of an infinite number of variables.

[1]See, W.F. Chen, *Lecture notes on Techniques and Method in Path Integral Quantization of Field Theory* [NCTS]. These notes are particularly clear and comprehensive.

The second possible definition rests on the fact that in many cases the function φ is an element of a separable Hilbert space and hence can be expanded on an orthonormal basis $\{\alpha_n\}_{n=0,1,\ldots}$:

$$\begin{cases} \varphi(x) = \sum_{n=1}^{\infty} \alpha_n(x) \int dx' \alpha_n(x')\varphi(x'), \\ \int dx\, \alpha_n(x)\alpha_m(x) = \delta_{nm}, \\ \sum_{n=1}^{\infty} \alpha_n(x)\alpha_n(x') = \delta(x - x'), \end{cases} \tag{C.3}$$

so that, finally, the functional is still expressed as a function of an infinite number of variables

$$\mathcal{E}(\{\varphi\}) = \mathcal{E}(\alpha_1, \alpha_2, \ldots). \tag{C.4}$$

C.1. Functional Differentiation

By analogy with ordinary differentiation, the functional derivative with respect to φ of a given functional $\mathcal{E}(\{\varphi\})$ is defined through

$$\frac{\delta}{\delta\varphi(y)}\mathcal{E}(\varphi(x)) = \lim_{\varepsilon\to 0} \frac{\mathcal{E}(\varphi(x) + \varepsilon\delta(x-y)) - \mathcal{E}(\varphi(x))}{\varepsilon}, \tag{C.5}$$

so that we have the useful relation

$$\frac{\delta\varphi(x)}{\delta\varphi(y)} = \lim_{\varepsilon\to 0} \frac{\varphi(x) + \varepsilon\delta(x-y) - \varphi(x)}{\varepsilon} = \delta(x-y). \tag{C.6}$$

For instance, the functional

$$\mathcal{E}(\{\varphi\}) = \int dx' K(x, x')\varphi(x') \tag{C.7}$$

has the following functional derivative with respect to $\varphi(y)$:

$$\frac{\delta}{\delta\varphi(y)} \int dx' K(x, x')\varphi(x') = K(x, y). \tag{C.8}$$

It can be checked that, *mutatis mutandis*, the ordinary chain rule for derivatives still applies for functional differentiation.

In statistical mechanics or in quantum field theory, one encounters expressions like

$$\begin{aligned}\mathcal{E}(\{\varphi\}) &= K_0(x) + \int dx' K_1(x;x')\varphi(x') \\ &\quad + \int dx'dx'' K_2(x;x',x'')\varphi(x')\varphi(x'') + \cdots \\ &\quad + \int dx' K_\ell(x,x_1,x_2,\ldots,x_\ell)\varphi(x_1)\varphi(x_2)\cdots\varphi(x_\ell) + \cdots \qquad \text{(C.9)}\end{aligned}$$

which can be called a functional power series in φ. It is not difficult to check that, when the kernels K_ℓ are *symmetric* in all their arguments $(x, x_1, x_2, \ldots, x_\ell)$, then one has

$$K_\ell\,(x,x_1,x_2,\ldots,x_\ell) = \frac{1}{\ell!}\,\frac{\delta}{\delta\varphi(x_1)\delta\varphi(x_2)\cdots\delta\varphi(x_\ell)}\mathcal{E}(\{\varphi\})\bigg|_{\varphi\equiv 0}, \qquad \text{(C.10)}$$

and in this case only can one speak of a functional Taylor expansion of the functional $\mathcal{E}(\{\varphi\})$. When this symmetry property is not satisfied, then K_ℓ is not given by this last expression. A sufficient condition for a functional to possess a Taylor expansion

$$\mathcal{E}(\{\varphi\}) = \sum_{\ell=0}^{\infty}\frac{1}{\ell!}\,\frac{\delta}{\delta\varphi(x_1)\delta\varphi(x_2)\cdots\delta\varphi(x_\ell)}\mathcal{E}\left(\{\varphi\}\right)\bigg|_{\varphi\equiv 0}\varphi(x_1)\varphi(x_2)\cdots\varphi(x_\ell) \qquad \text{(C.11)}$$

is that the function $\theta(\lambda)$ defined by

$$\theta\,(\lambda) \underset{\text{def}}{\equiv} \mathcal{E}\left(\{\varphi + \lambda\varphi'\}\right) \qquad \text{(C.12)}$$

possesses a Taylor expansion.[2] The proof, which is quite simple, uses the alternative definition of the functional differentiation

$$\frac{\delta}{\delta\varphi(y)}\mathcal{E}\left(\varphi(x)\right) = \lim_{\lambda\to 0}\frac{\mathcal{E}\left(\varphi(x)+\lambda\varphi'(x)\right) - \mathcal{E}\left(\varphi(x)\right)}{\lambda}. \qquad \text{(C.13)}$$

C.2. Functional Integration

The notion of functional (or path) integration goes back to 1923, with N. Wiener approach of the Brownian motion. Later R.P. Feynman invented the path integrals, now commonly used in quantum physics, with a then new interpretation of quantum mechanics.[3]

[2] See C. Nash, *op. cit.*

[3] R.P. Feynman, Rev. Mod. Phys. **20**, 367 (1948).

A functional integral is an expression of the form

$$\int \mathcal{D}\varphi\, \mathcal{E}(\{\varphi\}), \tag{C.14}$$

to which it is intended to make sense and where $\varphi(x)$ is supposed to be an element of a separable Hilbert space such as $\mathcal{L}^2(C^n)$. In general, one is interested in functional integrals of the form

$$\int \mathcal{D}\varphi\, \mathcal{E}(\{\varphi\}) \exp[-S(\{\varphi\})], \tag{C.15}$$

where $S(\{\varphi\})$ is quadratic in φ, and where $\mathcal{E}(\{\varphi\})$ can be expanded into a Taylor series. In particular, one is interested in Gaussian integrals, i.e. where $S[\{\varphi\}]$ is quadratic and where the functional $\mathcal{E}$ is independent of φ:

$$\int \mathcal{D}\varphi \exp\left[-\frac{1}{2}\int d^4x\, \varphi(x) A\varphi(x)\right], \tag{C.16}$$

where A is an operator acting on φ. In an ordinary finite-dimensional space, such a Gaussian integral would be equal to $[2\pi \det(A)]^{-1/2}$, provided the positive definite character of the matrix A is verified. Functional integrals require some simple generalizations. In the above integral, the functional

$$\mathcal{E}(\{\varphi\}) \equiv \left[-\frac{1}{2}\int d^4x\, \varphi(x) A\varphi(x)\right] \tag{C.17}$$

is then approximated as

$$\mathcal{E}(\{\varphi\}) \approx -\frac{1}{2}\sum_{i=1}^{i=n}\sum_{j=1}^{j=n} \varphi_i A_{ij} \varphi_j, \tag{C.18}$$

with

$$A_{ij} = \int dx\, \alpha_i(x) A \alpha_j(x). \tag{C.19}$$

The functional integral under study is now defined as

$$\int \mathcal{D}\varphi \exp\left[-\frac{1}{2}\int d^4x\, \varphi(x) A\varphi(x)\right] = \lim_{n\to\infty} \int \prod_{\ell=1}^{\ell=n} d^4x_\ell \exp\left[-\frac{1}{2}\sum_{i=1}^{i=n}\sum_{j=1}^{j=n} \varphi_i A_{ij}\varphi_j\right]; \tag{C.20}$$

in other words, it is defined (in fact, ill-defined!) as the limit of an ordinary n-dimensional Gaussian integral. Note that if A_{ij} were an ordinary diagonalizable matrix, the value of this integral would be

$$\int \prod_{\ell=1}^{\ell=n} d^4x_\ell \ \exp\left[-\frac{1}{2}\sum_{i=1}^{i=n}\sum_{j=1}^{j=n}\varphi_i A_{ij}\varphi_j\right] = \det\left[2\pi A\right]^{-1/2}, \tag{C.21}$$

with, obviously,

$$\det\left[2\pi A\right] = \exp(\mathrm{Tr}\ \ln 2\pi A). \tag{C.22}$$

This last relation is then extended to operators and hence

$$\int \mathcal{D}\varphi\ \exp\left[-\frac{1}{2}\int d^4x\,\varphi(x)A\varphi(x)\right] = \exp\left(-\frac{1}{2}\mathrm{Tr}\ \ln 2\pi A\right). \tag{C.23}$$

Let us calculate this last integral in the case of the free real scalar field

$$\begin{aligned}\mathcal{L}_E &= \frac{1}{2}\int_0^\beta d^4x\,\{\partial_\mu\varphi(x)\partial_\mu\varphi(x) + m^2\varphi^2(x)\}\\ &= \frac{1}{2}\int_0^\beta d^4x\,\{-(\partial/\partial\tau)\,\varphi(x)\cdot(\partial/\partial\tau)\,\varphi(x)\\ &\quad -\boldsymbol{\partial}\varphi(x)\cdot\boldsymbol{\partial}\varphi(x) + m^2\varphi^2(x)\},\end{aligned} \tag{C.24}$$

and remember that we are working in the Euclidean space so that

$$\int_0^\beta d^4x = \sum_n \int d^3x \to \sum_n\sum_{\mathbf{k}} = \sum_n \frac{1}{(2\pi)^3}\int d^3k, \tag{C.25}$$

where the sum over n is over the Matsubara frequencies $\omega_n = 2\pi nT$. An integration by parts in the expression of $\mathcal{L}_E$ yields

$$\mathcal{L}_E = -\frac{1}{2}\int_0^\beta d^4x\left\{\varphi(x)\left[(\partial/\partial\tau)^2 + \boldsymbol{\partial}^2\right]\varphi(x) - m^2\varphi^2(x)\right\}, \tag{C.26}$$

which, in Fourier space can be rewritten as

$$\mathcal{L}_E = \frac{1}{2}\sum_n\sum_{\mathbf{k}}\{\varphi(\omega_n,\mathbf{k})\left[\omega_n + \mathbf{k}^2 + m^2\right]\varphi(\omega,\mathbf{k})\}, \tag{C.27}$$

with $\omega_n = 2\pi nT$. In the preceding expressions for $\mathcal{L}_E$, the operator A is respectively given by

$$\begin{cases} A = -\left[(\partial/\partial\tau)^2 + \boldsymbol{\partial}^2\right] + m^2, \\ A = k^2 + m^2 = \omega^2 + \mathbf{k^2} + m^2. \end{cases} \tag{C.28}$$

Using the Fourier form of the operator A thus provides

$$\log Z = -\frac{1}{2}\sum_n \frac{1}{(2\pi)^3}\int d^3k \ln\left[\{\omega_n^2 + \mathbf{k}^2 + m^2\}\right], \tag{C.29}$$

where an irrelevant factor of 2π has been eliminated since it plays no role in the thermodynamics of the system; it only leads a constant term. In order to calculate log Z, we use the known formula

$$\frac{1}{2}\sum_{n=-\infty}^{n=+\infty} \frac{1}{\omega^2 + \omega_n^2} = -\frac{\beta}{4\omega}\left[1 + \frac{2}{\exp(\beta\omega) - 1}\right], \tag{C.30}$$

to arrive at

$$\log Z = -\int \frac{d^3k}{(2\pi)^3}\left\{\frac{1}{2}\beta\omega(\mathbf{k}) + \log\left[1 - \exp\left(-\beta\omega(\mathbf{k})\right)\right]\right\}, \tag{C.31}$$

where, once more, an irrelevant infinite constant term has been dropped.

With the above definition of a Gaussian functional integral, one can calculate integrals of the form

$$I = \int \mathcal{D}\varphi\, \mathcal{E}\left[\{\varphi\}\right] \exp\left(-\int d^4x\, \varphi(x) A \varphi(x)\right), \tag{C.32}$$

i.e. the ones to which one generally has to face, as

$$I = \sum_{n_1, n_2, \ldots, n_\ell} \int \mathcal{D}\varphi\, \mathcal{E}_{n_1 n_2 \ldots n_\ell} \varphi^{n_1} \varphi^{n_2} \ldots \varphi^{n_\ell} \exp\left(-\int d^4x\, \varphi(x) A \varphi(x)\right). \tag{C.33}$$

They can indeed be deduced by functional derivations with respect to $j(x)$ from the generating functional

$$I(j) = \int \mathcal{D}\varphi\, \exp\left(-\int_0^\beta d^4x \left[\varphi(x) A \varphi(x) + j(x)\varphi(x)\right]\right) \tag{C.34}$$

since the latter can also be cast into a Gaussian form by "completing the square."

Appendix D

Units+

D.1. Ordinary Units

$$c = 2.99792458 \times 10^{10}$$

$$h = 1.05457266 \times 10^{-27} = 6.582122 \times 10^{-22}\,\text{MeV s}$$

$$k_{\text{Boltz}} = 1.3806513 \times 10^{-16} = 8.617344 \times 10^{-11}\,\text{MeV/K}$$

$$e = 4.8032068 \times 10^{-10}$$

$$\alpha = e^2/\hbar c = 137.0359895^{-1}$$

$$G = 6.67259 \times 10^{-8}$$

$$g = 980.665$$

$$m_{\text{electron}} = 0.91093897 \times 10^{-27} = 0.51099906\,\text{MeV} = 5.92989 \times 10^{9}\,\text{K}$$

$$m_{\text{proton}} = 1.6726231 \times 10^{-24} = 938.27231\,\text{MeV} = 1.0888184 \times 10^{13}\,\text{K}$$

$$m_{\text{neutron}} = 1.6749286 \times 10^{-24} = 939.56563\,\text{MeV} = 1.0903193 \times 10^{13}\,\text{K}$$

$$m_{\pi}^{\pm} = 2.48801 \times 10^{-25} = 139.567\,\text{MeV}$$

$$m_{\pi}^{\text{o}} = 2.40594 \times 10^{-25} = 134.963\,\text{MeV}$$

$$\lambda_{\text{c}_{\text{electron}}} = 3.86159323 \times 10^{-11}$$

$$\lambda_{\text{c}_{\text{electron}}}^{-3} = 1.73660252 \times 10^{31}$$

$$\lambda_{\text{c}_{\text{proton}}} = 2.10308937 \times 10^{-14}$$

$$\lambda_{\text{c}_{\text{proton}}}^{-3} = 1.07504542 \times 10^{41}$$

$$\lambda_{\text{c}_{\text{neutron}}} = 2.10019445 \times 10^{-14}$$

$$\lambda_{\text{c}_{\text{neutron}}}^{-3} = 1.0794971 \times 10^{41}$$

$$\lambda_{c_{\pi^{\pm}}} = 1.41385 \times 10^{-13}$$

$$\lambda_{c_{\pi^{\text{o}}}} = 1.46208 \times 10^{-13}$$

+Compiled by H.D. Sivak.

$$\begin{aligned}
\text{MeV} &= 1.60217733 \times 10^{-6} = 5.0677289 \times 10^{-3}\,\text{fm}^{-1} \\
&= 1.16045 \times 10^{10}\,\text{K} \\
\text{gram} &= 5.6095862 \times 10^{26}\,\text{MeV} \\
\text{fermi}^{-1} &= 197.32705\,\text{MeV} \\
\text{second}^{-1} &= 1.5192669 \times 10^{21}\,\text{MeV} \\
\text{Ry} &= 1/2\,\alpha^2 m_e c^2 = 2.1798741 \times 10^{-11} = 13.6056981\,\text{eV} \\
r_e &= e^2/m_e c^2 = 2.81794092 \times 10^{-13} \\
\sigma_T &= 8\pi/3r_e^2 = 0.66524616 \times 10^{-24} \\
a_0 &= \hbar^2/m_e e^2 = 0.529177249 \times 10^{-8} \\
\mu_\text{B} &= e\hbar/2m_e c = 9.2740154 \times 10^{-21} \\
\mu_\text{N} &= e\hbar/2m_p c = 5.0507866 \times 10^{-24} \\
N_\text{A} &= 6.0221367 \times 10^{23} \\
\sigma &= \pi^2 k_B^4/60\hbar^3 c^2 = 5.670399 \times 10^{-5} \\
&= 6.418014 \times 10^{41}\,1/\text{cm}^2\,\text{s}\,\text{MeV}^3
\end{aligned}$$

D.2. Other Units of Interest

Absolute units

$$\begin{aligned}
\text{Energy} &= \sqrt{\hbar c^5/G} = 1.95633 \times 10^{16} \\
\text{Time} &= \sqrt{\hbar G/c^5} = 5.39056 \times 10^{-44} \\
\text{Length} &= \sqrt{\hbar G/c^3} = 1.61605 \times 10^{-33} \\
\text{Mass} &= \sqrt{\hbar c/G} = 2.17671 \times 10^{-5}
\end{aligned}$$

Degenerate electron Fermi gas at $p_F = m$

$$\begin{aligned}
\text{Pressure} &= 7.38231 \times 10^{22} = 4.60768 \times 10^{28}\,\text{MeV/cm}^3 \\
&= 3.54031 \times 10^{-4}\,\text{MeV}^4 = 82.1393\,\text{g/cm}^3 \\
\text{Energy density} &= 6.05263 \times 10^{23} = 3.77776 \times 10^{29}\,\text{MeV/cm}^3 \\
&= 2.90264 \times 10^{-3}\,\text{MeV}^4 = 673.446\,\text{g/cm}^3 \\
\text{Electron density} &= n = 5.86515 \times 10^{29} = 4.5065 \times 10^{-3}\,\text{MeV}^3
\end{aligned}$$

Critical magnetic field

$$H_\text{crit} = m_\text{electron}^2 c^3/e\hbar = 4.4140056 \times 10^{14}\,\text{Gauss}$$

Symmetric nuclear matter at the saturation density (as a degenerate Fermi gas)

$$
\begin{aligned}
n_{\text{saturation}} &= 0.16\,\text{fm}^{-3} = 1.6 \times 10^{38} = 1.23 \times 10^{6}\,\text{MeV}^3 \\
\text{Fermi momentum} &= 1.33\,\text{fm}^{-1} = 263\,\text{MeV} = 0.28\,m_{\text{neutron}} \\
\text{Pressure} &= 3.67 \times 10^{33} = 2.29 \times 10^{39}\,\text{MeV/cm}^3 \\
&= 1.76 \times 10^{7}\,\text{MeV}^4 = 4.09 \times 10^{12}\,\text{g/cm}^3 \\
\text{Energy density} &= 2.46 \times 10^{35} = 1.54 \times 10^{41}\,\text{MeV/cm}^3 \\
&= 1.18 \times 10^{7}\,\text{MeV}^4 = 2.74 \times 10^{14}\,\text{g/cm}^3
\end{aligned}
$$

Appendix E

Some Useful Formulae for Wigner Functions

E.1. Useful Formulae for Bosons

We now provide a few useful formulae for bosons and, the calculations being straightforward, no details are given. For simplicity, it is assumed that $\langle\phi\rangle = 0$. A first remark, trivial but important to avoid errors, deals with the Fourier transform of the fields, i.e.

$$\phi(k) = \int d^4x \, \exp(ik\cdot x)\phi(x); \tag{E.1}$$

one has

$$\begin{aligned}\phi^*(k) &= \int d^4x \exp(-ik\cdot x)\phi^*(x) \\ &\neq \int d^4x \exp(ik\cdot x)\phi^*(x).\end{aligned} \tag{E.2}$$

The Fourier transform of the Wigner operator

$$f_{\rm op}(x,p) = \frac{1}{(2\pi)^4}\int d^4R \, \exp\left(-ip\cdot R\right)\phi^*\left(x+\frac{1}{2}R\right)\phi\left(x-\frac{1}{2}R\right) \tag{E.3}$$

is given by

$$f_{\rm op}(k,p) = \frac{1}{(2\pi)^4}\phi^*\left(p-\frac{1}{2}k\right)\phi\left(p+\frac{1}{2}k\right), \tag{E.4}$$

or, equivalently,

$$\phi^*\left(k\right)\phi\left(k'\right) = \left(2\pi\right)^4 f_{\rm op}\left(k-k', \frac{k+k'}{2}\right), \tag{E.5}$$

while the product of fields reduces to

$$\begin{cases}\phi^*\left(x\right)\phi\left(y\right) = \displaystyle\int d^4\xi \exp\left(i\xi\cdot(x-y)\right) f_{\rm op}\left(\frac{x+y}{2},\xi\right), \\ \phi\left(x\right)\phi^*\left(y\right) = \displaystyle\int d^4\xi \exp\left(i\xi\cdot(x-y)\right)^* f_{\rm op}\left(\frac{x+y}{2},\xi\right).\end{cases} \tag{E.6}$$

Note also that $f^+ = {}^*f$. In the derivation of the energy–momentum tensor, use was made of the following useful relations:

$$\begin{cases} \partial^\mu_{(y)}\phi^*(y)\,\phi(x) = \displaystyle\int d^4p \exp(ip\cdot(y-x))\left[+ip^\mu + \frac{1}{2}\partial^\mu_{(w)}\right] \\ \qquad\qquad \times f_{\rm op}(w,p)\big|_{w=(y+x)/2}\,, \\ \phi^*(y)\,\partial^\mu_{(x)}\phi(x) = \displaystyle\int d^4p \exp(ip\cdot(y-x))\left[-ip^\mu + \frac{1}{2}\partial^\mu_{(w)}\right] \\ \qquad\qquad \times f_{\rm op}(w,p)\big|_{w=(y+x)/2}\,, \end{cases} \tag{E.7}$$

$$\begin{cases} \phi^*(y)\,\partial^2_{(x)}\phi(x) = \displaystyle\int d^4p \exp(ip\cdot(y-x))\left[-ip^2 - ip\cdot\partial_{(w)} + \frac{1}{4}\partial^2_{(w)}\right] \\ \qquad\qquad \times f_{\rm op}(w,p)\big|_{w=(y+x)/2}\,, \\ \partial^2_{(y)}\phi^*(y)\,\phi(x) = \displaystyle\int d^4p \exp(ip\cdot(y-x))\left[-p^2 + ip\cdot\partial_{(w)} + \frac{1}{4}\partial^2_{(w)}\right] \\ \qquad\qquad \times f_{\rm op}(w,p)\big|_{w=(y+x)/2}\,, \end{cases} \tag{E.8}$$

$$\partial^\mu_{(y)}\phi^*(y)\,\partial^\nu_{(x)}\phi(x) = \int d^4p \exp(ip\cdot(y-x))\Big[-p^\mu p^\nu + \frac{1}{2}ip^{(\mu}\partial^{\nu)}_{(w)} + \frac{1}{4}\partial^\mu_{(x)}\partial^\nu_{(x)}\Big]\, f_{\rm op}(w,p)\big|_{w=(y+x)/2}\,. \tag{E.9}$$

For completeness, the main equations derived from the equations

$$\left[k^2 - \Pi(k)\right] f_{\rm op}(k) = 0 \tag{E.10}$$

are repeated:

$$\begin{cases} \left\{p\cdot k - \frac{1}{2}\left[\Pi\left(p+\frac{1}{2}k\right) - \Pi\left(p-\frac{1}{2}k\right)\right]\right\} f_{\rm op}(k) = 0, \\ \left\{p^2 + \frac{1}{4}k^2 - \frac{1}{2}\left[\Pi\left(p+\frac{1}{2}k\right) + \Pi\left(p-\frac{1}{2}k\right)\right]\right\} f_{\rm op}(k) = 0. \end{cases} \tag{E.11}$$

The main observables then read

$$\langle J^\mu(k)\rangle = \frac{1}{2}\int d^4p f(k,p)\left[2p^\mu - \int_{-1/2}^{+1/2} ds\, \nabla^\mu \Pi\,(p+ks)\right] \tag{E.12}$$

(still with $\nabla^\mu \equiv \partial/\partial p_\mu$) for the four-current; and

$$\langle T^{\mu\nu}(k)\rangle = \int d^4p f(k,p)\left[2p^\mu p^\nu - \frac{1}{2}k^\mu k^\nu\right] f(k,p) - a^{\mu\nu}(k) - \eta^{\mu\nu}\mathcal{L}(k) \tag{E.13}$$

for the energy–momentum tensor; and where $a^{\mu\nu}(k)$ is given by

$$a^{\mu\nu}(k) = \int_0^{1/2} ds \int d^4p f(k,p)\left[\left(p^\nu - \frac{1}{2}k^\nu\right)\nabla^\mu \Pi\left(p + \frac{1}{2}k\right) + \left(p^\nu + \frac{1}{2}k^\nu\right)\nabla^\mu \Pi\left(p - \frac{1}{2}k\right)\right] \tag{E.14}$$

and where the Lagrangian is given by

$$\mathcal{L}(k) = \int d^4p \left[p^2 - \frac{1}{4}k^2 - \Pi(p)\right] f(k,p). \tag{E.15}$$

Finally, let us also note the very useful relations

$$k \cdot \nabla\Pi(p \pm ks) = \pm\frac{\partial}{\partial s}\Pi(p \pm ks). \tag{E.16}$$

E.2. Useful Formulae for Fermions

The equations of motion for the quasifermion, in the case of interacting quasiparticles,

$$\begin{cases} i\gamma \cdot \partial\psi(x) - \int d^4x'\, \Sigma(x,x')\,\psi(x') = 0, \\ \overline{\psi}(x')i\gamma \cdot \overleftarrow{\partial} + \int d^4x'\, \overline{\psi}(x')\Sigma(x',x) = 0, \end{cases} \tag{E.17}$$

are obtained from the Lagrangian

$$\mathcal{L} = \frac{i}{2}\overline{\psi}(x)\gamma \cdot \partial\psi(x) - \int d^4x'\, \overline{\psi}\left(x + \frac{1}{2}x'\right) \times \Sigma\left(x + \frac{1}{2}x', x - \frac{1}{2}x'\right)\psi\left(x + \frac{1}{2}x'\right). \tag{E.18}$$

Note the relation

$$\Sigma(x,x') = \gamma^0\Sigma^+(x',x)\,\gamma^0 \tag{E.19}$$

which arises from the consistency of the equations of motion, where the cross indicates the Hermitian conjugation. From the Lagrangian, the four-current is derived as

$$J^{\mu}_{\mathrm{op}}(x) = \overline{\psi}(x)\gamma^{\mu}\psi(x) + i\int d^4y \int_{-1/2}^{+1/2} ds\; y^{\mu}\overline{\psi}\left(x + y\left[s + \frac{1}{2}\right]\right) \times \Sigma\left[x + y\left[s + \frac{1}{2}\right], x + y\left[s - \frac{1}{2}\right]\right]\psi\left[x + y\left[s - \frac{1}{2}\right]\right] \tag{E.20}$$

and the energy–momentum tensor as

$$T^{\mu\nu} = \frac{i}{2}\overline{\psi}(x)\gamma^{\mu}\overleftrightarrow{\partial}^{\nu}\psi(x) - \int d^4y\, y^{\mu}\int_0^{1/2} ds\left\{\partial^{\nu}\overline{\psi}\left[x - y\left(s - \frac{1}{2}\right)\right] \times \Sigma\left[x - y\left(s - \frac{1}{2}\right), x - y\left(s + \frac{1}{2}\right)\right]\psi\left[x - y\left(s + \frac{1}{2}\right)\right] - \overline{\psi}\left[x + y\left(s + \frac{1}{2}\right)\right]\Sigma\left[x + y\left(s + \frac{1}{2}\right), x + y\left(s - \frac{1}{2}\right)\right] \times \partial^{\nu}\overline{\psi}\left[x - y\left(s - \frac{1}{2}\right)\right]\right\} - \eta^{\mu\nu}\mathcal{L}. \tag{E.21}$$

Let us note that this tensor is nonconservative since

$$\partial_{\mu}T^{\mu\nu} = \int d^4x'\,\overline{\psi}\left(x + \frac{1}{2}x'\right)\Sigma\left(x + \frac{1}{2}x', x - \frac{1}{2}x'\right)\psi\left(x - \frac{1}{2}x'\right). \tag{E.22}$$

In term of the covariant Wigner operator for the quasifermions and of the convenient definition

$$\Sigma\,(x,y) \equiv \widetilde{\Sigma}\left(\frac{x+y}{2}, x - y\right), \tag{E.23}$$

the four-current reads

$$J^{\mu} = \mathrm{Sp}\int d^4p\left\{\gamma^{\mu}F(x,p) + i\int d^4y\, y^{\mu}\int_{-1/2}^{+1/2} ds \times \exp(ip\cdot y)\Sigma(x + ys, y)\right\}F(x + ys, p)\}, \tag{E.24}$$

while the energy–momentum tensor has the form

$$T^{\mu\nu} = \mathrm{Sp}\int d^4p\left\{p^\mu\gamma^\nu - \eta^{\mu\nu}\left[\gamma\cdot p - \widetilde{\Sigma}(x,p)\right]\right\}F(x,p)$$
$$+\int d^4y d^4p\int_0^{1/2} ds y^\mu \exp(ip\cdot y)$$
$$\times\left[\widetilde{\Sigma}\,(x+ys,y)\left(p^\nu+\frac{1}{2}\partial^\nu\right)F(x+ys,p)\right.$$
$$\left.+\widetilde{\Sigma}\,(x-ys,y)\left(p^\nu-\frac{1}{2}\partial^\nu\right)F(x-ys,p)\right] \tag{E.25}$$

and the equations satisfied by the Wigner function in Fourier space read

$$\begin{cases}\gamma\cdot\left(p+\frac{1}{2}k\right)F(k,p) \\ \quad = -\int\frac{d^4k'}{(2\pi)^4}\widetilde{\Sigma}\left(k-k',p+\frac{1}{2}k'\right)F\left(k',p-\frac{1}{2}[k-k']\right) \\ F(k,p)\gamma\cdot\left(p-\frac{1}{2}k\right) \\ \quad = -\int\frac{d^4k'}{(2\pi)^4}F\left(k',p+\frac{1}{2}[k-k']\right)\widetilde{\Sigma}\left(k-k',p-\frac{1}{2}k'\right).\end{cases} \tag{E.26}$$

Bibliography

Abellàn J., A. Navarro and E. Alvarez, Thermodynamics and cosmology, *J. Phys.* **A10**, L129 (1977).

Abonyi I., Steady state solution of the relativistic Boltzmann equation, *Z. Angew. Math. Phys.* **XI**, 169 (1960).

Abonyi I., Le problème de l'équation de Boltzmann relativiste, *Cahiers Phys.* **18**, 461 (1964).

Abonyi I., Sur la solution de l'équation de Boltzmann relativiste sans collisions en présence de certains champs extérieurs, *Acta Phys. Acad. Sci. Hungar.* **23**, 247 (1967).

Acquista C. and J.L. Anderson, Radiative transfer of partially polarized light, *Astrophys. J.* **191**, 567 (1974).

Adam D. and H.D. Sivak, Screening potential due to relativistic degenerate electron gas embedded in a strong magnetic field, *Physica* **A133**, 307 (1985).

Adam D., Models of magnetic white dwarfs, *Astron. Astrophys.* **160**, 95 (1986).

Adam Ph. and R. Hakim, Covariant Wigner function approach to the relativistic charged gas in a strong magnetic field II: spin zero in thermal equilibrium, *Physica* **A113**, 491 (1982).

Adler S.L., J.N. Bahcall, C.G. Callan and M.N. Rosenbluth, Photon splitting in a strong magnetic field, *Phys. Rev. Lett.* **25**, 1061 (1970).

Aitken D.M., Relativistic irreversible thermodynamics. M.Sc. thesis, University of Alberta, Edmonton, 1964.

Akama H., Relativistic Boltzmann equation for plasmas, *J. Phys. Soc. Jpn.* **28**, 478 (1970).

Akhiezer A.I. and S.V. Peletminski, Use of the methods of quantum field theory for the investigation of the thermodynamical properties of a gas of electrons and photons, *Sov. Phys. JETP* **11**, 1316 (1960).

Akhiezer A.I. and A.G. Sitenko, Contribution to the theory of plasma fluctuations, *Sov. Phys. JETP* **14**, 462 (1960).

Aldrovandi R. and N.R. Teixeira filho, Relativistic phase-space for bosons and fermions, *Rev. Bras. Fis.* **5**, 383 (1975).

Aldrovandi R. and N.R. Teixeira filho, Recursion relation for the relativistic quantum phase-space, *Lett. Nuovo Cimento* **17**, 9 (1976).

Aldrovandi R. and N.R. Teixeira filho, The role of the symmetry group in the cluster decomposition of the relativistic phase-space, *Lett. Nuovo Cimento* **15**, 591 (1976).

Alexanian M., Covariant black-body and the 3 K radiation field, *Lett. Nuovo Cimento* **1**, 75 (1969).

Altherr T., Gauge field theory methods for ultra-degenerate and ultra-relativistic plasmas, *Astropart. Phys.* **1**, 133 (1992).

Altherr T., Introduction to thermal field theory, *Int. J. Mod. Phys.* **A8**, 5605 (1993).

Altherr T., Thermal field theory and its application to astrophysics. Prepared for Workshop on Quantum Infrared Physics (Paris, France; 6–10 Jun. 1994), in Paris 1994; Quantum Infrared Physics.

Altherr T., E. Petitgirard and T. del Rio Gaztelurrutia, Photon propagation in dense media, *Astropart. Phys.* **1**, 289 (1993).

Alvarez E., The relativistic scalar entropy, *Phys. Lett.* **A70**, 363 (1970).

Alvarez E., On the definition of entropy in relativistic kinetic theory, *Phys. Lett.* **A60**, 9 (1977).

Alvarez E., On the generalized relativistic Liouville equilibrium, *J. Phys.* **A9**, 687 (1976).

Alvarez E., Fluid limit of kinetic cosmology. Unpublished (1976).

Alvarez E., On the transport equations for a one-component relativistic gas, *J. Phys.* **A9**, 1861 (1976).

Alvarez E., Cosmological solutions of the Einstein–Liouville system. Unpublished.

Alvarez E., Quark matter in Hartree–Fock approximation, *Phys. Rev.* **D23**, 1715 (1981).

Alvarez E., The Hartree approximation as a high density limit, *Phys. Lett.* **B110**, 315 (1982).

Alvarez E., High-density limit of quantum chromodynamics, *Phys. Rev.* **D27**, 2085 (1983).

Alvarez E., Supersymmetric equilibrium, *Lett. Nuovo Cimento* **36**, 467 (1983).

Alvarez E. and J.M. Gracia-Bondia, Absorption lines of quasistellar objects: a kinematical description, *Astron. Astrophys.* **36**, 299 (1974).

Alvarez E. and L. Bel, Generalized angular-velocity formula and kinematical analysis of 3C 279, *Astrophys. J.* **179**, 687 (1973).

Alvarez E., Relativistic many-body physics (Unpublished; Madrid, 1984).

Alvarez E., L. Bel and J.M. Gracia-Bondia, Kinematical description of quasi-stellar objects, *Astron. Astrophys.* **40**, 381 (1975).

Alvarez E. and R. Hakim, Equation d'état pour le plasma de quarks: quelques applications astrophysiques (Journées Relativistes, Genève, 1978).

Amiranashvili Sh.G., L.S. Kuz'menkov and O.O. Trubachev, Applicability of the ideal fluid model in high-temperature plasma physics, *Sov. Phys. J.* **32**, 305 (1989).

Anderson J.L., Relativistic Boltzmann theory and the Grad method of moments, In eds. M. Carmeli, S.I. Fickler and L. Witten, *Relativity–Proc. Midwest Conf. on Relativity* (Cincinnati, USA 1969) (Plenum, New York, 1970), p. 109.

Anderson J.L. and E.A. Spiegel, The moment method in relativistic radiative transfer, *Astrophys. J.* **171**, 127 (1972).
Anderson J.L. and Witting H.R., Entropy production in an expanding gas-filled universe, *Trans. N.Y. Acad. Sci.* **35**, 636 (1973).
Anderson J.L., Relativistic Grad polynomials, *J. Math. Phys.* **15**, 1116 (1974).
Anderson J.L. and Witting H.R., A relativistic relaxation-time model for the Bolzmann equation, *Physica* **74**, 466 (1974).
Anderson J.L. and Witting H.R., Relativistic quantum transport coefficients for Fermi–Dirac and Bose–Einstein gases, *Physica* **74**, 489 (1974).
Anderson J.L., Variational principles for relativistic transport coefficients, *Physica* **A79**, 569 (1975).
Anderson J.L. and E.A. Spiegel, Radiative transfer through a flowing medium, *Astrophys. J.* **202**, 454 (1975).
Anderson J.L., The equations of radiative hydrodynamics, *Gen. Relativ. Gravit.* **7**, 53 (1976).
Anderson J.L., Variational principles for calculation of transport coefficients of relativistic multicomponent systems, *Physica* **A85**, 287 (1976).
Anderson J.L. and A.C. Payne Jr., The relativistic Burnett equations and sound propagation, *Physica* **A85**, 261 (1976).
Anderson J.L., Equations of hydrodynamics in general relativity in the slow motion approximation, *Gen. Relativ. Gravit.* **6**, 197 (1976).
Anderson J.L. and A.J. Kox, On the correct forms of the transport coefficients for a relativistic gas of hard spheres, *Physica* **A89**, 408 (1977).
Anderson J.L., Stochastic forces and fluctuation–dissipation theorems in a relativistic one-component gas, *Physica* **A163**, 501 (1990).
Anderson J.L. and E. Nowotny, Stochastic forces and fluctuation–dissipation theorem in a one-component gas, *Physica* **163**, 501 (1990).
Anile A.M., Evolution of Shock Waves in Relativistic Continuum Mechanics (1981).
Anile A.M., Gravitational Focussing of Shock Waves in General Relativity (Journées Relativistes, Torino, 1983).
Anile A.M., J.C. Miller and S. Motta, Formation and damping of relativistic strong shocks, *Phys. Fluids* **26**, 1450 (1983).
Anile A.M. and O. Muscato, Simple waves in relativistic magnetofluidodynamics, *Lett. Nuovo Cimento* **38**, 581 (1983).
Anile A.M. and G. Russo, Corrugation stability for plane relativistic shock waves, *Phys. Fluids* **29**, 2847 (1986).
Antonsen F., Quantum theory in curved space–time using the Wigner function, *Phys. Rev.* **D56**, 920 (1987).
Applequist T., Thermodynamics and finite temperature field theory, *Nucl. Phys.* **A418**, 35 (1984).
Arcidiacono G., Sulla propagazione delle onde nella termodinamica relativista, *Collect. Math.* **37**, 3 (1976).
Asseo E., D. Gerbal, J. Heyvaerts and M. Signore, General-relativistic kinetic theory of waves in a massive particle medium, *Phys. Rev.* **D13**, 2724 (1976).

Badziag P., Gauge independent Wigner function approach to spontaneous emission and to quantum kinetic equations, *Physica* **A130**, 565 (1985).

Bailin D. and A. Love, Superfluid quark matter, *J. Phys.* **A12**, L283 (1979).

Bailin D. and A. Love, The colour superconducting phase transition in quark matter, *Phys. Lett.* **107B**, 377 (1981).

Bailin D. and A. Love, Superfluid order in relativistic quark matter, *Nucl. Phys.* **B190**, 751 (1981).

Bailin D. and A. Love, Superfluidity in ultra-relativistic quark matter, *Nucl. Phys.* **B190**, 175 (1981).

Bailin D. and A. Love, Superconductivity in quark matter, *Nucl. Phys.* **B205**, 119 (1982).

Bakshi P., R. Cover and G. Kalman, Polarization and electromagnetic properties of the vacuum in the static limit for strong magnetic fields, In *Conf. Role of Magnetic Fields in Physics and Astrophysics* (1974).

Bakshi P., R. Cover and G. Kalman, Polarization and electromagnetic properties of the vacuum in the static limit for strong magnetic fields, *Ann. N.Y. Acad Sci.* **257**, 95 (1975).

Bakshi P., R. Cover and G. Kalman, Electromagnetic response in strong magnetic fields: general formalism and vacuum polarization for parallel propagation, *Phys. Rev.* **D14**, 2532 (1976).

Balescu R., On the statistical mechanics of a relativistic quantum plasma, *Physica* **31**, 1599 (1965).

Balescu R. and T. Kotera, On the covariant formulation of classical relativistic statistical mechanics, *Physica* **33**, 558 (1967).

Balescu R., T. Kotera and E. Pena, Lorentz transformations in phase space and in physical space, *Physica* **33**, 581 (1967).

Balescu R. and R. Sergysels, Evolution equations for the coherence function in a relativistic plasma, *Nuovo Cimento* **50**, 378 (1967).

Balescu R., M. Baus and A. Pytte, On some mathematical aspects of classical relativistic statistical mechanics, *Bull. Cl. Sci. Acad. R. Belg.* **53**, 1043 (1967).

Balescu R., A covariant formulation of quantum relativistic statistical mechanics, I: Phase space description of a relativistic quantum plasma, *Acta Phys. Austriaca* **28**, 309 (1968).

Balescu R., Kinetic equations and Lorentz transformations, *Physica* **38**, 119 (1968).

Balescu R. and A. Pytte, Geometrical and tensorial Lorentz transformations, *Nuovo Cimento* **B55**, 51 (1968).

Balescu R., Relativistic statistical thermodynamics, *Physica* **40**, 309 (1968).

Balescu R., A covariant formulation of quantum relativistic statistical mechanics, I: The Liouville equations for a relativistic quantum plasma, *Acta Phys. Austriaca* **29**, 313 (1969).

Balescu R., A covariant formulation of quantum relativistic statistical mechanics, II. The Liouville equations for a relativistic quantum plasma, *Acta Phys. Austriaca* **29**, 313 (1969).

Balescu R. and L. Brenig, Relativistic covariance of nonequilibrium statistical mechanics, *Physica* **54**, 504 (1971).

Balescu R., L. Brenig and I. Paiva-Veretennicoff, On the relativistic plasma-dynamical and hydrodynamical normal modes, *Physica* **74**, 447 (1974).

Balescu R., L. Brenig and I. Paiva-Veretennicoff, Kinetic theory of the plasma-dynamical modes and the transport coefficients of a relativistic plasma, *Physica* **A81**, 17 (1975).

Balescu R., L. Brenig and I. Paiva-Veretennicoff, On the relativistic plasma-dynamical and hydrodynamical normal modes, *Physica* **A81**, 1 (1975).

Balescu, R., L. Brenig and I. Paiva-Veretennicoff, Plasma-dynamical normal modes and transport coefficients of a relativistic plasma, in *Colloques Internationaux du Centre National de la Recherche Scientifique no. 236. Théories cinétiques classiques et relativistes*, 71 (Editions du Centre National de la Recherche Scientifique, Paris, 1975).

Bampi F. and A. Morro, Relativistic heat equation in causal nonstationary thermodynamics, *Phys. Lett.* **A79**, 156 (1980).

Bampi F. and A. Morro, Two approaches to nonstationary relativistic thermodynamics, *J. Math. Phys.* **21**, 1201 (1980).

Bancel D., Problème de Cauchy pour l'équation de Boltzmann non linéaire en relativité générale, *C. R. Acad. Sci.* **272**, 1685 (1971).

Bancel D., Problème de Cauchy pour l'équation de Boltzmann relativiste, *J. Math. Pure Appl.* **51**, 33 (1972).

Bancel D. and Y. Choquet-Bruhat, Existence, uniqueness and local stability for the Einstein–Maxwell–Boltzmann system, *Commun. Math. Phys.* **XVIII**, 263 (1973).

Bancel, D., Problème de Cauchy pour l'équation de Boltzmann en relativité générale, *Ann. Inst. H. Poincaré* **33**, 83 (1973).

Bancel D., Cauchy problem for the Einstein–Boltzmann system, In *Colloques Internationnaux du Centre National de la Recherche Scientifique no. 236.* Théories cinétiques classiques et relativistes, 223 (Editions du Centre National de la Recherche Scientifique, Paris, 1975).

Barbachoux C., F. Debbasch and J.P. Rivet, The spatially one-dimensional relativistic Ornstein–Uhlenbeck process in an arbitrary inertial frame, *Eur. Phys. J.* **B19**, 37 (2001).

Barbachoux C., F. Debbasch and J.P. Rivet, Covariant Kolmogorov equation and entropy current for the relativistic Ornstein–Uhlenbeck process, *Eur. Phys. J.* **B23**, 487 (2001).

Barcon X. and R. Lapiedra, Classical dilute relativistic plasmas in equilibrium II. Thermodynamic functions, *Phys. Rev.* **A28**, 3030 (1984).

Barcon X., Self-oscillations in classical dilute relativistic plasmas, *Physica* **A123**, 617 (1984).

Baring M.G., Synchrotron emission, pair production and annihilation in strongly magnetised relativistic plasmas, *Mon. Not. R. Astron. Soc.* **235**, 51 (1988).

Barker B.M., M.S. Bhatia and G. Szamosi, Excluded-volume model of a degenerate Fermi gas of hard spheres, *Nuovo Cimento* **B52**, 355 (1967).

Barrois B.C., Superconducting quark matter, *Nucl. Phys.* **B129**, 390 (1977).

Barrabes C., Contribution à l'étude des phénomènes de transport en relativité générale. Thesis, Tours (1984).

Barrabes C., Elastic and thermoeleastic media in general relativity, *Nuovo Cimento* **B28**, 377 (1975).
Barrabes C. and J. Henry, Relativistic kinetic theory of a system of particles with variable rest mass, *J. Phys.* **A9**, 1425 (1976).
Barrabes C., Equilibrium and nonequilibrium relativistic kinetic theory of particles with variable masses, *Nuovo Cimento* **B68**, 161 (1982).
Barrabes C., Transient relativistic thermodynamics of a system of particles with variable masses, *Physica* **A16**, 484 (1982).
Barrabes C., Transient thermodynamics of electromagnetic media in general relativity, *J. Math. Phys.* **26**, 798 (1985).
Barrabes C. and W. Israel, Relativistic kinetic theory of a system of cosmic strings: general formalism, *Ann. Phys.* (*N.Y.*) **175**, 213 (1987).
Barton G., On the finite temperature quantum electrodynamics of free electrons and photons, *Ann. Phys.* (*N.Y.*) **200**, 271 (1990).
Barut A.O., Covariant quantum statistics of fields, *Phys. Rev.* **109**, 1376 (1958).
Bayin S.S., Equation of state at ultrahigh densities and the speed of sound, *Phys. Rev.* **D21**, 1503 (1980).
Baym G. and S.A. Chin, Landau theory of relativistic Fermi liquids, *Nucl. Phys.* **A262**, 527 (1976).
Baym G., J.-P. Blaizot and B. Svetitsky, Emergence of new quasiparticles in quantum electrodynamics at finite temperature, *Phys. Rev.* **D46**, 4043 (1992).
Beaudet G. and M. Tassoul, Fitting formulae for a relativistic and/or degenerate electron gas, *Astron. Astrophys.* **13**, 209 (1971).
Bechler A., Polarization tensor and collective phenomena in the relativistic e^-e^+ plasma, *J. Phys.* **G13**, 1327 (1987).
Beckenstein J.D. and E. Oron, New conservation laws in general-relativistic magnetohydrodynamics, *Phys. Rev.* **D18**, 1809 (1978).
Beckmann R., F. Karsch and D.E. Miller, Bose–Einstein condensation of a relativistic gas in d dimensions, *Phys. Rev. Lett.* **43**, 1277 (1979).
Bel L., Kinetic theory of cosmology, *Astrophys. J.* **155**, 83 (1969).
Beliaev S.T. and G.I. Budker, The relativistic kinetic equation, *Sov. Phys. Dokl.* **1**, 218 (1956).
Ben-Yaakov U., Relativistic brownian motion and the spectrum of thermal radiation, *Phys. Rev.* **D23**, 1441 (1981).
Ben-Ya'acov U., Strings, superfluid vortices, and relativity, *Phys. Rev.* **D44**, 2452 (1991).
Ben-Ya'acov U., Extended dual transformation including quantum vortices, *Phys. Lett.* **B274**, 352 (1992).
Ben-Ya'acov U., Unified dynamics of quantum vortices, *Nucl. Phys.* **B382**, 597 (1992).
Ben-Ya'acov U., Non-linearity, the choice of Green function and self-regularization in quantum vortex dynamics, *Nucl. Phys.* **B382**, 616 (1992).
Ben-Ya'acov U., Relativistic superfluid vortices and Helmoltz's theorem, *J. Phys.* **A27**, 7165 (1994) (1993).

Berezdivin R. and R.K. Sachs, A lemma on the Einstein–Liouville equations, In eds. M. Carmeli, S.J. Fickler and L. Witten *Relativity–Proc. Midwest Conf. on Relativity* (Cincinnati, USA, 1969), (Plenum, New York, 1970), p. 25.

Berezdivin R. and R.K. Sachs, Matter symmetries in general relativistic kinetic theory, *J. Math. Phys.* **14**, 1254 (1973).

Berezhiani V.I., R.D. Hazeltine and S.M. Mahajan, Radiation reaction and relativistic hydrodynamics, *Phys. Rev.* **E69**, 056406 (2004).

Berezin V.A., Unusual hydrodynamics, *Int. J. Mod. Phys.* **A2**, 1591 (1987).

Bergmann P.G., Generalized statistical mechanics, *Phys. Rev.* **84**, 1026 (1951).

Bergmann P.G., The special theory of relativity, In ed. S. Flügge *Handbuch der Physik (Encyclopedia of Physics)* 4, Prinzipien der Elektrodynamik und Relativitätstheorie, (Springer-Verlag, Berlin, 1962), p. 159.

Bernard C., Feynman rules for gauge theories at finite temperature, *Phys. Rev.* **D9**, 3312 (1974).

Bernstein J., *Kinetic Theory in the Expanding Universe* (Cambridge University Press, 1988).

Bernstein J. and S. Dodelson, Relativistic Bose-Gas, *Phys. Rev. Lett.* **66**, 683 (1991).

Best C. and J.M. Eisenberg, Pair creation in the transport equations using the equal-time Wigner function, *Phys. Rev.* **D47**, 4639 (1993).

Bezchastnov V.G. and G.G. Pavlov, Quantum relativistic theory of synchrotron/cyclotron radiation in a strongly magnetised plasma: the case of longitudinal propagation, *Astrophy. Space Sci.* **148**, 257 (1988).

Bezchastnov V.G. and G.G. Pavlov, Quantum relativistic theory of cyclotron radiation in a strongly magnetised plasma with anisotropic temperature. Varenna School on Plasma Astrophysics (1988) (ESA, Noordwijk, 1989).

Bezchastnov V.G. and G.G. Pavlov, Quantum relativistic theory of cyclotron radiation in a strongly magnetised plasma with anisotropic temperature. Varenna School on Plasma Astrophysics (1988) (ESA, Noordwijk, 1989).

Bezzerides B. and D.F. DuBois, Quantum electrodynamics of nonthermal relativistic plasmas: kinetic theory, *Ann. Phys. (N.Y.)* **70**, 10 (1972).

Bialas A. and W. Czyz, Boost-invariant Boltzmann–Vlasov equation for relativistic quark–antiquark plasma, *Phys. Rev.* **D30**, 2371 (1984).

Bialynicki-Birula I., E.D. Davis and J. Rafelski, Evolution modes of the vacuum Wigner function in strong-field QED, *Phys. Lett.* **B311**, 329 (1993).

Bialynicki-Birula I., E.D. Davis and J. Rafelski, Evolution modes of the vacuum Wigner function in strong-field QED, *Phys. Lett.* **B311**, 329 (1993).

Bichteler K., Bemerkung über relativistische Stossinvarianten, *Z. Phys.* **182**, 521 (1965).

Bichteler K., On the Cauchy problem of the relativistic Boltzmann equation, *Commun. Math. Phys.* **4**, 352 (1967).

Bilaniuk O.M. and E.C.G. Sudarshan, Causality and spacelike signals, *Nature* **223**, 386 (1969).

Bildhauer S., Freely propagating polarized radiation in curved space–times, *Astron. Astrophys.* **219**, 25 (1989).

Biskamp D., Derivation of the quantum Vlasov equation and dispersion relation of a relativistic electron gas, *Z. Naturforschung,* **22a**, 869 (1967).

Blaizot J.-P., Quantum field theory at finite temperature and density, in High Density and High Temperature Physics (Namwon, Korea, 1991).

Blaizot J.-P. and E. Iancu, Kinetic theory and quantum electrodynamics at high temperature, *Nucl. Phys.* **B390**, 589 (1993).

Blaizot J.-P. and E. Iancu, Hard thermal loops excitations, in *Quantum Infrared Physics* (Paris, 1994).

Blaizot J.-P. and E. Iancu, Gauge structure and semi-classical aspects of hard thermal loops, *Nucl. Phys.* **B434**, 662 (1995).

Blaizot J.-P., E. Iancu and R.R. Parwani, On the screening of static electromagnetic fields in hot QED plasmas, *Phys. Rev.* **D52**, 2543 (1995).

Blaizot J.-P., J.F. Berger, J. Dechargé and M. Girod, Microscopic and macroscopic determinations of nuclear compressibility, *Nucl. Phys.* **A591**, 435 (1995).

Blaizot J.-P., Lifetime of quasiparticles excitations in hot gauge theories, in *Continuous Advances in QCD* (Minneapolis, 1996).

Blaizot J.-P. and E. Iancu, Lifetime of quasiparticles in hot QED plasmas, *Phys. Rev. Lett.* **76**, 3080 (1996).

Blaizot J.-P. and E. Iancu, Non-perturbative aspects of screening phenomena in Abelian and non-Abelian gauge theories, *Nucl. Phys.* **B459**, 559 (1996).

Blaizot J.-P., Quasiparticles in ultrarelativistic plasmas plasmas, *Nucl. Phys.* **A606**, 347 (1996).

Blaizot J.-P. and E. Iancu, Lifetime of quasiparticles and collective excitations in hot QED plasmas, *Phys. Rev.* **D55**, 973 (1997).

Blaizot J.-P. and E. Iancu, The Bloch–Nordsieck propagator at finite temperature, *Phys. Rev.* **D56**, 7877 (1997).

Blättel B., V. Koch, W. Cassing and U. Mosel, Covariant Boltzmann–Uehling–Uhlenbeck approach for heavy-ion collisions, *Phys. Rev.* **C38**, 1767 (1988).

Blättel B., V. Koch, K. Weber, W. Cassing and U. Mosel, Transverse momentum analysis in the relativistic BUU approach, *Nucl. Phys.* **A495**, 381c (1989).

Blättel B., V. Koch, W. Cassing and U. Mosel, Transport theoretical analysis of relativistic heavy-ion collisions, *Rep. Prog. Phys.* **56**, 1 (1993).

Blazhievskii L.F. and A.G. Zayats, Quasi-relativistic electron gas, *Sov. Phys. J.* **25**, 243 (1982).

Blazhievskii L.F. and S.S. Semak, Degenerated quasi-relativistic electron gas in electrostatic gas, *Ukr. Fiz. Zh.* **31**, 469 (1986).

Blinnikov S.I., Thermodynamics of a relativistic Fermi gas, *Sov. Astron. Lett.* **13**, 346 (1987).

Blokhintsev D.I., Relativistic density matrix in phase space, Preprint Dubna (SSSR) P2-7424 (1973).

Bludman S.A. and M.A. Ruderman, Noncausality and instability in ultradense matter, *Phys. Rev.* **D1**, 3243 (1970).

Bludman S.A. and K.A. van Riper, Equation of state of an ideal Fermi gas, *Astrophys. J.* **212**, 859 (1977).

Boisseau B. and W.A. van Leeuwen, Relativistic Boltzmann equation in $d + 1$ spacetime dimensions, *Ann. Phys. (N.Y.)* **195**, 376 (1989).

Boisseau B., Vortex in a weakly relativistic perfect isentropic fluid and Nambu–Goto hydrodynamics, *Phys. Rev.* **D61**, 083504 (2000).

Boisseau B., Vortex in a weakly relativistic Bose gas at zero temperature and relativistic fluid approximation, *J. Phys.* **A37**, 7923 (2002).

Bolsterli M., Relativistic second-order energy in infinite fermion matter, *Phys. Rev.* **D11**, 312 (1975).

Botermans W. and R. Malfliet, Quantum transport theory of nuclear matter, *Phys. Rep.* **198**, 115 (1990).

Bowers R.L., Asymptotic limit for the speed of sound in a system of relativistically interacting particles, *Phys. Rev. Lett.* **29**, 509 (1972).

Bowers R.L. and R.L. Zimmerman, Relativistic quantum many-body theory in a Riemannian space–time, *Phys. Rev.* **D7**, 296 (1973).

Bowers R.L., J.A. Campbell and R.L. Zimmerman, Relativistic many-body theory for strongly interacting matter, *Phys. Rev.* **D7**, 2278 (1973).

Bowers R.L., J.A. Campbell and R.L. Zimmerman, Model equation of state for strongly interacting superdense matter, *Phys. Rev.* **D7**, 2289 (1973).

Bowers R.L., J.A. Campbell and R.L. Zimmerman, Relativistic baryon effective mass and thresholds for strongly interacting superdense matter, *Phys. Rev.* **D8**, 1077 (1973).

Bowers R.L., A.M. Gleeson and R.D. Pedigo, Relativistic superdense matter in cold systems: theory, *Phys. Rev.* **D12**, 3043 (1975).

Bowers R.L., A.M. Gleeson and R.D. Pedigo, Relativistic superdense matter in cold systems: applications, *Phys. Rev.* **D12**, 3056 (1975).

Bowers R.L., A.M. Gleeson and R.D. Pedigo, Approaches to the analysis of relativistic superdense matter, *Nuovo Cimento* **B41**, 441 (1977).

Bowers R.L., A.M. Gleeson, R.D. Pedigo and J.W. Wheeler, A finite temperature system of strongly interacting baryons, *Phys. Rev.* **D15**, 2125 (1977).

Braaten E. and R.D. Pisarski, Simple effective Lagrangian for hard thermal loops, *Phys. Rev. Lett.* **D45**, R1827 (1992).

Brandt H.E., Symmetry of the complete second-order non-linear conductivity tensor for an unmagnetized relativistic turbulent plasma, *J. Math. Phys.* **24**, 1332 (1983).

Brandt H.E., Symmetry of the complete second-order non-linear conductivity tensor for an unmagnetized relativistic turbulent plasma, *J. Math. Phys.* **24**, 1332 (1983).

Braaten E., Neutrino emissivity of an ultrarelativistic plasma from positron and plasmino annihilation, *Astrophys. J.* **392**, 70 (1992).

Braaten E., Effective field theory for plasmas at all temperatures and densities, Preprint.

Brink D.M. and J. da Providencia, Landau damping in infinite nuclear matter, *Nucl. Phys.* **A500**, 242 (1989).

Brinkmann W., Temporal evolution of electron–positron plasma, In Workshop on Electron–positron Pairs in Astrophysics, *AIP Conf. Proc.* (Greenbelt, 1983).

Brittan M., Correlation corrections to the mean-field theory of nuclear matter, *Phys. Lett.* **B79**, 27 (1978).

Brown G.E., The equation of state of dense matter, *Phys. Rep.* **163**, 167 (1988).

Bugaev K.A., B. Kämpfer and V.I. Zhdanov, Generalized shock adiabatics and relativistic nuclear collisions, *Phys. Rev.* **D40**, 2903 (1989).

Bugaev K.A. and M.I. Gorenstein, Thermodynamically self-consistent class of nuclear matter EOS and compression shocks in relativistic nuclear collisions, *Z. Phys.* **C43**, 261 (1989).

Bunatan G.G., Thermodynamical theory of pion condensation, *Nucl. Phys.* **A404**, 525 (1983).

Buneman O., Covariant dispersion theory for a collisionless plasma in a uniform field, In *Relativistic Plasmas — The Coral Gable Conference*, eds. O. Buneman and W.B. Pardo (W.A. Benjamin, New York, 1968).

Buneman O., Experiments on relativistic plasma, In *Relativistic Plasmas — The Coral Gable Conference*, eds. O. Buneman and W.B. Pardo (W. Benjamin, New York, 1968).

Buneman O. and W.B. Pardo (eds.), *Relativistic Plasmas — The Coral Gable Conference* (W.A. Benjamin, New York, 1968).

Buneman O., Dispersion characteristics of "warm" and "hot" plasmas, *Astrophys. J.* **235**, 616 (1980).

Buot F.A., Magnetic susceptibility of a relativistic Dirac electron gas, *Phys. Rev.* **A8**, 1570 (1973).

Buot F.A., Erratum: Weyl transform and the magnetic susceptibility of a relativistic Dirac electron gas, *Phys. Rev.* **A9**, 2811 (1974).

Burakovsky L., Relativistic statistical mechanics and particle spectroscopy, *Found. Phys.* **28**, 1577 (1998).

Burakovsky L. and L.P. Horwitz, Equilibrium relativistic mass distribution. Preprint TAUP-2048-93.

Burakovsky L. and L.P. Horwitz, Galilean limit of equilibrium relativistic mass distribution. Preprint TAUP-2081-93.

Burakovsky L. and L.P. Horwitz, Equilibrium relativistic mass distribution for indistinguishable events. Preprint TAUP-2115-93.

Burakovsky L. and L.P. Horwitz, Galilean limit of equilibrium relativistic mass distribution in event–anti-event system and "realistic" equation of state for hot hadronic matter. Preprint TAUP-2136-94.

Burakovsky L. and L.P. Horwitz, Relativistic mass distribution for indistinguishable events. Preprint TAUP-2161-94.

Burakovsky L. and L.P. Horwitz, Equilibrium relativistic mass distribution, *Physica* **A201**, 666 (1993).

Burakovsky L. and L.P. Horwitz, Generalized Boltzmann equation in a manifestly covariant relativistic statistical mechanics [arXiv:hep-th/9508122 v1].

Burakovsky L., L.P. Horwitz and W.C. Schieve, On relativistic Bose–Einstein condensation [arXiv:hep/th/9508121 v1].

Burakovsky L., L.P. Horwitz and W.C. Schieve, A new relativistic high temperature Bose–Einstein condensation, *Phys. Rev.* **D54**, 4029 (1996).

Buti B., Plasma oscillations and Landau damping in a relativistic gas, *Phys. Fluids* **5**, 1 (1962).

Caianiello E.R. and G. di Genova, Some consequences of phase space geometry, In *Quantum Field Theory — Symp. in Honour of Hiroomi Umezawa*, ed. F. Mancini (North-Holland, Amsterdam, 1986).

Caillon J.C., P. Gabinski and J. Labarsouque, The Landau parameters of nuclear matter in relativistic Hartree approximation, *J. Phys.* **28**, 189 (2002).

Calkoen C.J. and Ch. G. van Weert, Thermal conductivity of a degenerate electron gas, *Phys. Lett.* **A106**, 271 (1984).

Calkoen C.J. and S.R. de Groot, Volume viscosity of a massive neutrino gas, *Physica* **A110**, 222 (1982).

Calkoen C.J. and S.R. de Groot, Volume viscosity of lepton systems: including massive neutrino, *Physica* **A111**, 531 (1984).

Calkoen C.J. and S.R. de Groot, The volume viscosity of a degenerate massive neutrino system, *Physica* **A127**, 627 (1984).

Callen H. and G. Horwitz, Relativistic thermodynamics, *Am. J. Phys.* **39**, 938 (1971).

Campbell D.K., R.F. Dashen and J.T. Manassah, Chiral symmetry and pion condensation I: Model-dependent results, *Phys. Rev.* **D12**, 979 (1975).

Campbell D.K., R.F. Dashen and J.T. Manassah, Chiral symmetry and pion condensation II: General formalism, *Phys. Rev.* **D12**, 1010 (1975).

Calzetta E. and B.L. Hu, Wigner function and quantum Liouville equation in curved spacetime, In *The Physics of Phase Space, Nonlinear Dynamics and Chaos, Geometric Quantization, and Wigner Function — Proc. First Int. Conf. Physics of Phase Space* (University of Maryland, College Park), Vol. 278, eds. Y.S. Kim and W.W. Zachary (Springer-Verlag, Berlin, 1987).

Calzetta E., S. Habib and B.L. Hu, Quantum kinetic field theory in curved spacetime: covariant Wigner function and Liouville–Vlasov equations, *Phys. Rev.* **D37**, 2001 (1988).

Calzetta E., Wigner functions in quantum gravity and quantum cosmology, In *Proc. Fourth Lake Louise Winter Institute*, eds. A. Astbury, B.A. Campbell, W. Israel, A.N. Kamal and F.C. Khanna (Frontiers in Physics, Singapore, 1989).

Calzetta E. and B.L. Hu, Wigner distribution function and phase-space formulation of quantum cosmology, *Phys. Rev.* **D40**, 380 (1989).

Canuto V. and H.-Y. Chiu, Magnetic moment of a magnetized Fermi gas, *Phys. Rev.* **173**, 1229 (1968).

Canuto V. and H.-Y. Chiu, Properties of high-density matter in intense magnetic fields, *Phys. Rev. Lett.* **21**, 110 (1968).

Canuto V. and H.-Y. Chiu, Quantum theory of an electron gas in intense magnetic fields, *Phys. Rev.* **173**, 1210 (1968).

Canuto V. and H.-Y. Chiu, Thermodynamic properties of a magnetized Fermi gas, *Phys. Rev.* **173**, 1220 (1968).

Canuto V., H.-Y. Chiu and L. Fassio-Canuto, Quantum theory of an electron gas with anomalous magnetic moments in intense magnetic fields, *Phys. Rev.* **176**, 1438 (1968).

Canuto V., H.-Y. Chiu and L. Fassio-Canuto, Electron bremsstrahlung in intense magnetic fields, *Phys. Rev.* **185**, 1607 (1969).
Canuto V., H.-Y. Chiu and L. Fassio-Canuto, Thermodynamic approach to the equation of state of a magnetized Fermi gas, *Astrophys. Space Sci.* **3**, 258 (1969).
Canuto V. and H.-Y. Chiu, Longitudinal electric conductivity of a relativistic gas in an intense magnetic field, *Phys. Rev.* **188**, 2446 (1969).
Canuto V., H.-Y. Chiu, C. Chiuderi and H.J. Lee, New state of ferromagnetism in degenerate electron gas and magnetic fields in collapsed bodies, *Phys. Rev. Lett.* **23**, 390 (1969).
Canuto V., Electrical conductivity and conductive opacity of a relativistic electron gas, *Astrophys. J.* **159**, 641 (1970).
Canuto V., H.-Y. Chiu, C. Chiuderi and H.J. Lee, New sources of intense magnetic fields in neutron stars, *Nature* **225**, 47 (1970).
Canuto V. and C. Chiuderi, Transverse electrical conductivity of a relativistic gas in an intense magnetic field, *Phys. Rev.* **D1**, 2219 (1970).
Canuto V., On a quantum mechanical relation for a magnetized electron gas, *Phys. Rev.* **A3**, 648 (1971).
Canuto V. and H.-Y. Chiu, Intense magnetic fields in astrophysics, *Space Sci. Rev.* **12**, 735 (1971).
Canuto V. and J. Ventura, Quantum theory of the dielectric constant of a magnetized plasma and astrophydical applications I: Theory, *Space Sci. Rev.* **18**, 104 (1972).
Canuto V., S. Kumar and H.J. Lee, Origin of strong magnetic fields, *Nature* **235**, 9 (1972).
Canuto V. and L. Fassio-Canuto, Photoneutrino energy losses in strong magnetic fields, *Phys. Rev.* **D7**, 1601 (1973).
Canuto V. and C.K. Chou, New dispersion relation for a strongly magnetized degenerate electron plasma with anisotropic pressure, *Phys. Fluids* **16**, 1273 (1973).
Canuto V., Equation of state at ultra-high densities I: Neutron stars, *Ann. Rev. Astron. Astrophys.* **12**, 167 (1974).
Canuto V., C. Chiuderi, C.K. Chou and L. Fassio-Canuto, Neutrino reactions in strong magnetic fields, *Astrophys. Space Sci.* **28**, 145 (1974).
Canuto V. and B. Datta, Meson–meson interactions and the equation of state for dense nucleon, *Phys. Lett.* **B56**, 148 (1975).
Canuto V. and J. Lodenquai, Behavior of matter at superhigh density, *Phys. Rev.* **D11**, 233 (1975).
Canuto V. and E. Tsiang, On the equation of state for an electron gas in an intense magnetic field, *Phys. Lett.* **B62**, 366 (1976).
Canuto V. and J. Ventura, Quantizing magnetic fields in astrophysics, *Fund. Cosmic Phys.* **2**, 203 (1977).
Canuto V. and E. Tsiang, On the hydrodynamic expansion of a relativistic gas, *Astrophys. J.* **213**, 27 (1977).
Canuto V., B. Datta and G. Kalman, Superdense neutron matter, *Astrophys. J.* **221**, 275 (1978).

Canuto V. and B. Datta, High density matter in the universe, *Nucl. Phys. J.* **A328**, 320 (1979).

Canuto V., J. Lodenquai and P. Olesen, The improved energy–momentum tensor for a scalar field, hydrodynamical dissipation, and the equation of state at high density. Preprint.

Carruthers P.A. and F. Zachariasen, Covariant transport theory approach to multiparticle production processes. Preprint.

Carruthers P.A. and F. Zachariasen, Transport equation approach to multiparticle production (American Physical Society meeting, Williamsburg, 1974).

Carruthers P.A. and F. Zachariasen, Unified approach to multiparticle production at high energy (Orbis Scientiae II, Coral Gable, 1975).

Carruthers P.A. and F. Zachariasen, Relativistic quantum transport theory approach to multiparticle production, *Phys. Rev.* **D13**, 950 (1976).

Carruthers P.A. and F. Zachariasen, Quantum collision theory with phase-space distribution functions, *Rev. Mod. Phys.* **55**, 245 (1983).

Carruthers P., Collective, stochastic and nonequilibrium behavior of highly excited hadronic matter, *Nucl. Phys.* **A418**, 501 (1984).

Carter B., (1) The canonical treatment of heat conduction and superfluidity in relativistic hydrodynamics, (2) Covariant theory of conductivity in ideal fluid or solid media, in Relativistic Hydrodynamics (CIME lectures, ed. M. Anile; Noto, Sicily, 1987), (Vaidya Raychaudhury Festschrift, 1983).

Carter B., Conductivity with causality in relativistic hydrodynamics: the regular solution to Eckart's problem, in Relativistic Hydrodynamics — *Proc. Int. Conf. Gravitation and Cosmology* (Goa, India, 1987), ed. C.V. Vishveshwara (Cambridge University Press, 1989).

Celenza L.S. and H.J. Pirner, A phenomenological approach to pion condensation, *Nucl. Phys.* **A294**, 357 (1978).

Celenza L.S. and C.M. Shakin, Relativistic many-body theory, *Phys. Rev.* **C24**, 2704 (1981).

Celik T. and D.E. Miller, Thermodynamic properties of a relativistic Fermi system with a Bose condensate at finite temperature, *Z. Phys.* **C24**, 59 (1984).

Cercignani C., Speed of propagation of infinitesimal disturbances in a relativistic gas, *Phys. Rev. Lett.* **50**, 1122 (1983).

Cercignani C. and A. Majorana, Analysis of thermal, sound, and shear waves according to a relativistic kinetic model, *Phys. Fluids* **28**, 1673 (1985).

Cercignani C. and A. Majorana, Analysis of thermal, sound and shear waves according to a relativistic kinetic model, *Phys. Fluids* **28**, 1673 (1985).

Cercignani C. and A. Majorana, Structure of shock waves in relativistic simple gases, *Phys. Fluids* **31**, 1064 (1988).

Cercignani C. and G.M. Kremer, On relativistic collisional invariants, *J. Stat. Phys.* **96**, 439 (1999).

Cercignani C. and G.M. Kremer, Trend to equilibrium of a degenerate relativistic gas, *J. Stat. Phys.* **98**, 441 (2000).

Cercignani C. and G.M. Kremer, Moment closure of the relativistic Anderson and Witting model equation, *Physica* **A290**, 192 (2001).

Cercignani C. and G.M. Kremer, The relativistic Boltzmann equation: theory and application (Birkhäuser; Basel, 2002).

Chaichian M., R. Hagedorn and M. Hayashi, Ideal relativistic quantum gas and invariant phase space, *Nucl. Phys.* **B92**, 445 (1975).

Chaichian M. and M. Hayashi, Application of the cluster decomposition for an ideal relativistic quantum gas, *Phys. Rev.* **D15**, 402 (1977).

Cheng B., D.N. Schramm and J.W. Truran, Interaction rates at high magnetic field strengths and high degeneracy. Preprint FERMILAB-Pub-93/260.

Chernikov N.A., The generalized stochastic problem of particle motion, *Sov. Phys. Dokl.* **1**, 103 (1956).

Chernikov N.A., The relativistic collision integral, *Sov. Phys. Dokl.* 2, 248 (1957).

Chernikov N.A., Reduction of the relativistic collision integral to the Boltzmann form, *Sov. Phys. Dokl.* **5**, 764 (1960).

Chernikov N.A., Relativistic kinetic equation and the equilibrium state of a gas in a static, spherically symmetric gravitational field, *Sov. Phys. Dokl.* **5**, 786 (1960).

Chernikov N.A., Flux vector and mass tensor of a relativistic ideal gas, *Sov. Phys. Dokl.* **7**, 414 (1962).

Chernikov N.A., Relativistic Maxwell–Boltzmann distribution and integral form of the conservation laws, *Sov. Phys. Dokl.* **7**, 428 (1962).

Chernikov N.A., Kinetic equation for a relativistic gas in an arbitrary gravitational field, *Sov. Phys. Dokl.* **7**, 397 (1962).

Chernikov N.A., The relativistic gas in the gravitational field, *Acta Phys. Pol.* **23**, 629 (1963).

Chernikov N.A., Derivation of the equations of relativistic hydrodynamics from the relativistic transport equation, *Phys. Lett.* **5**, 115 (1963).

Chernikov N.A., Equilibrium distribution of the relativistic gas, *Acta Phys. Pol.* **26**, 1069 (1964).

Chernikov N.A., Microscopic foundation of relativistic hydrodynamics, *Acta Phys. Pol.* **27**, 465 (1964).

Chin S.A. and J.D. Walecka, An equation of state for nuclear and higher-density matter based on a relativistic mean-field theory, *Phys. Lett.* **B52**, 24 (1974).

Chin S.A., A relativistic many-body theory of high density matter, *Ann. Phys. (N.Y.)* **108**, 301 (1977).

Chiu H.-Y. and L. Fassio-Canuto, Quantized synchroton radiation in intense magnetic fields, *Phys. Rev.* **185**, 1614 (1969).

Chiu H.-Y. and V. Canuto, Physics of high density matter in intense magnetic fields, *Phys. Rev. Lett.* **21**, 110 (1968).

Choquet-Bruhat Y., Theorem of uniqueness and local stability for Liouville–Einstein equations, *J. Math. Phys.* **11**, 3238 (1970).

Chou T.T., C.N. Yang and E. Yen, Single-particle momentum distribution at high energies and concept of partition temperature, *Phys. Rev. Lett.* **54**, 510 (1985).

Cissoko M., Relativistic radiation hydrodynamics: shock and deflagration waves, *Phys. Rev.* **D55**, 4555 (1997).

Ciubotariu C.D. and J. Gottlieb, The relativistic magnetohydrodynamics of a rotating plasma, *Physica* **63**, 393 (1973).
Ciubotariu C.D., D.I. Zoler and N.S. Thirer, The stability of the relativistic Alfvén waves, *Beitr. Plasma Phys.* **23**, 117 (1983).
Clemmow P.C. and A.J. Wilson, A relativistic form of Boltzmann's transport equation in the absence of collisions, *Proc. Cambridge Philos. Soc.* **53**, 222 (1957).
Comer G.L. and D. Langlois, Hamiltonian formulation for relativistic superfluids, *Class. Quantum Grav.* **11**, 709 (1994).
Constantinescu D.H., Electron self-energy in a magnetic field, *Nucl. Phys.* **B44**, 288 (1972).
Constantinescu D.H., Equation of state for condensed matter in a very strong magnetic field. Preprint THEP (May 1974).
Constantinescu D.H. and P. Rehàk, Condensed Matter in a Very Strong Magnetic Field at High Pressure and Zero Temperature — Meeting on the Role of Magnetic Fields in Physics and Astrophysics (Nordita, Copenhagen, 1974).
Constantinescu D.H. and P. Rehàk, Equation of state for condensed matter in a very strong magnetic field, *Ann. N.Y. Acad. Sci.* **257**, 85 (1975).
Constantinescu D.H., Vacuum polarization in a magnetic field, *Nucl. Phys.* **B36**, 121 (1976).
Constantinescu D.H., Condensed matter in a very strong magnetic field, at high pressure and zero temperature, *Nuovo Cimento* **B32**, 177 (1976).
Constantinescu D.H., Thermodynamics of the statistical model of atoms in very strong magnetic fields: an approach to the properties of condensed matter in the outer crust of pulsars, *Phys. Rev.* **D18**, 1820 (1978).
Cooper F., Landau's hydrodynamical model of particle production (American Physical Society meeting, Williamsburg, 1974).
Cooper F. and D.H. Sharp, Transport and hydrodynamic aspects of multiparticle production, *Phys. Rev.* **D8**, 194 (1975).
Cooper F. and D.H. Sharp, Pion production from a classical source: transport and hydrodynamical properties, *Phys. Rev.* **D8**, 194 (1975).
Cooper F., The hydrodynamical model of multiparticle production, in Ettore Majorana Conf. Proc. "Is There an Ultimate Temperature in Hadron Physics?" (Bielefeld, 1975).
Cooper F., G.S. Guralnik and S.H. Kasdan, Collective phenomena in $\lambda\phi^4$ field theory treated in the random-phase approximation, *Phys. Rev.* **D14**, 1607 (1976).
Cooper F. and M. Feigenbaum, Transport approach to multiparticle production: collective phenomena and renormalization, *Phys. Rev.* **D14**, 583 (1976).
Cover R., G. Kalman and P. Bakshi, Electromagnetic response in strong magnetic field II: Particle polarization and mode structure for parallel propagation, *Phys. Rev.* **D20**, 3015 (1979).
Cuccoli A., V. Tognetti, P. Verruchi and R. Vaia, Quantum thermodynamics in classical phase space, *Phys. Rev.* **A45**, 8418 (1992).
Daicic J., N.E. Frankel and V. Kowalenko, Magnetized pair Bose gas: relativistic superconductor, *Phys. Rev. Lett.* **71**, 1779 (1993).

Daicic J., N.E. Frankel, R.M. Gailis and V. Kowalenko, Statistical mechanics of the magnetized pair quantum gas, *Phys. Rep.* **237**, 63 (1994).

Daicic J. and N. E. Frankel, Superconductivity of the Bose gas, *Phys. Rev.* **D53**, 5745 (1996).

Das A., Extended phase space I: Classical fields, *J. Math. Phys.* **21**, 1506 (1980).

Das A., Extended phase space I: Unified spin 1/2 fields, *J. Math. Phys.* **21**, 1521 (1980).

Das A., Extended phase space II: Unified meson fields, *J. Math. Phys.* **21**, 1513 (1980).

Davis R.L., Ginsburg–Landau from field theory, *Mod. Phys. Lett.* **A6**, 73 (1991).

Daucourt G. and K. Rose, Polarized radiation in relativistic cosmology, *Astron. Nachr.* **299**, 11 (1978).

Dawson J.F. and J. Piekarewicz, Pion condensation in the Walecka model, *Phys. Rev.* **C43**, 2631 (1991).

de Boer W.P.H., Quantum kinetic theory of charged Klein–Gordon particles I: Quantum distribution function, *Proc. Kon. Nederl. Akad. Wetensch.* **82**, 337 (1979).

de Boer W.P.H., Quantum kinetic theory of charged Klein–Gordon particles II: Perturbation theory of non-equilibrium statistical Green function, *Proc. Kon. Nederl. Akad. Wetensch.* **82**, 353 (1979).

de Boer W.P.H., Quantum kinetic theory of charged Klein–Gordon particles IV: Linear kinetic equation, *Proc. Kon. Nederl. Akad. Wetensch.* **83**, 17 (1980).

de Groot S.R., W.A. van Leeuwen and C.G. van Weert, Relativistic Kinetic Theory (North-Holland, Amsterdam, 1980).

de Groot S.R. and L.G. Suttorp, The relativistic energy–momentum tensor in polarized media I: The atomic energy–momentum conservation laws, *Physica* **37**, 284 (1967).

de Groot S.R. and L.G. Suttorp, The relativistic energy–momentum tensor in polarized media II: The angular momentum laws and the symmetry of the energy–momentum tensor, *Physica* **37**, 297 (1967).

de Groot S.R. and L.G. Suttorp, The relativistic energy–momentum tensor in polarized media IV: The macroscopic material energy–momentum tensor, *Physica* **39**, 41 (1968).

de Groot S.R. and L.G. Suttorp, The relativistic energy–momentum tensor in polarized media V: Statistical thermodynamics of electromagnetic phenomena, *Physica* **39**, 61 (1968).

de Groot S.R. and L.G. Suttorp, The relativistic energy–momentum tensor in polarized media VI: The difference between the energy–momentum tensors in the presence and in the absence of external fields, *Physica* **39**, 77 (1968).

de Groot S.R. and L.G. Suttorp, The relativistic energy–momentum tensor in polarized media VII: Discussion of the results in connexion with previous works, *Physica* **39**, 84 (1968).

de Groot S.R. and L.G. Suttorp, The relativistic energy–momentum tensor in polarized media III: Statistical theory of the energy–momentum laws, *Physica* **39**, 28 (1968).

de Groot S.R., C.G. van Weert, W. Th. Hermens and W.A. van Leeuwen, Relativistic Onsager relations in kinetic gas theory, *Phys. Lett.* **A26**, 345 (1968).
de Groot S.R., C.G. van Weert, W. Th. Hermens and W.A. van Leeuwen, Relativistic entropy law for a gas mixture outside equilibrium, *Phys. Lett.* **A26**, 439 (1968).
de Groot S.R., C.G. van Weert, W. Th. Hermens and W.A. van Leeuwen, On relativistic kinetic gas theory I: The second law for a gas mixture outside equilibrium, *Physica* **40**, 257 (1968).
de Groot S.R., C.G. van Weert, W. Th. Hermens and W.A. van Leeuwen, On relativistic kinetic gas theory II: Reciprocal relations between transport phenomena, *Physica* **40**, 581 (1969).
de Groot S.R., C.G. van Weert, W. Th. Hermens and W.A. van Leeuwen, On relativistic kinetic gas theory III: The nonrelativistic limit and its range of validity, *Physica* **42**, 309 (1969).
de Groot S.R., C.G. van Weert, W. Th. Hermens and W.A. van Leeuwen, On the role of space and time symmetries in kinetic theory, *Physica* **43**, 109 (1969).
de Groot S.R., Relativistic Boltzmann theory, In *Int. Symp.* 100 Years Boltzmann Equation (Vienna, 4–8 Sept. 1972 (Springer-Verlag, Wien, 1973). Also in *Acta Phys. Austriaca Suppl.* **10**, 529 (1973).
de Groot S.R., Transportverschijnselen en geluidsvoortplanting in relativistische systemen van massive deeltjes en van neutrino's, *Versl. Afd. Nat. Kon. Nederl. Akad. Wetensch.* **82**, 72 (1973).
de Groot, S.R., Transport coefficients in special relativity, *Physica* **69**, 12 (1973).
de Groot S.R., Relativistische kinetische Theorie von Transporterscheinungen, *Z. Physik* **A262**, 349 (1973).
de Groot S.R., Théorie cinétique relativiste, *Revista Mexicana de Fisica* **23**, 405 (1974).
de Groot S.R., Théorie cinétique relativiste, In *Colloques Internationaux du Centre National de la Recherche Scientifique no. 236.* Théories cinétiques classiques et relativistes (Editions du Centre National de la Recherche Scientique, Paris, 1975).
de Groot S.R., The laws of relativistic thermodynamics I: The macroscopic quantities, *Physica* **A88**, 172 (1977).
de Groot S.R., The laws of relativistic thermodynamics II: The density, momentum and energy laws, *Physica* **A88**, 182 (1977).
de Groot S.R., The laws of relativistic thermodynamics III: The entropy law, *Physica* **A88**, 425 (1977).
de Boer W.P.H., Quantum kinetic theory of charged Klein–Gordon particles III: Feynman rules and thermal equilibrium, *Proc. Kon. Nederl. Akad. Wetensch.* **83**, 1 (1980).
de Jagher P.C. and F.W. Sluijter, A covariant hierarchy of kinetic equations for classical particles and fields, *Contrib. Plasma Phys.* **28**, 169 (1988).
Debbasch F., K. Mallick and J.P. Rivet, The relativistic Ornstein–Uhlenbeck process, *J. Stat. Phys.* **88**, 945 (1997).
Debbasch F. and J.P. Rivet, A diffusion equation for the relativistic Ornstein–Uhlenbeck process, *J. Stat. Phys.* **90**, 1179 (1998).

Debbasch F., J.P. Rivet and W.A. van Leeuwen, Invariance of the relativistic one-particle distribution function, *Physica* **A301**, 181 (2001).

Debbasch F., A diffusion process in curved spacetime, *J. Math. Phys.* **45**, 2744 (2004).

de Jong F. and R. Malfliet, Conserving relativistic many-body approach: equation of state, spectral function and occupation probabilities of nuclear matter, *Phys. Rev.* **C44**, 998 (1991).

Delsante A.E. and N.E. Frankel, Equation of state of ideal neutrons in an intense magnetic field, *Phys. Lett.* **A67**, 435 (1978).

Delsante A.E. and N.E. Frankel, Relativistic degenerate electron plasma, *Phys. Lett.* **A66**, 481 (1978).

Delsante A.E. and N.E. Frankel, Relativistic degenerate electron plasma in an intense magnetic field, *Phys. Lett.* **A67**, 279 (1978).

Delsante A.E. and N.E. Frankel, Relativistic paramagnetism: quantum statistics, *Phys. Rev.* **D20**, 1795 (1979).

Desjarlais M.P. and R.N. Sudan, A thermal distribution function for relativistic magnetically insulated electron flows, *Phys. Fluids* **29**, 1746 (1986).

Diaz Alonso J. and R. Hakim, Statistique Quantique Relativiste de la Matière Baryonique (Journées Relativistes, Genève, 1978).

Diaz Alonso J. and R. Hakim, Thermodynamical Properties of the Relativistic Quantum Plasma, *Phys. Lett.* **66A**, 476 (1978).

Diaz Alonso J. and R. Hakim, Quasiboson excitation spectrum of the relativistic quantum scalar plasma in the Hatree-Vlasov approximation, *Phys. Rev.* **D38**, 1780 (1988). *Phys. Rev.* **D29**, 2690 (1984).

Diaz Alonso J., Relativistic quantum analysis of a pion–nucleon system, *Ann. Rev. (N.Y.)* **160**, 1 (1985).

Diaz Alonso J. and R. Hakim, Quantum fluctuations of the relativistic scalar plasma, *Phys. Rev.* **D29**, 2690 (1984).

Diaz Alonso J., Field theoretical model for nuclear and neutron matter I: Equation of state, *Phys. Rev.* **D31**, 1315 (1985).

Diaz Alonso J., Relativistic quantum analysis of a pion–nucleon system, *Ann. Rev. (N.Y.)* **160**, 1 (1985).

Diaz Alonso J., Field theoretical model for nuclear and neutron matter I: Equation of state, *Phys. Rev.* **D31**, 1315 (1985).

Diaz Alonso J. and J.M. Ibanez Cabanell, Field theoretical model for nuclear and neutron matter II: Neutron stars, *Astrophys. J.* **291**, 308 (1985).

Diaz Alonso J. and J.M. Ibanez Cabanell, Field theoretical model for nuclear and neutron matter II: Neutron stars, *Astrophys. J.* **291**, 308 (1985).

Diaz Alonso J. and R. Hakim, Quasi-boson excitation spectrum of the relativistic quantum scalar plasma in the Hartree–Vlasov approximation, *Phys. Rev.* **D38**, 1780 (1988).

Diaz Alonso J., J.M. Ibanez Cabanell and H.D. Sivak, Field theoretical model for nuclear and neutron matter: thermal effects, *Phys. Rev.* **C39**, 671 (1989).

Diaz Alonso J., A. Perez Canyellas and H.D. Sivak, Linear response and Friedel oscillations of the pion field in relativistic nuclear matter, *Nucl. Rev.*, **A505**, 695 (1989).

Diaz Alonso J., J.M. Ibanez Cabanell and H.D. Sivak, Field theoretical model for nuclear and neutron matter: Thermal effects, *Phys. Rev.* **C39**, 671 (1989).

Diaz Alonso J. and A. Perez Canyellas, A study of meson dynamics in relativistic models of nuclear matter, *Nucl. Rev.* **A526**, 623 (1991).

Diaz Alonso J., E. Gallego and A. Perez Canyellas, Friedel oscillations in relativistic models of nuclear matter, *Phys. Rev. Lett.* **73**, 2536 (1994).

Dijkstra J.J., Mathematical aspects of relativistic kinetic theory II: The summational invariants, *Proc. Kon. Nederl. Akad. Wetensch.* **B81**, 265 (1978).

Dijkstra J.J., Mathematical aspects of relativistic kinetic theory II(B): The summational invariants, *Proc. Kon. Nederl. Akad. Wetensch.* **B81**, 277 (1978).

Dijkstra J.J. and W.A. van Leeuwen, Mathematical aspects of relativistic kinetic theory, *Physica* **A90**, 450 (1978).

Dixit V.V. and D. Miller, Model for relativistic Bose systems in d dimensions, *Phys. Rev.* **A28**, 1709 (1983).

Divine N., Numerical evaluation of the degenerate equation of state, *Astrophys. J.* **142**, 1652 (1965).

Dominguez Tenreiro R., The Liouville equation in general relativity: a particular kind of solutions. Preprint ETUAM/80-8 (1980).

Dominguez Tenreiro R. and R. Hakim, Bases de la Théorie Cinétique Relativiste du Gaz d'Electrons Quantiques (Journées Relativistes, Bruxelles, 1976).

Dominguez Tenreiro R. and R. Hakim, Théorie Cinétique et Propriétés de Transport du Gaz Dégénéré d'Electrons en Présence d'un Champ Magnétique Fort (Journées Relativistes, Vendôme, 1977).

Dominguez Tenreiro R. and R. Hakim, Transport coefficients of the relativistic degenerate electron gas in a strong magnetic field, *J. Phys.* **A10**, 1525 (1977).

Dominguez Tenreiro R. and R. Hakim, Transport properties of the relativistic degenerate electron gas in a strong magnetic field: covariant relaxation-time model, *Phys. Rev.* **D15**, 1435 (1977).

Dominguez Tenreiro R., R. Hakim and H.D. Sivak, Dispersion relations of relativistic plasmas from hydrodynamical equations, *Nuovo Cimento* **B57**, 45 (1980).

Dominguez Tenreiro R., R. Hakim and H.D. Sivak, Waves in a relativistic anisotropic plasma from hydrodynamical equations, *Phys. Rev.* **A24**, 1561 (1981).

Dominguez Tenreiro R. and R. Hakim, Covariant Wigner function approach to the relativistic charged gas in a strong magnetic field, I: Electron gas, *Physica* **A113**, 477 (1982).

Drake J.F. Y.C. Lee and N.L. Tsintsadze, Parametric instabiliies in strongly relativistic, plane-polarized electromagnetic waves, *Phys. Rev. Lett.* **36**, 31 (1976).

Dreyer W. and W. Weiss, The classical limit of relativistic extended thermodynamics, *Ann. Inst. H. Poincaré* **45**, 401 (1986).

Droz-Vincent Ph. and R. Hakim, Collective motions of the relativistic gravitational gas, *Ann. Inst. H. Poincaré* **9**, 17 (1967).

Droz-Vincent Ph., Direct interactions in relativistic statitical mechanics, *Found. Phys.* **27**, 363 (1997).

Dudynski M. and M.L. Ekiel-Jezewska, Causality of the linearized Boltzmann equation, *Phys. Rev. Lett.* **55**, 2831 (1985).
Dudley R.M., A note on Lorentz-invariant Markov processes, *Ark. Mat.* **6**, 575 (1967).
Dudley R.M., Lorentz-invariant Markov processes in relativistic phase-space, *Ark. Mat.* **6**, 241 (1965).
Dunkel J. and P. Haenggi, On the theory of the relativistic Brownian motion: the 1 + 1-dimensional case, *Phys. Rev.* **E71**, 016124 (20045).
Dunkel J. and P. Haenggi, Relativistic Brownian motion, *Phys. Rep.* **471**, 1 (2009).
Dunning-Davies J., Particle-number fluctuations, *Nuovo Cimento* **B57**, 315 (1968).
Dunning-Davies J., A comment on the ideal relativistic Bose gas, *J. Phys.* **A14**, 3013 (1981).
Dunning-Davies J., The *d*-dimensional Landsberg gas, *J. Phys.* **A14**, 3005 (1981).
Eberly J.H. and A. Kujawski, Relativistic statistical mechanics and black-body radiation, *Phys. Rev.* **155**, 10 (1967).
Eberly J.H. and A. Kujawski, Relativistic statistical mechanics and black-body radiation II: Spectral correlations, *Phys. Rev.* **166**, 197 (1968).
Eckart C., The thermodynamics of irreversible processes III: Relativistic theory of the simple fluid, *Phys. Rev.* **58**, 919 (1940).
Eggleton P.P., Faulkner J. and B.P. Flannery, An approximate equation of state for stellar material, *Astron. Astrophys.* **23**, 325 (1973).
Ehlers J., Beiträge zur relativistischen Mechanik kontinuerlicher Medien, (Akad. Wiss. Mainz, 1961), p. 791.
Ehlers J., P. Geren and R.K. Sachs, Isotropic solutions of the Einstein–Liouville equations, *J. Math. Phys.* **9**, 1344 (1968).
Ehlers J., Kinetic theory of gases in general relativity theory, In *General Relativity and Gravitation* (Haifa, 1969) (Gordon and Breach, London, 1971), p. 145.
Ehlers J. and R.K. Sachs, Kinetic theory and cosmology, In *Astrophysics and General Relativity*, Vol. 2, eds. M. Chretien, S. Deser and J. Goldstein (Gordon and Breach, New York, 1971).
Ehlers J., Survey of general relativity theory, In ed. W. Israel, *Relativity, Astrophysics and Cosmology* (W. Israel Reidel, Dordrecht, 1973), p. 1.
Ehlers J. and A. Peres, Kinetic theory of gases in general relativity, In eds. W.C. Schieve and J.S. Turner, *Lectures in Statistical Physics* (Springer-Verlag, Berlin, 1974), p. 78.
Ehlers J., Progress in relativistic statistical mechanics, thermodynamics and continuum mechanics, In *General Relativity and Gravitation*, In eds. G. Shaviv and J. Rosn (Tel Aviv, 1974) (Wiley, Chichester, 1975), p. 145.
Ehlers J., Contributions to the relativistic mechanics of continuous media, *General Relativity Gravitation* **25**, 1225 (1993).
Eliezer S. and R.M. Weiner, SU(3) symmetry breaking and superfluidity of hadronic matter, *Phys. Lett.* **B50**, 463 (1974).

Eliezer S. and R.M. Weiner, Phase transitions of second and zero kind in high-energy physics: a phenomenological field-theoretical approach, s-model and superfluidity of hadronic matter, *Phys. Rev.* **D13**, 87 (1976).

Ellis G.F.R., D.R. Matravers and R. Treciokas, An exact anisotropic solution of the Einstein–Liouville equations, *General Relativity Gravitiation* **15**, 931 (1983).

Ellis G.F.R., D.R. Matravers and R. Treciokas, Anisotropic solutions of the Einstein–Boltzmann equations: I. General formalism, *Ann. Phys. (N.Y.)* **150**, 455 (1983).

Ellis G.F.R., R. Treciokas and D.R. Matravers, Anisotropic solutions of the Einstein–Boltzmann equations: II. Some exact properties of the equations, *Ann. Phys. (N.Y.)* **150**, 487 (1983).

Elmfors P., P. Liljenberg, D. Persson and Bo-Sture Skagerstam, Thermal versus vacuum magnetization, in QED, *Phys. Rev.* **D51**, 5885 (1994).

Elmfors P., P. Liljenberg, D. Persson and Bo-Sture Skagerstam, Condensation and magnetization of the relativistic Bose gas, *Phys. Lett.* **B348**, 462 (1995).

Elmfors P., P. Liljenberg, D. Persson and Bo-Sture Skagerstam, Comment on "Magnetized pair bose gas: relativistic superconductor", *Phys. Rev. Lett.* **75**, 2067 (1995).

Eltgroth P.G., Nonplanar relativistic flow, *Phys. Fluids* **15**, 2140 (1972).

Eltgroth P.G., Similarity analysis for relativistic flow in one dimension, *Phys. Fluids* **14**, 2631 (1971).

Elze H.-T., W. Greiner and J. Rafelski, The relativistic ideal Fermi gas revisited, *J. Phys.* **G6**, L149 (1980).

Elze H.-T. and W. Greiner, Quantum statistics with internal symmetry, *Phys. Rev.* **A33**, 1879 (1986).

Elze H.-T., M. Gyulassy and D. Vasak, Transport equations for the QCD quark Wigner operator, *Nucl. Phys.* **B276**, 706 (1986).

Elze H.-T., M. Gyulassy and D. Vasak, Transport equations for the QCD gluon Wigner operator, *Phys. Lett.* **B177**, 402 (1986).

Elze H.-T. and U. Heinz, Quark–gluon transport theory, *Phys. Rev.* **A33**, 1879 (1986).

Elze H.-T., D.E. Miller and K. Redlich, Gauge theories at finite temperatures and chemical potential, *Phys. Rev.* **D35**, 748 (1987).

Elze H.-Th., Gluon transport equations, plasma oscillations and screening, *Z. Phys.* **C38**, 211 (1987).

Elze H.-T., M. Gyulassy, D. Vasak, H. Heinz, H. Stöcker and W. Greiner, Towards a relativistic self-consistent quantum transport theory of hadronic matter, *Mod. Phys. Lett.* **A2**, 451 (1987).

Elze H.-T., K. Kajantie and T. Toimela, Chromomagnetic screening at high temperature, *Z. Phys.* **C37**, 601 (1988).

Elze H.-T., U. Heinz, K. Kajantie and T. Toimela, High temperature gluon matter in the background gauge, *Z. Phys.* **C37**, 305 (1988).

Elze H.-T., Gluon transport equations, plasma oscillations, and screening, *Z. Phys.* **C38**, 211 (1988).

Escobedo M., Mischler S. and M.1A. Valle, Homogeneous Boltzmann equation in quantum relativistic kinetic theory, *Electron. J. Diff. Eq.* (Monograph 04, 2003).

Esposito F.P. and E.N. Glass, General relativistic magnetofluids, *J. Math. Phys.* **18**, 705 (1977).

Evans R.W. and P.B. Shaw, Pressure and composition of a weakly degenerate nucleon gas, *Phys. Rev.* **D14**, 2526 (1976).

Evans T.E., What is being calculated with thermal field theory, In *Proc. Ninth Louise Winter Institute: Particle Physics and Cosmology*, eds. A. Astbury, B.A. Campbell, W. Israel, F.C. Khanna, D. Page and J.L. Pinfold (World Scientific, Singapore, 1995).

Evtimova E.N., Phase space representation of a spin-1/2 particle interacting with an external electromagnetic field: I. General examination, *J. Phys.* **A27**, 993 (1994).

Ezawa H., Y. Tomozawa and H. Umezawa, Quantum statistics of fields and multiple production of mesons, *Nuovo Cimento* **V**, 810 (1957).

Feldman Y. and G.E. Tauber, The internal state of a gas of particles with spin, *General Relativity Gravitity* **12**, 837 (1980).

Feldmeier H., M. Schönhofen and M. Cubero, The relativistic Vlasov equation with non-radiating meson fields, *Nucl. Phys.* **A495**, 337c (1989).

Fernandez Velicia F.J., New functional expansions for the Fermi–Dirac functions. Preprint.

Ferrarese G., Sulla termomeccanica dei continui relativistici, *Fis. Mat.* **4**, 151 (1985).

Ferrer E.J., V. de la Incera and A.E. Shabad, Bose–Einstein condensation in many-particle gauge theories and external charge, *Nuovo Cimento* **A98**, 245 (1987).

Ferrer Obeso E.J. and H. Perez-Rojas, Mass operator of an e^+e^--plasma in a constant and uniform magnetic field, *Cienc. Tec. Fis. Mat.* **4**, 63 (1984).

Flowers E., N. Itoh, Transport properties of dense matter I, *Astrophys. J.* **206**, 218 (1976).

Flowers E. and N. Itoh, Transport properties of dense matter II, *Astrophys. J.* **230**, 847 (1979).

Fonarev O.A., Wigner function and quantum kinetic theory in curved space–time and external fields [arXiv: gr-qc/9309005 v1].

Fowler G.N., N. Stelte and R.M. Weiner, New functional expansions for the Fermi–Dirac functions, *Nucl. Rev.* **A319**, 349 (1979).

Fox R., C.G. Kuper and S.G. Lipson, Faster-than-light group velocities and causality violation, *Proc. R. Soc. London* **A316**, 515 (1970).

Fradkin E.S., The Green's function method in quantum statistics, *Nucl. Phys.* **12**, 465 (1959).

Fradkin E.S., Some general relations in statistical quantum electrodynamics, *Sov. Phys. JETP* **11**, 114 (1960).

Frankel N.E., K.C. Hines and R.L. Dewar, Energy loss due to binary collisions in a relativistic plasma, *Phys. Rev.* **A20**, 2120 (1979).

Frankel N.E., K.C. Hines and V. Kowalenko, Relativistic boson pair plasma, Laser Part. *Beams*, **3**, 251 (1985).

Frezzotti A., The propagation of small disturbances in an ultrarelativistic gas according to the method of elementary solutions, *J. Stat. Phys.* **46**, 255 (1987).

Fristrom C.C., The equation of state for degenerate dwarf stars, *Fund. Cos. Phys.* **12**, 119 (1987).

Fu K.Y., Energy conversion between longitudinal and transverse waves by mode–mode coupling in a relativistic plasma, *Plasma Phys.* **15**, 1221 (1973).

Fujimura F.S. and C.F. Kennel, Numerical solutions of the trans-relativistic shock relations, *Astron. Astrophys.* **79**, 299 (1979).

Furnstahl R.J. and B.D. Serot, Covariant mean-field calculation of finite-temperature nuclear matter, *Phys. Rev.* **C41**, 262 (1990).

Furnstahl R.J. and B.D. Serot, Covariant Feynman rules at finite temperature: application to nuclear matter, *Phys. Rev.* **C43**, 105 (1991).

Galitskivi V.M., Yu. B. Ivanov and V.A. Khangulyan, Kinetic coefficients of nuclear matter, *Sov. J. Nucl. Phys.* **30**, 401 (1980).

Garbaczewski P., Relativistic problem of random flights and Nelson's stochastic mechanics, *Phys. Lett.* **A164**, 6 (1992).

Garpman S.I.A., D. Sperber and M. Zielinska-Pfabe, The temperature dependence of nuclear transport coefficients, *Nuovo Cimento* **B57**, 11 (1980).

Glaser W., ZZ, *Z. Phys.* **94**, 317 (1935).

Glaser W., ZZ, *Z. Phys.* **94**, 677 (1935).

Glaser W., ZZ, *Z. Phys.* **118**, 727 (1942).

Gleeson A.M. and S. Raha, Shock waves in relativistic nuclear matter I, *Phys. Rev.* **C21**, 1065 (1980).

Gleeson A.M. and S. Raha, Shock waves in relativistic nuclear matter II: Angular anisotropy in multiparticle spectra, *Phys. Rev.* **C26**, 1521 (1982).

Gedalin M., D.B. Melrose and E. Grunman, Long waves in a relativistic pair plasma in a strong magnetic field, *Phys. Rev.* **E57**, 3399 (1998).

Genga R.O., High-frequency expansion of relativistic classical plasma dielectric tensor in a magnetic field, *Int. J. Theor. Phys.* **28**, 71 (1989).

Georgiou A., A virial theorem for relativistic plasmas, *Phys. Lett.* **A62**, 484 (1977).

Georgiou A., Relativistic theory of Bernstein waves, in XVI Int. Conf. Phenomena, In *Ionized Gases* (Dusseldorf, 1983) eds. W. Botticher, H. Wenk and E. Schulz-Gulde (University of Dusseldorf, 1984).

Georgiou A., Relativistic transport equation for plasmas, *Class Quantum Grav.* **12**, 1491 (1995).

Gerbal D., Relativistic dispersive media, *Physica* **47**, 227 (1970).

Gerbal D., Potentiel gravitationnel de Debye–Hückel dans l'univers de Lemaître, *C. R. Acad. Sci.* **272**, 701 (1971).

Gerbal D. and M. Prud'Homme, Some remarks about relativistic line profiles, *J. Quant. Spectros. Rad. Transf.* **14**, 351 (1974).

Gerbal D., On the relativistic kinetic theory of a gas, mean free 4-path, mean free 3-path and collision time, *Physica* **71**, 124 (1974).

Ghosh S., S. Mandal and S. Chakrabarty, Relativistic version of Landau theory of Fermi liquid in presence of strong quantizing magnetic field: an exact formalism, *Ann. Phys.* **312**, 398 (2004).

Glassey R.T. and W.A. Strauss, Large velocities in the relativistic Maxwell–Vlasov equations, *J. Fac. Sci. Tokyo* **A1**, 36, 615 (1989).

Goble D.F., Pion superfluidity in nuclear matter, *Ann. Phys. (N.Y.)* **90**, 295 (1975).

Gonzales Felipe R. and H. Perez Rojas, Structure of the polarization operator for relativistic plasma in external electric field, *Ciencias Tec. Fis. Mat.* **4**, 53 (1984).

Gonzales Felipe R., A. Perez Martinez and H. Perez Rojas, Relativistic quantum Hall effect for 3D and 2D electron plasma in an external magnetic field, *Mod. Phys. Lett.* **B4**, 1103 (1990).

Gordeev A.N., The power energy spectrum of cosmic rays as a solution of the general covariant kinetic equation, *Sov. Astron.* **10**, 826 (1967).

Gordeev A.N., On the relativistic Fokker–Planck equation. Preprint (1967).

Gorenstein M.I., The nuclear equation of state and shock phenomena in relativistic heavy-ion collision, In *The Nuclear Equation of State*, ed. W. Greiner. Preprint UFTP 232/1989 (1989).

Gorenstein M.I. and S.N. Yang, Quasi-pions with temperature-dependent dispersion relation, *J. Phys.* **G21**, 1053 (1995).

Gorenstein M.I. and S.N. Yang, Gluon plasma with a medium-dependent dispersion relation, *Phys. Rev.* **D52**, 5206 (1995).

Goto K., Relativistic magnetohydrodynamics, *Prog. Theor. Phys.* **20**, 1 (1958).

Gould R.J., Kinetic theory of relativistic plasmas, *Phys. Fluids* **24**, 102 (1981).

Granik A., Taylor–Maccoll problem in relativistic gas dynamics, *Phys. Fluids* **29**, 1449 (1986).

Grassi F., R. Hakim and H.D. Sivak, Statistical mechanics of a relativistic magnetized plasma and its radiation field, *Phys. Lett.* **A116**, 30 (1986).

Grassi F., R. Hakim and H.D. Sivak, The Gaussian approximation for the $\lambda\varphi^4$-theory revisited, *Int. J. Mod. Phys.* **A6**, 4579 (1991).

Greco A., On the general relativistic magnetohydrodynamics with a generalized thermodynamical differential equation, *Ann. Inst. H. Poincaré* **20**, 245 (1974).

Guess A.W., The relativistic degenerate gas, *Adv. Astron. Astrophys.* **4**, 153 (1966).

Guichelaar J., W.A. van Leeuwen and S.R. de Groot, On relativistic kinetic gas theory VII: The propagation and absorption of sound, *Physica* **59**, 97 (1972).

Guichelaar J., W.A. van Leeuwen and S.R. de Groot, On relativistic kinetic gas theory XII: A kinetic description of the dispersion and absorption of sound, *Physica* **68**, 342 (1973).

Guichelaar J., W.A. van Leeuwen and S.R. de Groot, Sound dispersion and absorption derived from relativistic theory, *Phys. Lett.* **A43**, 323 (1973).

Guichelaar J., On relativistic kinetic gas theory XIII: Sound propagation in a multicomponent mixture, *Physica* **75**, 593 (1974).

Guichelaar J., Sound propagation in a relativistic gas mixture, *Physica* **74**, 330 (1974).
Haber H.E. and H.A. Weldon, Thermodynamics of an ultra-relativistic Bose gas, *Phys. Rev. Lett.* **46**, 1497 (1981).
Haber H.E. and H.A. Weldon, Finite temperature symmetry breaking as Bose-Einstein condensation, *Phys. Rev.* **D25**, 502 (1982).
Haber H.E. and H.A. Weldon, On the relativistic Bose–Einstein integrals, *J. Math. Phys.* **23**, 1852 (1982).
Habib S. and H.E. Kandrup, Gravitational Laudau damping for an isotropic cluster of stars, *Astrophys. J.* **309**, 176 (1986).
Hafizi M., The Gaussian approximation for interacting charged scalar fields, *Eur. Phys. J.* **C11**, 181 (1999).
Hagedorn R. and K. Redlich, Statistical thermodynamics in relativistic particles and ion physics: canonical or grand-canonical?, *Z. Phys.* **C27**, 541 (1985).
Hagedorn R., Statistical thermodynamics of strong interactions at high energies, *Nuovo Cimento Suppl.* **3**, 147 (1965).
Hagedorn R., Cours sur la Thermodynamique des Interactions Fortes à Hautes Energies (Cours à la 2ème Ecole d'Eté de Gif-sur-Yvette, 1970).
Hagedorn R., Thermodynamics of strong interactions, Academic Training Program of CERN, 1970–1971; CERN Report 71-12 (1971).
Hagedorn R., Lectures Given at the Meeting on Hadron Thermodynamics, (Torino, 1972).
Hagedorn R., The pressure ensemble as a tool for describing the hadron–quark phase transition, *Z. Phys.* **C17**, 265 (1983).
Hagedorn R., The long way to the statistical bootstrap model, (hot hadronic matter; Divonne, 1995).
Hakim R., Sur les équations de Vlasov d'un plasma relativiste, *C. R. Acad. Sci.* **260** (1965).
Hakim R., A covariant theory of relativistic Brownian motion, *J. Math. Phys.* **6**, 1482 (1965).
Hakim R., Quelques méthodes en mécanique statistique relativiste, *C. R. Acad. Sci.* **260** (1965).
Hakim R., Relativistic perfect gas in an external force field, *Phys. Rev.* **162**, 128 (1967).
Hakim R., Sur la mécanique statistique relativiste, *Ann. Inst. H. Poincaré* **6**, 225 (1967).
Hakim R., Remarks on relativistic statistical mechanics I: Reduced densities, *J. Math. Phys.* **8**, 1315 (1967).
Hakim R., Remarks on relativistic statistical mechanics II: Hierarchies for the reduced densities, *J. Math. Phys.* **8**, 1379 (1967).
Hakim R., Relativistic stochastic processes, *J. Math. Phys.* **9**, 1805 (1968).
Hakim R. and A. Mangeney, Kinetic equations including radiation effects I: Vlasov approximation, *J. Math. Phys.* **9**, 116 (1968).
Hakim R., On the relativistic scalar plasma, *Phys. Rev.* **166**, 175 (1968).
Hakim R., Self-consistent homogeneous cosmological models I: Generalities, *Phys. Rev.* **173**, 1235 (1968).

Hakim R. and A. Mangeney, Remarks on Relativistic Thermodynamics, *Lett. Nuovo Cimento* **1**, 429 (1969).

Hakim R. and A. Mangeney, Collective oscillations of a relativistic radiating electron plasma, *Phys. Fluids* **14**, 2751 (1971).

Hakim R. and A. Mangeney, Relativistic kinetic equations with radiation effects, In *Kinetic Equations*, eds. R.L. Liboff and N. Rostoker (Gordon and Breach, New York, 1971).

Hakim R. and Ch. Vilain, The applications of relativistic kinetic theory to cosmological models: some observational consequences, *Astron. Astrophys.* **25**, 211 (1973).

Hakim R., Equilibrium statistical mechanics of relativistic particles with variable masses I: Nonquantal theory, *J. Math. Phys.* **15**, 1310 (1974).

Hakim R., Statistical mechanics of relativistic dense matter, *Riv. Nuovo Cimento* **1**, 1 (1978).

Hakim R. and J. Heyvaerts, Covariant Wigner function approach to relativistic quantum plasmas, *Phys. Rev.* **A18**, 1250 (1978).

Hakim R., Infinite diffusion velocity and causality in special relativity, *Lett. Nuovo Cimento* **25**, 108 (1979).

Hakim R., Fonction de Wigner d'Equilibre du Plasma Relativiste Plongé dans un Champ Magnétique Intense (Colloque Soc. Fr. Phys. "Plasmas Denses et Champs Magnétiques Intenses"; Meudon, 1979).

Hakim R. and J. Heyvaerts, Excitation spectrum of the relativistic quantum plasma, *J. Phys.* **A13**, 2001 (1980).

Hakim R., Plasma Physics Techniques for QED and QCD Systems — Int. Symp. Statistical Mechanics of Quarks and Gluons (Bielefeld, Aug. 1980), In *Statistical Mechanics of Quarks and Hadrons*, ed. H. Satz (North-Holland, Amsterdam, 1981).

Hakim R. and H.D. Sivak, Covariant Wigner function approach to the relativistic quantum electron gas in a strong magnetic field, *Ann. Phys. (N.Y.)* **139**, 230 (1982).

Hakim R., L. Mornas, P. Peter and H.D. Sivak, Relaxation time approximation for relativistic dense matter, *Phys. Rev.* **D46**, 4603 (1992).

Hakim R. and L. Mornas, Transport coefficients of relativistic dense matter — collective effects, *Phys. Rev.* **C47**, 2846 (1993).

Hakim R. and H.D. Sivak, Relativistic quasi-particles and their statistics, *Class. Quantum Grav.* **10**(Suppl.), 223 (1993).

Hakim R. and H.D. Sivak, Transport properties of relativistic quasiparticles, in *Hot Hadronic Matter*, eds. J. Rafelski and J. Letessier (Plenum, New York, 1995).

Hamada T. and Y. Nakamura, Specific heat of relativistic degenerate electron gas, *Prog. Theor. Phys.* **36**, 465 (1966).

Hamity V.H., Relativistic thermodynamics, *Phys. Rev.* **187**, 1745 (1969).

Harding A.K., One photon pair annihilation in magnetised relativistic plasmas, *Astrophys. J.* **300**, 167 (1986).

Harrington B.J. and H.K. Shepard, Superconducting properties of pion condensates, *Phys. Rev.* **D16**, 3437 (1977).

Harrington B.J. and H.K. Shepard, Superconducting properties of pion condensates II, *Phys. Rev.* **D19**, 1713 (1979).
Harris E.G., Relativistic magnetohydrodynamics, *Phys. Rev.* **108**, 1357 (1957).
Harta H. and T. Kugo, Operator formalism of statistical mechanics of gauge theory in covariant gauges, *Phys. Rev.* **D21**, 3333 (1980).
Havas P., Some basic problems in the formulation of a relativistic statistical mechanics of interacting particles, In Statistical Mechanics of Equilibrium and Non-equilibrium, ed. J. Meixner (North-Holland, Amsterdam, 1965).
Havas P. and J. Swenson, Relativistic thermodynamics of fluids I, *Ann. Phys. (N.Y.)* **118**, 259 (1979).
Hayes L.M. and D.B. Melrose, Dispersion in a relativistic quantum electron gas I: General distribution functions, *Austr. J. Phys.* **37**, 615 (1984).
Haymaker R.W. and J. Perez-Mercader, Relativistic thermodynamics I, *Phys. Rev.* **D27**, 1353 (1983).
Hazlehurst J. and W.L.W. Sargent, Hydrodynamics in a radiation field — a covariant treatment, *Astrophys. J.* **130**, 276 (1959).
Hazletine R.D. and S.M. Mahajan, Fluid description of relativistic magnetized plasma, *Astrophys. J.* **567**, 1262 (2002).
Hazletine R.D. and S.M. Mahajan, Radiation reaction in fusion plasmas, *Phys. Rev.* **E70**, 046407 (2004).
Hazletine R.D. and S.M. Mahajan, Closed fluid description of relativistic magnetized plasma interacting with radiation field, *Phys. Rev.* **E69**, 07904 (2004).
Hazletine R.D. and S.M. Mahajan, Relativistic magnetohydrodynamics. Preprint.
Hebeler K. and Friman B., Relativistic theory of dense matter. Preprint.
Hebenstreit H., Balance equation for a relativistic plasma I: Differential term, *Physica* **A117**, 631 (1983).
Hebenstreit H., Balance equation for a relativistic plasma II: Collision term, *Physica* **A117**, 653 (1983).
Heckler A.F., Astrophysical applications of quantum corrections to the equation of state of a plasma, *Phys. Rev.* **D49**, 611 (1994).
Heintzmann H. and M. Novello, Action principle for a hot plasma in curved spacetime, *Phys. Rev.* **A27**, 2671 (1983).
Heinz U., Landau damping: new aspects of an old story, *Phys. Lett.* **A109**, 385 (1985).
Heinz U., Quark–gluon transport theory I: The classical theory, *Ann. Phys. (N.Y.)* **161**, 48 (1985).
Heinz U., P.R. Subramanian, H. Stöcker and W. Greiner, Formation of antimatter clusters in the hadronization phase transition. Preprint BNL-37016 (1985).
Heinz U., Quark–gluon transport theory II: Color response and color correlations in a quark–gluon plasma, *Ann. Phys. (N.Y.)* **168**, 148 (1986).
Heinz U., Gauge covariant linear response analysis of QCD plasma oscillations, *Physica* **A158**, 111 (1989).
Heinz U., P. Koch, K.S. Lee, E. Schenedermann and H. Weigert, Dynamical evolution and particle production in relativistic nuclear collisions, In *Phase Structure of Strongly Interacting Matter*, ed. J. Cleymans (Springer, Berlin, 1990).

Heinz U., New developpements in quark–gluon transport theory, In *Berkeley Workshop on Pre-equilibrium Parton Dynamics*, ed. Xin Nian Wang (World Scientific, Singapore, 1993).

Hénin F., Liouville equation for a relativistic charged particle, *Physica* **29**, 1233 (1963).

Henneaux M., The Gibbs entropy production in general relativity, *Lett. Nuovo Cimento* **38**, 609 (1983).

Hermens W. Th., Relativistic kinetic theory of chemical reactions and other transport phenomena. Part I: General discussion, *Proc. Kon. Nederl. Akad. Wetensch.* **B74**, 376 (1971).

Hermens W. Th., Relativistic kinetic theory of chemical reactions and other transport phenomena. Part II: Normally reactive mixture, *Proc. Kon. Nederl. Akad. Wetensch.* **B74**, 461 (1971).

Hermens W. Th., Relativistic kinetic theory of chemical reactions and other transport phenomena. Part III: Highly reactive mixture, *Proc. Kon. Nederl. Akad. Wetensch.* **B74**, 478 (1971).

Hermens W. Th., On relativistic kinetic gas theory VIII: Reciprocal relations for a reactive mixture, *Physica* **60**, 472 (1972).

Hermens W. Th., W.A. van Leeuwen, Ch. G. van Weert and S.R. de Groot, *Physica* **A161**, 300 (1989).

Hernquist L., Relativistic electron transport in a quantizing magnetic field, *Astrophys. J.* **56**(Suppl.), 325 (1984).

Herold H., H. Ruder and G. Wunner, Relativistic effects on the photon propagation in a strongly magnetised plasma, *Plasma Phys.* **23**, 775 (1981).

Herold H., H. Ruder and G. Wunner, Cyclotron emission in strongly magnetised plasmas, *Astron. Astrophys.* **115**, 90 (1982).

Hinata S., H. Kishi, Collisional relaxation of relativistic and degenerate electrons, *J. Plasma Phys.* **43**, 311 (1990).

Hines D.F. and N.E. Frankel, Relativistic charged Bose gas, *Phys. Lett.* **A69**, 301 (1978).

Hiscock W.A. and L. Lindblom, Stability and causality in dissipative relativistic fluids, *Ann. Phys. (N.Y.)* **151**, 466 (1983).

Hiscock W.A. and L. Lindblom, Generic instabilities in first-order dissipative relativistic fluids theories, *Phys. Rev.* **D31**, 725 (1985).

Hiscock W.A., Generic instabilities in first-order dissipative relativistic fluids theories II: Havas–Swenson-type theories, *Phys. Rev. Lett.* **33**, 1527 (1986).

Hiscock W.A. and L. Lindblom, Linear plane waves in dissipative relativistic fluids, *Phys. Rev.* **D35**, 3723 (1987).

Hiscock W.A. and J.B. Lail, Relativistic fluid flows aroud cosmic strings, *Phys. Rev.* **D37**, 869 (1988).

Hiscock W.A. and L. Lindblom, Generic instabilities in first-order dissipative relativistic fluids theories, *Phys. Lett.* **A131**, 509 (1988).

Hiscock W.A. and J. Salmonson, Dissipative Boltzmann–Robertson–Walker cosmologies, *Phys. Rev.* **D43**, 3249 (1991).

Hiscock W.A. and L. Lindblom, Stability in dissipative relativistic fluids theories, *Contemp. Math.* **71**, 181 (1988).

Hiscock W.A. and T.S. Olson, Effects of frame choice on nonlinear dynamic in relativistic heat conducting fluid theories, *Phys. Rev.* **A141**, 125 (1989).

Hogan C.J., Nucleation of cosmological phase transitions, *Phys. Lett.* **B133**, 172 (1983).

Holcomb K.A. and T. Tajima, General-relativistic plasma physics in the early universe, *Phys. Rev.* **D40**, 3809 (1989).

Holland P.R., A. Kyprianidis, Z. Maric and J.P. Vigier, Relativistic generalization of the Wigner function and its interpretation in the causal stochastic formulation of quantum mechanics, *Phys. Rev.* **A33**, 480 (1986).

A. Holl, V.G. Morozov and G. Röpke, Kinetic theory of QED plasmas in a strong electromagnetic field II: The mean-field description. Preprint (2001).

Holm D.D. and B.A. Kupershmidt, Hamiltonian theory of relativistic magneto-hydrodynmics with anisotropic pressure, *Phys. Fluids* **29**, 3889 (1986).

Holm D.D. and B.A. Kupershmidt, Lyapunov stability conditions for relativistic multifluid plasmas, *Physica* **D18**, 405 (1986).

Hoodbhoy P., A variational approach to dense relativistic matter using functional techniques, *Ann. Phys. (N.Y.)* **139**, 68 (1982).

Hora H., Strong coupling of electrons to black-body radiation at high temperatures, *Phys. Stat. Sol.* **B686**, 685 (1978).

Hore S.R. and N.E. Frankel, Thermodynamic functions in the intermediate region of an ideal Fermi gas in a weak magnetic field, *Phys. Rev.* **A12**, 2177 (1975).

Hore S.R. and N.E. Frankel, Thermodynamic functions of an ideal Fermi gas in the intermediate quantum region, *Phys. Rev.* **A12**, 1617 (1975).

Horvath J.I., Contributions to the relativistic generalization of the kinetic theory of gases and statistical mechanics, ZZ, **ZZ**, 47 (ZZ).

Horowitz C.J. and B.D. Serot, Self-consistent Hartree description of a finite nuclei in a relativistic quantum field theory, *Nucl. Phys.* **A368**, 503 (1981).

Horowitz C.J. and B.D. Serot, The pion–nucleon interaction in a relativistic Hartree–Fock theory of high-density nuclear matter, *Phys. Lett.* **B108**, 377 (1982).

Horowitz C.J. and B.D. Serot, Relativistic Hartree–Fock description of high-density nuclear matter, *Phys. Lett.* **B109**, 341 (1982).

Horowitz C.J. and B.D. Serot, Properties of nuclear and neutron matter a relativistic Hartree–Fock theory, *Nucl. Phys.* **A399**, 529 (1983).

Horowitz C.J. and B.D. Serot, Two-nucleon correlations in a relativistic theory of nuclear matter, *Phys. Lett.* **B137**, 287 (1984).

Horowitz C.J. and B.D. Serot, The relativistic two-nucleon problem in nuclear matter, *Nucl. Phys.* **A464**, 613 (1987).

Horowitz C.J. and B.D. Serot, Nuclear currents in a relativistic mean-field theory, *Nucl. Phys.* **A468**, 539 (1987).

Horwitz L.P., Rest frames in relativistic thermodynamics, *Phys. Rev.* **D4**, 3812 (1971).

Horwitz L.P. and E. Dagan, The covariant Boltzmann–Fokker–Planck equation and its associated short-time transition probability, *J. Phys.* **A21**, 1017 (1988).

Horwitz L.P., S. Shashoua and W.C. Schieve, A manifestly covariant relativistic Boltzmann equation for the evolution of a system of events, *Physica* **A161**, 300 (1989).

Hosoya A., M.-A. Sakagami and M. Takao, Nonequilibrium thermodynamics in field theory: Transport coefficients, *Ann. Phys. (N.Y.)* **154**, 229 (1984).

Hosoya A. and K. Kajantie, Transport coefficient of QCD matter, *Nucl. Phys.* **B250**, 666 (1985).

Hosotani Y., Magnetic superconductivity and a linear response theory for non-abelian gauge fields, *Phys. Rev.* **D23**, 2318 (1981).

Hsieh S.-H. and V. Canuto, Scale covariant theory of gravitation V: Kinetic theory, *Astrophys. J.* **248**, 790 (1981).

Hu B., Relativistic superconductivity and superfluidity: a unified view. Preprint (1977).

Hsu J.P., Four-dimensional symmetry from a broad viewpoint, 4-covariant statistical mechanics with common time, *Nuovo Cimento* **B80**, 183 (1983).

Hsu J.P., Four-dimensional symmetry from a broad viewpoint, 4-covariant statistical mechanics with common time, *Nuovo Cimento* **B78**, 85 (1983).

Hsu J.P., Four-dimensional symmetry from a broad viewpoint, 4-covariant statistical mechanics with common time, *Nuovo Cimento* **B78**, 85 (1984).

Hwa R.C., Statistical description of hadron constituents as a basis for the fluid model of high-energy collisions, *Phys. Rev.* **D10**, 2260 (1974).

I.A. and R.V. Polovin, Theory of relativistic magnetohydrodynamic waves, *Sov. Phys. JETP* **36**, 1316 (1959).

Ignat'ev Yu. G., Relativistic kinetics and cosmology I, *Sov. Phys. J.* **23**, 682 (1981).

Ignat'ev Yu. G. Relativistic kinetics and cosmology II, *Sov. Phys. J.* **23**, 771 (1981).

Ignat'ev Yu. G., Relativistic kinetic theory and conformal transformations, *Sov. Phys. J.* **25**, 372 (1982).

Ignat'ev Yu. G., Relativistic kinetics and theory of gravitation, In *General Relativity and Gravitation*, Vol. 1, *Classical Relativity — Proc. 10th Int. Conf. General Relativity and Gravitation* 4–9, July 1983; Padova, Italy, eds. B. Bertotti, F. de Felice and A. Pascolini (Consiglio Nazionale delle Ricerche, Rome, 1983), p. 63.

Ignat'ev Yu. G., Conservation laws and thermodynamic equilibrium in the general relativistic kinetic theory of inelastically interacting particles, *Sov. Phys. J.* **26**, 1068 (1984).

Ignat'ev Yu. G., Relativistic canonical formalism and the invariant single-particle distribution function in the general theory of relativity, *Sov. Phys. J.* **26**, 686 (1984).

Ignat'ev Yu. G., Relativistic kinetic equations for inelastically interacting particles in a gravitational field, *Sov. Phys. J.* **26**, 690 (1984).

Ignat'ev Yu. G., Relativistic kinetics of baryon production in the big bang, *Sov. Astron.* **29**, 371 (1985).

Ikeuchi S. and K. Tomisaka, The structure and expansion law of a shock wave in an expanding universe, *Astrophys. J.* **265**, 583 (1983).

Ilgenfritz E.M. and J. Kripfganz, Quantum Liouville equation and non-equilibrium processes in quantum field theory, *Phys. Lett.* **A108**, 133 (1985).

Imaeda K., Statistical thermodynamics of hadron gas and liquid drops, *Lett. Nuovo Cimento* **1**, 290 (1971).

Imre K., Oscillations in a relativistic plasma, *Phys. Fluids* **5**, 459 (1962).

Imshennik V.S. and Yu. L. Morozov, Relativistically covariant equations for the coupling of radiation with matter, *Sov. Astron.* **13**, 628 (1970).

Imshennik V.S. and Yu. L. Morozov, Radiative relativistic gas dynamics of high-temperature phenomena (Atomizdat, Moscow, 1981) (in Russian).

Imshennik V.S. and Yu. L. Morozov, Dissipative relativistic radiative hydrodynamics, *Sov. Astron.* **33**, 619 (1990).

Israel W., Relativistic theory of shock waves, *Proc. R. Soc. London* **A259**, 129 (1960).

Israel W., Relativistic kinetic theory of a simple gas, *J. Math. Phys.* **4**, 1163 (1963).

Israel W. and J.N. Vardalas, Transport coefficients of a relativistic quantum gas, *Lett. Nuovo Cimento* **4**, 887 (1970).

Israel W., The relativistic Boltzmann equation, In General Relativity: Papers in Honour of J.L. Synge, ed. L.O'. Raifeartaigh (Clarendon, Oxford, 1972), p. 201.

Israel W., Electrodynamics, gravitation and spin, *Lett. Nuovo Cimento* **7**, 860 (1973).

Israel W., Foundations of a relativistic kinetic theory of spinning particles, In *Colloques Internationaux du Centre National de la Recherche Scientifique no. 236.* Théories cinétiques classiques et relativistes (Editions du Centre National de la Recherche Scientifique, Paris, 1975), p. 169.

Israel W., Thermo-field dynamics of black holes, *Phys. Lett.* **A57**, 107 (1976).

Israel W., Nonstationary irreversible thermodynamics: a causal relativistic theory, *Ann. Phys. (N.Y.)* **100**, 310 (1976).

Israel W. and J.M. Stewart, Thermodynamics of nonstationary and transient effects in a relativistic gas, *Phys. Lett.* **A58**, 213 (1976).

Israel W., The dynamics of polarization, *Gen. Relativ. Gravit.* **9**, 451 (1978).

Israel W. and J.M. Stewart, Transient relativistic thermodynamics and kinetic theory, *Ann. Phys. (N.Y.)* **118**, 341 (1979).

Israel W. and J.M. Stewart, On transient relativistic thermodynamics and kinetic theory II, *Proc. R. Soc.* **A365**, 43 (1979).

Israel W. and J.M. Stewart, Progress in relativistic thermodynamics and electrodynamics of continuous media, In *General Relativity and Gravitation 2*, ed. A. Held (Plenum, Oxford, 1980).

Israel W., Thermodynamics of relativistic systems, *Physica* **A106**, 204 (1981).

Israel W., Covariant superfluid mechanics, *Phys. Lett.* **A86**, 79 (1981).

Israel W., Equivalence of two theories of relativistic superfluid mechanics, *Phys. Lett.* **A92**, 77 (1982).

Israel W. and H.E. Kandrup, Non-equilibrium statistical mechanics in the general theory of relativity I: General formalism, *Ann. Phys. (N.Y.)* **152**, 30 (1984).

Israel W., Thermodynamics and field statistics of a relativistic superfluid, In *Proc Third Seminar on Quantum Gravity*, eds. M.A. Markov, V.A. Berezin and V.P. Frolov (World Scientific, Singapore, 1985), p. 246.

Israel W., Relativistic thermodynamics, thermofield statistics and superfluids, *J. Nonequil. Therm.* **11**, 295 (1986).

Ito D. and H. Tanaka, Theory of meson multiple production, *Nuovo Cimento* **VII**(Suppl.), 91 (1958).

Itoh N., S. Mitake, H. Iyetomo and S. Ichimaru, Transport properties of dense matter, In *Proc Third Marcel Grossman Meeting* (Shanghai, 1982), ed. Hu Ning (North-Holland, 1983).

Ivanov M.A. and G.A. Schul'man, Equilibrium in relation to pair formation in a quantizing magnetic field, *Sov. Phys. J.* **20**, 1220 (1977).

Ivanov M.A. and G.A. Schul'man, Thermodynamic functions of equilibrium radiation and electron–positron pairs in quantizing magnetic fields, *Sov. Phys. J.* (1978).

Ivanov M.A. and G.A. Schul'man, On the physical conditions in the interior of superdense, magnetized, physical objects, *Sov. Astron.* **23**, 197 (1979).

Ivanov M.A. and G.A. Schul'man, Neutrino energy emission in *b*-interaction of electrons and poditrons with nuclei of hot matter in the presence of a quantizing magnetic field, *Sov. Astron.* **25**, 76 (1981).

Ivanov M.A. and G.A. Schul'man, Kinetic equilibrium of beta-processes in high-temperature, superdense matter with a strong frozen-in magnetic field, *Sov. Astron.* **24**, 311 (1981).

Ivanov Yu. B., Kinetic description of hadron plasma, *Sov. J. Nucl. Phys.* **46**, 63 (1987).

Ivanov Yu. B., Relativistic mean field kinetic approach to hadron plasma and three-fluid dynamics, *Nucl. Phys.* **A474**, 669 (1987).

Ivanov M.A., S.S. Lipovetskii and V.S. Sekerzhitskii, Ideality of Fermi gases in a strong magnetic field, *Astrophys.* **31**, 532 (1989).

Ivanov, M.A., S.S. Lipovetskij and V.S. Sekerzhitskij, Neutronization of strongly magnetized matter, *Astron. Zh.* **70**, 531 (1993).

Jackson A.D., Nuclear matter theory: a status report, *Annu. Rev. Nucl. Part. Sci.* **33**, 105 (1983).

Jaminon M. and C. Mahaux, Relativistic quantum field approach to nuclear matter and nuclei, In *Recent Progress in Many-Body Theories*, eds. J.G. Zhabolitzky, M. de Llano, M. Fortes and J.W. Clark (Springer, Berlin, 1981).

Jancovici B., On the relativistic degenerate electron gas, *Nuovo Cimento* **25**, 428 (1962).

Jancovici B., Comportement des électrons dans un champ magnétique de 10^{13} gauss ou plus et implications astrohysiques, *Nuovo Cimento* **30**(suppl.), C3 (1969).

Jancovici B., Radiative correction to the ground-state energy of an electron in an intense magnetic field, *Phys. Rev.* **187**, 2275 (1969).

Janicke L., Non-linear electromagnetic waves in a relativistic plasma, *Plasma Phys.* **19**, 209 (1977).

Jannussis A. and N. Patargias, Eigenfunctions and eigenvalues of the Wigner oerator, *Phys. Lett.* **A53**, 357 (1975).

Jannussis A., A. Streklas, D. Sourlas and K. Vlachos, Relativistic Wigner operator and its distribution, *Lett. Nuovo Cimento* **18**, 349 (1977).

Jannussis A., A. Streklas, D. Sourlas and K. Vlachos, Wigner representation of Bloch electrons in uniform fields, *Physica* **A107**, 587 (1981).

Jauch J.M. and K.M. Watson, Phenomenological quantum electrodynamics, *Phys. Rev.* **74**, 950 (1948).

Jauch J.M. and K.M. Watson, Phenomenological quantum electrodynamics, Part II: Interaction of the field with charges, *Phys. Rev.* **74**, 1485 (1948).

Jauch J.M. and K.M. Watson, Phenomenological quantum electrodynamics, Part III. Dispersion, *Phys. Rev.* **75**, 1249 (1949).

K.A. Johns and P.T. Landsberg, Maximum entropy arguments in relativistic statistical mechanics, *J. Phys.* **A3**, 121 (1970).

Johnson M.H. and C.F. McKee, Relativistic hydrodynamics in one dimension, *Phys. Rev.* **D3**, 858 (1971).

Jones R.D. and A. Pytte, Hamiltonian of a slightly relativistic plasma, *Phys. Fluids* **23**, 269 (1980).

Jones R.D. and A. Pytte, Kinetic equation of a slightly relativistic plasma, *Phys. Fluids* **23**, 273 (1980).

Jordan P., J. Ehlers, W. Kundt and R.K. Sachs, Beiträge zur relativistischen Mechanik kontinuierlicher Medien: strenge Lösungen der Feldgleichungen der allgemeinen Relativitäts-theorie IV, *Akad. Wiss. Lit. Mainz, Abh. Math. Naturw. Kl.* **11**, 793 (1961).

Joyce M., K. Kainulainen and T. Prokopec, The quantum Bolltzmann equation in a non-trivial background, In *3rd Int. Conf. Particle Physics and the Early Universe* (Trieste, 1999).

Joyce M., K. Kainulainen and T. Prokopec, Quantum transport equations for a scalar field, *Phys. Lett.* **B474**, 402 (2000).

Jüttner F., Das Maxwellsche Gesetz der Geschwindigkeitsverteilung in der Relativitätheorie, *Ann. Physik* **34**, 856 (1911).

Jüttner F., Die Dynamik eines bewegten Gases in der Relativitätheorie, *Ann. Physik* **35**, 145 (1912).

Jüttner F., Die relativistische Quantentheorie des idealen Gases, *Z. Physik.* **47**, 542 (1928).

Kaiser G., Phase-space approach to relativistic quantum mechanics I: Coherent-state representation for massive scalar particles, *J. Math. Phys.* **18**, 952 (1977).

Kaiser G., Phase-space approach to relativistic quantum mechanics II: Geometrical aspects, *J. Math. Phys.* **19**, 502 (1978).

Kaiser G., Phase-space approach to relativistic quantum mechanics III: Quantization, relativity, localization and gauge freedom, *J. Math. Phys.* **22**, 705 (1981).

Kalman G., Equilibrium and linear response of a classical scalar plasma, *Phys. Rev.* **161**, 156 (1967).

Kalman G., Zero sound in a relativistic Fermi gas and Zeldovich's model, *Phys. Rev.* **158**, 144 (1967).

Kalman G. and S.T. Lai, Ferromagnetic transitions in a short range vector field model, *Phys. Lett.* **A34**, 75 (1971).

Kalman G., P. Bakshi and R. Cover, Strong magnetic fields in the pulsar cruts and atmosphere, In *Cosmic Plasma Physics*, ed. K. Schindler (Plenum, New York, 1972).

Kalman G., The concept of relativistic exchange energy, in *Second Windsor Conf. Relativistic Physics* (1972).

Kalman G., A vector plasma model for superdense matter and superluminal sound propagation, *Ann. Phys. (N.Y.)* **73**, 19 (1972).

Kalman G., Relativistic fermion gas interacting through a scalar field I: Hartree approximation, *Phys. Rev.* **D9**, 1656 (1974).

Kalman G. and S.T. Lai, Superluminal sound and ferromagnetic transition in the Zeldovich's model, AIU Symposium No. 53, *Physics of Dense Matter*, ed. C.J. Hansen (1972) (Reidel, Dordrecht, 1975).

Kalman G., V. Canuto and B. Datta, Self-consistency condition and high-density virial theorem in relativistic many-particle systems, *Phys. Rev.* **D13**, 3493 (1976).

Kanai T., Pion condensation I: Nonrelativistic treatment for π-condensation, *Prog. Theor. Phys.* **56**, 1802 (1976).

Kanai T., Pion condensation II: Relativistic Klein–Gordon treatment, *Prog. Theor. Phys.* **57**, 1320 (1977).

Kandrup H.E., Stochastic gravitational fluctuations in a self-consistent mean field theory, *Phys. Rep.* **63**, 1 (1980).

Kandrup H.E., A generalized Landau equation for a system with a self-consistent mean field, *Astrophys. J.* **244**, 316 (1981).

Kandrup H.E., Thermodynamic fluctuations and gravitational instabilities, *Mon. Not. R. Astron. Soc.* **202**, 717 (1983).

Kandrup H.E., A covariant kinetic equation for a self-gravitating system, *Physica* **A126**, 461 (1984).

Kandrup H.E., On the stability of a new relativistic kinetic equation, *Astrophys. J.* **282**, 361 (1984).

Kandrup H.E., Role of Maxwellian distribution in an expanding universe, *Phys. Rev.* **D30**, 2067 (1984).

Kandrup H.E., Statistical mechanics of the gravitational in a conformally static setting, *J. Math. Phys.* **25**, 3286 (1984).

Kandrup H.E. and S. Hill Kandrup, An exact closed equation for the cosmological pair correlations, *Astrophys. J.* **277**, 1 (1984).

Kandrup H.E., Non-equilibrium statistical mechanics in the general theory of relativity II: Linear fields in a kinetic approximation. Preprint (1984).

Kandrup H.E., Non-equilibrium statistical mechanics in the general theory of relativity III: Collisional stellar dynamics. Preprint (1985).

Kandrup H.E., Relativistic stellar dynamics and dense galactic nuclei, In *Fourth Marcel Grossman Meeting on General Relativity*, ed. R. Ruffini (North-Holland, Amsterdam, 1985).

Kandrup H.E., Statistical mechanics in a conformally static setting II, *J. Math. Phys.* **26**, 2850 (1985).

Kandrup H.E., Collisional stellar dynamics in general relativity: an overview. IAU Symposium No. 113, Dynamics of Star Clusters (1985).

Kandrup H.E. and G. Severne, Mean field fluctuations and the evolution of self-gravitating systems, *Astrophys. Space Sci.* **126**, 177 (1986).

Kandrup H.E., Relativistic stellar dynamics as an example of relativistic statistical mechanics, *Astrophys. Space Sci.* **124**, 359 (1986).

Kandrup H.E., On the entropy of a quantum field, *Phys. Lett.* **B185**, 382 (1987).

Kandrup H.E., Conditional probabilities and entropy in (minisuperspace) quantum cosmology, *Class. Quantum Grav.* **5**, 903 (1988).

Kandrup H.E., Generalized Wigner functions in curved spaces: a new approach, *Phys. Rev.* **D37**, 2165 (1988).

Kandrup H.E., A covariant Landau equation for discreteness fluctuations, *Class. Quantum Grav.* **6**, 519 (1989).

Kapusta J. and T. Toimela, Friedel oscillations in relativistic QED and QCD, *Phys. Rev.* **D37**, 3731 (1988).

Kapusta J. and T. Toimela, Gauge-invariant correlation functions, *Phys. Rev.* **D39**, 3197 (1989).

Kapusta J. and P.V. Landshoff, Finite-temperature field theory *J. Phys.* **C15**, 267 (1989).

Karsch F. and D.E. Miller, Equation of state for hadronic matter, *Phys. Lett.* **B79**, 259 (1978).

Karsch F. and H. Satz, On the thermodynamics of extended hadrons, *Phys. Rev.* **D2**, 1168 (1980).

Katsouleas T. and W.B. Mori, Wave-breaking amplitude of relativistic oscillations in a thermal plasma, *Phys. Rev. Lett.* **61**, 90 (1988).

Katz J. and G. Horwitz, Heat flow in rotating relativistic perfect fluids and global thermodynamic parameters, In Modern Developments in Thermodynamics, ed. B. Gal-Or (Israel Universities Press and Wiley, New York, 1974), p. 237.

Kaufman A.N. and D.D. Holm, The Lie-transformed Vlasov action principle: relativistically covariant wave propagation and self-consistent ponderomotive effects, *Phys. Lett.* **A105**, 277 (1984).

Kelly, D.C., Diffusion: a relativistic appraisal, *Am. J. Phys.* **36**, 585 (1968).

Kelly D.C., Collision operator for a relativistic Lorentz gas, *Phys. Fluids* **12**, 799 (1969).

Kelly D.C., Electrical and thermal conductivities of a relativistic degenerate plasma, *Astrophys. J.* **179**, 599 (1973).

Kennel C.F. and R. Pellat, Relativistic nonlinear plasma waves in a magnetic field, *J. Plasma Phys.* **15**, 335 (1976).

Kennel C.F., M.E. Gedalin and J.G. Lominadze, Relativistic MHD from two-fluid hydrodynamics (TFHD), In *Varenna Summer School and Workshop on Plasma Astrophysics (1988)*, eds. T.D. Guyenne and J.J. Hunt (ESA, Noordwijk, 1989).

Khakimova M. and F. Kh. Khakimov, On a second sound existence possibility in a relativistic plasma, In *XVI Int. Conf. Phenomena in Ionized Gases*, (Dusseldorf, 1983), W. Botticher, H. Wenk and E. Schulz-Gulde (University of Dusseldorf, 1984).

Khalilov V.R., Equilibrium of a degenerate gas of nucleons and electrons in a strong magnetic field, *Theor. Math. Phys.* **130**, 74 (2001).

Khlebnikov V.I., Long gravitational waves in an ultrarelativistic two-component universe, *Sov. Phys. JETP* **59**, 6 (1984).

Kirzhnits D.A. and N. Zh. Takibaev, The thermodynamic description of hadronic matter, *Sov. Phys. JETP* **48**, 397 (1978).

Kjellmert B., A new type of coherent oscillations in a relativistic electron plasma, *Plasma Phys.* **17**, 649 (1975).

Klein O., On the thermodynamical equilibrium of fluids in gravitational fields, *Rev. Mod. Phys.* **21**, 531 (1949).

Klimas A. and G. Sandri, Relativistic charged particles in an electromagnetic field I: Uniform, stationary fields. Preprint (1967).

Klimontovich Yu. L., Relativistic transport equations for a plasma I, *Sov. Phys. JETP* **37**, 524 (1960).

Klimontovich Yu. L., Relativistic transport equations for a plasma II, *Sov. Phys. JETP* **37**, 876 (1960).

Klimov V.V., Spectrum of elementary Fermi excitations in quark–gluon plasma, *Sov. J. Nucl. Ener.* **33**, 934 (1981).

Klimov V.V., Collective oscillations in a hot quark–gluon plasma, *Sov. Phys. JETP* **55**, 199 (1982).

Kluitenberg G.A., S.R. de Groot and P. Mazur, Relativistic thermodynamics of irreversible processes I: Heat conduction, diffusion, viscous flow and chemical reactions; formal part, *Physica* **19**, 689 (1953).

Kluitenberg G.A., S.R. de Groot and P. Mazur, Relativistic thermodynamics of irreversible processes II: Heat conduction and diffusion; physical part, *Physica* **19**, 1079 (1953).

Kluitenberg G.A. and S.R. de Groot, Relativistic thermodynamics of irreversible processes III: Systems without polarization and magnetization in an electromagnetic field, *Physica* **20**, 199 (1955).

Kluitenberg G.A. and S.R. de Groot, Relativistic thermodynamics of irreversible processes IV: Systems with polarization and magnetization in an electromagnetic field, *Physica* **21**, 148 (1955).

Kluitenberg G.A. and S.R. de Groot, Relativistic thermodynamics of irreversible processes V: The energy–momentum tensor of the macroscopic electromagnetic field, the macroscopic forces acting on the matter and the first and second law of thermodynamics, *Physica* **21**, 169 (1955).

Ko C.M., Q. Li and R. Wang, Relativistic Vlasov equation for heavy-ion collisions, *Phys. Rev. Lett.* **59**, 1084 (1987).

Ko C.M. and Q. Li, Relativistic Vlasov–Uehling–Uhlenbeck model for heavy-ion collisions, *Phys. Rev.* **C37**, 2270 (1988).

Ko C.M., Relativistic Vlasov–Uehling–Uhlenbeck model for high energy heavy-ion collisions, *Nucl. Phys.* **A495**, 321c (1989).

Ko C.M. and Q. Li, Relativistic Vlasov–Uehling–Uhlenbeck equation for nucleus–nucleus collisions, *Phys. Rev.* **C39**, 849 (1989).

Koch P., J. Rafelski and W. Greiner, Covariant quantum statistics of fields, *Phys. Lett.* **B123**, 151 (1983).

Kosowski S., Effective determination of dissipative parameters in a relativistic gas within the framework of kinetic theory, *Fluid Dyn. Trans.* **4**, 547 (1969).

Kosowski S., The structure of a shock wave in an ultra-relativistic gas, *Arch. Mech. Stos.* **1**, 21, 33 (1969).

Kosowski S., Transport coefficients of an ultrarelativistic gas, *Arch. Mech. Stos.* **4**, 21, 391 (1969).

Kosowski S., The balance and production of entropy in a relativistic medium, *Arch. Mech.* **23**, 65 (1971).

Kothari D.S. and B.N. Singh, ZZ, *Proc. R. Soc.* **178**, 135 (1941).

Kothari D.S. and B.N. Singh, ZZ, *Proc. R. Soc.* **189**, 414 (1942).

Kowalenko V., N.E. Frankel and K.C. Hines, Response theory of particle–antiparticle plasmas, *Phys. Rep.* **126**, 109 (1985).

Kowalski K.L., Thermal averages in covariant gauges, *Phys. Rev.* **D32**, 2835 (1985).

Kox A.J., W.A. van Leeuwen and S.R. de Groot, Diffusion and thermal diffusion coefficients of a relativistic gas mixture, *Phys. Lett.* **A47**, 221 (1974).

Kox, A.J., Diffusion coefficients of a relativistic Lorentz gas, *Phys. Lett.* **A53**, 173 (1975).

Kox A.J., J. Guichelaar and S.R. de Groot, Théorie relativiste de l'influence d'un phénomène de relaxation sur la propagation du son, *C. R. Acad. Sci.* **A283**, 959 (1976).

Kox A.J., On relativistic kinetic gas theory XVIII: Diffusion in a Lorentz gas of hard spheres, *Physica* **A84**, 603 (1976).

Kox A.J., S.R. de Groot and W.A. van Leeuwen, On relativistic kinetic gas theory XVI: The temperature dependence of the transport coefficients for a simple gas of hard spheres, *Physica* **A84**, 155 (1976).

Kox A.J., S.R. de Groot and W.A. van Leeuwen, On relativistic kinetic gas theory XVII: Diffusion and thermal diffusion in a binary mixture of hard spheres, *Physica* **A84**, 165 (1976).

Kranys M., Relativistic hydrodynamics with irreversible thermodynamics without the paradox of inifinite velocity of heat conduction, *Nuovo Cimento* **B42**, 51 (1966).

Kranys M., Non-truncated relativistic Chernikov–Grad 13-moment equations as an approach to nonstationary thermodynamics, *Phys. Lett.* **A33**, 77 (1970).

Kranys M., Kinetic derivation of nonstationary general relativistic thermodynamics, *Nuovo Cimento* **B8**, 417 (1972).

Kranys M., Phase and signal velocities of waves in dissipative media: special relativistic theory, *Arch. Rat. Mech. Anal.* **48**, 274 (1972).

Kranys M., Relativistic theory of waves in dissipative media, In *Modern Developments in Thermodynamics*, ed. B. Gal-Or (Israel Universities Press and Wiley, New York, 1974), p. 259.

Kranys M. and J. Teichmann, Phase and signal velocities of eigenwaves in a binary dissipative plasma, *J. Plasma Phys.* **11**, 269 (1974).

Kranys M., Eigenwaves in a relativistic gas according to Marle's 14-moment description, *Ann. Inst. H. Poincaré* **AXXV**, 197 (1976).

Kranys M., Hyperbolic elasticity of dissipative media and its wave propagation modes, *J. Phys.* **A10**, 689 (1977).

Kranys M., Extended thermodynamics of ideal gases with 14 fields, *Ann. Inst. H. Poincaré* **45**, 419 (1986).

Kranys M., Relativistic theory of waves in dissipative media, In *Colloques Internationaux du Centre National de la Recherche Scientifique no. 236. Théories cinétiques classiques et relativistes*, (Editions du Centre National de la Recherche Scientifique, Paris, 1975), p. 315.

Kraus K. and P.T. Landsberg, Elementary statistical mechanics of a relativistic gas in thermal equilibrium, *J. Phys.* **A7**, 2130 (1974).

Krizan J.E. and P. Havas, Relativistic corrections in the statistical mechanics of interacting particles, *Phys. Rev.* **128**, 2916 (1962).

Krizan J.E., Kinetic equation with relativistic interaction corrections for a plasma, *Phys. Rev.* **A140**, 1155 (1965).

Krizan J.E., Relation of the plasma kinetic equation for the Darwin Hamiltonian in the relativistic Landau equation, *Phys. Rev.* **152**, 136 (1966).

Krizan J.E., Time-symmetric, approximately relativistic particle interactions and radiation, *Phys. Rev.* **D22**, 3017 (1980).

Krizan J.E., Matter and radiation in equilibrium, *Phys. Rev.* **D31**, 1968 (1985).

Krizan J.E., Relativistic Doppler-shift effects, *Phys. Rev.* **D31**, 3140 (1985).

Kudashev V.R., A.B. Mikhailovskii and G.I. Suramlishvili, The low-frequency mode of electromagnetic parametric instabilities occurring in relativistic plasma, *Phys. Scripta* **29**, 490 (1984).

Kulikov I.K. and P.I. Pronin, Bose gas in gravitational field, *Int. J. Mod. Phys.* **32**, 1261 (1993).

Kulikov I.K. and P.I. Pronin, Low temperature properties of a quantum Fermi gas in curved space-time, *Int. J. Mod. Phys.* **34**, 1843 (1995).

Kunuose M., Relativistic thermal plasmas: time development of electron–positron pair concentration, *Astrophys. J.* **321**, 186 (1987).

Kursunoglu B., Relativistic plasma, *Nucl. Fus.* **1**, 213 (1960).

Kursunoglu B., Relativistic plasma, In *Proc. Fifth Int. Conf. Ionization Phenomena in Gases* (Munich, 1961) (North-Holland, Amsterdam, 1962).

Kursunoglu B., Relativistic plasma in a magnetic field, *Nuovo Cimento* **43**, 209 (1966).

Kursunoglu B., Symmetries of a relativistic plasma, In *Relativistic Plasmas — The Coral Gable Conference*, eds. O. Buneman and W.B. Pardo (W. Benjamin, New York, 1968).

Kusunose M. and F. Takahara, Pair equilibrium in a relativistic plasma with magnetic fields, *Prog. Theor. Phys.* **69**, 1443 (1983).

Kusunose M. and F. Takahara, Electron–positron pair equilibrium in a mildly relativistic plasma: effects of the Comptonization of soft photons, *Prog. Theor. Phys.* **73**, 41 (1985).
Kuz'menkov L.S., A.A. Sokolov and O.O. Trubachev, Nonlinear Langmuir waves in a relativistic plasma, *Sov. Phys. J.* **26**, 1076 (1983).
Kuz'menkov L.S., L.A. Polyakov, M.I. Sitnov and O.O. Trubachev, Representing the longitudinal dielectric permittivity of relativistic plasma by a convenient series in the long wavelength region, *Moscow Univ. Phys. Bull.* **37**, 92 (1982).
Kuz'menkov L.S. and M.I. Sitnov, Dielectric constant of an ultrarelativistic plasma, *Sov. Phys. J.* **25**, 817 (1982).
Landsberg P.T. and J. Dunning-Davies, Ideal relativistic Bose gas condensation, *Phys. Rev.* **A138**, 1049 (1965).
Landsberg P.T. and J. Dunning-Davies, Statistical thermodynamics of the ideal relativistic quantum gas, In *Statistical Mechanics of Equilibrium and Non-equilibrium*, ed. J. Meixner (North-Holland, Amsterdam, 1965), p. 36.
Landsberg P.T. and K.A. Johns, The problem of moving thermometers, *Proc. R. Soc.* **A306**, 477 (1968).
Landsberg P.T., Special relativistic thermodynamics. Review in *A Critical Review of Thermodynamics*, eds. E.B. Stuart, B. Gal-Or and A.J. Brainard (Mono Book, Baltimore, 1970), p. 253.
Landsberg P.T. and K.A. Johns, Ergodicity and the Lorentz transformation of time-based probabilities, *J. Phys.* **A3**, 113 (1970).
Landsberg P.T., W.C. Saslaw, A.J. Haggett, On the adiabatic expansion of a relativistic gas, *Mon. Not. R. Astron. Soc.* **154**, 7P (1971).
Landsberg P.T., Thermodynamic inequalities in special relativity, *Ann. Phys. (N.Y.)* **70**, 1 (1972).
Landsberg P.T. and K.A. Johns, Elementary statistical mechanics of a relativistic gas in thermal equilibrium II, *J. Phys.* **A10**, 953 (1977).
Landsberg P.T., Thought experiment to determine the special relativistic temperature transformation, *Phys. Rev. Lett.* **45**, 149 (1980).
Landsman N.P. and Ch. G. van Weert, Real- and imaginary-time field theory at finite temperature and density, *Phys. Rep.* **145**, 141 (1987).
Langdon A.B., J. Arons and C.E. Max, Structure of magnetosonic shocks in electron–positron plasmas, *Phys. Rev. Lett.* **61**, 779 (1988).
Lanza A., J.C. Miller and S. Motta, Relativistic shocks in a Synge gas, *Lett. Nuovo Cimento* **35**, 309 (1982).
Lanza A., J.C. Miller and S. Motta, Formation and damping of relativistic strong shocks in a Synge gas, *Phys. Fluids* **28**, 97 (1985).
Lanza A., J.C. Miller and S. Motta, Relativistic shocks in a gas of interacting particles and radiation, *Nuovo Cimento*, **98**, 119 (1987).
Lapedes A. and E. Mottola, Complex path integrals and finite temperature, *Nucl. Phys.* **B203**, 58 (1982).
Lapiedra R. and E. Santos, Classical relativistic statistical mechanics: the case of a hot dilute plasma, *Phys. Rev.* **D23**, 2181 (1981).
Lazar M., Relativistic kinetic theory of waves in equilibrium magnetized pair plasma: general dispersion relations, *Phys. Scripta* **18**, 146 (2003).

Lee H.J., V. Canuto, H.-Y. Chiu and C. Chiuderi, New state of ferromgnetism in degenerate electron gas and magnetic fields in collapsed bodies, *Phys. Rev. Lett.* **23**, 390 (1969).

Lee T.D., Hydrogen content and energy-productive mechanism of white dwarfs, *Astrophys. J.* **111**, 625 (1950).

Leermakers M.C.J. and Ch. G. van Weert, Effective electron mass in finite temperature QED, *Phys. Lett.* **B135**, 118 (1984).

Leermakers M.C.J. and Ch. G. van Weert, Effective electron mass in finite temperature QED, *Nucl. Phys.* **B248**, 671 (1985).

Leermakers M.C.J. and Ch. G. van Weert, Energy–momentum tensor of a QED plasma at finite temperature and density, *Ann. Phys. (N.Y.)* **160**, 264 (1985).

Leermakers R., Charged and colored Fermi liquids, *Phys. Rev.* **D44**, 3687 (1991).

Leermakers R., Unexpected phenomena in covariant Landau theory, *Phys. Rev.* **B49**, 475 (1994).

Lemoine D., Fluctuations in the relativistic plasma and primordial magnetic fields, *Phys. Rev.* **D51**, 2677 (1995).

Lemoine M., A quasi-particle approach to the QED plasma. Unpublished (1994).

Lemoine M., Etude du plasma relativiste quantique de paires e^+e^-. Thesis, Paris (1995).

Lerche I., Initial-value problem for relativistic plasma oscillations, *J. Math. Phys.* **8**, 1838 (1967).

Lerche I., Supraluminous plasma waves and radiation from density inhomogeneities, In *Relativistic Plasmas — The Coral Gable Conferences*, eds. O. Buneman and W.B. Pardo (W. Benjamin, New York, 1968).

Lerche I., On the passage of radiation through inhomogeneous, moving media. XIII: Causality, analyticity, group speed, transparency, and allied phenomena with emphasis on some recent calculations by Ko and Chuang, *Astrophys. J.* **222**, 1020 (1978).

Li Q. and C.M. Ko, Covariant Vlasov equation based on the Walecka model, *Mod. Phys. Lett.* **3**, 465 (1988).

Liang E.P.T., Relativistic simple waves: shock damping and entropy production, *Astrophys. J.* **211**, 361 (1977).

Lichnerowicz A. and R. Marrot, Propriétés statistiques des ensembles de particules en relativité restreinte, *C. R. Acad. Sci.* **210**, 759 (1940).

Lichnerowicz A., Sur le théorème d'hydrodynamique relativiste, *C. R. Acad. Sci.* **211**, 117 (1940).

Lichnerowicz A., Sur les équations de l'hydrodynamique des fluides visqueux et la notion de fluide incompressible en relativité générale, *C. R. Acad. Sci.* **219**, 270 (1944).

Lichnerowicz A., Etude mathématique des fluides thermodynamiques relativistes, *Commun. Math. Phys.* **1**, 328 (1966).

Lichnerowicz A., Sur les ondes de choc en magnétohydrodynamique relativiste, *C. R. Acad. Sci.* **262**, 153 (1966).

Lightmann A.P., Fundamental processes in pair plasmas, In *Workshop on Electron–positron Pairs in Astrophysics* (AIP Conf. Proc., Greenbelt, 1983).

Lightmann A.P., Relativistic plasmas, *Adv. Space Res.* **8**, 547 (1988).

Linde A.D., Phase transitions in gauge theories and cosmology, *Rep. Prog. Phys.* **42**, 389 (1979).

Lindblom L. and W.A. Hiscock, A one-fluid model of superfluids, *Phys. Lett.* **A131**, 280 (1988).

Lindquist R.W., Relativistic transport theory, *Ann. Phys. (N.Y.)* **37**, 487 (1966).

Liu I.-S., I. Müller and T. Ruggeri, Relativistic thermodynamics of gases, *Ann. Phys. (N.Y.)* **169**, 191 (1986).

Lopuszanski J., Relativisierung der Theorie des stochastischen Prozesse, *Acta Phys. Pol.* **XII**, 87 (1953).

Love A. and S.J. Stow, Relativistic effects selecting the superfluid phase of neutron star matter, *J. Phys.* **A16**, L41 (1983).

Lucquiaud J.C., Analyse harmonique sur l'hyperboloide de Minkowski, In *Colloques Internationaux du Centre National de la Recherche Scientifique no. 236. Théories cinétiques classiques et relativistes* (Editions du Centre National de la Recherche Scientifique, Paris, 1975), p. 183.

Lucquiaud J.L., Sur l'équation de Boltzmann relativiste et applications. Thèse de doctorat d'Etat, Dijon (1975).

Lucquiaud J.C., On the relativistic Boltzmann equation, *J. Math. Phys.* **19**, 2659 (1978).

Lucquiaud J.C., Polynômes orthogonaux sur l'hyperboloïde de Minkowski, *C. R. Acad. Sci.* **A282**, 809 (1976).

Luke S.K. and G. Szamosi, On the general relativistic Boltzmann equation, *Nuovo Cimento* **B70**, 100 (1970).

Lurçat F. and P. Mazur, Statistical mechanical evaluation of phase-space integrals, *Nuovo Cimento* **31** (1964) 140.

Lyutikov M., Waves in a one-dimensional magnetized relativistic pair plasma, *Mon. Not. R. Soc. London*, **293**, 447 (1998).

Maartens R. and S.D. Maharaj, Collision-free gases in spatially homogeneous space–times, *Commun. Math. Phys.* **17**, 343 (1970).

Maartens R. and S.D. Maharaj, Collision-free gases in spatially homogeneous space–times, *J. Math. Phys.* **26**, 2869 (1985).

Maartens R. and S.D. Maharaj, Invariant solutions of Liouville's equation in Robertson–Walker space–times, *Class. Quantity Grav.* **3**, 1005 (1986).

Maartens R. and S.D. Maharaj, Collision-free gases in static space–times, *J. Math. Phys.* **27**, 2514 (1986).

Maartens R. and S.D. Maharaj, Conformal Killing vectors in Robertson–Walker space–times, *Gen. Relativ. Gravit.* **19**, 1223 (1987).

Maartens R. and S.D. Maharaj, Collision-free gases in Bianchi space–times, *Gen. Relativ. Gravit.* **22**, 595 (1990).

MacCallum M.A.H., J.M. Stewart and B.G. Schmidt, Anisotropic stresses in homogeneous cosmologies, *Proc. R. Soc.* **A365**, 43 (1979).

Macedo P.G. and A.H. Nelson, Propagation of gravitational waves in a magnetized plasma, *Phys. Rev.* **D28**, 2382 (1983).

Magneville A., Plasma waves in hot relativistic beam-plasma systems I: Dispersion relations, *J. Plasma Phys.* **44**, 191 (1990).

Magneville A., Plasma waves in hot relativistic beam-plasma systems II: Kinetic and hydrodynamic instabilities, *J. Plasma Phys.* **44**, 213 (1990).

Mahajan S.M. and R.D. Hazletine, Fluid description of relativistic magnetized plasma, *Astrophys. J.* **567**, 1262 (2002).

Maharaj S.D. and R. Maartens, Collision-free gases in static space–times, *J. Math. Phys.* **27**, 2514 (1986).

Maharaj S.D. and R. Maartens, Exact inhomogeneous Einstein–Liouville solutions in Robertson–Walker space-times, *Class. Quantum Grav.* **19**, 499 (1987).

Maharaj S.D. and R. Maartens, General solution of Liouville's equation in Robertson–Walker space–times, *Class. Quantum Grav.* **19**, 1217 (1987).

Majorana A. and A.M. Anile, Magnetoacoustic shock waves in a relativistic gas, *Phys. Fluids* **30**, 3045 (1968).

Majorana A. Analytical solutions of the Rankine–Hugoniot relations for a relativistic simple gas, *Nuovo Cimento* **B98**, 111 (1987).

Malfliet R., Relativistic transport theory, *Nucl. Phys.* **A553**, 763c (1993).

Mangeney A., Contribution à l'interaction entre un champ de rayonnement et un plasma, *Ann. Phys. (Paris)* **10**, 191 (1965).

Mangeney A. and M. Signore, Relativistic anisotropic plasma. Preprint (1974).

Mangeney A. and M. Signore, Sur une magnétohydrodynamique relativiste anisotrope. Rapport Observatoire de Meudon (France) (1974).

Mangeney A. and M. Signore, The motion, in the drift approximation, of a relativistic charged particle in a strong magnetic field, *Phys. Lett.* **A48**, 65 (1974).

Mann A. and R.M. Weiner, Superfluidity of hadronic matter, *Nuovo Cimento.* **A10**, 625 (1972).

Mann A. and R.M. Weiner, Derivation of the critical temperature for the superfluid phase of hadronic matter, *Phys. Lett.* **B40**, 383 (1972).

Mansouri R. and A. Rakei, Distribution function of a collisionless gas in the de Sitter spacetime, *Class. Quantum Grav.* **5**, 321 (1988).

Manzke G. and D. Kremp, Kinetic equation of a slightly relativistic plasma, *Physica* **A97**, 153 (1979).

Mark D.W., The unobservability of higher-order moments in the cosmic background radiation (1983).

Marle C., Modèle cinétique pour l'établissement des lois de la conduction de la chaleur et de la viscosité en théorie de la relativité, *C. R. Acad. Sci.* **260**, 6539 (1965).

Marle C., Modèle cinétique pour un gaz relativiste à deux constituants, *C. R. Acad. Sci.* **A262**, 100 (1966).

Marle C., Modèle cinétique relativiste pour un gaz avec conduction thermique, viscosité et diffusion. Publ. Institut Français du Pétrole (f'évrier, 1966).

Marle C., Sur une nouvelle famille de polynômes généralisant, dans l'espace–temps de la relativité restreinte, les polynômes de Hermite–Grad, *C. R. Acad. Sci.* **263**, 485 (1966).

Marle C., Sur l'établissement des équations de l'hydrodynamique des fluides relativistes dissipatifs I. L'équation de Boltzmann relativiste, *Ann. Inst. Henri Poincaré* **A10**, 67 (1969).

Marle C., Sur l'établissement des équations de l'hydrodynamique des fluides relativistes dissipatifs II. Méthodes de résolution approchée de l'équation de Boltzmann relativiste, *Ann. Inst. Henri Poincaré* **A10**, 127 (1969).

Marle C., Extension de la méthode de Grad à la résolution de l'équation de Boltzmann relativiste, In *Colloques Internationaux du Centre National de la Recherche Scientifique nr. 184. La magnétohydrodynamique classique et relativiste* (Editions du Centre National de la Recherche Scientifique, Paris, 1970), p. 129.

Marle C.M., Le développement de Hilbert en théorie cinétique relativiste, In *Colloques Internationaux du Centre National de la Recherche Scientifique no. 236. Théories cinétiques classiques et relativistes* (Editions du Centre National de la Recherche Scientifique, Paris, 1975), p. 255.

Marshall T.W., Random electrodynamics, *Proc. R. Soc.* **A276**, 475 (1963).

Marshall T.W., Statistical electrodynamics, *Proc. Cambridge Philos. Soc.* **61**, 537 (1963).

Marshall T.W., Statistical electrodynamics, In *Statistical Mechanics of Equilibrium and Non-equilibrium*, ed. J. Meixner (North-Holland, Amsterdam, 1965).

Masaki Isao, Energy–momentum tensor for radiation and radiative viscosity, *Publ. Astr. Soc. Jpn.* **23**, 425 (1971).

Matarrese S., Phonons in a relativistic perfect fluid, In *Fourth Marcel Grossman Meeting on General Relativity*, ed. R. Ruffini (North-Holland, Amsterdam, 1986).

Matravers D.R., Relativistic kinetic theory and cosmology, In *Rotating Objects and Relativistic Physics*, eds. F.J. Chinea and L.M. Gonzales-Romero (Springer, Berlin, 1993).

Matsui T., Fermi-liquid properties of nuclear matter in a relativistic mean-field theory, *Nucl. Phys.* **A370**, 365 (1981).

Matzner R.A., Closed rotating cosmologies containing matter described by the kinetic theory formalism, *Ann. Phys. (N.Y.)* **65**, 438 (1971).

Matzner R.A. and C.W. Misner, Dissipative effects in the expansion of the universe, *Astrophys. J.* **171**, 415 (1972).

Matzner R.A., Closed rotating cosmologies containing matter described by the kinetic theory entropy production in the collision time approximation, *J. Math. Phys.* **13**, 931 (1972).

Maugin G.A., Magnetized deformable media in general relativity, *Ann. Inst. Henri Poincaré* **15**, 275 (1971).

Maugin G.A. and A.C. Eringen, Polarized elastic materials with electronic spin — a relativistic approach, *J. Math. Phys.* **13**, 1777 (1972).

Maugin G.A. and A.C. Eringen, Deformable magnetically saturated media-II-constitutive theory, *J. Math. Phys.* **13**, 1334 (1972).

Maugin G.A., An action principle in general relativistic magnetohydrodynamics, *Ann. Inst. Henri Poincaré* **16**, 133 (1972).

Maugin G.A., Remarks on dissipative processes in the continuum theory of micromagnetics, *J. Phys.* **A5**, 1550 (1972).

Maugin G.A., Relativistic theory of magnetoelastic interactions II: Constitutive Theory, *J. Phys.* **A6**, 306 (1973).

Maugin G.A., Relativistic theory of magnetoelastic interactions III: Isotropic media, *J. Phys.* **A6**, 1647 (1973).

Maugin G.A., Relativistic theory of magnetoelastic interactions IV: Hereditary processes, *J. Phys.* **A7**, 818 (1974).

Maugin G.A., On relativistic deformable solids and the detection of gravitational waves, *Gen. Relativ. Gravit.* **5**, 13 (1974).

Maugin G.A., Exact relativistic theory of wave propagation in pre-stressed non-linear elastic solids, *Ann. Inst. Henri Poincaré* **38**, 155 (1978).

Maugin G.A., On Maxwell's covariant equations in matter, *J. Franklin Inst.* **305**, 11 (1978).

Maugin G.A., On the covariant equations of relativistic electrodynamics of continua I: General equations, *J. Math. Phys.* **19**, 1198 (1978).

Maugin G.A., On the covariant equations of relativistic electrodynamics of continua II: Fluids, *J. Math. Phys.* **19**, 1206 (1978).

Maugin G.A., On the covariant equations of relativistic electrodynamics of continua III: Elastic solids, *J. Math. Phys.* **19**, 1212 (1978).

Maugin G.A., Relativistic effective interactions and nucleon–nucleon scattering. Preprint CTP 1144 (1984).

Max C.E., Steady state solutions for relativistically strong electromagnetic waves in plasma, *Phys. Fluids* **16**, 1277 (1973).

Mazumber B.G., Plasma neutrino emission in a strong magnetic field: transverse plasma, *Indi. J. Phys.* **B59**, 77 (1985).

Mazumber B.G., Plasma neutrino emission in a strong magnetic field and neutron star models, *Indi. J. Phys.* **B59**, 88 (1985).

McKee C.R. and S.A. Colgate, Relativistic shock hydrodynamics, *Astrophys. J.* **181**, 903 (1973).

Meixner J. and H.G. Reik, Thermodynamik der irreversiblen Prozesse: relativistische Thermodynamik der irreversiblen Prozesse, In *Handbuch der Physik III/2*, ed. S. Flügge (Springer-Verlag, Berlin, 1959), p. 494.

Melrose D.B., Symmetry properties of nonlinear responses in a plasma, *Plasma Phys.* **14**, 1035 (1972).

Melrose D.B., A covariant formulation of wave dispersion, *Plasma Phys.* **15**, 99 (1973).

Melrose D.B., A relativistic quantum theory for processes in collisionless plasmas, *Plasma Phys.* **16**, 845 (1974).

Melrose D.B. and R.J. Stoneham, Vacuum polarization and photon propagation in a magnetic field, *Nuovo Cimento* **A32**, 435 (1976).

Melrose D.B. and R.J. Stoneham, The polarization tensor for a magnetized vacuum, *J. Phys.* **A10**, 1211 (1977).

Melrose D.B., The transformation properties of the energy–momentum tensor for dispersive media, *Aust. J. Phys.* **34**, 563 (1981).

Melrose D.B., Covariant description of dispersion in a relativistic thermal electron gas, *Aust. J. Phys.* **35**, 41 (1982).

Melrose D.B., Quantum electrodynamics in strong magnetic fields II: Photon fields and interactions, *Aust. J. Phys.* **36**, 775 (1983).

Melrose D.B. and A.J. Parle, Quantum electrodynamics in strong magnetic fields I: Electron states, *Aust. J. Phys.* **36**, 755 (1983).

Melrose D.B. and A.J. Parle, Quantum electrodynamics in strong magnetic fields III: Electron–photon interactions, *Aust. J. Phys.* **36**, 799 (1983).

Melrose D.B. and J. Kuijpers, Resonant parts of nonlinear response tensors, *J. Plasma Phys.* **32**, 239 (1984).

Melrose D.B., The cubic response tensor for a magnetized plasma: a covariant forward-scattering approach, *Aust. J. Phys.* **40**, 139 (1987).

Melrose D.B., Covariant form of Trubnikov's response tensor for a relativistic magnetized thermal plasma, *J. Plasma Phys.* **57**, 479 (1995).

Melrose D.B., The response tensor for a highly relativistic magnetized thermal plasma, *J. Plasma Phys.* **58**, 735 (1997).

Melrose D.B., Response of a relativistic anisotropic thermal plasma. Part I: Unmagnetized particles, *J. Plasma Phys.* **58**, 705 (1997).

Melrose D.B., Response of a relativistic anisotropic thermal plasma. Part II: Magnetized particles, *J. Plasma Phys.* **58**, 721 (1997).

Merches I., A relativistically covariant formalism in the theory of plasma stability, *J. Plasma Phys.* **23**, 259 (1980).

Messer J., Isotropic solutions of the Einstein–Vlasov equations with lowest order corrections, *Class. Quantum. Grav.* **3**, 589 (1986).

Messer J., On the moment hierarchy of the first-order quantum corrected Einstein–Vlasov equations, *Class. Quantum. Grav.* **4**, 1383 (1987).

Mészàros P. and J. Ventura, Vacuum-polarization effects on Thomson cross sections in a magnetized plasma, *Phys. Rev. Lett.* **41**, 1544 (1978).

Mészàros P. and J. Ventura, Vacuum-polarization effects on radiative opacities in a strong magnetic field, *Phys. Rev.* **D19**, 3565 (1979).

Mezzasalma S.A., Relativistic Brownian motion and its geometrical analogy. From scaling laws in polymer physics to statistical thermodynamics at quantum level, In *The Encyclopedia of Surface and Colloid Science* (Hubbard A. Marcel Dekker, New York, 2002).

Midorikawa S. and Y. Yamaguchi, Screening effect in relativistic QED plasma, *Prog. Theor. Phys.* **76**, 311 (1986).

Mies S. and B. Schürmann, A linearized Uehling–Uhlenbeck equation for high energy nucleus–nucleus collisions, *Ann. Phys. (N.Y.)*, **166**, 189 (1986).

Migdal A.B., Pion fields in nuclear matter, *Rev. Mod. Phys.* **50**, 107 (1978).

Mikhailovskii A.B., O.G. Onishchenko and A.I. Smolyakov, Kinetic description of a relativistic plasma with a one-dimensional momentum distribution of initial states, *Phys. Lett.* **A97**, 103 (1983).

Mikhailovskii A.B., G.I. Suramlishvili, V.R. Kudashev and E.G. Tatarinov, Stabilization of the magneto-modulational instability in relativistic plasmas, *Phys. Lett.* **A101**, 137 (1984).

Mikhailovskii A.B., G.I. Suramlishvili, V.R. Kudashev and E.G. Tatarinov, Magnetic field generation in a relativistic plasma with high frequency waves, *Plasma Phys. Cont. Fus.* **26**, 493 (1984).

Mikhailovskii A.B. and O.G. Onishchenko, Drift instabilities of a relativistic plasma I: Kinetic description of drift effects in a relativistic plasma, *J. Plasma Phys.* **37**, 15 (1987).

Mikhailovskii A.B. and O.G. Onishchenko, Drift instabilities of a relativistic plasma II: Kinetic theory of low-frequency drift instabilities of a relativistic finite-pressure plasma, *J. Plasma Phys.* **37**, 29 (1987).

Miller D.E., Thermodynamics of a pion gas, *Phys. Lett.* **A61**, 291 (1977).

Miller D.E., Solvable thermodynamical models for hadronic systems, ZZ, ZZ, ZZ (1977) peut-être dans Gal-Or?

Miller D.E. and F. Karsch, On the theory of relativistic quantum gases I: The covariant formulation for equilibrium systems. Preprint BI-TP 79/08.

Miller D.E. and F. Karsch, On the theory of relativistic quantum gases II: The exact equation of state for ideal systems. Preprint BI-TP 79/09.

Miller D.E. and F. Karsch, Exact equation of state for ideal relativistic quantum gases, *Phys. Rev.* **A22**, 1210 (1980).

Miller D.E. and F. Karsch, Covariant structure of relativistic gases in equilibrium, *Phys. Rev.* **D24**, 2564 (1981).

Miller D.E. and E. Suhonen, Relativistic ensembles and the mass spectrum, *Phys. Rev.* **D26**, 2944 (1982).

Miller D.E. and P.S. Ray, Thermodynamics of a relativistic Fermi gas in a strong magnetic field, *Helv. Phys. Acta* **57**, 704 (1984).

Miller D.E. and P.S. Ray, Condensation of a charged relativistic Bose gas in the presence of a magnetic field, *Phys. Rev.* **A33**, 1990 (1986).

Miller J.C. and O. Pantano, Hydrodynamics of the cosmological quark–hadron transition, *Phys. Rev.* **D40**, 1789 (1989).

Miller J.C. and O. Pantano, Computations of the growth of bubbles during the cosmological quark–hadron transition, *Phys. Rev.* **D42**, 3334 (1990).

Minich R.W. and L.R. Ram Mohan, Elementary-particle symmetries in relativistic many-body theory at finite temperature and density, *Phys. Rev.* **D19**, 1582 (1979).

Miralles J.A. and K.A. van Riper, *Accurate evaluation of Fermi–Dirac integrals and their derivatives For arbitrary degeneracy and relativity* [arXiv:astro-ph/9509124 v1].

Mofiz U.A. and U. de Angelis, Nonlinear propagation and localization of intense electromagnetic waves in relativistic plasmas, *J. Plasma Phys.* **33**, 107 (1985).

Moiseev S.S., On the distribution function for dissipative processes in a rarefiled relativistic gas, *Sov. Phys. JETP* ZZ, 391 (ZZ).

Morgenstern B. and W. Nörenberg, Dissipative fluid dynamics, *Nucl. Phys.* **A492**, 93 (1989).

Morley P.D., Thermodynamic potential in quantum electrodynamics, *Phys. Rev.* **D17**, 598 (1978).

Morley P.D. and M.B. Kislinger, Relativistic many-body theory, quantum chomodynamics and neutron stars/supernova, *Phys. Rep.* **51**, 63 (1979).

Mornas L., Propriétés de transport de la matière dense relativiste. Thesis, Grenoble (1992).

Mornas L., Collective effects on transport coefficients of relativistic nuclear matter II: BUU collision term. Preprint GSI-93-21 (1993) ZZ.

Mornas L., Transport coefficients of relativistic nuclear and neutron matter with in-medium effects, *Nucl. Phys.* **A573**, 554 (1994).

Mornas L. and K. Morawetz, Relativistic transport coefficients with generalized mass-shell constraints [arXiv:nucl-th/9909073 v2].

Morozov V.G., G. Röpke and A. Holl, Kinetic theory of quantum electrodynamic plasma in a strong electromagnetic field II: The covariant mean-field approximation, *Theor. Math. Phys.* **132**, 1029 ((2002).

Mourad J., The Wigner–Weyl formalism and the relativistic semi-classical approximation, *Phys. Lett.* **A179**, 231 (1993).

Mrowczynski S. and U. Heinz, Towards transport theory of hadron gases, *Ann. Phys.* (*N.Y.*) **169**, 48 (1986).

Mrowczynski S. and U. Heinz, Relativistic transport equations from quantum field theory, Walecka model case, In *Statistical Description of Transport in Plasma, Astro- and Nuclear Physics* (Les Houches) eds. J. Misquich, G. Pelletier and P. Schuck (Nova Science, 1993).

Mrowczynski S. and U. Heinz, Towards a relativistic transport theory of nuclear matter, *Ann. Phys.* (*N.Y.*) **229**, 1 (1994).

Mrowczynski S. and B. Müller, Wigner functional approach to quantum field dynamics, *Phys. Rev.* **D50**, 7542 (1994).

Müller P., Reduzierte Dichten in der relativistischen statistischen Mechanik (Diplomearbeit; Hamburg, 1970).

Namyslowski J.M., Relativistic two-body forces in many-body systems, *Phys. Rev.* **C20**, 2356 (1979).

Naumov N.D., Kinetic theory of a relativistic plasma, *Fizika* **24**, 78 (1981).

Naumov N.D., Statistical mechanics of a relativistic plasma. Moskovskii Universitet, Vestnik, Seriia 3 — *Fiz. Astron.* **22**, 80 (1981).

Naumov N.D., A kinetic theory of relativistic plasma II, *Sov. Phys. J.* **24**, 409 (1981).

Naumov N.D., Particle correlations in a relativistic plasma, *Moscow Univ. Phys. Bull.* **37**, 74 (1982).

Naumov N.D., A relativistic gas in a gravitational field, *Fizika* **25**, 17 (1982).

Naumov N.D., Statistical mechanics of relativistic charges and radiation, *Sov. Phys. J.* **26**, 764 (1983).

Newcomb W.A., Warm relativistic electron fluid, *Phys. Fluids* **25**, 846 (1982).

Newcomb W.A., Warm relativistic electron fluid II: Small m/e limit, *Phys. Fluids* **29**, 881 (1986).

Nicholson D.R. and M.J. Schwartz, Energy conversion between longitudinal and transverse waves by mode–mode coupling in a relativistic plasma, *Plasma Phys.* **16**, 275 (1974).

Nieto M.M., Exact state and fugacity equations for the ideal quantum gases, *J. Math. Phys.* **11**, 1346 (1970).

Nightingale J.D., Independent investigations concerning bulk viscosity in relativistic homogeneous isotropic cosmologies, *Astrophys. J.* **185**, 105 (1973).

Nishimura K., Polynomial expansion method for problems involving a relativistic Marxwellian distribution, *J. Math. Phys.* **12,** 1840 (1971).
Noble R.J., Semi-classical probability distribution function for finite-temperature field theory, *Phys. Rev.* **D23**, 1392 (1981).
Norton R.E., Elementary particle scattering and statistical quasiparticles in quantum statistical mechanics, *Ann. Phys.* (*N.Y.*) **170**, 18 (1986).
Norton R.E. and J.M. Cornwall, On the formalism of relativistic many-body theory, *Ann. Phys.* (*N.Y.*) **91**, 106 (1975).
O'Connell R.F., Magnetic moment of a magnetized electron gas and magnetic field in white dwarfs and neutron stars, *Lett. Nuovo Cimento* **III**, 218 (1970).
O'Connell R.F. and K.M. Roussel, Origin of magnetic field in white dwarfs and neutron stars, *Nature* **231**, 32 (1971).
O'Connell R.F. and K.M. Roussel, On the origin of magnetic field in white dwarfs and neutron stars, *Lett. Nuovo Cimento* **2**, 55 (1971).
O'Connell R.F. and K.M. Roussel, Magnetic properties of degenerate electron gas and implications for metals, white dwarfs, and neutron stars, *Astron. Astrophys.* **18**, 198 (1972).
Olson T.S., Stability and causality in the Israel–Stewart energy frame theory, *Ann. Phys.* (*N.Y.*) **199**, 18 (1990).
Olson T.S. and W.A. Hiscock, Stability, causality and hyperbolicity in Carter's "regular" theory of relativistic heat-conducting fluids, *Phys. Rev.* **D41**, 3687 (1990).
Olson T.S. and W.A. Hiscock, Relativistic dissipative hydrodynamics and the nuclear equation of state, *Phys. Rev.* **C39**, 1818 (1991).
Orlov Yu. N. and I.P. Pavlotsky, BBGKY-hierarchies and Vlasov's equations in postgalilean approximation, *Physica* **A151**, 318 (1988).
Oron O. and L.P. Horwitz, Relativistic Brownian motion. Report No. TAUO 2728-2002.
Oron O. and L.P. Horwitz, Relativistic Brownian motion in 3+1 dimensions [arxiv: math-ph/0312003].
Ortega Blanco L.C. and H. Perez Rojas, Behavior of polarised electromagnetic waves in a relativistic plasma in an intense magnetic field, *Ciencias Tec. Fis. Mat.* **2**, 115 (1983).
Osnes E. and D. Stottman, Causal constraints on the nuclear equation of state, *Phys. Lett.* **B166**, 5 (1986).
Ozernoy L.M. and A. Aharonian, Nuclear gamma-ray lines from CEN A: evidence for their origin in relativistic plasma, *Astrophys. Space Sci.* **66**, 497 (1979).
Paczynski B., Models of X-ray bursters with radius expansion, *Astrophys. J.* **267**, 315 (1983).
Pauli W., Theory of Relativity (1921) (reprinted by Dover, New York, 1981).
Pavlov G.G., Yu. A. Shibanov and Yu. N. Gnedin, Effects of vacuum polarization by a strong magnetic field on the cyclotron radiation of a thermal plasma, *JETP Lett.* **30**, 137 (1979).
Pavlov G.G. and V.G. Bezchastnov, Cyclotron and synchrotron radiation in superstrong magnetic field of neutron stars, *Adv. Space Res.* **8**, 563 (1988).

Pavon D., D. Jou and J. Casas-Vasquez, About the relativistic temperature gradient, *Phys. Lett.* **A78**, 317 (1980).

Pavon D., D. Jou and J. Casas-Vasquez, Heat conduction in relativistic extended thermodynamics, *J. Phys.* **A13**, L77 (1980).

Pavon D., D. Jou and J. Casas-Vasquez, Electric conduction in relativistic thermodynamics, *J. Non-equil. Thermodyn.* **5**, 191 (1980).

Pavon D., D. Jou and J. Casas-Vasquez, On a covariant formulation of dissipative phenomena, *Ann. Inst. Henri. Poincaré* **A36**, 79 (1982).

Pavon D., D. Jou and J. Casas-Vasquez, Equilibrium and non-equilibrium fluctuations in relativistic fluids, *J. Phys.* **A16**, 775 (1983).

Peacock J.A., Fermi acceleration by relativistic shock waves, *Mon. Not. R. Astron. Soc.* **196**, 135 (1981).

Pegoraro F. and F. Porcelli, Equation of state for relativistic plasma waves, *Phys. Fluids* **27**, 1665 (1984).

Pengfei Z. and U. Heinz, Relativistic kinetic equations for electromagnetic, scalar and pseudo-scalar interactions, *Phys. Rev.* **D53**, 2096 (1996).

Perez Rojas H. and A.E. Shabad, Thermodynamic potential in electron–positron system in a magnetic field, *Sov. Phys.–Lebedev Inst.* **7**, 13 (1976).

Perez Rojas H., One-loop calculations for polarization operator in a strong magnetic field: finite temperature and chemical potential. Preprint *Acad. Cien. Cuba 71* (1978).

Perez Rojas H., Polarization of a relativistic electron and positron gas in a strong magnetic field: propagation of electromagnetic waves. Preprint *Acad. Cien. Cuba No. 70* (1978).

Perez Rojas H., Quantum statistics of an electron gas in an external constant uniform magnetic field I: the Green's function and thermodynamic relations. Preprint *Acad. Cien. Cuba No.* 1 (1978).

Perez Rojas H., Potencial termodinamico de un gas electr'on–positron en un campo magnetico constante y uniforme, *Ciencias Tec. Fis. Mat. No.* 1, **59** (1981).

Perez Rojas H. and A.E. Shabad, Absorption and dispersion of electromagnetic eigenwaves of electron–positron plasma in a strong magnetic field, *Ann. Phys. (N.Y.)* **138**, 1 (1982).

Perez Rojas H., Interaction of the electron–positron field with a uniform, constant magnetic field, *Ciencias Tec. Fis. Mat. No.* 2, **93** (1983).

Perez Rojas H., Electroweak plasma in a magnetic field. Preprint GFT-IMACC-1-1985.

Perez Rojas H., Electroweak plasma in a magnetic field, *Acta Phys. Pol.* **B17**, 861 (1986).

Perez Rojas H., Explicit evaluation of the thermodynamical potential for a gas of charged bosons at high temperatures, *Ciencias Tec. Fis. Mat. No.* 7, **60** (1987).

Perez Rojas H. and R. Torres Rivero, Bose–Einstein condensation with increasing temperature? *Phys. Lett.* **A137**, 13 (1989).

Perez Rojas H. and R. Torres Rivero, Bose–Einstein condensation may occur in a constant magnetic field, *Phys. Lett.* **B379**, 148 (1996).

Perez Rojas H. and L. Villegas-Lelovski, Bose–Einstein condensation in a constant magnetic field, *Braz. J. Phys.* **30**, 410 (2000).

Persson D. and V. Zeitlin, A note on QED with magnetic field and chemical potential, *Phys. Rev.* **D51**, 2026 (1995).

Pham M.Q., Aspect géométrique de l'équation de Boltzmann relativiste, *C. R. Acad. Sci.* **263**, 106 (1966).

Pham M.Q., Sur une théorie relativiste des fluides thermodynamiques. Thesis, Paris (1954).

Pham M.Q., ZZ, *C. R. Acad. Sci.* **236**, 2299 (1953).

Pham M.Q., ZZ, *C. R. Acad. Sci.* **237**, 22 (1953).

Pham M.Q., ZZ, *C. R. Acad. Sci.* **238**, 324 (1954).

Pichon G., Les varétés caractéristiques des fluides visqueux en relativité générale, *C. R. Acad. Sci.* **256**, 1675 (1963).

Pichon G., Diffusion du rotationnel dans les fluides visqueux incompressibles, *C. R. Acad. Sci.* **259**, 2363 (1964).

Pichon G., Etude relativiste des fluides visqueux et chargés. Thesis, Paris (1965).

Pichon G., Equation de Boltzmann relativiste et applications. Preprint (février 1966).

Pichon G., Calcul et utilisation de la fonction de distribution d'un fluide parfait, *C. R. Acad. Sci.* **A264**, 544 (1967).

Pichon G., Problèmes mathématiques de l'équation de Boltzmann relativiste, *Commun. Math. Phys.* **19**, 189 (1970).

Pichon B., Numerical calculation of the generalized Fermi–Dirac integrals, *Comput. Phys. Commun.* **55**, 127 (1989).

Pina E. and R. Balescu, Some aspects of relativistic statistical mechanics, *Acta Phys. Austriaca* **28**, 309 (1968).

Pirk K. and G. Börner, Wigner functions and explicit computation of quantum corrections in Robertson–Walker spacetimes, *Class. Quantum Grav.* **6**, 1855 (1989).

Plinio J.B. and D. Gerbal, Dispersion relations for zero-helicity gravitational waves in a relativistic expanding medium, *J. Phys.* **A75**, 3351 (1982).

Ploszaczak M. and M.J. Rhoades-Brown, Approximation scheme for the quantum Liouville equation using phase-space distribution functions, *Phys. Rev. Lett.* **55**, 147 (1985).

PloszaJczak M. and M.J. Rhoades-Brown, Nonequilibrium truncation scheme for the statistical mechanics of relativistic matter, *Phys. Rev.* **D33**, 3686 (1986).

Ploszaczak M. and M.J. Rhoades-Brown, Self-consistent truncation scheme for the quantum kinetic theory of relativistic nuclear matter, In *Physics and High-Energy Heavy-Ion Reactions*, eds. M. Kutschera and M. Ploszaczak (World Scientific, Singapore, 1987).

Polak P.H., W.A. van Leeuwen and S.R. de Groot, On relativistic kinetic gas theory X: Transport coefficients in the intermediate relativistic regime values for special models, *Physica* **66**, 455 (1973).

Prakash M., T.L. Ainsworth, J.P. Blaisot and H. Wolter, Features of hot dense nuclear matter, In *Symp. Unified Concepts of Many-Body Problems* (Stony Brook), eds. T.T.S. Kuo and J. Speth (North-Holland, Amsterdam, 1987).

Posilicano A. Poincaré invariant Markov processes and Gaussian random fields on relativistic phase space, *Letts. Math. Phys.* **42**, 85 (1997).

Prigogine I., Some aspects of relativistic statistical mechanics, In *Statistical Mechanics of Equilibrium and Non-equilibrium*, ed. J. Meixner (North-Holland, Amsterdam, 1965).

Prigogine I. and P. de Gottal, Relativistic effects in statistical hydrodynamics, *Physica* **31**, 677 (1965).

Prozorkevich A.V. and S.A. Smolyanskii, Derivation of relativistic transport equations of a plasma in a strong electromagnetic field, *Theor. Math. Phys.* **23**, 608 (1975).

Prozorkevich A.V. and S.A. Smolyanskii, Theory of quantum relativistic kinetic equations, *Theor. Math. Phys.* **28**, 773 (1976).

Prozorkevich A.V. and S.A. Smolyanskii, Covariant formalism in the theory of quantum relativistic kinetic equations, *Theor. Math. Phys.* **31**, 357 (1977).

Prozorkevich A.V. and S.A. Smolyanskii, A generalized kinetic equation and its application to models of relativistic nuclear dynamics (1995).

Prozorkevich A.V., S.A. Smolyanskii and V.D. Toneev, Kinetic description of the baryon subsystem of nuclear matter at intermediate energies, *Phys. Atom. Nucl.* **59**, 766 (1996).

Pulsifer P. and G. Kalman, Pair-creation collective modes in an electron gas, *Phys. Rev.* **A45**, 5820 (1992).

Ramanathan R., Relativistic diffusion equation (1986).

Ray J.R. and J.C. Zimmerman, General-relativistic kinetic theory of the Robertson–Walker and Petrov G-4 VIII cosmologies, *Nuovo Cimento* **B42**, 183 (1977).

Ray P.S., Energy loss of relativistic electrons in plasma in accordance with covariant Fokker–Planck equation, *J. Plasma Phys.* **24**, 75 (1980).

Redkoborodyi Yu. N., Quantum theory of the screening effect on thermonuclear reactions I: Relativistic electron plasma, *Astrophysika* **12**, 495 (1976).

Redlich K. and L. Turko, Phase transitions in the statistical bootstrap model with an internal symmetry, *Z. Phys.* **C5**, 201 (1980).

Remler E.A., Use of the Wigner representation in scattering problems, *Ann. Phys. (N.Y.)* **95**, 455 (1975).

Remler E.A., Some connections between relativistic classical mechanics, statistical mechanics and quantum field theory, *Phys. Rev.* **D16**, 3464 (1977).

Remler E.A., Particle propagation in quasiparticle models of relativistic heavy-ion collisions, *Phys. Rev. Lett.* **54**, 989 (1985).

Robinson P.A., Relativistic plasma dispersion functions, *J. Math. Phys.* **27**, 1206 (1986).

Robnik M., K.D. Fritz and E. Schrufer, Propagation of relativistic plasma waves: global nonlinear aspects, *Astron. Astrophys.* **165**, 254 (1986).

Roth G., Oscillations of a relativistic electron plasma in an external magnetic field, *Acta Phys.* **27**, 309 (1969).

Rudberg H., On the theory of relativistic diffusion (Almqvist & Wicksells Boktryckeri AB, Uppqala, 1957).

Ruderman M., Causes of sound faster than light in classical models of ultradense matter, *Phys. Rev.* **172**, 1286 (1968).
Ruderman M., Matter in superstrong magnetic fields, In *IAU Symp. 53 — The Physics of Dense Matter* (1972).
Rukhadze A.A. and V.P. Silin, On the magnetic susceptibility of a relativistic electron gas, *Sov. Phys. JETP* **11**, 610 (1960).
Russo G. and A.M. Anile, Stability properties of relativistic shock waves: basic results, *Phys. Fluids* **30**, 2406 (1968).
Rylov Yu. A., Quantum mechanics as a theory of relativistic Brownian motion, *Ann. Phys. (Leipzig)* **27**, 1 (1971).
Saad D., L.P. Horwitz and R.I. Arshansky, Off-shell electromagnetism in manifestly covariant relativistic quantum mechanics. Preprint TAUP 1665-88 (1989).
Sabbata V. and M. Gasperini, On the possibility of speed higher than c inside hadronic matter, *Lett. Nuovo Cimento* **34**, 337 (1982).
Sachs R.K. and J. Ehlers, Kinetic theory and cosmology, In *Brandeis University Summer Institute in Theoretical Physics, 1968: Astrophysics and General Relativity*, eds. M. Chrétien, S. Deser and J. Goldstein (Gordon and Breach, New York, 1971), pp. 2, 333.
Sakai J. and T. Kawata, Nonlinear Alfvén wave in an ultra-relativistic electron–positron plasma, *J. Phys. Soc. Jpn.* **49**, 753 (1980).
Salasnich L., Particles and anti-particles in a relativistic Bose condensate [arXiv:math-phys/0207030 v1].
Santamato E., Gauge-invariant statistical mechanics and average action principle for the Klein–Gordon particle in geometric quantum mechanics, *Phys. Rev.* **D32**, 2615 (1985).
Santilli R.M., Causality and relativistic plasmas, In *Relativistic Plasmas — The Coral Gable Conference*, eds. O. Buneman and W.B. Pardo (W. Benjamin, New York, 1968).
Santini F. and G. Szamosi, Linear oscillations in relativistic plasma, *Nuovo Cimento* **37**, 685 (1965).
Sasaki M., On the relativistic gases, In *Max Planck Festschrift 1958*, eds. B. Kockel, W. Macke and A. Papapetrou (VEB Deutscher Verlag der Wissenschaften, Berlin, 1958).
Schaaf M.E., Longitudinal and transverse heat conduction in quantizing magnetic fields of neutron stars, *Astron. Astrophys.* **205**, 335 (1988).
Schaeffer J., The classical limit of the relativistic Vlasov–Maxwell system, *Commun. Math. Phys.* **104**, 403 (1986).
Scheidegger A.E. and C.D. McKay, Quantum statistics of fields, *Phys. Rev.* **83**, 125 (1951).
Scheidegger A.E. and R.V. Krotkov, Relativistic statistical thermodynamics, *Phys. Rev.* **89**, 1096 (1952).
Schimd L.A., Heat transfer in relativistic charged-fluid flow, *Nuovo Cimento* **47**, 1 (1967).
Schlickeiser, R., Relativistic kinetic theory of plasma waves, *Phys. Scripta* **T75**, 33 (1998).

Schmidt-Burgk J., Temperature aspects of Landau orbital ferromagnetism in white dwarfs and neutron stars, *Astron. Astrophys.* **26**, 335 (1973).
Schneidermann E. and U. Heinz, Relativistic hydrodynamics in a global fashion, *Phys. Rev.* **C47**, 1738 (1993).
Schnitzer H.J., Hartree approximation in relativistic field theory, *Phys. Rev.* **D10**, 2041 (1974).
Schulman G.A., Neutronization at cold hydrogen in the presence of extra powerful magnetic fields, *Astrofiz.* **10**, 542 (1974).
Schulman G.A., *Astrofiz.* **11**, 89 (1975).
Schulman G.A., *Sov. Astron.* **19**, 698 (1976).
Schulman G.A., *Sov. Astron.* **20**, 425 (1976).
Schulman G.A., *Sov. Astron.* **20**, 689 (1976).
Schulman G.A., *Sov. Astron.* **21**, 590 (1977).
Schulman G.A., *Astrofiz.* **13**, 657 (1977).
Schulman G.A., On the physical conditions in the interiors of superdense, magnetized, astrophysical objects, *Sov. Astron.* **23**, 197 (1979).
Schwartz D.J., Exact solutions for cosmological perturbations with collisionless matter, In *Proc. Birth of the Universe and Fondamental Physics* (Roma, 1994).
Schweizer M.A., Relativistic radiative hydrodynamics, *Ann. Phys.* (*N.Y.*) **183**, 80 (1988).
Selikhov A.V., Transformation of covariant quark Wigner operator to noncovariant one. Preprint IAE-4725/1 (1988).
Selikhov A.V. and M.V. Gyulassy, QCD Fokker–Planck equations with color diffusion, *Phys. Rev.* **C49**, 1726 (1994).
Serot B.D., Pion dynamics in the relativistic nuclear many-body problem, In *Workshop on the Interaction Between Medium Energy Nucleons in Nuclei* (Bloomington, USA, 1982).
Serot B.D., Quantum hadrodynamics and nuclear matter, In *Conf. Intersections Between Particle and Nuclear Physics* (Steamboat Springs, USA, 1984).
Serot, B.D. and J.D. Walecka, The relativistic nuclear many-body problem, In *Advances in Nuclear Physics*, Vol. 16, eds. J.W. Negele and E. Vogt (Plenum, New York, 1986).
Serot B.D., A relativistic theory of nuclear matter, In *Nuclear Matter and Heavy Ion Collisions*, eds. M. Soyeur, H. Flocard, B. Tamain and M. Porneuf (Plenum, New York, 1989).
Serot B.D., Quantum hadrodynamics, *Rep. Prog. Phys.* **55**, 1855 (1992).
Serot B.D., Covariant calculations of finite-temperature nuclear matter, In *Relativistic Many-Body Physics*, eds. B.C. Clark, R.J. Perry and J.P. Vary (World Scientific, Singapore, 1989).
Service A.T., Fitting formulae for the equation of state of a perfect, semirelativistic Boltzmann gas, *Astrophys. J.* **307**, 60 (1986).
Shikin I.S., Wave theory in Chew–Goldberger–Low relativistic magnetohydrodynamics, *Sov. J. Plasma Phys.* **2**, 13 (1976).
Shin G.R., I. Bialynicki-Birula and J. Rafelski, Wigner functions for relativistic spin 1/2 particles, *Phys. Rev.* **A46**, 645 (1992).

Shivamoggi B.K., Relativistic modulational instability of an electromagnetic wave in a magnetized plasma, *Phys. Rev.* **A31**, 1728 (1985).

Shivamoggi B.K., Relativistic nonlinear waves in a magnetic plasma, *Czech. J. Phys.* **B39**, 1007 (1989).

Shukla P.K., R. Bharuthram and N.L. Tsintsadze, Fully relativistic filamentation instability of strong electromagnetic radiation in unmagnetized plasmas, *Phys. Scripta* **38**, 578 (1988).

Shun-Jin W., B.-A. Li, W. Bauer and J. Randrup, Relativistic transport theory for hadronic matter, *Ann. Phys. (N.Y.)* **209**, 251 (1991).

Siemens P.J., Relativistic transport theory of fluctuating fields for hadrons, In *Nuclear Matter and Heavy Ion Collisions*, eds. M. Soyeur, H. Flocard, B. Tamain and M. Porneuf (Les Houches, 1989) (Plenum, New York, 1989).

Siemens P.J., M. Soyeur, G.D. White, L.J. Lantto and K.T.R. Davies, Relativistic transport theory of fluctuating fields for hadrons, *Phys. Rev.* **C40**, 2641 (1989).

Silin V.P., On the electromagnetic properties of a relativistic plasma, *Sov. Phys. JETP* **11**, 1136 (1960).

Silin V.P., Collision integral for charged particles, *Physica* **A81**, 597 (1975).

Silin V.P. and V.N. Ursov, Kinetic theory of the quark–gluon plasma, *Sov. Phys. Dokl.* **30**, 594 (1985).

Signore M., Sur les conditions supplémentaires en théorie cinétique relativiste, *C.R. Acad. Sci.* **274**, 1131 (1972).

Similton P.L., Conservation laws for relativistic guiding-center plasma, *Phys. Lett.* **A112**, 33 (1985).

Singh S. and R.K. Pathria, Bose–Einstein condensation in an Einstein universe, *J. Phys.* **A17**, 2983 (1984).

Sinyukov Yu. M., On the relativistic statistical physics and thermodynamics of ideal gases. Preprint ITP-82-94E (1982).

Sinyukov Yu. M., On the relativistic partition function of ideal gases, *Phys. Lett.* **B127**, 443 (1983).

Sinyukov Yu. M., On the relativistic partition function of ideal gases, *Phys. Lett.* **B127**, 443 (1983).

Sisir Roy, Relativistic Brownian motion and the space–time approach to quantum mechanics, *J. Math. Phys.* **21**, 71 (1980).

Siskens Th. J. and Ch. G. van Weert, Relativistic kinetic theory of quantum systems IV: Transport equation for the neutrino component of a neutrino/antineutrino mixture, *Physica* **A86**, 80 (1977).

Siskens Th. J. and Ch. G. van Weert, Relativistic kinetic theory of quantum systems V: Transport equations for the massive components of a neutrino/antineutrino mixture, *Physica* **A87**, 369 (1977).

Siskens Th. J. and Ch. G. van Weert, Transition amplitudes and probabilities for the Weinberg–Salam model, *Physica* **A89**, 163 (1977).

Siskens Th. J., Ch. G. van Weert and W.P.H. de Boer, Influence of the spin and the Weinberg angle on the bulk viscosity of a neutrino–antineutrino mixture, *Phys. Lett.* **A61**, 167 (1977).

Siskens Th. J. and L.J. van den Horn, Transport coefficients for the cosmological fluid during the lepton era, *Phys. Lett.* **A64**, 489 (1978).

Siskens Th. J. and L.J. van den Horn, Weak neutral currents and neutrino transport, *Phys. Lett.* **B76**, 374 (1978).

Siskens Th. J. and A.J. Kox, Recent results in relativistic quantum kinetic theory, *Statphys.* **13**, 53 (1978).

Siskens Th. J., Ch. G. van Weert, W.P.H. de Boer and A.J. Kox, Recent results in relativistic quantum kinetic theory, *Statphys.* **13**, 704 (1978).

Siskens Th. J. and Ch. G. van Weert, Relativistic kinetic theory of quantum systems VI: Bulk viscosity of a reactive ZZ, *Physica* **A91**, 303 (1978).

Sivak H.D., Fluctuations in the relativistic quantum plasma, *Ann. Phys.* (*N.Y.*), **159**, 351 (1985).

Sivak H.D., Friedel oscillations in a relativistic quantum plasma, *Physica* **A129**, 408 (1985).

Sivak H.D., Electromagnetic perturbations of the relativistic quantum plasma, *Phys. Rev.* **A34**, 653 (1986).

Skjervold J.E. and E. Ostgaard, Matter in supersrong magnetic fields, *Can. J. Phys.* **64**, 256 (1986).

Smolyansky, S.A., A.V. Prozorkevich, S. Schmidt, D. Blaschke, G. Roepke and V.D. Toneev, Relativistic quantum kinetic equation of the Vlasov type for systems with internal degrees of freedom [arXiv:nucl-th/9709057].

Sofel M., U. Heinz, B. Müller and W. Greiner, A self-consistent solution of the coupled Dirac–Einstein equations in the mean field approximation, *Lett. Nuovo Cimento* **22**, 391 (1978).

Song C., Dense nuclear matter: Landau Fermi-liquid theory and chiral Lagrangian with scaling, *Phys. Rep.* **347**, 289 (2001).

Sorge H., H. Stocker and W. Greiner, Poincaré invariant Hamiltonian dynamics: modelling multi-hadronic interactions in a phase space approach, *Ann. Phys.* (*N.Y.*) **192**, 266 (1989).

Standen G.B. and D.J. Toms, Bose–Einstein condensation of the magnetized ideal Bose gas [arXiv:cond-mat/9705180 v2].

Stewart J.M., Fundamental problems in relativistic kinetic theory, In *Colloques Internationaux du Centre National de la Recherche Scientifique no. 236. Théories cinétiques classiques et relativistes* (Editions du Centre National de la Recherche Scientifique, Paris, 1975), p. 151.

Stewart J.M., Neutrino viscosity in cosmological models, *Astrophys. Lett.* **2**, 133 (1968).

Stewart J.M., Non-equilibrium processes in the early universe, *Mon. Not. R. Astron. Soc.* **145**, 347 (1969).

Stewart J.M., M.A.H. MacCallum and D.W. Sciama, Thermodynamics and cosmology, *Comments Astrophys. Space Phys.* **11**, 206 (1970).

Stewart J.M., Non-equilibrium Relativistic Kinetic Theory (Springer, Berlin, 1971).

Stewart J.M., Perturbations in an expanding universe of free particles, *Astrophys. J.* **176**, 323 (1972).

Stewart J.M., Relativistic thermodynamics and kinetic theory, with applications to cosmology, In *Cargèse Lectures in Physics*, Vol. 6, ed. E. Schatzman (Gordon and Breach, New York, 1973), p. 165.

Stewart J.M., On transient relativistic thermodynamics and kinetic theory, *Proc. R. Soc.* **A357**, 59 (1977).

Stratonovich R.L., *Sov. Phy. Dokl.* **1**, 72 (1956).

Strumai A., Damping of relativistic shocks in an expanding universe, *Nuovo Cimento* **B92**, 91 (1986).

Sugawara H., Relativistic collective motion, *Phys. Rev.* **D12**, 3272 (1975).

Sutapa G., S. Mandal and S. Chakrabarty, Relativistic version of Landau theory of Fermi liquid in presence of strong quantizing magnetic field — an exact formalism, *Ann. Phys. (N.Y.)* **312**, 398 (2004).

Synge J.L., The energy tensor of continuous medium, *Proc. R. Soc. Can.* **28**, 127 (1934).

Synge J.L., The Relativistic Gas (North-Holland, Amsterdam, 1957).

Szekeres P. and A.S. Barnes, Cosmologies with a Synge gas and radiation, *Mon. Not. R. Soc.* **189**, 767 (1979).

Szell K., ZZ, Hungar. *Acta Phys.* **1**, 19 (1949).

Takahara F. and M. Kusunose, Equilibrium pair density in a relativistic plasma with magnetic fields, In *Workshop on Electron–Positron Pairs in Astrophysics* (AIP Conf. Proc., Greenbelt, 1983).

Takahashi Y. and H. Umezawa, Thermo field dynamics, *Coll. Phenom.* **2**, 55 (1975).

Tam K.-K., A covariant treatment of relativistic plasma oscillations, *Can. J. Phys.* **46**, 1763 (1968).

Tam K.-K., J. O'Hanlon and K.-K. Tam, Radiation damping of relativistic plasma oscillations, *Nuovo Cimento* **LXIIIB**, 241 (1969).

Tam K.-K. and D. Kiang, The relativistic Boltzmann equation and the equations of magnetohydrodynamics with radiative reaction, *Prog. Theor. Phys.* **62**, 1245 (1979).

Tang C.L. and J. Meixner, Relativistic theory of the propagation of plane electromagnetic waves, *Phys. Fluids* **4**, 148 (1961).

Taub A.H., Relativistic Rankine–Hugoniot equations, *Phys. Rev.* **74** , 328 (1948).

Taub A.H., General relativistic variational principle for perfect fluids, *Phys. Rev.* **94**, 1468 (1954).

Tauber G.E. and J.W. Weinberg, Internal state of a gravitating gas, *Phys. Rev.* **122**, 1342 (1961).

Tauber G.E., On a general relativistic statistical mechanics of interacting particles, In *Modern Developments in Thermodynamics*, ed. B. Gal-Or (Israel Universities Press and Wiley, New York, 1974), p. 247.

Tauber G.E., Relativistic magnetohydrodynamics in a conformal flat space. Lecture given at Second Advanced Seminar of International School of Relativistic Astrophysics (1976). Preprint TAUP 565–76.

Tauber G.E., Plasma kinetic equation in an expanding universe. Preprint TAUP 1004-82.

Tezuka H., Instability in relativistic nuclear matter, *Phys. Rev.* **C22**, 2585 (1980).

Thacker H.B., Statistical mechanics of a (1+1)-dimensional quantum field theory at finite density and temperature, *Phys. Rev.* **D16**, 2515 (1977).
Theimer O. and E.H. Dirk, Relativistic corrections for the conventional, classical Nyquist theorem, *Nuovo Cimento* **B76**, 73 (1983).
Theis J., G. Graebner, G. Buchwald, J. Marhun, W. Greiner, H. Stocker and J. Polonyi, Phase transition of the nucleon–antinucleon plasma in a relativistic mean field theory, *Phys. Rev.* **D28**, 2286 (1983).
Thomas L.H., The radiation field in a fluid in motion, *Quart. J. Math., Oxford Ser.* **1**, 239 (1930).
Thompson W.B., The self-consistent test-particle approach to relativistic kinetic theory, In *General Relativity — Papers in Honour of J.L. Synge*, ed. O'Raifeartaigh, Clarendon, Oxford, (1972).
Thorne K.S., Relativistic radiative transfer: moment formalism, *Mon. Not. R. Astron. Soc.* **194**, 439 (1981).
Titeica S., O complementare statistica a teoriei electronice clasice, *Studii si Cercetari de fizica* **7**, 7 (1956).
Toms D.J., Bose–Einstein condensation of a charged relativistic ideal gas in a general homogeneous magnetic field, *Astrophys. J.* **D50**, 6457 (1994).
Toms D.J., Weak field superconductivity for relativistic charged gases at high temperature, *Phys. Rev.* **D51**, 1895 (1995).
Tooper R.F., Entropy of a relativistic electron gas, *Astrophys. J.* **157**, 1391 (1969).
Tooper R.F., On the equation of state of a relativistic Fermi–Dirac gas at high temperature, *Phys. Rev.* **156**, 1075 (1969).
Touscheck B., Covariant statistical mechanics, *Nuovo Cimento* **B58**, 295 (1968).
Touscheck B., On the measurement of temperature in covariant statistical mechanics, *Symp. Math.* **12**, 35 (1973).
Treciokas R. and G.F.R. Ellis, Isotropic solutions of the Einstein–Liouville equations, *Commun. Math. Phys.* **8**, 1344 (1968).
Treciokas R. and G.F.R. Ellis, Isotropic solutions of the Einstein–Boltzmann equations, *Commun. Math. Phys.* **23**, 1 (1971).
Treciokas R. and G.F.R. Ellis, Anisotropic solutions of the Einstein–Boltzmann equations. Preprint (1980).
Tretyak V., On relativistic models in the equilibrium statistical mechanics, *Cond. Matter Phys.* **1**, 553 (1998).
Tsintsadze N.L. and E.G. Tsikarishvili, Parametric instabilities in relativistic plasma, *Astrophys. Space Sci.* **39**, 191 (1976).
Tsytovich V.N., Spatial dispersion in a relativistic plasma, *Sov. Phys. JETP* 13, 1249 (1961).
Tsytovich V.N., Quantum fluctuations and relativistic radiative quantum kinetics of charged particles, In *Int. Conf. Plasma Physics* (Kiev, 1987), ed. A.G. Sitenko (World Scientific, Singapore, 1987).
Tsytovich V.N., Radiative corrections to the Landau damping of intense plasma waves, *Sov. Phys. JETP* **66**, 959 (1987).
Turko L., Quantum gases with internal symmetry, *Phys. Lett.* **B104**, 153 (1981).
Turko L., Quantum gases with internal symmetry revisited — condensation of hadronic matter, *Eur. J. Phys.* **61**, 297 (1994).

Twareque Ali S. and G. Emch, Covariant conditional expectations in quantum probability theory. Preprint ICTP IC/74/72 (1974).

Tznanakis C., On the validity of Liouville's theorem in general relativity, *Lett. Nuovo Cimento* **38**, 606 (1983).

Udey N. and W. Israel, General relativistic radiative transfer: the 14-moment approximation, *Mon. Not. R. Astron. Soc.* **199**, 1137 (1981).

Uechi H., Fermi-liquid properties of nuclear matter in a Dirac–Hartree–Fock approximation, *Nucl. Phys.* **A501**, 813 (1989).

Urban P., ZZ, *S.B. Akad. Wiss. Wien* **152**, 111 (1943).

Vainer B.V. and P.D. Nasel'skii, Quasilinear relaxation of the distribution function for collisionless relativistic gravitating particles in an expanding universe, *Sov. Astron.* **28**, 264 (1984).

van Dantzig D., On relativistic gas theory, *Proc. Sect. Sci. Kon. Nederl. Akad. Wetensch.* **42**, 608 (1939).

van Dantzig D., On relativistic thermodynamics, *Proc. Sect. Sci. Kon. Nederl. Akad. Wetensch.* **42**, 601 (1939).

van Dantzig D., On the phenomenological thermodynamics of moving matter, *Physica* **6**, 673 (1939).

van Dantzig D., Stress tensor and particle density in special relativity theory, *Nature* **143**, 855 (1939).

van den Horn L.J. and Th. J. Siskens, Relativistic kinetic theory of quantum systems VII: Transport coefficients for (ene) and (enm) mixtures, *Physica* **A95**, 141 (1979).

van den Horn L.J. and Ch. G. van Weert, Bulk viscosity of a reactive ν system, *Phys. Lett.* **A84**, 226 (1981).

van den Horn L.J. and Ch. G. van Weert, Transport properties of degenerate neutrinos and photons, *Phys. Lett.* **B101**, 447 (1981).

van den Horn L.J. and Ch. G. van Weert, Effects of neutrino degeneracy on the transport properties of presupernovae, *Phys. Lett.* **A83**, 88 (1981).

van den Horn L.J. and Ch. G. van Weert, Kinetic theory of neutrinos, *Z. Phys.* **C26**, 53 (1984).

van den Horn L.J. and Ch. G. van Weert, Transport properties of neutrinos in stellar collapse II: Shear viscosity, heat conduction and diffusion, *Astron. Astrophys.* **136**, 74 (1984).

van Eijck M.A., R. Kobes and Ch. G. van Weert, Transformations of real-time finite-temperature Feynman rules, *Phys. Rev.* **D50**, 4097 (1994).

van Erkelens H. and W.A. Leeuwen, Chapman–Enskog method for a relativistic ionized gas, *Phys. Lett.* **A57**, 197 (1976).

van Erkelens H. and W.A. Leeuwen, Relativistic Boltzmann-theory for a plasma I: The entropy production, *Physica* **A89**, 113 (1977).

van Erkelens H. and W.A. Leeuwen, Relativistic Boltzmann-theory for a plasma II: Reciprocal relations, *Physica* **A89**, 225 (1977).

van Erkelens H. and W.A. Leeuwen, Relativistic Boltzmann-theory for a plasma III: Viscous phenomena, *Physica* **A90**, 97 (1978).

van Erkelens H. and W.A. Leeuwen, Relativistic Boltzmann-theory for a plasma IV: A variational method, *Physica* **A91**, 88 (1978).

van Erkelens H. and W.A. Leeuwen, Relativistic Boltzmann-theory for a plasma V: Reduction of the collision integrals, *Physica* **A101**, 205 (1980).

van Erkelens H. and W.A. Leeuwen, Relativistic Boltzmann-theory for a plasma VI: The relativistic Landau equation, *Physica* **A107**, 48 (1981).

van Erkelens H. and W.A. Leeuwen, Relativistic Boltzmann-theory for a plasma VII: Propagation of electromagnetic waves, *Physica* **A111**, 462 (1982).

van Erkelens H. and W.A. Leeuwen, Relativistic Boltzmann-theory for a plasma VIII: Energy transport in hydromagnetic waves, *Physica* **A112**, 331 (1982).

van Erkelens H. and W.A. Leeuwen, Relativistic Boltzmann-theory for a plasma IX: Attenuation of hydromagnetic waves, *Physica* **A116**, 499 (1982).

van Erkelens H. and W.A. Leeuwen, Relativistic Boltzmann-theory for a plasma X: Electrical conduction of the cosmological fluid, *Physica* **A123**, 72 (1984).

van Kampen N.G., Lorentz-invariance of the distribution function in phase space, *Physica* **43**, 244 (1969).

van Kampen N.G., Relativistic thermodynamics, *J. Phys. Jpn.* **26**(Suppl.), 316 (1969).

van Kampen N.G., A model for relativistic heat transport, *Physica* **46**, 315 (1970).

van Leeuwen W.A. and S.R. de Groot, On relativistic kinetic gas theory IV: The linearized transport equation and its solution for a mixture of relativistic Maxwellian molecules, *Physica* **51**, 1 (1971).

van Leeuwen W.A. and S.R. de Groot, On relativistic kinetic gas theory V: The viscosity for a gas of relativistic Maxwellian molecules, *Physica* **51**, 16 (1971).

van Leeuwen W.A. and S.R. de Groot, On relativistic kinetic gas theory VI: The coefficients of heat conductivity, diffusion and thermal diffusion for a binary mixture of relativistic Maxwellian molecules, *Physica* **51**, 32 (1971).

van Leeuwen W.A., On the relativistic kinetic theory of transport phenomena I: General expressions for the transport coefficients, *Proc. Kon. Nederl. Akad. Wetensch.* **B74**, 122 (1971).

van Leeuwen W.A., On the relativistic kinetic theory of transport phenomena II(A), *Proc. Kon. Nederl. Akad. Wetensch.* **B74**, 139 (1971).

van Leeuwen W.A., P.H. Polak and S.R. de Groot, Transport coefficients calculated by means of a relativistic extension of Enskog's method, *Phys. Lett.* **A37**, 323 (1971).

van Leeuwen W.A., On the relativistic kinetic theory of transport phenomena II(B), *Proc. Kon. Nederl. Akad. Wetensch.* **B74**, 150 (1971).

van Leeuwen W.A., On the relativistic kinetic theory of transport phenomena III: The solution of the linearized transport equations for Maxwellian particles, *Proc. Kon. Nederl. Akad. Wetensch.* **B74**, 269 (1971).

van Leeuwen W.A., On the relativistic kinetic theory of transport phenomena IV: The calculation of the transport coefficients for Maxwellian particles, *Proc. Kon. Nederl. Akad. Wetensch.* **B74**, 276 (1971).

van Leeuwen W.A., P.H. Polak and S.R. de Groot, Coefficients de transport calculés à l'aide d'une extension relativiste de la méthode d'Enskog, *C. R. Acad. Sci.* **274**, 431 (1972).

van Leeuwen W.A. and S.R. de Groot, Transport coefficients of a neutrino gas, *Lett. Nuovo Cimento* **6**, 470 (1973).

van Leeuwen W.A., P.H. Polak and S.R. de Groot, On relativistic kinetic gas theory IX: Transport coefficients for systems of particles with arbitrary interaction, *Physica* **63**, 65 (1973).

van Leeuwen W.A., P.H. Meltzer and S.R. de Groot, Coefficients de transport d'un gaz de neutrinos, *C. R. Acad. Sci.* **B279**, 45 (1974).

van Leeuwen W.A., A.J. Kox and S.R. de Groot, Kinetic theory of transport coefficients of a relativistic binary mixture, *Phys. Lett.* **A47**, 31 (1974).

van Leeuwen W.A., A.J. Kox and S.R. de Groot, On relativistic kinetic gas theory XIV: Transport coefficients of a binary mixture, general expressions, *Physica* **A79**, 233 (1975).

van Leeuwen W.A., On relativistic kinetic gas theory XV: The Ritz method in relativistic Boltzmann theory, *Physica* **A81**, 249 (1975).

van Weert Ch. G., On the relativistic kinetic theory of particles with a magnetic dipole moment in an external electro-magnetic field I: The relativistic transport equation for a simple gas of magnetic dipole particles, *Proc. Kon. Nederl. Akad. Wetensch.* **B73**, 381 (1970).

van Weert Ch. G., On the relativistic kinetic theory of particles with a magnetic dipole moment in an external electromagnetic field II: Conservation laws and equilibrium properties for a gas mixture magnetic dipole particles, *Proc. Kon. Nederl. Akad. Wetensch.* **B73**, 397 (1970).

van Weert Ch. G., On the relativistic kinetic theory of particles with a magnetic dipole moment in an external electromagnetic field III: The first Chapman–Enskog approximation, *Proc. Kon. Nederl. Akad. Wetensch.* **B73**, 500 (1970).

van Weert Ch. G., On the relativistic kinetic theory of particles with a magnetic dipole moment in an external electromagnetic field IV: Transport coefficients and space-time invariance principles, *Proc. Kon. Nederl. Akad. Wetensch.* **B73**, 517 (1970).

van Weert Ch. G., P.C. de Jagher and S.R. de Groot, On relativistic kinetic gas theory XI: Volume viscosity in a model consisting of particles with excited states, *Physica* **66**, 474 (1973).

van Weert Ch. G., Relativistic treatment of multipole radiation from an extended charge-current distribution, *Physica* **65**, 452 (1973).

van Weert Ch. G., W.A. van Leeuwen and S.R. de Groot, Elements of relativistic kinetic theory, *Physica* **69**, 441 (1973).

van Weert Ch. G. and W.P.H. de Boer, Relativistic kinetic theory of quantum systems I: Wigner functions for a relativistic spin 1/2 system, *Physica* **A81**, 597 (1975).

van Weert Ch. G., Wigner function and normal ordering, *Physica* **A100**, 641 (1980).

van Weert Ch. G., Generalized hydrodynamics from relativistic kinetic theory, *Physica* **A111**, 537 (1982).

van Weert Ch. G., Maximum entropy principle and relativistic hydrodynamics, *Ann. Phys. (N.Y.)* **140**, 133 (1982).

van Weert Ch. G. and L.J. van den Horn, Transport properties of neutrinos in stellar collapse I: Bulk viscosity of collapsing stellar cores, *Astron. Astrophys.* **125**, 93 (1983).

van Weert Ch. G. and M.C.J. Leermakers, Transverse modes of a relativistic quantum plasma, *Phys. Lett.* **A97**, 159 (1983).
van Weert Ch. G., M.C.J. Leermakers, A.J.W. Hooiveld and F. Mensonides, Normal modes of a relativistic quantum plasma; the one-component plasma, *Physica* **A127**, 388 (1984).
van Weert Ch. G. and M.C.J. Leermakers, Relativistic Landau theory, *Phys. Lett.* **A101**, 195 (1984).
van Weert Ch. G. and M.C.J. Leermakers, Statistical quasi-particles and the thermodynamics of hot dense matter, *Physica* **A131**, 465 (1985).
van Weert Ch. G. and M.C.J. Leermakers, Landau theory of a QED plasma, *Acta Phys. Pol.* **B16**, 459 (1985).
van Weert Ch. G., M.C.J. Leermakers and A.M.J. Schackel, Landau kinetic theory of quark–gluon matter, *Physica* **A134**, 577 (1986).
Varro S. and J. Javainen, Gauge-invariant relativistic Wigner functions, *J. Opt.* **B5**, S402 (2003).
Vasak D., M. Gyulassy and H.-T. Elze, Quantum transport theory for an abelian plasma, *Ann. Phys. (N.Y.)* **173**, 462 (1987).
Ventura J., Scattering of light in a strongly magnetized plasma, *Phys. Rev.* **D19**, 1684 (1979).
Viello P. and Salvati M., On the relativistic plasma-dynamical and hydrodynamical normal modes, *Phys. Fluids* **19**, 1523 (1976).
Vignon B., Sur une méthode nouvelle de résoluion approché de l'équation de Boltzmann relativiste, *C. R. Acad. Sci.* **266**, 594 (1968).
Vignon B., Une méthode de résolution approchée de l'équation de Boltzmann relativiste, *Ann. Inst. Henri Poincaré* **A10**, 31 (1969).
Vignon B., Sur l'utilisation d'un principe variationnel dans l'étude de l'équation de Boltzmann relativiste, *C. R. Acad. Sci.* **273**, 1331 (1971).
Vinnik D.V., A.V. Prozorkevich, S.A. Smolyansky, V.D. Toneev, M.B. Hecht, C.D. Roberts and S.M. Schmidt, Plasma production and thermalisation in a strong field, *Eur. Phys. J.* **C22**, 341 (2001).
Vishniac E.T., Relativistic collisionless particles and the evolution of cosmological perturbations, *Astrophys. J.* **257**, 456 (1982).
Visvanathan S., Magnetic properties of a relativistic electron gas, *Phys. Fluids* **5**, 701 (1962).
Vshivtsev A.S., B.V. Magnitskii, I.N. Maslov, V.R. Khalikov and V.K. Perez-Fernandez, A relativistic electron gas in a strong magnetic field, *Sov. Astron.* **33**, 249 (1989).
Walecka J.D., A theory of highly condensed matter, *Ann. Phys. (N.Y.)* **83**, 491 (1974).
Walecka J.D., Equation of state for neutron matter at finite T in a relativistic mean-field theory, *Phys. Lett.* **B59**, 109 (1975).
Walecka J.D., The nuclear sum rule in a relativistic model, *Nucl. Phys.* **A399**, 387 (1983).
Walker A.G., The Boltzmann equations in general relativity, *Proc. Edinburgh Math. Soc.* **2**, 238 (1934).

Wandel A. and A. Yahil, Thermodynamics of pair production in a non-degenerate Fermi gas, *Astron. Astrophys.* **72**, 367 (1979).
Wang R.-C., D. Zhu, C.-Y. Wong and C.-M. Ko, Liouville equation of relativistic charged fermion, *Acta Phys. Sin.* **40**, 14 (1991).
Watanabe S., Symmetry of physical laws. Part I: Symmetry of space–time and balance theorems, *Rev. Mod. Phys.* **27**, 26 (1955).
Watanabe S., Symmetry of physical laws. Part II: Q-number theory of space–time, inversions and charge-conjugation, *Rev. Mod. Phys.* **27**, 40 (1955).
Watson K.M., S.A. Bludman and M.N. Rosenbluth S., Statistical mechanics of relativistic streams I, *Phys. Fluids* **3**, 741 (1960).
Wayland, J.R., Relativistic Stochastic Processes, In *Lectures in High-Energy Astrophysics*, (NASA SP-199), eds. H. Ögelman and J.R. Wayland (Scientific and Technical Information Division, Office of Technology Utilization, NASA, Washington, DC, 1969).
Webb G.M., G.P. Zank and J.F. McKenzie, Relativistic oblique magnetohydrodynamic shocks, *J. Plasma Phys.* **37**, 117 (1987).
Weber F. and M.K. Weigel, Relativistic nuclear and neutron matter at finite temperatures, *Z. Phys.* **A330**, 249 (1988).
Weber F. and M.K. Weigel, Equation of state of dense baryonic matter, *J. Phys.* **G15**, 765 (1989).
Weber F. and M.K. Weigel, Neutron stars properties and the relativistic nuclear equation of state of many-baryon matter, *Nucl. Phys.* **A493**, 549 (1989).
Weber F. and M.K. Weigel, Relativistic many-baryon matter and neutron stars, *Nucl. Phys.* **A495**, 363c (1989).
Weigel M.K. and J. Winter, On the general theory of nuclear non-equilibrium systems (statistical equations and nuclear hydrodynamics), *J. Phys.* **G9**, 913 (1983).
Weldon H.A., Covariant calculations at finite temperature: the relativistic plasma, *Phys. Rev.* **D26**, 1394 (1982).
Weldon H.A., Effective fermion masses of order gT in high-temperature gauge theories with exact chiral invariance, *Phys. Rev.* **D26**, 2789 (1982).
Weldon H.A., Simple rules for discontinuities in finite-temperature field theory, *Phys. Rev.* **D28**, 2007 (1983).
Weldon H.A., $\lambda\varphi^4$ coupling at high temperature, *Phys. Lett.* **B174**, 427 (1986).
Weldon H.A., Proof of zeta-function regularization of high-temperature expansions, *Nucl. Phys.* **B270**, 79 (1986).
Weldon H.A., Comparison of symmetry breaking in the canonical and microcanonical ensembles, *Ann. Phys. (N.Y.)* **193**, 177 (1989).
Weldon H.A., Particles and holes, *Physica* **A158**, 169 (1989).
Weldon H.A., Scalar field propagators in the microcanonical ensemble, *Ann. Phys. (N.Y.)* **193**, 166 (1989).
Weldon H.A., Quasi-particle in finite-temperature field theory [arXiv:hep-ph/9809330 v2].
Wehrberger K. and F. Beck, Effect of vacuum fluctuations on the quasi-elastic response functions in quantum hadrodynamics, *Nucl. Phys.* **A491**, 587 (1989).

Weinberg S., Entropy generation and the survival of protogalaxies in an expanding universe, *Astrophys. J.* **168**, 175 (1971).

Weinberg S., Gauge and global symmetries at high temperature, *Phys. Rev.* **D9**, 3357 (1974).

Wienke B.R., Mean, mean-square, and most probable momentum for a relativistic Maxwellian ensemble, *Am. J. Phys.* **43**, 317 (1975).

Wilets L., Green's functions method for the relativistic field theory many-body problem, In *Mesons in Nuclei*, Vol. III eds. M. Rho and D. Wilkinson (North-Holland, Amsterdam, 1979).

Williams D.R.M. and D.B. Melrose, Covariant response tensors for spin zero and spin one boson/antiboson plasmas, *Aust. J. Phys.* **42**, 59 (1989).

Wilson J.W., Transport methods for relativistic ions, *Trans. Am. Nucl. Soc.* **52**, 389 (1986).

Winter J., Covariant extension of the Wigner transformation to non-abelian Yang–Mills symmetries for a Vlasov equation approach to the quark–gluon plasma, *J. Phys. (Paris)* **45**, C6 (1984).

Winter J., Wigner transformation in curved space–time and the curvature correction of the Vlasov equation for semi-classical gravitating systems, *Phys. Rev.* **D32**, 1871 (1985).

Winterberg F., Radiative collapse of a relativistic electron–positron plasma to ultrahigh densities, *Phys. Rev.* **A19**, 1356 (1979).

Witte N.S., R.L. Dawe and K.C. Hines, Relativistic charged bosons in a magnetic field I: Wave functions and matrix elements, *J. Math. Phys.* **28**, 1864 (1987).

Witte N.S., V. Kowalenko and K.C. Hines, Relativistic charged bosons in a magnetic field II, *Phys. Rev.* **D38**, 3667 (1988).

Wright T.P. and G.R. Hadley, Relativistic distribution functions and applications to electron beams, *Phys. Rev.* **A12**, 686 (1975).

Yadavalli S., A note on the relativistic Boltzmann equation and some applications, 368 (1961).

Yakovlev D.G., Transport properties of the degenerate electron gas of neutron stars along the quantizing magnetic field, *Astrophys. Space Sci.* **98**, 37 (1984).

Yoon P.H., Electromagnetic Weibel instability in a fully relativistic bi-Maxwellian plasma, *Phys. Fluids* **B1**, 1336 (1989).

Yoon P.H. and R.C. Davidson, Exact analytical model of the classical Weibel instability in a relativistic anisotropic plasma, *Phys. Rev.* **A35**, 2718 (1987).

Yoon P.H. and R.C. Davidson, Alternative representation of the dielectric tensor for a relativistic magnetized plasma in thermal equilibrium, *J. Plasma Phys.* **43**, 269 (1990).

Zakharov A.V., A kinetic theory for the growth of perturbations in an isotropic cosmological model, and the ultrarelativistic limit, *Sov. Astron.* **22**, 528 (1978).

Zakharov A.V., Influence of collisionless particles on the growth of gravitational perturbations in an isotropic universe, *Sov. Phys. JETP* **50**, 221 (1979).

Zakharov A.V., Collision integral in a Friedmann universe, *Sov. Phys. JETP* **59**, 1 (1984).

Zakharov A.V., Relativistic kinetic equation for gravitating particles in a Friedmann universe, *Sov. Astron.* **33**, 624 (1989).

Index

www.ingramcontent.com/pod-product-compliance
Ingram Content Group UK Ltd.
Pitfield, Milton Keynes, MK11 3LW, UK
UKHW022109170926
8624IPUK00024B/348

9 789814 322430